CW00945942

Complex Analysis with MATHEMATICA®

Complex Analysis with MATHEMATICA®

William T. Shaw
St Catherine's College, Oxford and
Oxford Centre for Industrial and Applied Mathematics

CAMBRIDGE
UNIVERSITY PRESS

University Printing House, Cambridge CB2 8BS, United Kingdom

Cambridge University Press is part of the University of Cambridge.

It furthers the University's mission by disseminating knowledge in the pursuit of education, learning and research at the highest international levels of excellence.

www.cambridge.org
Information on this title: www.cambridge.org/9780521836265

First published 2006
Reprinted with corrections 2008

A catalogue record for this publication is available from the British Library

ISBN 978-0-521-83626-5 Hardback

Additional resources for this publication at www.cambridge.org/9780521836265

For Helen and Benjamin

A CD-ROM formerly accompanied this publication:
the contents of the CD-ROM are now located at
www.cambridge.org/9780521836265

Contents

Preface **xv**
Why this book? xv
How this text is organized xvi
Some suggestions on how to use this text xxi
About the enclosed CD xxii
Exercises and solutions xxiv
Acknowledgements xxiv

1 Why you need complex numbers **1**
Introduction 1
1.1 First analysis of quadratic equations 1
1.2 *Mathematica* investigation: quadratic equations 3
Exercises 8

2 Complex algebra and geometry **10**
Introduction 10
2.1 Informal approach to 'real' numbers 10
2.2 Definition of a complex number and notation 12
2.3 Basic algebraic properties of complex numbers 13
2.4 Complex conjugation and modulus 14
2.5 The Wessel–Argand plane 14
2.6 Cartesian and polar forms 15
2.7 DeMoivre's theorem 21
2.8 Complex roots 25
2.9 The exponential form for complex numbers 29
2.10 The triangle inequalities 32
2.11 *Mathematica* visualization of complex roots and logs 33
2.12 Multiplication and spacing in *Mathematica* 35
Exercises 35

3 Cubics, quartics and visualization of complex roots **41**
Introduction 41
3.1 *Mathematica* investigation of cubic equations 42
3.2 *Mathematica* investigation of quartic equations 45

3.3 The quintic 51
3.4 Root movies and root locus plots 51
Exercises 53

4 Newton–Raphson iteration and complex fractals **56**
Introduction 56
4.1 Newton–Raphson methods 56
4.2 *Mathematica* visualization of real Newton–Raphson 57
4.3 Cayley's problem: complex global basins of attraction 59
4.4 Basins of attraction for a simple cubic 62
4.5 More general cubics 67
4.6 Higher-order simple polynomials 71
4.7 Fractal planets: Riemann sphere constructions 73
Exercises 76

5 A complex view of the real logistic map **78**
Introduction 78
5.1 Cobwebbing theory 79
5.2 Definition of the quadratic and cubic logistic maps 80
5.3 The logistic map: an analytical approach 81
5.4 What about n=3,4,...? 89
5.5 Summary of our root-finding investigations 91
5.6 The logistic map: an experimental approach 91
5.7 Experiment one: $0 < \lambda < 1$ 92
5.8 Experiment two: $1 < \lambda < 2$ 93
5.9 Experiment three: $2 < \lambda < \sqrt{5}$ 93
5.10 Experiment four: $2.45044 < \lambda < 2.46083$ 95
5.11 Experiment five: $\sqrt{5} < \lambda < \sqrt{5} + \epsilon$ 96
5.12 Experiment six: $\sqrt{5} < \lambda$ 96
5.13 Bifurcation diagrams 98
5.14 Symmetry-related bifurcation 100
5.15 Remarks 102
Exercises 103

6 The Mandelbrot set **105**
Introduction 105
6.1 From the logistic map to the Mandelbrot map 105
6.2 Stable fixed points: complex regions 107
6.3 Periodic orbits 110
6.4 Escape-time algorithm for the Mandelbrot set 114
6.5 *MathLink* versions of the escape-time algorithm 120
6.6 Diving into the Mandelbrot set: fractal movies 126
6.7 Computing and drawing *the* Mandelbrot set 129
Exercises 135
Appendix: C Code listings 136

7 Symmetric chaos in the complex plane **138**
Introduction 138
7.1 Creating and iterating complex non-linear maps 139
7.2 A movie of a symmetry-increasing bifurcation 143
7.3 Visitation density plots 145
7.4 High-resolution plots 146
7.5 Some colour functions to try 146
7.6 Hit the turbos with *MathLink*! 148
7.7 Billion iterations picture gallery 149
Exercises 154
Appendix: C code listings 155

8 Complex functions **159**
Introduction 159
8.1 Complex functions: definitions and terminology 159
8.2 Neighbourhoods, open sets and continuity 163
8.3 Elementary vs. series approach to simple functions 165
8.4 Simple inverse functions 169
8.5 Branch points and cuts 171
8.6 The Riemann sphere and infinity 175
8.7 Visualization of complex functions 176
8.8 Three-dimensional views of a complex function 183
8.9 Holey and checkerboard plots 187
8.10 Fractals everywhere? 189
Exercises 192

9 Sequences, series and power series **194**
Introduction 194
9.1 Sequences, series and uniform convergence 194
9.2 Theorems about series and tests for convergence 196
9.3 Convergence of power series 202
9.4 Functions defined by power series 205
9.5 Visualization of series and functions 205
Exercises 207

10 Complex differentiation **208**
Introduction 208
10.1 Complex differentiability at a point 209
10.2 Real differentiability of complex functions 211
10.3 Complex differentiability of complex functions 212
10.4 Definition via quotient formula 213
10.5 Holomorphic, analytic and regular functions 214
10.6 Simple consequences of the Cauchy–Riemann equations 214
10.7 Standard differentiation rules 215
10.8 Polynomials and power series 217
10.9 A point of notation and spotting non-analytic functions 220

10.10 The Ahlfors–Struble(?) theorem 221
Exercises 233

11 Paths and complex integration 237
Introduction 237
11.1 Paths 237
11.2 Contour integration 240
11.3 The fundamental theorem of calculus 241
11.4 The value and length inequalities 242
11.5 Uniform convergence and integration 243
11.6 Contour integration and its perils in *Mathematica!* 244
Exercises 245

12 Cauchy's theorem 248
Introduction 248
12.1 Green's theorem and the weak Cauchy theorem 248
12.2 The Cauchy–Goursat theorem for a triangle 250
12.3 The Cauchy–Goursat theorem for star-shaped sets 254
12.4 Consequences of Cauchy's theorem 255
12.5 *Mathematica* pictures of the triangle subdivision 259
Exercises 261

13 Cauchy's integral formula and its remarkable consequences 263
Introduction 263
13.1 The Cauchy integral formula 263
13.2 Taylor's theorem 265
13.3 The Cauchy inequalities 271
13.4 Liouville's theorem 271
13.5 The fundamental theorem of algebra 272
13.6 Morera's theorem 274
13.7 The mean-value and maximum modulus theorems 275
Exercises 275

14 Laurent series, zeroes, singularities and residues 278
Introduction 278
14.1 The Laurent series 278
14.2 Definition of the residue 282
14.3 Calculation of the Laurent series 282
14.4 Definitions and properties of zeroes 286
14.5 Singularities 287
14.6 Computing residues 292
14.7 Examples of residue computations 293
Exercises 299

15 Residue calculus: integration, summation and the argument principle **302**
Introduction 302
15.1 The residue theorem 302
15.2 Applying the residue theorem 304
15.3 Trigonometric integrals 305
15.4 Semicircular contours 313
15.5 Semicircular contour: easy combinations of trigonometric functions and polynomials 316
15.6 Mousehole contours 318
15.7 Dealing with functions with branch points 320
15.8 Infinitely many poles and series summation 324
15.9 The argument principle and Rouché's theorem 328
Exercises 335

16 Conformal mapping I: simple mappings and Möbius transforms **338**
Introduction 338
16.1 Recall of visualization tools 338
16.2 A quick tour of mappings in *Mathematica* 340
16.3 The conformality property 347
16.4 The area-scaling property 348
16.5 The fundamental family of transformations 348
16.6 Group properties of the Möbius transform 349
16.7 Other properties of the Möbius transform 350
16.8 More about `ComplexInequalityPlot` 354
Exercises 355

17 Fourier transforms **357**
Introduction 357
17.1 Definition of the Fourier transform 358
17.2 An informal look at the delta-function 359
17.3 Inversion, convolution, shifting and differentiation 363
17.4 Jordan's lemma: semicircle theorem II 366
17.5 Examples of transforms 368
17.6 Expanding the setting to a fully complex picture 372
17.7 Applications to differential equations 373
17.8 Specialist applications and other *Mathematica* functions and packages 376
Appendix 17: older versions of *Mathematica* 377
Exercises 379

18 Laplace transforms **381**
Introduction 381
18.1 Definition of the Laplace transform 381
18.2 Properties of the Laplace transform 383

18.3 The Bromwich integral and inversion 387
18.4 Inversion by contour integration 387
18.5 Convolutions and applications to ODEs and PDEs 390
18.6 Conformal maps and Efros's theorem 395
Exercises 398

19 Elementary applications to two-dimensional physics 401
Introduction 401
19.1 The universality of Laplace's equation 401
19.2 The role of holomorphic functions 403
19.3 Integral formulae for the half-plane and disk 406
19.4 Fundamental solutions 408
19.5 The method of images 413
19.6 Further applications to fluid dynamics 415
19.7 The Navier–Stokes equations and viscous flow 425
Exercises 430

20 Numerical transform techniques 433
Introduction 433
20.1 The discrete Fourier transform 433
20.2 Applying the discrete Fourier transform in one dimension 435
20.3 Applying the discrete Fourier transform in two dimensions 437
20.4 Numerical methods for Laplace transform inversion 439
20.5 Inversion of an elementary transform 440
20.6 Two applications to 'rocket science' 441
Exercises 448

21 Conformal mapping II: the Schwarz–Christoffel mapping 451
Introduction 451
21.1 The Riemann mapping theorem 452
21.2 The Schwarz–Christoffel transformation 452
21.3 Analytical examples with two vertices 454
21.4 Triangular and rectangular boundaries 456
21.5 Higher-order hypergeometric mappings 463
21.6 Circle mappings and regular polygons 465
21.7 Detailed numerical treatments 470
Exercises 470

22 Tiling the Euclidean and hyperbolic planes 473
Introduction 473
22.1 Background 473
22.2 Tiling the Eudlidean plane with triangles 475
22.3 Tiling the Eudlidean plane with other shapes 481
22.4 Triangle tilings of the Poincaré disc 485
22.5 Ghosts and birdies tiling of the Poincaré disc 490
22.6 The projective representation 497

22.7 Tiling the Poincaré disc with hyperbolic squares 499
22.8 Heptagon tilings 507
22.9 The upper half-plane representation 510
Exercises 512

23 Physics in three and four dimensions I 513
Introduction 513
23.1 Minkowski space and the celestial sphere 514
23.2 Stereographic projection revisited 515
23.3 Projective coordinates 515
23.4 Möbius and Lorentz transformations 517
23.5 The invisibility of the Lorentz contraction 518
23.6 Outline classification of Lorentz transformations 520
23.7 Warping with *Mathematica* 524
23.8 From null directions to points: twistors 529
23.9 Minimal surfaces and null curves I: holomorphic parametrizations 531
23.10 Minimal surfaces and null curves II: minimal surfaces and visualization in three dimensions 535
Exercises 538

24 Physics in three and four dimensions II 540
Introduction 540
24.1 Laplace's equation in dimension three 540
24.2 Solutions with an axial symmetry 541
24.3 Translational quasi-symmetry 543
24.4 From three to four dimensions and back again 544
24.5 Translational symmetry: reduction to 2-D 548
24.6 Comments 550
Exercises 551

Bibliograpy 553

Index 558

Preface

Why this book?

Since 1985, I have been fortunate to have taught the theory of complex variables for several courses in both the USA and the UK. In the USA I lectured a course on advanced calculus for engineers and scientists at MIT, and in the UK I have given tutorials on the subject to undergraduate students in mathematics at both Cambridge and Oxford. Indeed, draft versions of this text have been inflicted on my students at Balliol and, more recently, at St. Catherine's over the last fourteen years. Few topics have given me such pleasure to teach, given the rich yet highly accessible structure of the subject, and it has at times formed the subject of my research, notably in its development into twistor theory, and latterly in its applications to financial mathematics. A parallel thread of my work has been in the applications of computer algebra and calculus systems, and in particular *Mathematica*®, to diverse topics in applied mathematics. This book is in part an attempt to use *Mathematica* to illuminate the topic of complex analysis, and draws on both these threads of my experience.

The book attempts also to inject some new mathematical themes into the topic and the teaching of it. These themes I feel are, if not actually missing, under-emphasized in most traditional treatments. It is perfectly possible for students to have had a formal training in mathematics that leaves them unaware of many key and/or beautiful topics. If we take the beginning of the historical time-line to supply our first example, many students will not be aware of how to solve a cubic equation, despite this procedure being one of the key early developments in this field. Having invented complex numbers to cope with a general quadratic, early algebraists found that the cubic could also be solved. This is of paramount importance, not just for the elegance of the solution, but also because it is the first indication that the fundamental theorem of algebra might be a possible theorem! If we take a more recent example, those students that do consider the applications to basic physics will almost always emerge with the entirely mistaken notion that complex variable methods are limited to a few very special problems in two-dimensional electrostatics or fluid dynamics. Similarly, the Möbius transform will often be presented only as a neat trick for mapping shapes around the complex plane, and its profound links to relativistic physics, via its equivalence to the Lorentz transformation, are ignored.

So in addition to providing illuminations and visualizations with *Mathematica*, I have tried to put back and indeed add some of the topics that I feel students ought to know. In particular, and unusually for a text targeted at undergraduate mathematicians or graduate students in other disciplines, this book includes a friendly introduction to the theory of spinors and twistors, thereby unlocking the applications of complex functions to problems in three and four dimensions.

But you do not have to accept this particular set of views to make good use of this book. It is perfectly possible to use this text to teach a standard course in complex analysis, ignore my idiosyncratic additions, and take the *Mathematica* elements as purely an embedded tool that has been used to generate some of the pictures!

Mathematica makes its appearance in many different ways. In several chapters it is there purely to provide, literally, illustrations. In some places it is used as a checking tool, for example when calculating residues and integrals. In other chapters it is fundamental, and indeed in Part II, it is the centre of a set of investigations into the solving of equations by iteration. In places its rich library of special functions, the ability to evaluate them over the complex plane, to do calculus with them, come to the fore. It is particularly valuable when applied to topics in conformal mapping. Finally, *Mathematica's* wonderful graphics are universally useful.

How this text is organized

It is best to think of the material of this book as being grouped informally into five parts. These are as follows:

Part I Basic complex number theory and history

Attention will be focused on three topics, each of which constitutes one chapter:

- Chapter 1: Why you need complex numbers;
- Chapter 2: Complex algebra and geometry;
- Chapter 3: Cubics, quartics and visualization of complex roots.

The idea of this part of the book is to explain how and why complex numbers were introduced, and then to go on to discuss elementary properties of the complex number system. This material is at a level normally to be found in final year high school programs or introductory college level. Chapter 3 should be regarded as optional, but is highly recommended for any students with an interest in the history of the subject. It covers the treatment of cubics and quartics, which is not usually taught in modern courses, and also includes some material on the visualization of roots of polynomials.

Part II Iterated mappings

Part I showed how to define complex numbers and how to use them to solve low-order polynomial equations. The methods used to treat the quadratic, cubic and quartic equations are the classical techniques that have been known for some time – many hundreds of years in some cases.

Now that computer systems are available, you can explore, either through this text, or directly yourselves through the use of these *Mathematica* notebooks, the rich structure that is obtained by the application of *iterative* equation-solving techniques to these same simple polynomial systems. This idea originates with A. Cayley in the 19th century, who although able to understand quickly the complex structure of Newton–Raphson methods when applied to a quadratic, was frustrated by the corresponding problem with a cubic. In getting to grips with Cayley's problem, we shall quickly encounter some of the most beautiful objects in modern mathematics – chaotic systems and fractals.

This part of the text consists of material that is not part of a *traditional* course on complex analysis. It may be skipped by those using this text to pursue such a traditional route, who should proceed to Part III. Part III does not rely on any of the material in Part II.

In Part II, all of the systems that you will see can be regarded as special cases of the general first-order iterated map:

$$z_{n+1} = f(z_n)$$

You will be able to explore how this works for various choices of the function f. One way or other, f is to be associated with the solution of a low-order polynomial equation. The association of the iterated map with the polynomial equation can take place in several ways, and two will be considered here.

The first approach will involve polynomial (or even transcendental) equations of the form

$$g(z) = 0$$

and you will explore the Newton–Raphson iteration scheme given by the choice

$$f(z) = z - \frac{g(z)}{g'(z)}$$

The second scheme will involve a polynomial equation that is already written (for example, by simply isolating the linear term, if there is one) in the form

$$z = f(z)$$

and you will explore the 'cobwebbing' solution scheme based on iteration of this representation.

Attention will be focused on four topics, each of which constitutes one chapter. Of these four topics, the first is specifically Newton–Raphson. The next two may be regarded as being associated with the cobwebbing method. The fourth topic is a complex extension of the cobwebbing method with symmetry. In order of presentation, the topics are:

- Chapter 4: Newton–Raphson iteration and complex fractals;
- Chapter 5: A complex view of the real logistic map;
- Chapter 6: The Mandelbrot set;
- Chapter 7: Symmetric chaos in the complex plane.

In Chapter 4 the solution of a low-order polynomial equation will be reconsidered in the complex plane using Newton–Raphson iteration, as part of an investigation of Cayley's problem (Cayley, 1879). This is a standard technique for solving real non-linear equations – our purpose here is to explore what happens when Newton–Raphson is applied in the complex plane, and to use the computer to understand why Cayley was defeated by the cubic!

In this part of the text we will engage occasionally engage in 'fashionable number crunching', as chaos theory was once famously described. The desire to produce some spectacular pictures is never far from one's mind. But not all mathematics has to be useful, and the uncovering of beauty is a worthwhile goal in itself. So in this part of the book I shall indulge shamelessly in some fashionable number crunching - this is sometimes referred to somewhat pompously as 'experimental mathematics'. But *good* experiments should be designed to test some theory about what should happen, and we can use complex numbers, to some extent, to provide a framework for first formulating a hypothesis regarding what may happen in a simple real non-linear system.

The logistic map, as developed by May (1976), is the place where this experimentation will commence for real, with Chapter 5. This is usually regarded as a real mapping, so what is it doing here? The point is that we shall not just indulge in computation, but shall attempt to predict, through the machinery of complex numbers, what should happen in a certain experiment. It will turn out that the period-doubling behaviour of the logistic map is in fact a simple and predictable result that requires nothing more than 'end of high school' mathematics. The experimentation will serve to confirm our hypotheses about it. What is surprising and fascinating is the transition to chaos that follows, and there are indeed many properties of the logistic map that are still not properly understood.

It is admittedly very hard to extend this approach to more complicated non-linear systems, so we shall then rely more substantially on numerical experiments for our other chapters. In Chapter 6 we shall extend the cobwebbing concept to the complex plane using the simple quadratic (Mandelbrot) map. Finally, in Chapter 7, we shall revisit the logistic map again, constructing complex versions of it possessing various types of symmetry, leading to the recently developed concepts of symmetric chaos. This leads to some stunning imagery, discovered by Field and Golubitsky. Their text (Field and Golubitsky, 1992) is one of the most beautiful books I have ever seen. Here we will see how some of their work can be readily investigated using *Mathematica*.

Part III Traditional complex analysis

By the beginning of Part III you will have seen how to define complex numbers and how to use them to solve simple polynomial equations by both classical solution methods (and by modern iterative techniques if you have worked through Part II). In Part III you begin the study of complex functions from a formal point of view. Your goal is to understand the calculus of complex functions – differentiation, integration, series (just as in the real case) – and the very *special* results that apply to *complex* differentiable functions, in manifest distinction to the real case. The plan of this part of the text is as follows:

- Chapter 8: Complex functions;
- Chapter 9: Sequences, series and power series;
- Chapter 10: Complex differentiation;
- Chapter 11: Paths and complex integration;
- Chapter 12: Cauchy's theorem;
- Chapter 13: Cauchy's integral formula and its remarkable consequences;
- Chapter 14: Laurent series, zeroes, singularities and residues;
- Chapter 15: Residue calculus: integration, summation and the argument principle.

There are various ways of presenting and ordering this material, and it is worth explaining the particular approach taken here. Our approach is to give a first introduction to standard functions in Chapter 8, by extension of their definitions for real variables. Next, in Chapter 9, we assume some basic results from real analysis related to sequences and series. A summary of results about sequences and series are presented without formal proof. Students of pure mathematics should consult a good calculus or basic real analysis text for background on this (a comprehensive text is the book by Rudin, 1976). Then we define power series for complex functions, and establish their convergence within a circle of convergence. Then, in Chapter 10, differentiability is introduced. The approach to complex differentiability is based on the notion of a local linear approximation to a function – equivalent to the notion that there is a tangent to a complex curve. The definition quite frequently given, based on the quotient formula, is given as an aside. There are several very good reasons for this approach. First, *the quotient formula for the derivative does not work for functions of two or more real or complex variables*, so if we were to take this approach we could not sensibly relate complex differentiation to differentiation of functions of two real variables, nor can we make a generalization to functions of several complex (or real) variables without starting again with the linear approximation approach. I think it is better to do it properly in the first place. Second, the standard properties of derivatives such as the product, ratio and chain rules are really easy to write down within the linear approximation framework. Once differentiability has been defined, the differentiability of a power series within the circle of convergence is then established immediately. We then redefine our standard basic functions in terms of power series – their differentiability properties are then obvious.

Chapter 10 also includes a discussion of the theorem that the author has tentatively called the 'Ahlfors-Struble' theorem. This is the means by which one can recover a holomorphic function from its real part alone (or from just the imaginary part) by a purely *algebraic* method. This idea seems to have been rediscovered several times over the years. It is a very powerful technique when linked to the symbolic power of *Mathematica* and I have also given a discussion of the history, to justify crediting the result to Ahlfors and Struble, in Section 10.10.

Next, in Chapter 11, paths and integrals along paths are defined. Chapter 12 introduces the key theorem of this section – Cauchy's theorem – that certain integrals vanish identically. This is the key to the magic that follows, and a standard approach to the consequences of Cauchy's theorem is given in Chapters 13–15, culminating in the evaluation of certain integrals and series by the calculus of residues. Some of the material here can be augmented by other texts and I would recommend Rudin (1976), particularly as it also proceeds in a manner that makes the multi-variable case straightforward. I also suggest that geometrically-minded students look at Needham's (1997) beautiful book, *Visual Complex Analysis*.

Part IV Standard applications

In this part of the book you explore the basic applications of the material. Most first courses in complex variable theory include at least some of these topics, though the transform material may also find its way into other applied mathematics courses, and the basic applications to two-dimensional physics could also be useful in courses on potential theory and/or fluid dynamics. This part begins with basic conformal mapping – more advanced conformal maps are revisited in Chapter 21. Similarly, numerical issues with transforms are deferred to Chapter 20. You should note that Chapters 17–18 also discuss more advanced topics in contour integration, including the development and application of Jordan's lemma for semicircles. The plan of this part of the book is as follows:

- Chapter 16: Conformal mapping I: simple mappings and Möbius transforms;
- Chapter 17: Fourier transforms;
- Chapter 18: Laplace transforms;
- Chapter 19: Elementary applications to two-dimensional physics.

The novel features in this part of the book include the use of *Mathematica* to visualize conformal maps and their applications to potential flow. The generalization of the convolution theorem for Laplace transforms due to Efros is also presented, and the discussion of fluid dynamics includes a discussion of viscous flow and the biharmonic equation in complex form.

Part V Advanced applications

In this part of the book you may explore material that is not so frequently presented in introductory complex variable texts. There are five topics:

- Chapter 20: Numerical transform techniques;
- Chapter 21: Conformal mapping II: the Schwarz–Christoffel transformation;
- Chapter 22: Tiling the Euclidean and hyperbolic planes;
- Chapter 23: Physics in three and four dimensions I;
- Chapter 24: Physics in three and four dimensions II;.

The first three of these chapters have been added because the integration of the presentation with *Mathematica* allows a full treatment of some issues that require a combination of numerical/advanced analytical and graphical methods on a computer. With a computer one can explore the numerical treatment of transforms, the beautiful applications of the Schwarz–Christoffel transformation, and produce stunning hyperbolic tilings! Finally, in the last two chapters, you can see how complex numbers are very useful for doing physics and geometry in more than two dimensions. For example, you will discover that the Möbius transformation is not just a dry device for mapping circles and lines, but is really the mapping at the heart of Einstein's theory of special relativity. You will discover how complex numbers may be used to solve non-linear partial differential equations such as arise for the shape of a soap bubble, in a formalism – Penrose's theory of twistors – that links the nineteenth century work of Weierstrass to modern minimal surface and string theory. In the last chapter you will see at last the true power of holomorphic functions in solving the 3-D Laplace equation and the wave equation in four dimensions, again through Penrose's theory of twistors.

Some suggestions on how to use this text

In the end this is up to you, the reader, whether you are student or teacher. But in writing this material I have had several possible course threads in my mind. Let's look at a few possibilities.

A basic computer-enhanced course on complex numbers and solving equations

This might consist of Chapters 1–7. The unifying theme is the solution of equations. In the first few chapters the emphasis is on solving polynomial equations by traditional attempts at factorization, whereas in Chapters 4–7 we look at iterative methods of solution and the consequences.

A traditional mathematics course on complex analysis

As a minimum this would consist of Chapters 8–15, with parts of Chapters 1–3 for less well prepared students, and some portions of Chapters 16–19, 21–22 depending on the scope of the presentation.

For physics and engineering

Students taking a serious mathematics component could use Chapters 8–15 together with material from 16–19 and 23–24.

For a numerical programme

Students studying numerical methods could draw on material from Chapters 4–6, with 7 for fun, a review of 17–18 and then 20–21.

Material for specific courses in physics and engineering

It is hopefully evident that some topics may be useful for parts of other programmes. Obvious cases include courses on potential theory, whether in electrostatics, gravity or fluids, which frequently dip into complex variable theory. This material is available here, notably in Chapter 19, but it is to be hoped that those who dip into Chapter 19 also take a good look at Chapters 23 and 24!

Motivational mathematics

Many of the topics developed here could also be used as motivational material, perhaps for students not taking specialist mathematics, physics or engineering programs, but on more general courses. In my view, having an appreciation of the beauty, indeed the art, of mathematics is a vital component of an advanced education. The material in Part II could be extensively drawn on for such a program, in addition to snippets from other chapters.

Playing

Everybody should play! You can have fun just trying out the *Mathematica* implementations in many of the chapters. You can have even more fun by coming up with better ways of doing things than the author has done here and letting the author know.

*About the enclosed CD

The enclosed CD contains three directories, entitled 'Notebooks', 'MathLink' and 'Goodies'.

*A CD-ROM formerly accompanied this publication: the contents of the CD-ROM are now located at www.cambridge.org/9780521836265

The Notebooks directory contains electronic copies of all the chapters of the text, in the form of *Mathematica* notebooks. These have been finalized in *Mathematica* 5.1 and therefore should open directly in V5.1 or later. If you are using an earlier version you will get a warning that you can ignore and open the file anyway. If you are using version 4.x or even 3.x you may find that some things do not quite work as in the text. The results from `Integrate` have now stabilized but differed in earlier versions, so you should watch out for that, particularly in the sections where you check a contour integral or work out a Schwarz–Christoffel map. There are other minor stylistic issues, such as `Conjugate` appearing in output form as the whole word 'Conjugate' in older versions, whereas now the output form is a simple star!

The MathLink directory contains *MathLink* code in the form of (a) source for any system, (b) binaries for some systems. The source consists of `.tm` files and `.c` files. A lack of resources prevents me from making immediately useful binaries for every operating system.

The Goodies directory contains encrypted information pertinent to *Mathematica* technologies beyond version 5.2 that will be made available once such technology is officially released. See below for more details on this. First I need to remark on *kernel* compatibility issues in general.

The author is unable to offer support on the code or *MathLink* issues. But I do wish to receive bug reports on kernel operation. This code started off as working in *Mathematica* 2.2, and has been updated for compatibility with 3.x, 4.x, 5.x. As *Mathematica* has been updated it has become increasingly difficult to retain total compatibility with older versions, as noted above. The evolution of the software has in fact resulted in better code for this book, as I have been driven to write code that relies less on a trick that might work in one version, and more towards code that uses the fundamental principles of *Mathematica* .

Please let me know if you find anything that does NOT work under versions 5.1 or 5.2. You are strongly encouraged to use version 5.x or later, until a new version is released. I have made an effort to explain where there is a significant different between the way this book works between major versions. As for *Mathematica* technologies beyond Version 5.2, I cannot comment on any of the details of unreleased software, but you should see the author's web site at King's College London:

`www.mth.kcl.ac.uk/staff/w_shaw.html`

and the CUP website for the book at

`www.cambridge.org/0521836263`

for updates when a new version is released, including a key to unlock the encrypted material in the 'Goodies' section of the CD. If you are using a *Mathematica* technology beyond version 5.2, please do not send me bug reports until you have first checked the CD and *then* the on-line information, as the author will do his best to ensure that the book as distributed together with the CD is future-proof.

Exercises and solutions

Each chapter ends with a collection of exercises. These consist of some requiring traditional thought and pen and paper analysis, others where you can additionally check the results with *Mathematica* and others that are entirely based on *Mathematica*. Questions entirely based on *Mathematica* are indicated by a polyhedral *Mathematica* icon, while those having some partial or optional involvement of *Mathematica* have the icon bracketed. Similarly sections of the book based primarily on the software are prefixed with the same icon.

Some problems are elementary exercises based on the material, while others are more open-ended investigations that do do not have a 'correct answer'. The author intends to make a 'Solutions to Selected Exercises' available on-line to educators at the earliest opportunity, and information of the progress on this will be available from the web sites noted above.

Acknowledgements

I need to start with the mathematics staff of Manshead School who, during the 1970s, inspired me with a love of mathematics. Then the inspirational teaching of A. F. Beardon and T. W. Körner at Cambridge got me hooked on complex analysis, and J. M. Stewart taught me how to apply it. Sir R. Penrose, F.R.S. showed me what could be done with complex variables and relativistic physics, and has provided me with more inspiration than anyone has a right to have. M. Perry set me on the course of looking at twistor models of string theory, and some of the material in Chapter 23, especially the twistor solution of the relativistic string (minimal surface) equations, arose from these studies.

Particular topics presented here have benefited from contributions from particular people. I am particularly grateful to L.N. Trefethen, F.R.S. for providing me with background material on modern approaches to conformal mapping, and to Vanessa Thomas for allowing me to use her work on tiling the hyperbolic plane. In addition, N. Hitchin, F.R.S. taught me how properly to use twistor methods in 3-D. The material of Chapter 7 was inspired by the beautiful book by M. Field and M. Golubitsky. The material in the final two chapters (23 and 24) owes a great deal to R. Penrose, who has educated me in all sorts of other matters including the hyperbolic tilings of Chapter 22, the fine details of fractals (Chapter 6) as well as everything to do with twistors. I am also grateful to J. Ockendon, F.R.S. for making me aware of the advanced applications of complex variables to fluids, so that this text, unusually for a first course, includes a discussion of viscous (biharmonic) flow. He also provoked me in several ways, most constructively by suggesting I look into ways of deducing the full structure of holomorphic functions from a purely algebraic treatment of their real (or imaginary) parts. I published a short educational note on this method (Shaw, 2004) and requested information on the history of the method. I am grateful to H. Boas, B. Margolis and others for responding to this request, and also providing some suggestions that lead to the improved *Mathematica* implementation given in Section 10.10.

Particular thanks go to N. Woodhouse and S. Howison for professional support during a difficult time coinciding with the latter stages of preparing this book. S. Howison, B. Hambly, J. Dewynne, A. Ilhan and C. Reisinger have all kept me sane with their considerable support on my real job of running a graduate programme in mathematical finance in Oxford.

On the *Mathematica* side I am indebted to S. Wolfram and the staff of Wolfram Research for providing this wonderful software, and to C. Wolfram, T. Wickham Jones and the UK team for particular support on numerous activities. I have used various programming devices borrowed from several other people over the years, and I am sure that ideas developed by P. Abbott and R. Maeder have found their way into several bits of code here and there! I also thank numerous unnamed individuals in technical support for help over the years, in patiently answering my questions. Gratitude is expressed to T. Gayley for compiling my *MathLink* binaries to run under a certain operating system whose name I shall not speak.

I also have to thank many students of Balliol and St Catherine's Colleges, Oxford for using and correcting various versions of this material. There were too many contributions to name everybody, but special thanks are due to Rosie Bailey, Robin Oliver-Jones and Elizabeth Lang. I also thank Frances Kirwan and Keith Hannabuss for allowing me to teach this topic at Balliol.

Several others are mentioned at specific places in the text. I apologize to anyone I have left out.

There are others I need to thank whose names I do not know. These are the anonymous reviewers who have patiently and thoroughly commented on this text during various revisions, and made numerous helpful comments. I think I agreed with all they said, and were it not for them a few important things would have been left out, and chapters presented in a more haphazard order.

This text was initially inspired by some discussions with Karen Mosman about ten years ago. Since then I have taken far too long to finish it. But it is better for the wait, in part because computer speed has rather changed since I started. First drafts of chapters were worked out on a 66 MHz machine, and the final draft was edited on a 1.4 GHz machine (already 'one-third speed' by standards as of late 2005). This means that many ideas for interactive work now work, well, interactively, rather than over a coffee break. I also learnt a lot from others about how to do things better in the meantime. *Mathematica* also got much better over time, particularly in regard to typesetting and the consistency of symbolic integration. These comments will hopefully go some way to appeasing my CUP overall editor, David Tranah, whose patience has been tested, and to whom I am indebted for his support and views. David Hemsley took on the task of editing the manuscript, and set me numerous challenges in formatting the book. I have not managed to do all the things he wished – typesetting glitches that remain are all my fault. Production was taken on by Jayne Aldhouse's team at CUP.

To finish this book I have neglected many people, notably my patient wife Helen and my impatient son Benjamin, who is worryingly interested in numbers. It is to them that this text is dedicated.

1 Why you need complex numbers

Introduction

The complex number system is now such an accepted part of mathematical analysis that it requires some adjustment of your point of view just to ask *why* you need so-called imaginary or complex numbers. But you should understand that there was, at first, considerable resistance to their introduction, even amongst those who felt compelled to invent them! Probably the first person to discuss them was Girolamo Cardano, in his text *Ars magna* (*The Great Art*), published in 1545 (Cardano, 1993). Cardano also was one of the first Western algebraists to cope with the concept of negative numbers, and to introduce negative roots. The additional headache involved in dealing with imaginary numbers was such that he largely kept them out of his book, with the exception of a brief discussion of the solution of the quadratic equation :

$$x(10 - x) = 40 \tag{1.1}$$

Cardano did not cope terribly well with the processes involved in managing this equation – as he put it: 'putting aside the mental tortures involved, multiply $5 + \sqrt{(-15)}$ by $5 - \sqrt{(-15)}$, making $25 - (-15)$, whence the product is 40'. He went on to add: 'So progresses arithmetic subtlety the end of which, as is said, is as refined as it is useless.'

One view of this book is therefore that it is devoted to a useless topic, but it is likely that Cardano might have shifted his opinions if he could only have experienced the outcome of his mental tortures. If you wish to experience Cardano's reservations personally, his text is available in translation. Also, an excellent exposition of the history of algebra in the sixteenth century is given by Burton (1995). Cardano pushed his luck in other ways, and through his fascination for astrology he managed to get himself imprisoned for heresy. The author hastens to add that it was not so much the introduction of imaginary numbers that got him into trouble – rather more offence was taken at his publishing a horoscope of Jesus Christ. Controversy surrounded Cardano and others interested in the solution of polynomial equations – the intense competition to understand quadratic, cubic and quartic equations generated considerable rivalry!

This book is devoted to explaining why complex numbers and complex analysis are two of the most useful topics in pure and applied mathematics, physics and engineering. You need them.

1.1 First analysis of quadratic equations

If you wish to understand how complex numbers arise from simple polynomial equations with real coefficients, it is sufficient to analyse the following quadratic equation:

$$ax^2 + bx + c = 0 \tag{1.2}$$

You can rewrite this in the form

$$a(x^2 + (2\,xb)/(2\,a)) + c = 0 \tag{1.3}$$

If you then 'complete the square', you obtain

$$a(x + b/(2\,a))^2 - b^2/(4\,a) + c = 0 \tag{1.4}$$

Hence, you arrive at:

$$(x + b/(2\,a))^2 = (b^2 - 4\,ac)/(2\,a)^2 \tag{1.5}$$

As long as x is real, the left side of this equation is the square of a real number, and is therefore non-negative. The right side is non-negative if and only if

$$b^2 - 4\,ac \geq 0 \tag{1.6}$$

You can, in this case, take the ordinary square root, to obtain

$$x + b/(2\,a) = \pm\left(\sqrt{(b^2 - 4\,ac)}\right)/(2\,a) \tag{1.7}$$

$$x = \left(-b \pm \sqrt{(b^2 - 4\,ac)}\right)/(2\,a) \tag{1.8}$$

If, on the other hand,

$$b^2 - 4\,ac < 0 \tag{1.9}$$

the square root cannot be taken in the usual way. The introduction of a quantity i (*Mathematica*® uses a double-struck character for the standard mathematical representation) satisfying

$$i^2 = -1 \tag{1.10}$$

resolves the matter, since then you can write

$$x = \left(-b \pm i\sqrt{(4\,a\,c - b^2)}\right)/(2\,a) \tag{1.11}$$

Thus you obtain complex roots of an equation with real coefficients. It is another matter to understand what happens when the coefficients themselves are complex. At first you might wonder if you have to extend the number system still further to cope. It is one of the important results of complex analysis that this is unnecessary. You will discover that all polynomial equations of degree n, with coefficients that are complex numbers, have n roots that are complex numbers. In other words: *Complex numbers are enough.*

Now is a good time to get a grip on the use of *Mathematica* to solve simple equations. If you are completely new to *Mathematica* you may first wish to explore the booklet *Getting Started with Mathematica…* (see your *Mathematica* documentation kit). If you do not wish to explore the use of *Mathematica* to solve equations, skip to the next chapter.

1.2 *Mathematica* investigation: quadratic equations

If you are using this text with a computer, start the *Mathematica* system by clicking (e.g. in the MacOS X 'dock') or double-clicking (on most other systems) on the appropriate icon on your computer system. This section will give you a brief introduction on using *Mathematica* to solve simple quadratic equations. In each case you can just type in the commands given in bold-face **Courier** ('typewriter') font, and enter the command using either Shift Return, RET; Enter, ENTER; or Insert (depending on your operating system). If you are viewing this notebook from a CD-ROM or other electronic source, you can of course just browse the existing material entering some or all of the commands.

Since the solution of equations is one of your main goals, and is a running theme of this book, now is a good time to explore the **Solve** function that is built into *Mathematica*. Although the material of this section is basic, it illustrates an important and basic point about how *Mathematica* 'solves' equations by *returning replacement rules*. You should try the following examples:

```
mysolution = Solve[x^2 - 1 == 0, x]
```

$\{\{x \to -1\}, \{x \to 1\}\}$

Note that you use a double equal sign when you wish to denote *equality* in a *Mathematica* expression representing an equation. (The use of a single equal sign is reserved for assignment.) The solution to this quadratic is now contained in the *Mathematica* expression **mysolution**, and we can ask about this expression:

```
mysolution
```

$\{\{x \to -1\}, \{x \to 1\}\}$

What we really want are the values of x *given* the result contained in **mysolution**. In *Mathematica* the English word 'given' is expressed by the combination **/.** (slash dot):

```
myx = x /. mysolution
```

$\{-1, 1\}$

This gets you a list of the solutions expressed in the variable **myx**. This list contains two elements, and we can extract each one by referring to the position in the list. This is achieved by the use of double square brackets, containing the position in the list of the result of interest:

```
myx[[2]]
```

1

■ Some other real equations with purely real roots

In the following example you go directly to the list of solutions:

```
myxTwo = x /. Solve[x^2 - 2 == 0, x]
```

$\{-\sqrt{2}, \sqrt{2}\}$

We can now ask for any particular result:

```
myxTwo[[1]]
```

$-\sqrt{2}$

```
myxThree = x /. Solve[x^2 + 2 x - 2 == 0, x]
```

$\{-1-\sqrt{3}, -1+\sqrt{3}\}$

```
myxFour = x /. Solve[x^2 - x - 1 == 0, x]
```

$\{\frac{1}{2}(1-\sqrt{5}), \frac{1}{2}(1+\sqrt{5})\}$

If you wish to extract the numerical values of the solution, you use the **N[ ]** function. This can be applied by placing the expression to be numericalized in single square brackets – this is how *Mathematica* expects to see all arguments to functions, including **N[ ]**:

```
N[myxFour]
```

{−0.618034, 1.61803}

You can also apply **N[ ]** 'afterwards', using the **//N** construction:

```
myxFour // N
```

{−0.618034, 1.61803}

■ Real equations with purely imaginary roots

In the following examples, *Mathematica* extracts the purely imaginary roots from an equation with real coefficients:

```
myxFive = x /. Solve[x^2 + 1 == 0, x]
```

$\{-i, i\}$

```
myxSix = x /. Solve[x^2 + 2 == 0, x]
```

$\{-i\sqrt{2}, i\sqrt{2}\}$

```
myxSeven = x /. Solve[x^2 + 12 == 0, x]
```

$\{-2i\sqrt{3}, 2i\sqrt{3}\}$

Note that in *Mathematica*'s standard form and traditional form for expressions, the square root of -1 is denoted by i. You can always ask *Mathematica* for the traditional mathematical form of an expression by using explicit conversion to **TraditionalForm**. In most of this book the traditional form is used for output – it is more elegant than **StandardForm**.

```
StandardForm[myxSix]
```

$\{-\text{i}\sqrt{2}, \text{i}\sqrt{2}\}$

```
TraditionalForm[%]
```

$\{-i\sqrt{2}, i\sqrt{2}\}$

Note that **StandardForm** uses an upright font and **TraditionalForm** an italic font. If you are new to *Mathematica*, note that the "per cent" symbol % is a useful shortcut to the last output. In this case there is just some basic tidying up of the spacing. In the second example there is more tidying to present the output essentially as you would write it on paper. If you look under the *Mathematica* Cell Menu you will see that there are menu commands and keyboard shortcuts for converting between:

InputForm – *Mathematica*'s standard pure text representation of expressions;
StandardForm – A compromise between traditional mathematical notation and an unambiguous computer representation;
TraditionalForm – Traditional mathematical notation.

■ Real equations with complex roots

Try the following examples:

```
myxEight = x /. Solve[x^2 + x / 2 + 1 == 0, x]
```

$\{\frac{1}{4}(-1 - i\sqrt{15}), \frac{1}{4}(-1 + i\sqrt{15})\}$

```
N[myxEight]
```

$\{-0.25 - 0.968246\, i, -0.25 + 0.968246\, i\}$

It is always a good idea to check that your answers regenerate the original quadratic:

```
Expand[(x - (myxEight[[1]])) (x - (myxEight[[2]]))]
```

$$x^2 + \frac{x}{2} + 1$$

■ Complex equations with complex roots

Although you may not yet know how to treat equations that have complex coefficients, it is perfectly possible (you will read how to deal with this later, in Chapter 2); *Mathematica* does not need any further instruction. Here is a good place to note that the ‘Input Form’ of the square root of -1 is the capital **I**:

```
myxNine =
 x /. Solve[(1 + I) x^2 + (2 + I) x + 3 - 2 I == 0, x]
```

$$\left\{\left(\frac{1}{4} + \frac{i}{4}\right)\left((-1 + 2\,i) + \sqrt{17}\right), \left(-\frac{1}{4} - \frac{i}{4}\right)\left((1 - 2\,i) + \sqrt{17}\right)\right\}$$

Finally, you can consider the case we started with:

```
myxTen = x /. Solve[a x^2 + b x + c == 0, x]
```

$$\left\{\frac{-b - \sqrt{b^2 - 4\,a\,c}}{2\,a}, \frac{\sqrt{b^2 - 4\,a\,c} - b}{2\,a}\right\}$$

■ Using **Factor** (more advanced)

You have seen how to use **Solve** to solve the equation. You can also use **Factor** to work on the quadratic (or indeed any polynomial) itself, if you are interested in looking at factorizations over only the integers:

```
Factor[2 x^2 - 2]
```

$$2\,(x - 1)\,(x + 1)$$

You can get a list of the factors together with their powers as follows:

```
FactorList[2 x^2 - 2]
```

$$\begin{pmatrix} 2 & 1 \\ x - 1 & 1 \\ x + 1 & 1 \end{pmatrix}$$

However, look what happens when you try the following:

```
Factor[x^2 + 1]
```

$$x^2 + 1$$

You can examine the effect of resetting one of the options:

```
Options[Factor]
```

{Extension → None, GaussianIntegers → False, Modulus → 0, Trig → False}

If we allow Gaussian integers, that is, complex numbers whose real and imaginary parts are both integers, then a different result is obtained:

```
Factor[x^2 + 1, GaussianIntegers -> True]
```

$(x - i)(x + i)$

You can obtain further information about **Factor**, or indeed any *Mathematica* function, by prefixing the function name with a question mark:

```
? Factor
```

```
Factor[poly] factors a polynomial over the integers. Factor[poly, Modulus->
   p] factors a polynomial modulo a prime p. Factor[poly, Extension->
   {a1, a2, ... }] factors a polynomial allowing coefficients
   that are rational combinations of the algebraic numbers ai. ...
```

So, for example, if you try the following, you get nowhere:

```
Factor[x^2 + 2, GaussianIntegers -> True]
```

$x^2 + 2$

But if you allow an extension, the factorization proceeds:

```
Factor[x^2 + 2, GaussianIntegers -> True,
    Extension -> {Sqrt[2]}]
```

$\left(\sqrt{2} - i\,x\right)\left(i\,x + \sqrt{2}\right)$

You can build up lists of suitable extensions very easily. Here you make a list of the square roots of the first three primes:

```
mylist = Table[Sqrt[Prime[k]], {k, 3}]
```

$\{\sqrt{2}, \sqrt{3}, \sqrt{5}\}$

This allows you to play with higher-order polynomials:

```
Factor[(x^2 - 3) (x^2 - 5) (x^2 - 6), Extension -> mylist]
```

$-\left(\sqrt{3} - x\right)\left(\sqrt{5} - x\right)\left(\sqrt{6} - x\right)\left(x + \sqrt{3}\right)\left(x + \sqrt{5}\right)\left(x + \sqrt{6}\right)$

But now we have jumped to higher-order polynomials. It is time to develop some more theory. In the next chapter you will see how to define complex numbers properly, and you will soon be able to explore the solution of cubic, quartic and other polynomial equations.

Exercises

1.1 Using pen and paper only (i.e. not using *Mathematica)* find the solutions of the following quadratic equations:

$$x^2 - 3 = 0$$

$$x^2 + 2x + 1 = 0$$

$$x^2 + 3 = 0$$

$$x^2 + x + 4 = 0$$

1.2 If $a > 0$, what is the minimum value of the expression

$$ax^2 + bx + c$$

and for what value of x does it occur? When is this minumum value negative, and what does this imply about the roots of the quadratic equation

$$ax^2 + bx + c = 0?$$

What happens when this minumum value is zero? What happens when it is positive? Interpret your results graphically.

1.3 Suppose that x_1 and x_2 are the roots of the quadratic equation

$$ax^2 + bx + c = 0$$

By writing this in the form

$$a(x - x_1)(x - x_2) = 0$$

show that

$$x_1 x_2 = c/a$$
$$x_1 + x_2 = -b/a$$

Hence write down a quadratic equation with roots $3 + 2i$ and $1 - i$.

1.4 Let f be the function given by:

$$f(x) = \lambda x(1 - x)$$

Without using *Mathematica*, find both solutions of the quadratic equation:

$$x = f(x)$$

Hence, without using *Mathematica*, find all the solutions of the quartic equation:

$$x = f(f(x))$$

Hint: the solutions of the quadratic equation are necessarily solutions of the quartic equation.

1.5 ❀ Use *Mathematica*'s **`Solve`** function to solve the quadratic equations in Exercise 1.1, and compare your solutions.

1.6 ❀ Plot the functions x^2, $x^2 - 1$, and $x^2 + 1$, using the *Mathematica* **`Plot`** function, For example, the first of these may be plotted with:

```
Plot[x^2, {x, -2, 2}]
```

1.7 ❀ Repeat the exercise of problem 1.6, but parametrize the constant by a value **`c`**. Explore the use of an animation to see the results, by trying the following:

```
Do[Plot[x^2 + c, {x, -2, 2}, PlotRange -> {-1, 3}],
  {c, -1, 1, 0.2}]
```

What is the relationship between the quantity $b^2 - 4ac$ in this case, and the location of the curve?

1.8 ❀ Use **`Factor`** to find the real linear and quadratic factors of the expression $x^4 - 1$, and hence find all solutions of the quartic equation $x^4 - 1 = 0$.

1.9 ❀ Using the methods of Exercise 1.8, find all six solutions to the equation $x^6 - 1 = 0$.

2 Complex algebra and geometry

Introduction

In the first chapter you saw why you need imaginary and complex numbers, by considering the solution of simple quadratic equations. In this chapter you will see how we set up complex numbers in general, and establish their basic algebraic and geometrical properties.

We shall assume that you have some understanding of what is meant by a real number. The exact nature and depth of this understanding will not materially affect the discussion thoughout most of this book, and this is not a book about the fundamentals of real analysis. We should, however, take a moment to remind ourselves what a 'real' number is, before we start defining 'imaginary' and 'complex' numbers. Students of pure mathematics should remind themselves of the details of these matters – there is really nothing for it but to go for a proper mathematical definition, and experience has shown that one needs to be slightly abstract in order to get it right, in the sense that the resulting definition contains all the numbers 'we need'. For a full exposition, complete with proofs, you should consult a text on real analysis, such as that by Rudin (1976). For our purposes it will mostly be sufficient to regard real numbers as being all the points on a line (which we call the real axis) extending to infinity in both directions. This contains positive and negative integers, rational numbers, such as 1/2 and 17/15, simple square roots such as the square root of 2, and other numbers such as π and e. When we come to consider certain results about limits of infinite sequences and series, it will be necessary to call on more formal results from analysis.

2.1 Informal approach to 'real' numbers

For completeness we begin by briefly exploring the necessity of the 'real' number system. In fact, one reason for having 'real' numbers at all is the inadequacy of integers and rationals (ratios of integers) for solving equations. This parallels our need for introducing complex numbers, but at a more basic level. So we begin by considering even simpler equations than those considered in Chapter 1. For example, consider the solution of the following two equations. Although this can be done with 'pen and paper', we carry this out in *Mathematica*, asking for x given (`/.`) the results of the **`Solve`** function. First note that this function returns a list of replacement rules:

```
Solve[2 x == 16, x]
```

$\{\{x \to 8\}\}$

To get the value of x given this result we ask for:

```
x /. Solve[2 x == 16, x]
```

{8}

The first, and here only, element of this list can be picked out using the double square brackets to indicate position in the list:

```
x /. Solve[2 x == 16, x][[1]]
```

8

```
x /. Solve[2 x == 15, x][[1]]
```

$$\frac{15}{2}$$

Both of these equations have whole number, or integer, coefficients. In the first case we obtain an integer solution. In the second case we do not obtain an integer, but we get a rational. Clearly we need to consider rational numbers if we are to solve simple linear equations with either integer or rational coeffiecients. Now consider the following two equations containing rational numbers as coefficients. Here we keep the whole list rather than asking for the first element:

```
x /. Solve[x^2 == 9/4, x]
```

$$\{-\frac{3}{2}, \frac{3}{2}\}$$

```
x /. Solve[x^2 == 2, x]
```

$$\{-\sqrt{2}, \sqrt{2}\}$$

The first equation with rational coefficients has a solution that is rational, but for the second, we have obtained an answer from *Mathematica* that remains expressed with square roots. There is in fact no rational number whose square is 2. So clearly we need to extend the system of rationals to cope with such equations. How do we do so? One might imagine that one could work through various equations, generating numbers such as the square roots of 2, 3, etc., to add to our number system. Unfortunately, it is no use just trying to add, one at a time, various fractional powers such as the square and higher order roots of the rationals, for we shall miss important irrational numbers such as e and π, though these numbers can be obtained in terms of limiting processes involving sequences of rationals. This results in the need for a more abstract approach, as described, for example, by Rudin (1976).

However, although this process of ‘adding new numbers’ is not adequate for giving a proper definition of the real numbers, it serves perfectly well as a motivation for defining complex numbers, and this is the route we shall take.

2.2 Definition of a complex number and notation

Traditionally there have been several different notations for complex numbers. They are all equivalent and you should get used to using the three different traditional forms introduced below. Since we are working with *Mathematica*, which has a convention of using capital letters for special symbols (including the square root of -1) in **`InputForm`**, we need to introduce a fourth notation for our *Mathematica* investigations. Throughout this book the normal mathematical notation will be used, except for our *Mathematica* expressions, where the *Mathematica* form will be used.

■ Ordered pair notation (formal mathematical)

The formal mathematical definition treats a complex number z as an ordered pair (x, y), where x and y are ordinary real numbers. We call x the 'real part' and y the 'imaginary part'. Note that this slightly awkward terminology implies that the imaginary part is actually a real number! This is also a convenient representation for when we wish to treat complex numbers as being vectors in two-dimensional space – then x and y are just the Cartesian components of the vector.

■ $x + iy$ notation ('normal' mathematics) and *Mathematica* standard and traditional forms

In this notation a complex number is represented by an expression of the form $x + i y$, where x and y are ordinary real numbers. The quantity x is called the 'real part' and y is the 'imaginary part'. Complex numbers thus expressed are related to the ordered pair notation by the relations that x is equivalent to $(x, 0)$ and $i y$ is equivalent to $(0, y)$. This is similar to the use of **`TraditionalForm`** notation in *Mathematica*, except that *Mathematica* uses a double-struck letter, $\mathbb{i}$, instead of an italic i, or as in many texts, a standard i.

■ $x + jy$ notation (engineering)

In some branches of engineering, particularly electrical engineering, the symbol j or j is used instead of i. It can usually be used completely interchangeably, though sometimes matters of convention (usually in the treatment of time-dependent complex functions) make it useful to think of j as being $-i$, in order to make sensible comparisons between different texts. This slightly odd point arises as the world of applied mathematics, physics and engineering is divided in many ways. One division arises from whether time-harmonic functions have a time-dependence of the form $\mathbb{e}^{-i\omega t}$ or $\mathbb{e}^{+j\omega t}$. Don't blame this author! Send e-mail to anyone but me, such as to anyone writing a book about applied electromagnetics that uses a convention opposite to the one you have been taught.

■ x + Iy notation (*Mathematica* **InputForm**)

In *Mathematica* the capital letter **I** is used instead of i or $\mathbb{i}$, in **InputForm**, in order to conform to *Mathematica*'s conventions of representing all special symbols with capitalized names. Similarly π becomes **Pi** and e or $\mathbb{e}$, becomes **E**. This applies in *Mathematica* when using **InputForm**, though in **StandardForm** or **TraditionalForm** it reverts to the notation of 'normal' mathematics, but with double-struck characters. Here is an example illustrating many of the issues:

```
expr = Exp[-I 2 Pi f t];

InputForm[expr]

  E^((-2*I)*f*Pi*t)

StandardForm[expr]
```

$\mathbb{e}^{-2\,\mathbb{i}\,\mathrm{f}\,\pi\,\mathrm{t}}$

```
TraditionalForm[expr]
```

$\mathbb{e}^{-2 \mathbb{i} f \pi t}$

Note that in some older versions of *Mathematica* the capital **I** of **InputForm** is also used in **StandardForm**. Another way of converting *Mathematica* cells between the different formats is to use the commands under the 'Cell Menu'. You might want to explore this now.

2.3 Basic algebraic properties of complex numbers

We define the operations of addition, subtraction, multiplication and division in a completely natural way. Formally, we obtain a *field*, though it is important to note that it is not an ordered field, in contrast to the real field. So addition is defined by

$$(a + b\mathbb{i}) + (c + d\mathbb{i}) = a + c + (b + d)\mathbb{i} \tag{2.1}$$

Multiplication is defined in a similarly straightforward way, bearing in mind that $\mathbb{i} \times \mathbb{i} = -1$:

$$(a + b\mathbb{i}) \times (c + d\mathbb{i}) = ac - bd + (bc + ad)\mathbb{i} \tag{2.2}$$

Division of one complex number by another complex number is defined by:

$$\frac{a + bi}{c + di} = \frac{(a + b\mathbb{i})(c - d\mathbb{i})}{(c + d\mathbb{i})(c - d\mathbb{i})} = \frac{(ac + bd) + (bc - ad)\mathbb{i}}{c^2 + d^2} \tag{2.3}$$

Note that for this to work, it cannot be the case that $c = d = 0$ – in other words, division is possible by any complex number except zero. Bearing in mind these definitions, it is straightforward to verify that standard algebraic properties such as associativity and commutativity of addition and multiplication apply.

2.4 Complex conjugation and modulus

Given a complex number

$$z = x + i y \tag{2.4}$$

we define its complex conjugate $\bar{z}$, also denoted by z^*

$$\bar{z} = x - i y \tag{2.5}$$

The modulus $|z|$ of a complex number is defined by

$$|z|^2 = z\bar{z} = x^2 + y^2 \tag{2.6}$$

$$|z| = \sqrt{z\bar{z}} \tag{2.7}$$

From these definitions a number of simple relations follow. First, complex conjugation is involutory – if we do it twice we arrive back at our original complex number:

$$\bar{\bar{z}} = z \tag{2.8}$$

Furthermore, we can extract the real and imaginary parts of a complex number by the use of conjugation:

$$2\mathrm{Re}(z) = z + \bar{z} \qquad 2i\mathrm{Im}(z) = z - \bar{z} \tag{2.9}$$

It is important to realize that the imaginary part of a complex number is real – we refer to the imaginary part as y rather than as iy. Finally, we can see that the operation of taking complex conjugates commutes with addition and multiplication:

$$\overline{(w + z)} = \bar{w} + \bar{z} \qquad \overline{(z w)} = \bar{z}\,\bar{w} \tag{2.10}$$

2.5 The Wessel–Argand plane

We can regard the real and imaginary parts of a complex number as giving the Cartesian coordinates of a point in a two-dimensional plane – the Wessel–Argand plane. This concept was developed independently by C. Wessel in 1797 and J. R. Argand in 1806 (see Penrose, 2004), but has become more commonly known as the *Argand plane*. We shall use this terminology from here on. So the complex number $z = x + i y$ is regarded as the point with coordinates (x, y). We identify the set of all complex numbers $\mathbb{C}$ with the two-dimensional real plane $\mathbb{R}^2$. Furthermore, we can give natural geometrical interpretations to some of the operations on complex numbers that we have introduced. We can look at some of these right away – others we shall defer until later.

First, the modulus $|z|$ gives the length of the vector (x, y). Second, the complex conjugate $\bar{z}$ is associated with the point with coordinates $(x, -y)$. In the following diagram the point $z = 3 + 2i$ is shown, together with its complex conjugate. The *Mathematica* code for generating the diagram is also shown – you may ignore it or modify it to show other complex numbers in Argand form, if you are using the electronic form of this text with *Mathematica*.

This is a good point to note how to write the double-struck i in Input Form. It is obtained by typing: ESC ii ESC

```
argandata = {PointSize[0.03], Point[{3, 2}],
   Point[{3, -2}], Line[{{0, 0}, {3, 2}}],
   Line[{{-5, 0}, {5, 0}}], Line[{{0, -5}, {0, 5}}],
Text["3+2i", {4.5, 2}], Text["3-2i", {4.5, -2}],
Text["x", {5, -1/2}], Text["y", {-1/2, 5}]};
Show[Graphics[argandata],
 PlotRange -> {{-6, 6}, {-6, 6}}, AspectRatio -> 1]
```

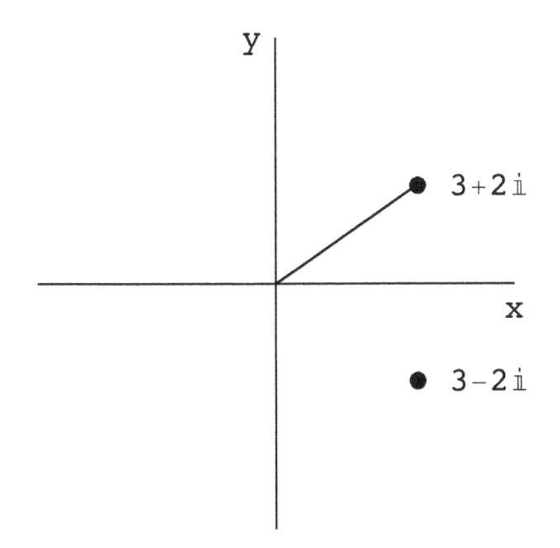

2.6 Cartesian and polar forms

Given a complex number in the form

$$z = x + iy \tag{2.11}$$

you have already seen how to define the modulus,

$$r = |z| \tag{2.12}$$

Given such a point (x, y) in the Argand plane, you can describe it in various coordinate systems. Instead of Cartesian (x, y) coordinates, you can use polar coordinates, (r, θ), related to Cartesian coordinates by the relations

$$x = r\cos(\theta)$$
$$y = r\sin(\theta) \tag{2.13}$$

In these coordinates, the complex number may be written

$$z = r\cos(\theta) + i\, r\sin(\theta) \;=\; r(\cos(\theta) + i\sin(\theta)) \tag{2.14}$$

This is the *polar form* of a complex number. The quantity θ is called the *argument* of z.

```
argandata = {PointSize[0.03], Point[{3, 2}],
Line[{{0, 0}, {3, 2}}],
   Line[{{-5, 0}, {5, 0}}], Line[{{0, -5}, {0, 5}}],
Circle[{0, 0}, 2, {0, ArcTan[2 / 3]}],
Text["3+2ⅈ", {4.2, 2}],
Text["θ", {3 / 2, 1 / 2}],
   Text["x", {5, -1 / 2}], Text["y", {-1 / 2, 5}]};
Show[Graphics[argandata], PlotRange -> {{-6, 6}, {-6, 6}},
 AspectRatio -> 1]
```

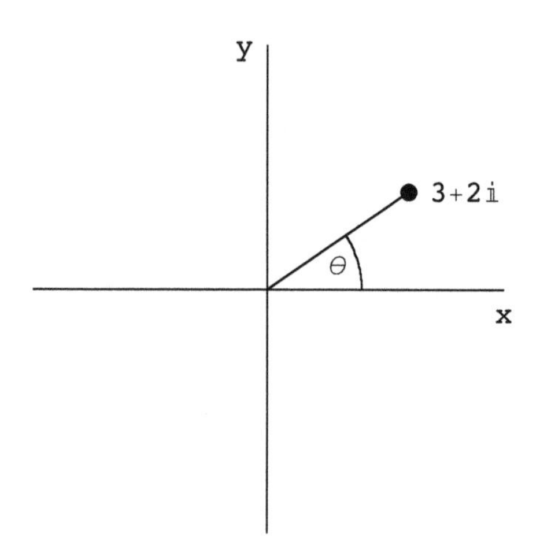

■ Issues surrounding the definition of the argument

The argument, θ, of a complex number is not uniquely defined by the formula

$$z = r(\cos(\theta) + i\sin(\theta)) \tag{2.15}$$

since you can add any multiple of 2π to θ and preserve this relation. This is the simplest example of a phenomenon that affects functions of complex variables in many important ways. In the present situation you just need to know how to define the argument unambiguously. The first step in doing so is to set a *principal range* for the argument. Two principal ranges are in common use, but it is important to understand that others may be defined. The two most commonly employed are to demand that either

$$0 \leqslant \theta < 2\pi \tag{2.16}$$

or that

$$-\pi < \theta \leqslant \pi \tag{2.17}$$

A value of θ satisfying one of these inequalities is called a *principal value* of the argument. Such a value can be denoted by a subscript P – you write θ_P for such a value. If such a value has been defined through the choice of a principal range, then any other valid value may be written

$$\theta = \theta_P + 2\pi k \tag{2.18}$$

for a positive or negative integer k.

■ Arguments of products and ratios

Suppose you have two complex numbers

$$z_1 = r_1(\cos(\theta) + i\sin(\theta)) \tag{2.19}$$

$$z_2 = r_2(\cos(\phi) + i\sin(\phi)) \tag{2.20}$$

Consider first the product:

$$z_1 z_2 = r_1(\cos(\theta) + i\sin(\theta))\, r_2(\cos(\phi) + i\sin(\phi)) \tag{2.21}$$

On expansion of the product, you obtain:

$$z_1 z_2 = r_1 r_2(\cos(\theta)\cos(\phi) - \sin(\theta)\sin(\phi) + i(\cos(\theta)\sin(\phi) + \sin(\theta)\cos(\phi))) \tag{2.22}$$

or, using a trigonometric formula:

$$z_1 z_2 = r_1 r_2(\cos(\theta + \phi) + i\sin(\theta + \phi)) \tag{2.23}$$

So the modulus of the product is the product of the moduli, and the argument of the product is the *sum* of the arguments. A similar calculation gives

$$\frac{z_1}{z_2} = \frac{r_1}{r_2}(\cos(\theta - \phi) + i\sin(\theta - \phi)) \tag{2.24}$$

so that the argument of the ratio is the difference of the arguments.

■ *Mathematica* functions for Cartesian and polar forms

If you are using *Mathematica* to manipulate simple complex numbers, you will find it useful to note the following functions, which are quite obvious in the way they work. To find the argument of a complex number, you use the **`Arg`** function:

```
Arg[3 + 2 I]
```

$$\tan^{-1}\left(\frac{2}{3}\right)$$

This leaves the answer in exact trigonometric form, which will be simplified if possible:

```
Arg[2 + 2 I]
```

$$\frac{\pi}{4}$$

This function uses the range $-\pi < \text{Arg} \leqslant \pi$ as the principal range. You can check this by asking for the argument of some values scattered near the negative real axis:

```
{Arg[-1 - 0.01 I], Arg[-1], Arg[-1 + 0.01 I]}
```

$\{-3.13159, \pi, 3.13159\}$

This is a good place to illustrate the nice way in which many *Mathematica* functions behave when acting on lists. A very useful shorthand is to write

```
Arg[{-1 - 0.01 I, -1, -1 + 0.01 I}]
```

$\{-3.13159, \pi, 3.13159\}$

This works for any function that has the attribute of 'being listable', and this can be checked by applying the **Attributes** function:

```
Attributes[Arg]
```

{Listable, NumericFunction, Protected}

The modulus of a complex number is found by using the **Abs** function, an abbreviation for absolute value:

```
Abs[3 + 2 I]
```

$\sqrt{13}$

The real and imaginary parts are extracted by use of the functions **Re** and **Im**, while complex conjugates are taken by the use of the **Conjugate** function:

```
Re[3 + 2 I]
```

3

```
Im[3 + 2 I]
```

2

```
Conjugate[3 + 2 I]
```

$3 - 2i$

Note that these are the representation of the functions in **StandardForm**. If you convert them to **TraditionalForm**, you obtain a representation that looks more familiar.

```
mylist = {Re[z], Im[z], Abs[z], Arg[z], Conjugate[z]};
TraditionalForm[mylist]
```

$\{\mathrm{Re}(z), \mathrm{Im}(z), |z|, \mathrm{Arg}(z), z^*\}$

Note that older versions of *Mathematica* do not use the star superscript for complex conjugates – rather the word 'Conjugate' is used in full. (The star was introduced relatively recently in version 5.1 of the software.)

■ The ComplexExpand function

There is a very useful general purpose function in *Mathematica* that extracts real and imaginary parts of expressions on the assumption that all symbols in the expression represent real variables. You can query *Mathematica* about its operation in the usual way:

```
? ComplexExpand
```

```
ComplexExpand[expr] expands expr assuming that all variables
   are real. ComplexExpand[expr, {x1, x2, ... }] expands expr
   assuming that variables matching any of the xi are complex. ...
```

At first sight, it gives results that are barely distinguishable from the ordinary **Expand** function:

```
ComplexExpand[(x + I y)^3]
```

$x^3 - 3y^2 x + i(3x^2 y - y^3)$

```
Expand[(x + I y)^3]
```

$x^3 + 3iyx^2 - 3y^2 x - iy^3$

However, the ability to retain some symbols as complex, and to set options to change the way the function works, gives some useful capabilities. First, we keep z as complex, and work out the real part of the cube of z:

```
ComplexExpand[Re[z^3], {z}]
```

$\mathrm{Re}(z)^3 - 3\,\mathrm{Im}(z)^2\,\mathrm{Re}(z)$

It is a good idea to explore the options of this function:

```
Options[ComplexExpand]
```

{TargetFunctions → {Re, Im, Abs, Arg, Conjugate, Sign}}

By changing the target function, you can convert to polar coordinates:

```
ComplexExpand[Re[z^3], {z},
 TargetFunctions -> {Abs, Arg}]
```

$|z|^3 \cos^3(\text{Arg}(z)) - 3\,|z|^3 \cos(\text{Arg}(z)) \sin^2(\text{Arg}(z))$

Alternatively, the expression can be simplified to a function of z and its complex conjugate:

```
ComplexExpand[Re[z^3],
 {z}, TargetFunctions -> Conjugate]
```

$$\frac{z^3}{2} + \frac{(z^*)^3}{2}$$

We shall revisit **ComplexExpand** later - it is particularly powerful when dealing with functions of complex variables, as the following simple example illustrates:

```
ComplexExpand[Sin[x + I y]]
```

$\cosh(y)\sin(x) + i\cos(x)\sinh(y)$

▪ *Mathematica*'s internal forms

A technical point that is sometimes useful is to note that *Mathematica* uses different names to represent various types of number, or, more generally, expressions. The name is called the **Head**:

```
Head[3 + 2 I]
```

Complex

```
Head[Re[3 + 2 I]]
```

Integer

```
Head[3. + 2. I]
```

Complex

```
Head[Re[3. + 2. I]]
```

Real

So *Mathematica* uses different forms of **Head** for different types of number. This is related to the way *Mathematica* represents such quantities internally. The internal representation may be extracted by the use of **FullForm**, which reveals a complex number as an object 'Complex[x, y]':

```
FullForm[2 + 3 I]
```

Complex[2, 3]

If you work with lists of complex numbers, **FullForm** reveals still more internal structure:

```
FullForm[{2 + 3 I, 4 + 5 I}]
```

List[Complex[2, 3], Complex[4, 5]]

2.7 DeMoivre's theorem

DeMoivre's theorem is fundamental to understanding how to take powers and roots of complex numbers. By implication it is central to solving quadratic equations with complex coefficients, since it is then required to take square roots of a complex number. The theorem states that for any real number α,

$$(\cos(\theta) + i\sin(\theta))^{\alpha} = \cos(\alpha\,\theta) + i\sin(\alpha\,\theta) \tag{2.25}$$

Once you understand the exponential form of complex numbers, this theorem will become completely obvious. However, it is useful to see how to prove and apply it without such knowledge, in various cases. The simplest and most important case is for positive integers, where $\alpha = n > 1$. Once you see the case $n = 2$ it becomes clear how to generalize via induction

$$\begin{aligned}(\cos(\theta) + i\sin(\theta))^2 \\ &= \cos^2(\theta) - \sin^2(\theta) + 2\,i\sin(\theta)\cos(\theta) \\ &= \cos(2\,\theta) + i\sin(2\,\theta)\end{aligned} \tag{2.26}$$

where the last line is obtained by application of the elementary trigonometric identities.

■ Inductive proof for positive integers

Suppose that the theorem is true for $n = k$:

$$(\cos(\theta) + i\sin(\theta))^k = \cos(k\theta) + i\sin(k\theta) \tag{2.27}$$

So you deduce that

$$(\cos(\theta) + i\sin(\theta))^{k+1} = (\cos(\theta) + i\sin(\theta))\,(\cos(k\theta) + i\sin(k\theta)) \tag{2.28}$$

$$= \cos(k\theta)\cos(\theta) - \sin(k\theta)\sin(\theta) + i\,(\cos(\theta)\sin(k\theta) + \cos(k\theta)\sin(\theta)) \tag{2.29}$$

$$= \cos((k+1)\theta) + i\sin((k+1)\theta) \tag{2.30}$$

by the application of trigonometric addition formulae. So the result is established for $k+1$. Since the result holds for $k = 2$, it holds for all positive integers, by induction.

■ Proof for negative integers

The corresponding result for negative integers follows by some elementary manipulations. First, let k be a positive integer, and note that

$$(\cos(\theta) + i\sin(\theta))^{-k} = \frac{1}{(\cos(\theta) + i\sin(\theta))^k} = \frac{1}{\cos(k\theta) + i\sin(k\theta)} \tag{2.31}$$

Now note that

$$\frac{1}{\cos(k\theta) + i\sin(k\theta)} = \frac{1}{\cos(k\theta) + i\sin(k\theta)}\cos(k\theta) - \frac{i\sin(k\theta)}{\cos(k\theta) - i\sin(k\theta)} \tag{2.32}$$

and this, on expanding the denominator and noting that

$$\cos^2(k\theta) + \sin^2(k\theta) = 1 \tag{2.33}$$

becomes

$$\cos(k\theta) - i\sin(k\theta) = \cos(-k\theta) + i\sin(-k\theta) \tag{2.34}$$

So it has been shown that:

$$(\cos(\theta) + i\sin(\theta))^{-k} = \cos(-k\theta) + i\sin(-k\theta) \tag{2.35}$$

which establishes the desired result.

■ Proof for rational powers

Now let p and q be integers, and consider:

$$\left(\cos\left(\frac{p\theta}{q}\right) + i\sin\left(\frac{p\theta}{q}\right)\right)^q = \cos(p\theta) + i\sin(p\theta) = (\cos(\theta) + i\sin(\theta))^p \tag{2.36}$$

This is true by DeMoivre's theorem for a positive integer, applied twice, first with q, then with p. So, taking the qth root of both sides,

$$\cos\left(\frac{p\theta}{q}\right) + i\sin\left(\frac{p\theta}{q}\right) = (\cos(\theta) + i\sin(\theta))^{p/q} \tag{2.37}$$

You will see shortly that there are other values of the qth roots of a complex number.

■ Classical trigonometric formulae

There are many applications of DeMoivre's theorem. Two of the most straightforward concern the development of trigonometric formulae relating trigonometric functions of multiple angles to powers of trigonometric functions. The arguments can be developed in two directions.

■ Multiple angles to powers

An example will suffice to illustrate the general principle. How do you express $\cos(4\,\theta)$ in terms of the pair of functions $\cos(\theta)$ and $\sin(\theta)$, or indeed just in terms of $\cos(\theta)$? Now, by DeMoivre's theorem:

$$\cos(4\theta) + i\sin(4\theta) = (\cos(\theta) + i\sin(\theta))^4 \tag{2.38}$$

Expanding the right side by the binomial theorem gives:

$$\cos^4(\theta) + 4\,i\sin(\theta)\cos^3(\theta) - 6\cos^2(\theta)\sin^2(\theta) - 4\,i\cos(\theta)\sin^3(\theta) + \sin^4(\theta) \tag{2.39}$$

Taking the real parts of this gives

$$\cos(4\theta) = \cos^4(\theta) + \sin^4(\theta) - 6\cos^2(\theta)\sin^2(\theta) \tag{2.40}$$

You can now write $\sin^2(\theta) = 1 - \cos^2(\theta)$ and expand the result, to obtain

$$\cos(4\theta) = 1 - 8\cos^2(\theta) + 8\cos^4(\theta) \tag{2.41}$$

■ Powers to multiple angles

The question here is how to express a power of $\cos(\theta)$ or $\sin(\theta)$, or indeed, a polynomial involving both, to sums of terms of the form $\cos(k\theta)$ and $\sin(k\theta)$, involving no powers, and just multiple angles. It is useful to be able to do this, for example, to compute integrals of polynomials involving trigonometric functions. Some of the manipulations involved in this will appear later also, when you see how to do integration using contour integrals and holomorphic functions. For now it is just an algebraic trick.

$$z = \cos(\theta) + i\sin(\theta) \tag{2.42}$$

$$\frac{1}{z} = \frac{1}{\cos(\theta) + i\sin(\theta)} = \cos(\theta) - i\sin(\theta) \tag{2.43}$$

If you take nth powers of both of these relations, and apply DeMoivre's theorem, you obtain:

$$z^n = \cos(n\,\theta) + i\sin(n\,\theta) \tag{2.44}$$

$$\frac{1}{z^n} = \cos(n\,\theta) - i\sin(n\,\theta) \tag{2.45}$$

Now, adding and subtracting these results gives, first,

$$z + \frac{1}{z} = 2\cos(\theta) \tag{2.46}$$

$$z - \frac{1}{z} = 2\,i\sin(\theta) \tag{2.47}$$

Secondly, you get, for general n:

$$z^n + \frac{1}{z^n} = 2\cos(n\theta) \tag{2.48}$$

$$z^n - \frac{1}{z^n} = 2\,i\sin(n\theta) \tag{2.49}$$

Now consider an example. On the one hand, we have

$$\left(z + \frac{1}{z}\right)^4 = (2\cos(\theta))^4 = 16\cos^4(\theta) \tag{2.50}$$

By the binomial theorem, we also have

$$\left(z + \frac{1}{z}\right)^4 = z^4 + 4z^2 + \frac{4}{z^2} + \frac{1}{z^4} + 6 = \left(z^4 + \frac{1}{z^4}\right) + 4\left(z^2 + \frac{1}{z^2}\right) + 6 \tag{2.51}$$

Now, using the results obtain so far, the right side can be written as

$$2\cos(4\theta) + 8\cos(2\theta) + 6 \tag{2.52}$$

Hence, equating the two forms, we have

$$16\cos^4(\theta) = 2\cos(4\theta) + 8\cos(2\theta) + 6 \tag{2.53}$$

or

$$\cos^4(\theta) = \frac{1}{8}(4\cos(2\theta) + \cos(4\theta) + 3) \tag{2.54}$$

This makes it easy to integrate a function like $\cos^4(\theta)$, because the right side is a sum of terms, each of which gives a simple integration problem.

■ *Mathematica* development of trigonometric formulae

There are several functions built into *Mathematica* that perform operations related to those just discussed. The conversion of multiple angles to powers can be partially accomplished using either **TrigExpand**, or **Expand** with the option **Trig -> True**:

```
TrigExpand[Cos[4 θ]]
```

$\cos^4(\theta) - 6\sin^2(\theta)\cos^2(\theta) + \sin^4(\theta)$

The operation can be completed by asking *Mathematica* to expand the result, given a supplementary rule about how powers of $\sin(\theta)$ are to be treated:

```
Expand[% /. Sin[θ]^k_ -> (1 - Cos[θ]^2)^(k/2)]
```

$$8\cos^4(\theta) - 8\cos^2(\theta) + 1$$

```
Expand[Cos[4 θ], Trig -> True]
```

$$\cos^4(\theta) - 6\sin^2(\theta)\cos^2(\theta) + \sin^4(\theta)$$

The opposite operation can be effected with the **TrigReduce** function. You can apply this to the two examples considered above:

```
TrigReduce[1 - 8 Cos[θ]^2 + 8 Cos[θ]^4]
```

$$\cos(4\,\theta)$$

```
TrigReduce[Cos[θ]^4]
```

$$\frac{1}{8}(4\cos(2\,\theta) + \cos(4\,\theta) + 3)$$

Another function that is sometimes useful is **TrigFactor**:

```
TrigFactor[Cos[4 θ]]
```

$$(\cos(2\,\theta) - \sin(2\,\theta))(\cos(2\,\theta) + \sin(2\,\theta))$$

```
? TrigFactor
```

```
TrigFactor[expr] factors trigonometric functions in expr. ...
```

2.8 Complex roots

We now come to one of the most important basic topics in the theory of complex numbers – how to find the n nth roots of a complex number. That is, the task is, given a complex number w, to find all n solutions for z of the equation

$$z^n = w \tag{2.55}$$

Let's begin by starting with a simple case, where $n = 2$.

■ Square roots

Suppose that w is a given complex number and we wish to solve the equation

$$z^2 = w \tag{2.56}$$

for z. First, we write w in polar form:

$$w = r(\cos(\theta) + i\sin(\theta)) \tag{2.57}$$

DeMoivre's theorem allows us to find one solution for z right away:

$$z = \sqrt{r}\left(\cos\left(\frac{\theta}{2}\right) + i\sin\left(\frac{\theta}{2}\right)\right) \tag{2.58}$$

In the case of the square root, you can probably guess how to find the other root:

$$z = -\sqrt{r}\left(\cos\left(\frac{\theta}{2}\right) + i\sin\left(\frac{\theta}{2}\right)\right) \tag{2.59}$$

But there is another way of thinking about this second solution that allows you to generalize to nth roots. For any integer value of k, we can use the ambiguity in the definition of the argument to write

$$w = r(\cos(\theta + 2k\pi) + i\sin(\theta + 2k\pi)) \tag{2.60}$$

By applying DeMoivre's theorem to this expression, you obtain

$$z = \sqrt{r}\left(\cos\left(\frac{\theta}{2} + k\pi\right) + i\sin\left(\frac{\theta}{2} + k\pi\right)\right) \tag{2.61}$$

where $\sqrt{r}$ is the ordinary positive square root of the positive real number r. Now, if k is even, we get the same value of z, whereas if it is odd we get the other root. To see this, note, for example, that

$$\cos(\alpha + \pi) = -\cos(\alpha) \tag{2.62}$$

and

$$\sin(\alpha + \pi) = -\sin(\alpha) \tag{2.63}$$

So it is sufficient to consider the cases $k = 0, 1$ only, to extract both square roots. This is the pattern we shall generalize to nth roots.

■ *n*th roots

Suppose that w is a given complex number and we wish to solve the equation

$$z^n = w \tag{2.64}$$

for z. First, we write w in polar form:

$$w = r(\cos(\theta + 2k\pi) + i\sin(\theta + 2k\pi)) \tag{2.65}$$

where k is any integer. By applying DeMoivre's theorem to this expression, you obtain:

$$z = r^{1/n}\left(\cos\left(\frac{\theta}{n} + \frac{2k\pi}{n}\right) + i\sin\left(\frac{\theta}{n} + \frac{2k\pi}{n}\right)\right) \tag{2.66}$$

This time, as you vary k, you can find n distinct complex numbers, all of which are roots of the number w. It is sufficient to consider

$$k = \{0, 1, 2, \ldots, n-1\} \tag{2.67}$$

To get a better feel for how this works, it is a good idea to consider the simplest case beyond square roots, and where $w = 1$.

■ Cube and *n*th roots of unity

The cube roots of unity are given by the list of possible values of:

$$z = \left(\cos\left(\frac{2k\pi}{3}\right) + i\sin\left(\frac{2k\pi}{3}\right)\right), \quad k = 0,\ 1,\ 2 \tag{2.68}$$

Working this out gives

$$\left\{1,\ -\frac{1}{2} + \frac{i\sqrt{3}}{2},\ -\frac{1}{2} - \frac{i\sqrt{3}}{2}\right\} \tag{2.69}$$

The first *complex* element of this list is sometimes treated as special, and given the name ω:

$$\omega = \frac{1}{2}\left(-1 + i\sqrt{3}\right) \tag{2.70}$$

Note that

$$\omega^3 = 1 \tag{2.71}$$

and that, by virtue of the identity

$$(1-\omega)(1+\omega+\omega^2) = 1 - \omega^3 = 0 \tag{2.72}$$

we have the result that the sum of the three cube roots of unity is zero:

$$1 + \omega + \omega^2 = 0 \tag{2.73}$$

Note also that the modulus of ω is unity. The quantity ω and its powers lie at the corners of a triangle in the complex plane, with the vertices of the triangle on the unit circle. You can easily visualize this, using the following plot:

```
CPlot[z_List] :=
Module[{r}, r = Map[{Re[#], Im[#]} &, z];
    ParametricPlot[{Cos[θ], Sin[θ]}, {θ, 0, 2 π},
        AspectRatio -> 1,
   PlotRange -> {{-1.1, 1.1}, {-1.1, 1.1}},
        PlotRegion -> {{0.05, 0.95}, {0.05, 0.95}},
        Epilog -> {PointSize[0.05], Map[Point, r]}]]
```

```
ω = 1/2 (-1 + ⅈ √3);
```

```
CPlot[{1, ω, ω^2}]
```

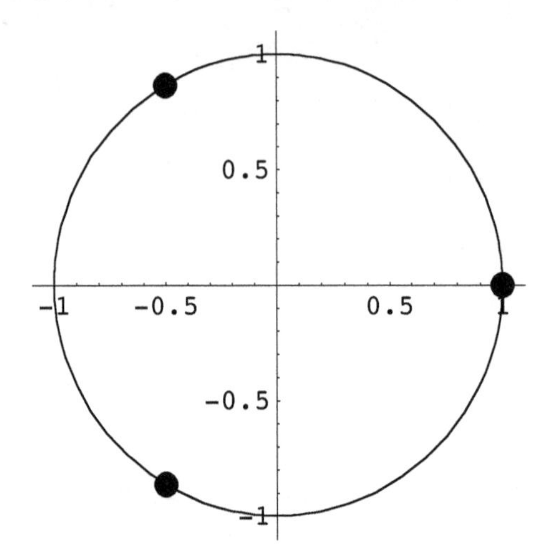

The behaviour of higher-order roots is similar. Here is a visualization of the n nth roots of unity, implemented in *Mathematica* :

```
ShowNthRoots[n_] := Module[{ω = Cos[2*Pi/n] +
I*Sin[2*Pi/n]},
CPlot[Table[ω^k, {k, 0, n - 1}]]];
```

For example, here are the seven 7th roots of unity:

```
ShowNthRoots[7]
```

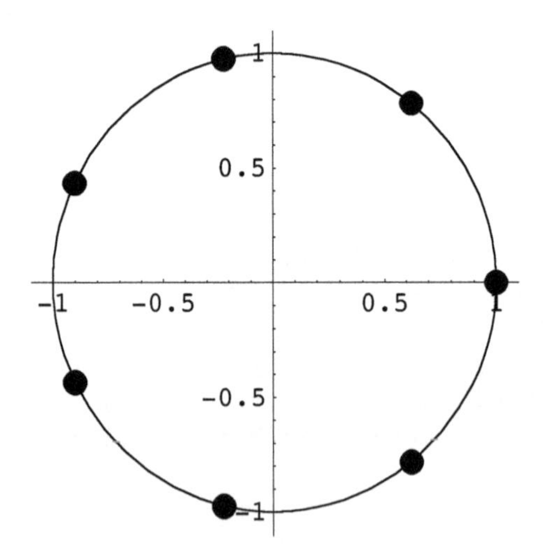

Here are the twenty-seven 27th roots of unity!

```
ShowNthRoots[27]
```

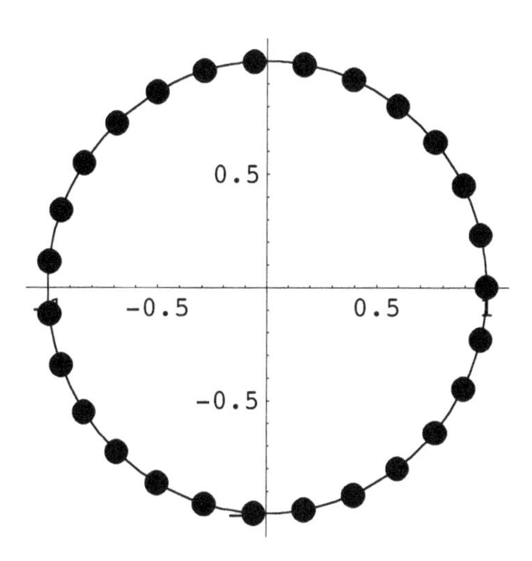

2.9 The exponential form for complex numbers

The purpose of this section is to establish the relationship:

$$e^{i\theta} = \cos(\theta) + i\sin(\theta) \tag{2.74}$$

There are a few different ways of understanding how one can arrive at this equation. Each of them relies on some knowledge from other areas of mathematics. Perhaps the simplest is to write down the Taylor series for the three functions involved. We may as well ask *Mathematica* to generate the relevant series for us, using the **Series** command. You might like to use the general Taylor series formula to check the following. First, let's take a look at the series for the exponential function:

```
Series[Exp[x], {x, 0, 10}]
```

$$1 + x + \frac{x^2}{2} + \frac{x^3}{6} + \frac{x^4}{24} + \frac{x^5}{120} + \frac{x^6}{720} + \frac{x^7}{5040} + \frac{x^8}{40320} + \frac{x^9}{362880} + \frac{x^{10}}{3628800} + O(x^{11})$$

Next, the series for the trig functions:

```
Series[Cos[x], {x, 0, 10}]
```

$$1 - \frac{x^2}{2} + \frac{x^4}{24} - \frac{x^6}{720} + \frac{x^8}{40320} - \frac{x^{10}}{3628800} + O(x^{11})$$

```
Series[Sin[x], {x, 0, 10}]
```

$$x - \frac{x^3}{6} + \frac{x^5}{120} - \frac{x^7}{5040} + \frac{x^9}{362880} + O(x^{11})$$

Now consider the series for the complex combination:

```
Series[Cos[x] + I*Sin[x], {x, 0, 10}]
```

$$1 + i x - \frac{x^2}{2} - \frac{i x^3}{6} + \frac{x^4}{24} + \frac{i x^5}{120} - \frac{x^6}{720} - \frac{i x^7}{5040} + \frac{x^8}{40320} + \frac{i x^9}{362880} - \frac{x^{10}}{3628800} + O(x^{11})$$

This is compared with the series for the exponential function with an imaginary argument:

```
Series[Exp[I*x], {x, 0, 10}]
```

$$1 + i x - \frac{x^2}{2} - \frac{i x^3}{6} + \frac{x^4}{24} + \frac{i x^5}{120} - \frac{x^6}{720} - \frac{i x^7}{5040} + \frac{x^8}{40320} + \frac{i x^9}{362880} - \frac{x^{10}}{3628800} + O(x^{11})$$

The terms displayed are identical. Once the pattern of the denominators is seen as factorials, it becomes clear that the relationship is exact. Another approach is to develop some differential equations. If we use the rule

$$\frac{de^y}{dy} = e^y \tag{2.75}$$

together with the chain rule, it follows that

$$\frac{de^{i\theta}}{d\theta} = i e^{i\theta} \tag{2.76}$$

Differentiating again, if follows that

$$\frac{de^{i\theta}}{d\theta^2} = -e^{i\theta} \tag{2.77}$$

Now the *general* solution of the equation

$$\frac{d^2 g(\theta)}{d\theta^2} = -g(\theta) \tag{2.78}$$

is given by

$$g(\theta) = a\cos(\theta) + b\sin(\theta) \tag{2.79}$$

for some constants a and b. But if we also impose initial conditions of the form

$$g(0) = 1,\ g'(0) = i \tag{2.80}$$

it follows that $a = 1$ and $b = i$. In summary, however you derive it, the relationship leads to the statement that

$$\cos(\theta) + i\sin(\theta) = e^{i\theta} \tag{2.81}$$

and so a general complex number can be written as

$$z = r(\cos(\theta) + i\sin(\theta)) = r\,e^{i\theta} = (|z|)\,e^{i\,\text{Arg}(z)} \tag{2.82}$$

This has several consequences. In particular

$$1 = e^{i\,0} = e^{2\pi i} = e^{-2\pi i} = e^{4\pi\,i} \tag{2.83}$$

and so on. Furthermore, we see that

$$e^{i\pi} = -1 \Longleftrightarrow e^{i\pi} + 1 = 0 \tag{2.84}$$

This relationship ties together, in one simple and rather mysterious equation, the five quantities e, i, π, 0, 1!

■ DeMoivre and roots revisited

Many formulae involving the arguments of complex numbers become 'obvious' when the exponential form is used. You should review, for example, the formulae for the products and ratios of complex numbers. DeMoivre's theorem bcomes particularly transparent in the exponential form:

$$(\cos(\theta) + i\sin(\theta))^n = (e^{i\theta})^n = e^{i\,n\theta} = \cos(n\theta) + i\sin(n\theta) \tag{2.85}$$

Next, suppose that

$$w = r\,e^{i\theta} = r\,e^{i(\theta+2k\pi)} \tag{2.86}$$

Then the n nth roots of w can be constructed in the form:

$$w^{1/n} = r^{1/n}\,e^{i(\theta/n+2k\pi/n)} \tag{2.87}$$

With k varying from 0 to $n-1$, all n roots are generated.

■ A first look at the exponential and logarithm functions

You can now see how to define the exponential function for a general complex number, by requiring that the property

$$e^{z_1+z_2} = e^{z_1}\,e^{z_2} \tag{2.88}$$

must hold for complex numbers, as it does for real ones. If we write

$$z = x + i y \tag{2.89}$$

then

$$e^z = e^{x+i\,y} = e^x\, e^{i\,y} = e^{x(\cos(y)+i\sin(y))} \tag{2.90}$$

You will see a more extensive discussion of the logarithm function and its properties later. For now, note that we can reorganize the exponential form for a complex number into the relation

$$z = |z|\, e^{i\,\mathrm{Arg}(z)} = e^{\log(|z|)}\, e^{i\,\mathrm{Arg}(z)} = e^{\log(|z|)+i\,\mathrm{Arg}(z)} \tag{2.91}$$

So, given z, if we seek a number w such that

$$z = e^w \tag{2.92}$$

one solution for w is

$$w = \log(|z|) + i\,\mathrm{Arg}(z) \tag{2.93}$$

This is identified as the logarithm of the complex number z:

$$\log(z) = \log(|z|) + i\,\mathrm{Arg}(z) \tag{2.94}$$

Note that the logarithm inherits the same ambiguity as the argument. This is the source of several interesting and useful properties of the log function that you will revisit many times.

2.10 The triangle inequalities

There are various properties relating to complex numbers that carry over from two-dimensional vector algebra, given the realization of a complex number as a point in the Argand plane. An important result is the triangle inequality. Geometrically, this is the result that the length of any one side of a triangle is less than or equal to the sums of the lengths of the other two sides. We can prove this inequality directly in complex number form. There are two versions. Both stem from the observation that for any complex number:

$$\mathrm{Re}(z) \leqslant |z| \tag{2.95}$$

Now apply this to a pair of complex numbers:

$$\mathrm{Re}(z_1\, \overline{z_2}) \leqslant |z_1\, \overline{z_2}| = |z_1|\,|z_2| \tag{2.96}$$

Now consider

$$\begin{gathered}|z_1 + z_2|^2 = (z_1 + z_2)\,\overline{(z_1 + z_2)} = |z_1|^2 + |z_2|^2 + \overline{z_2}\, z_1 + \overline{z_1}\, z_2 = \\ |z_1|^2 + |z_2|^2 + 2\,\mathrm{Re}(z_1\, \overline{z_2}) \leqslant |z_1|^2 + 2\,|z_2|\,|z_1| + |z_2|^2 = (|z_1| + |z_2|)^2\end{gathered} \tag{2.97}$$

That is,

$$|z_1 + z_2|^2 \leqslant (|z_1| + |z_2|)^2 \tag{2.98}$$

so that, since both sides are positive:

$$|z_1 + z_2| \leqslant |z_1| + |z_2| \tag{2.99}$$

This is the fundamental triangle inequality. You can extract another related inequality by repeating the argument with a minus sign:

$$\begin{aligned} |z_1 - z_2|^2 &= (z_1 - z_2)\overline{(z_1 - z_2)} = |z_1|^2 + |z_2|^2 - \overline{z_2}\, z_1 + \overline{z_1}\, z_2 = \\ &|z_1|^2 + |z_2|^2 - 2\,\mathrm{Re}(z_1\, \overline{z_2}) \geqslant |z_1|^2 + |z_2|^2 - 2\,|z_2|\,|z_1| = (|z_1| - |z_2|)^2 \end{aligned} \tag{2.100}$$

Now we do not know which modulus is bigger, so the strongest statement we can make is that

$$||z_1| - |z_2|| \leqslant |z_1 - z_2| \tag{2.101}$$

2.11 *Mathematica* visualization of complex roots and logs

An elegant way of understanding the behaviour of roots is to consider plotting a root of z as z wanders through the complex plane. We shall do this by just plotting the real part of an nth root of z as z varies in a disc around the origin. In polar coordinates, we have a function

$$f(r, \theta) = r^{1/n} \cos\left(\frac{\theta}{n}\right) \tag{2.102}$$

We let θ vary from 0 to $2\pi n$ to get all values of f. These plots show how a multi-valued surface is generated. As one loops around the origin in units of 2π, one jumps from one root to the next.

```
ViewRootSurface[n_Integer, resolution_Integer] :=
ParametricPlot3D[{r * Cos[theta],
   r * Sin[theta], r^(1 / n) * Cos[theta / n]},
{r, 0, 2}, {theta, 0, 2 * n * Pi},
PlotPoints -> {resolution, resolution * n},
Boxed -> False, Axes -> False,
  AspectRatio → 1, ViewPoint -> {-3, -3, 0}]
```

Here is the plot for a square root ($n = 2$), showing the two-valued nature of the square-root function:

```
ViewRootSurface[2, 20]
```

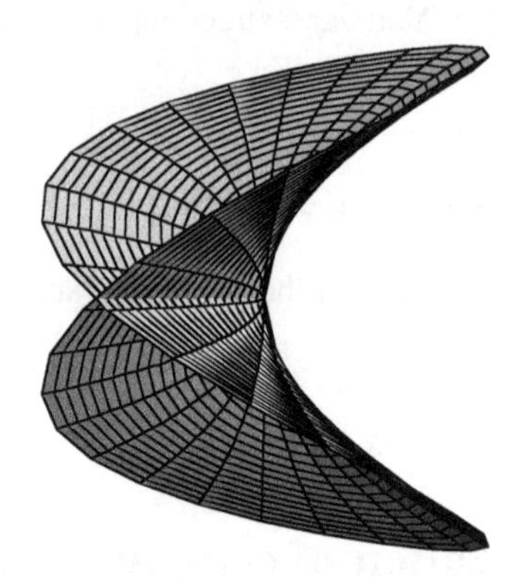

A similar view of the logarithm function reveals how the function never joins up again. Here we plot the imaginary part, which is just the argument:

```
ViewLogSurface[n_Integer, resolution_Integer] :=
ParametricPlot3D[
{r * Cos[theta], r * Sin[theta], theta},
{r, 0, 2}, {theta, 0, 2 * n * Pi},
PlotPoints -> {resolution, resolution * n},
Boxed -> False, Axes -> False, AspectRatio -> 1 / 2,
ViewPoint -> {-3, -2, 3}]

ViewLogSurface[3, 30]
```

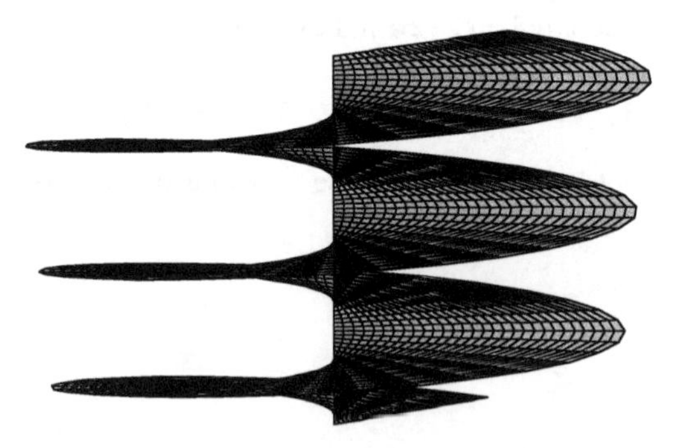

You might like to explore some other functions this way. If you are using *Mathematica* technology beyond version 5.2, there may be other options for displaying and viewing 3D graphics. Be sure to check the Front End menus and your documentation for options available in your current version. Also see the on-line supplement and CD.

2.12 Multiplication and spacing in *Mathematica*

Now, before you attempt the *Mathematica* exercises below, is a good time to note *Mathematica*'s conventions on spacing and multiplication. The quantity $x \times y$, giving the explicit multiplication of x by y, can be denoted in *Mathematica* explicitly by **x*y** or implicitly by **x y**, where there is a single space between x and y. Both of these are different from **xy** with no space, as the following example shows:

```
{TraditionalForm[x * y],
 TraditionalForm[x y], TraditionalForm[xy]}
```

$\{x\,y, x\,y, \mathrm{xy}\}$

The quantity **xy** with no space has a meaning all of its own as a new symbol! In this respect *Mathematica*'s conventions are rather more careful than those of ordinary mathematics. However, there is a downside to this in that TraditionalForm expressions can result in less than satisfactory layout for typesetting purposes. For example, the output of

```
TraditionalForm[x * y * z * p * q * r]
```

$p\;q\;r\;x\;y\;z$

is too widely spaced for most purposes, and the first example above means that omitting spaces altogether is wrong. Instead, the default spacing can be adjusted by surrounding each space in the output by a pair of ESC characters, which creates a thin space as follows:

$pqrxyz$

Exercises

The reader should note that familiarity with the material exemplified here will be assumed throughout the rest of this text – it is vital that you work through this entire set, otherwise later material involving many basic calculations with complex numbers, in their various forms, will be incomprehensible.

■ Mathematical, grouped by section topic

■ Algebra

2.1 Express in the form $x + i y$ each of the following numbers

$$\frac{1}{i}$$

$$\frac{1}{3 - 6i}$$

$(2+3i)+(3-6i)$

$(2+3i)-(3-6i)$

$(2+3i)*(3-6i)$

$$\frac{2+3i}{3-6i}$$

2.2 If $z = -2+5i$, express in the form $x+iy$ the quantity

$$3z^2 + z + \frac{2}{z}$$

2.3 Express in the form $x+iy$ each of the following:

i^2, i^3, i^4

Hence find in their simplest forms the values of

i^9, i^{43}, i^{4002}

2.4 Express in the form $x+iy$ each of the following:

$(1+i)^2, (1+i)^3, (1+i)^4$

▪ Conjugate and modulus

2.5 If $z = 3+2i$, what are the values (in $x+iy$ terms where appropriate) of

$$\bar{z},\ |z|,\ \frac{1}{z},\ \frac{1}{\bar{z}}.$$

2.6 If $|z| = 0$ and $z = x+iy$, show that $x = y = 0$.

2.7 If $z = x+iy$ and $z = \bar{z}$, show that $y = 0$. If instead $z = -\bar{z}$, show that $x = 0$. If instead you have

$$z^2 = \bar{z}^2$$

show that either $x = 0$, or $y = 0$, or both.

2.8 Let α be any real number and z be any complex number. Show that

$$|z-\alpha| = |\bar{z}-\alpha|$$

Hint: write $z = x+iy$ and consider the square of both sides.

▪ The Argand plane

2.9 If $z = 3+2i$, show the following complex numbers on the Argand plane:

$$z,\ \bar{z},\ \frac{1}{z},\ \frac{1}{\bar{z}}$$

2.10 If $z_1 = 3 + 2i$ and $z_2 = 1 - i$, show on a drawing of the Argand plane the following complex numbers:

$$z_1, z_2, z_1 - z_2, z_1 + z_2$$

What is the geometrical meaning of the complex number $z_1 - z_2$ and what is the interpretation of the quantity $|z_1 - z_2|$?

2.11 Show on a drawing of the Argand plane the following loci:

$$\{z \in \mathbb{C} : |z| = 2\}$$

$$\{z \in \mathbb{C} : |z - 3| = 1\}$$

$$\{z \in \mathbb{C} : |z + 1 + i| = 1\}$$

2.12 Show on a drawing of the Argand plane the following loci:

$$\{z \in \mathbb{C} : |z - 1| = |z + 1|\}$$

$$\{z \in \mathbb{C} : 2\,|z - 1| = |z + 1|\}$$

$$\{z \in \mathbb{C} : |z - 1| = 2\,|z + 1|\}$$

2.13 What is the geometrical interpretation of the result of exercise 2.8?

▪ Polar forms

2.14 For each of the following complex numbers
(i) find the modulus;
(ii) find the principal value of the argument;
(iii) express each in the form $r(\cos(\theta) + i\sin(\theta))$.

$$z = i$$

$$z = -i$$

$$z = 1 + i$$

$$z = -1 - i$$

$$z = \sqrt{3} + i$$

$$z = 1 + i\sqrt{3}$$

$$z = -1 - i\sqrt{3}$$

2.15 For each of the following pairs of complex numbers, z_1, z_2, convert each into the modulus–argument form, and use this to calculate both $z_1 z_2$ and z_1 / z_2.

$$z_1 = \frac{1+i}{\sqrt{2}},\ z_2 = \frac{1-i}{\sqrt{2}}$$

$$z_1 = \frac{\sqrt{3}+i}{2},\ z_2 = i$$

■ DeMoivre and trigonometry

2.16 Simplify the following expressions to the form $a + bi$:

$$(\cos(2\,\theta) + i\sin(2\,\theta))^{17}$$

$$\left(\cos\left(\frac{\pi}{6}\right) + i\sin\left(\frac{\pi}{6}\right)\right)^3$$

$$(\cos(3\,\theta) + i\sin(3\,\theta))^4\,(\cos(2\,\theta) - i\sin(2\,\theta))^6$$

2.17 Express each of $\cos(3\,\theta)$, $\cos(4\,\theta)$, $\cos(5\,\theta)$, in powers of $\cos(\theta)$.

2.18 Express each of $\sin(3\,\theta)$, $\sin(5\,\theta)$, in powers of $\sin(\theta)$.

2.19 Show that

$$\tan(3\,\theta) = \frac{3\tan(\theta) - \tan^3(\theta)}{1 - 3\tan^2(\theta)}$$

$$\tan(4\,\theta) = \frac{4\tan(\theta) - 4\tan^3(\theta)}{1 - 6\tan^2(\theta) + \tan^4(\theta)}$$

2.20 Show that

$$\cos^5(\theta) = \frac{1}{16}\,(10\cos(\theta) + 5\cos(3\,\theta) + \cos(5\,\theta))$$

$$\sin^5(\theta) = \frac{1}{16}\,(10\sin(\theta) - 5\sin(3\,\theta) + \sin(5\,\theta))$$

2.21 Calculate the indefinite integral

$$\int \sin^5(\theta)\,d\theta$$

2.22 Calculate the definite integral

$$\int_0^{\frac{\pi}{4}} \cos^4(\theta)\,d\theta$$

■ Roots

2.23 The three cube roots of unity are: $1,\ \omega,\ \omega^2$, where

$$\omega = \frac{1}{2}\left(-1 + i\sqrt{3}\right)$$

Show that $1 + \omega + \omega^2 = 0$.

2.24 Show that the sum of the n nth roots of unity is zero.

2.25 Find the three cube roots of -1 in the form $a + ib$.

2.26 Find the three cube roots of 3, and the four fourth roots of 4, in the form $a + i\,b$.

■ Exponential and log functions

2.27 Find the three cube roots of -1 in the form $r\,e^{i\theta}$.

2.28 Find the three cube roots of 3, and the four fourth roots of 4, in the form $re^{i\theta}$.

2.29 Calculate each of the following in the form $a + ib$:

$$e^{3+\frac{i\pi}{4}}$$

$$e^{\frac{1}{2}-\frac{3i\pi}{4}}$$

$$e^{\log(2)+\frac{\pi i}{4}}$$

2.30 Calculate all possible values of

$$\log i$$

$$\log\left(e^{3+\frac{7\pi i}{4}}\right)$$

$$\log(-4)$$

$$\log\left(i + \sqrt{3}\right)$$

in the form $a + ib$.

■ ✱ *Mathematica* exercises

2.31 ✱ Apply the format conversion operations **InputForm**, **StandardForm** and **TraditionalForm** to each of the following expressions:

```
Pi, E, I, Exp[I Pi], Exp[I Pi Cos[Pi]]
```

2.32 ✱ Use *Mathematica* to display on the Argand plane the locations of the four points equivalent to the complex numbers

$$1 + i,\ -1 + i,\ -1 - i,\ 1 - i$$

2.33 ✱ Use *Mathematica*'s built-in functions to calculate the modulus, argument and complex conjugate of each of the following:

$$3 + 2i,\ \frac{1-i}{\sqrt{2}},\ 3 + 4i$$

2.34 ✵ Use the *Mathematica* function **ComplexExpand** to work out the real and imaginary parts, with x and y real, of each of the following functions:

$$(x + iy)^4$$

$$(x - iy)^5$$

$$\cos(x + iy)$$

$$\cosh(x + iy)$$

$$\sinh(x + iy)$$

2.35 ✵ Use the *Mathematica* function **TrigExpand** to re-work Exercise 2.17 (also see the comment in Section 2.7 about supplementary rules).

2.36 ✵ Use the *Mathematica* function **TrigReduce** to re-work Exercise 2/18.

2.37 ✵ Show the eight eighth roots of unity in the Argand plane.

2.38 ✵ Find out what value *Mathematica* returns for the square and cube roots of $1,\ -1,\ i,\ -i$. (Use the **N** function to get a firm grip on the choice made.)

2.39 ✵ Use *Mathematica* to check your results for Exercise 2.29.

2.40 ✵ What does *Mathematica* return for the values of the log function you were asked to compute in Exercise 2.30?

2.41 ✵ Use the function **ViewRootSurface** to investigate the behaviour of the cube and fourth-root functions. If you are using *Mathematica* technology beyond version 5.2, be sure to investigate both Front End and kernel options for graphics in your current version of *Mathematica*. See the on-line supplement and CD.

3 Cubics, quartics and visualization of complex roots

Introduction

The solution of general quadratic equations becomes possible, in terms of simple square roots, once one has access to the machinery of complex numbers. The question naturally arises as to whether it is possible to solve higher-order equations in the same way. In fact, we must be careful to pose this question properly. We might be interested in whether we need to extend the number system still further. For example, if we write down a cubic equation with coefficients that are complex numbers, can we find all the roots in terms of complex numbers? We can ask similar questions for higher-order polynomial equations. The investigation of the solution of cubic and quartic equations is a topic that used to be popular in basic courses on complex numbers, but has become less fashionable recently, probably because of the extensive manipulations that are required. Armed with *Mathematica*, however, such manipulations become routine, and we can revisit some of the classic developments in algebra quite straightforwardly. These topics have become so unfashionable, in fact, that the author received some suggestions from readers of early drafts of this book that this material should be, if not removed altogether, relocated to an appendix! I have left this material here quite deliberately, having found numerous applications for the solutions of cubics, at least, in applied mathematics. You may feel free to skip this part of the material if you have no interest in cubics and higher order systems.

A further topic that is placed is here is the demonstration of some important techniques for visualizing the behaviour of roots of equations in the complex plane – root movies and root locus plots. This will use *Mathematica* functions for the numerical solution of polynomial equations.

The formula we shall develop for a cubic is a modern view of one commonly attributed to one 'Tartaglia', more correctly known as N. Fontana, and developed sometime before 1539. It was a special case for cubics with no quadratic term. A solution for such cubics was first *published* by Cardano in the *Ars Magna* (see Cardano, 1993), having allegedly being obtained confidentially from Tartaglia, in 1539, with a promise that it would not be revealed. This led to something of a feud. It is possible that Cardano was the first to realise that any cubic can be transformed to one where the quadratic term is absent, but the rest of the work (if not more) appears to be creditable to S. del Ferro. Cardano published his work mentioning the work by del Ferro and Tartaglia, having investigated del Ferro's posthumous papers and satisfied himself that del Ferro had a solution by 1526 (see Penrose, 2004). Tartaglia was not at all pleased. The management of quartic equations led to an escalation of the arguments between Cardano and Tartaglia, as the problem of the quartic was solved by L. Ferrari, a pupil of Cardano, using the solution of the cubic along the way. It all became terribly heated. As Ferrari said to

Tartaglia: 'You have written things that falsely and unworthily slander Signor Cardan, compared with whom you are hardly worth mentioning.' In return, Tartaglia referred to Ferrari as 'Cardan's creature'. We must of course always be grateful that modern day scientists *never* resort to this type of dreadful bickering, or have childish disputes about who established what result first!

We can approach the management of cubic equations by proceeding historically and traditionally, or by seeing what *Mathematica* does and then trying to understand it. Either way we come to the same conclusion, so we shall in fact begin by seeing what *Mathematica* does when presented with a cubic in symbolic form.

3.1 ❁ *Mathematica* investigation of cubic equations

Let's take a look a very simple cubic and a general one:

```
solOne = x /. Solve[x^3 - 1 == 0, x]
```

$$\left\{1, -\sqrt[3]{-1}, (-1)^{2/3}\right\}$$

```
solTwo = x /. Solve[x^3 - c x^2 - 3 a x - b == 0, x]
```

$$\{\frac{c}{3} + \frac{1}{3\sqrt[3]{2}}((2c^3 + 27ac + 27b + 3\sqrt{3}\sqrt{-108a^3 - 9c^2a^2 + 54bca + 4bc^3 + 27b^2})\wedge(1/3)) -$$
$$(\sqrt[3]{2}(-c^2 - 9a))/(3(2c^3 + 27ac + 27b + 3\sqrt{3}\sqrt{-108a^3 - 9c^2a^2 + 54bca + 4bc^3 + 27b^2})\wedge(1/3)),$$
$$\frac{c}{3} - \frac{1}{6\sqrt[3]{2}}((1 - i\sqrt{3})(2c^3 + 27ac + 27b + 3\sqrt{3}\sqrt{-108a^3 - 9c^2a^2 + 54bca + 4bc^3 + 27b^2})\wedge(1/3)) +$$
$$((1 + i\sqrt{3})(-c^2 - 9a))/$$
$$(3\,2^{2/3}(2c^3 + 27ac + 27b + 3\sqrt{3}\sqrt{-108a^3 - 9c^2a^2 + 54bca + 4bc^3 + 27b^2})\wedge(1/3)),$$
$$\frac{c}{3} - \frac{1}{6\sqrt[3]{2}}((1 + i\sqrt{3})(2c^3 + 27ac + 27b + 3\sqrt{3}\sqrt{-108a^3 - 9c^2a^2 + 54bca + 4bc^3 + 27b^2})\wedge(1/3)) +$$
$$((1 - i\sqrt{3})(-c^2 - 9a))/$$
$$(3\,2^{2/3}(2c^3 + 27ac + 27b + 3\sqrt{3}\sqrt{-108a^3 - 9c^2a^2 + 54bca + 4bc^3 + 27b^2})\wedge(1/3))\}$$

The latter is somewhat hard to interpret. However, a pattern appers to emerge if we restrict attention to cubics containing no quadratic term, and simplify the results:

```
solThree = Simplify[x /. Solve[x^3 - 3 a x - b == 0, x]]
```

$$\left\{\frac{\sqrt[3]{2}\,a}{\sqrt[3]{b + \sqrt{b^2 - 4a^3}}} + \frac{\sqrt[3]{b + \sqrt{b^2 - 4a^3}}}{\sqrt[3]{2}}, \frac{i\left(\sqrt[3]{2}\,(i + \sqrt{3})\left(b + \sqrt{b^2 - 4a^3}\right)^{2/3} - 2\left(-i + \sqrt{3}\right)a\right)}{2\,2^{2/3}\sqrt[3]{b + \sqrt{b^2 - 4a^3}}},\right.$$
$$\left.\frac{2i\left(i + \sqrt{3}\right)a + \sqrt[3]{2}\left(-1 - i\sqrt{3}\right)\left(b + \sqrt{b^2 - 4a^3}\right)^{2/3}}{2\,2^{2/3}\sqrt[3]{b + \sqrt{b^2 - 4a^3}}}\right\}$$

To get a better grip on this pattern you will need to investigate the cube roots of unity in some detail. First, however, note that any cubic can be reduced to a cubic with no quadratic term, by making a simple translation of the coordinates. You can get *Mathematica* to do this for you:

```
x^3 - c x^2 - 3 a x - b /. x -> X + A
```

$(A+X)^3 - c(A+X)^2 - 3a(A+X) - b$

```
Expand[%]
```

$A^3 - cA^2 + 3XA^2 + 3X^2A - 3aA - 2cXA + X^3 - cX^2 - b - 3aX$

```
Collect[%, X]
```

$A^3 - cA^2 - 3aA + X^3 + (3A - c)X^2 - b + (3A^2 - 2cA - 3a)X$

```
Coefficient[%, X^2]
```

$3A - c$

So if you set $A = c/3$ the new cubic has no quadratic term.

■ The basic cube root of unity

The basic cube root of unity is the quantity:

```
ω = 1/2 (-1 + I Sqrt[3]);
```

```
Expand[ω^3]
```

1

As a consequence of this we also have the following (which can be stated as the fact that the sum of the three cube roots of unity is zero):

```
Expand[ω^2 + ω + 1]
```

0

The quantity ω and its powers lie at the corners of a triangle in the complex plane. You can easily visualize this, using the following plot routine, which we recall from Chapter 2:

```
CPlot[z_List] :=
Module[{r}, r = Map[{Re[#], Im[#]} &, z];
ListPlot[r, PlotStyle -> PointSize[0.1],
AspectRatio -> 1,
   PlotRange -> {{-1.1, 1.1}, {-1.1, 1.1}},
PlotRegion -> {{0.05, 0.95}, {0.05, 0.95}}]]
```

```
CPlot[{1, ω, ω^2}]
```

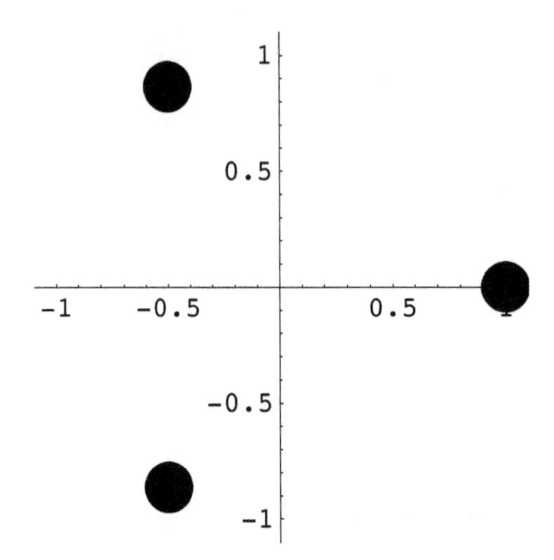

■ Solving the cubic

Consider the product of the following three quantities:

```
z1 = x - α - β;
z2 = x - α ω - β ω^2;
z3 = x - α ω^2 - β ω;
```

■ Exercise

Using pen and paper, you should work out for yourself, and simplify, the product of these three numbers. Here is what *Mathematica* gives as a check (the **Collect** function groups the terms multiplying each power of x):

```
Collect[Expand[z1 z2 z3], x]
```

$$x^3 - 3\,\alpha\,\beta\,x - \alpha^3 - \beta^3$$

Evidently the factorization of the polynomial

$$x^3 - 3ax - b \tag{3.1}$$

can be done explicitly if α and β can be found such that

$$\alpha\beta = a; \quad \alpha^3 + \beta^3 = b \tag{3.2}$$

We now show how this can be done, using *Mathematica* to develop the argument. Again, it is a useful exercise to work through this yourself with pen and paper.

```
eqn = α³ + β³ - b /. β -> a / α
```

$$\frac{a^3}{\alpha^3} + \alpha^3 - b$$

This is now an equation involving α^3, so we substitute accordingly:

```
sub = eqn /. {α³ -> λ, 1 / α³ -> 1 / λ}
```

$$\frac{a^3}{\lambda} - b + \lambda$$

This gives a quadratic equation for λ, which is readily solved:

```
Solve[sub == 0, λ]
```

$$\left\{\left\{\lambda \to \frac{1}{2}\left(b - \sqrt{b^2 - 4a^3}\right)\right\}, \left\{\lambda \to \frac{1}{2}\left(b + \sqrt{b^2 - 4a^3}\right)\right\}\right\}$$

```
alpharoots = α /. Solve[eqn == 0, α]
```

$$\left\{-\sqrt[3]{-\frac{1}{2}}\,\sqrt[3]{b - \sqrt{b^2 - 4a^3}},\ \frac{\sqrt[3]{b - \sqrt{b^2 - 4a^3}}}{\sqrt[3]{2}},\right.$$
$$\frac{(-1)^{2/3}\sqrt[3]{b - \sqrt{b^2 - 4a^3}}}{\sqrt[3]{2}},\ \sqrt[3]{\frac{b}{2} + \frac{1}{2}\sqrt{b^2 - 4a^3}},$$
$$\left.-\sqrt[3]{-1}\,\sqrt[3]{\frac{b}{2} + \frac{1}{2}\sqrt{b^2 - 4a^3}},\ (-1)^{2/3}\sqrt[3]{\frac{b}{2} + \frac{1}{2}\sqrt{b^2 - 4a^3}}\right\}$$

Superficially, we get six roots, but as our equations are symmetric under the swapping of α and β, three of the roots are the βs associated with three of the αs.The three roots of the cubic are then just:

$$\begin{aligned} &\alpha + \beta \\ &\beta\,\omega^2 + \alpha\,\omega \\ &\alpha\,\omega^2 + \beta\,\omega \end{aligned} \tag{3.3}$$

This, with a little reorganization, results in the formula already given. As noted in the introduction, the formula given is essentially equivalent to one developed by Tartaglia in about 1530 for cubics with no quadratic term.

3.2 *Mathematica* investigation of quartic equations

In what follows we shall pursue the general solution of the quartic using *Mathematica*. First we shall use *Mathematica*'s **Solve** function as a blunt tool. Then we shall explore the cunning intricacies of Ferrari's solution, using *Mathematica* to do all the algebra.

If you want to see what the full quartic solution is like, execute the following code. The output is not given here for reasons that will be obvious if you try it!

```
solTwo = x /. Solve[x^4 - d x^3 - c x^2 - 3 a x - b == 0, x]
```

The output of this is rather hard to interpret. This time, matters are not much simpler if you restrict attention to quartics containing no cubic term, which can be effected by a simple translation, as before. This time we shall show the result, suitably reduced:

```
solThree = x /. Solve[x^4 - c x^2 - 3 a x - b == 0, x]
```

$$\Big\{-\frac{1}{2}\sqrt{\left(\frac{2c}{3}+\frac{1}{3\sqrt[3]{2}}\left(\left(-2c^3-72bc+243a^2+\sqrt{(-2c^3-72bc+243a^2)^2-4(c^2-12b)^3}\right)\wedge(1/3)\right)+\left(\sqrt[3]{2}\,(c^2-12b)\right)\Big/\left(3\left(-2c^3-72bc+243a^2+\sqrt{(-2c^3-72bc+243a^2)^2-4(c^2-12b)^3}\right)\wedge(1/3)\right)\right)} - \frac{1}{2}\sqrt{\left(-(6a)\Big/\left(\sqrt{\left(\frac{2c}{3}+\frac{1}{3\sqrt[3]{2}}\left(\left(-2c^3-72bc+243a^2+\sqrt{(-2c^3-72bc+243a^2)^2-4(c^2-12b)^3}\right)\wedge(1/3)\right)+\left(\sqrt[3]{2}\,(c^2-12b)\right)\Big/\left(3\left(-2c^3-72bc+243a^2+\sqrt{(-2c^3-72bc+243a^2)^2-4(c^2-12b)^3}\right)\wedge(1/3)\right)\right)}\right)+\frac{4c}{3}-\frac{1}{3\sqrt[3]{2}}\left(\left(-2c^3-72bc+243a^2+\sqrt{(-2c^3-72bc+243a^2)^2-4(c^2-12b)^3}\right)\wedge(1/3)\right)-\left(\sqrt[3]{2}\,(c^2-12b)\right)\Big/\left(3\left(-2c^3-72bc+243a^2+\sqrt{(-2c^3-72bc+243a^2)^2-4(c^2-12b)^3}\right)\wedge(1/3)\right)\right)},$$

$$\frac{1}{2}\sqrt{\left(-(6a)\Big/\left(\sqrt{\left(\frac{2c}{3}+\frac{1}{3\sqrt[3]{2}}\left(\left(-2c^3-72bc+243a^2+\sqrt{(-2c^3-72bc+243a^2)^2-4(c^2-12b)^3}\right)\wedge(1/3)\right)+\left(\sqrt[3]{2}\,(c^2-12b)\right)\Big/\left(3\left(-2c^3-72bc+243a^2+\sqrt{(-2c^3-72bc+243a^2)^2-4(c^2-12b)^3}\right)\wedge(1/3)\right)\right)}\right)+\frac{4c}{3}-\frac{1}{3\sqrt[3]{2}}\left(\left(-2c^3-72bc+243a^2+\sqrt{(-2c^3-72bc+243a^2)^2-4(c^2-12b)^3}\right)\wedge(1/3)\right)-\left(\sqrt[3]{2}\,(c^2-12b)\right)\Big/\left(3\left(-2c^3-72bc+243a^2+\sqrt{(-2c^3-72bc+243a^2)^2-4(c^2-12b)^3}\right)\wedge(1/3)\right)\right)} - \frac{1}{2}\sqrt{\left(\frac{2c}{3}+\frac{1}{3\sqrt[3]{2}}\left(\left(-2c^3-72bc+243a^2+\sqrt{(-2c^3-72bc+243a^2)^2-4(c^2-12b)^3}\right)\wedge(1/3)\right)+\left(\sqrt[3]{2}\,(c^2-12b)\right)\Big/\left(3\left(-2c^3-72bc+243a^2+\sqrt{(-2c^3-72bc+243a^2)^2-4(c^2-12b)^3}\right)\wedge(1/3)\right)\right)},$$

$$\frac{1}{2}\sqrt{\left(\frac{2c}{3}+\frac{1}{3\sqrt[3]{2}}\left(\left(-2c^3-72bc+243a^2+\sqrt{(-2c^3-72bc+243a^2)^2-4(c^2-12b)^3}\right)\wedge(1/3)\right)+\left(\sqrt[3]{2}\,(c^2-12b)\right)\Big/\left(3\left(-2c^3-72bc+243a^2+\sqrt{(-2c^3-72bc+243a^2)^2-4(c^2-12b)^3}\right)\wedge(1/3)\right)\right)} - \frac{1}{2}\sqrt{\left((6a)\Big/\left(\sqrt{\left(\frac{2c}{3}+\frac{1}{3\sqrt[3]{2}}\left(\left(-2c^3-72bc+243a^2+\sqrt{(-2c^3-72bc+243a^2)^2-4(c^2-12b)^3}\right)\wedge(1/3)\right)+\left(\sqrt[3]{2}\,(c^2-12b)\right)\Big/\left(3\left(-2c^3-72bc+243a^2+\sqrt{(-2c^3-72bc+243a^2)^2-4(c^2-12b)^3}\right)\wedge(1/3)\right)\right)}\right)+\frac{4c}{3}-\frac{1}{3\sqrt[3]{2}}\left(\left(-2c^3-72bc+243a^2+\sqrt{(-2c^3-72bc+243a^2)^2-4(c^2-12b)^3}\right)\wedge(1/3)\right)-\left(\sqrt[3]{2}\,(c^2-12b)\right)\Big/\left(3\left(-2c^3-72bc+243a^2+\sqrt{(-2c^3-72bc+243a^2)^2-4(c^2-12b)^3}\right)\wedge(1/3)\right)\right)},$$

$$\frac{1}{2}\sqrt{\left(\frac{2c}{3}+\frac{1}{3\sqrt[3]{2}}\left(\left(-2c^3-72bc+243a^2+\sqrt{(-2c^3-72bc+243a^2)^2-4(c^2-12b)^3}\right)\wedge(1/3)\right)+\left(\sqrt[3]{2}\,(c^2-12b)\right)\Big/\left(3\left(-2c^3-72bc+243a^2+\sqrt{(-2c^3-72bc+243a^2)^2-4(c^2-12b)^3}\right)\wedge(1/3)\right)\right)} + \frac{1}{2}\sqrt{\left((6a)\Big/\left(\sqrt{\left(\frac{2c}{3}+\frac{1}{3\sqrt[3]{2}}\left(\left(-2c^3-72bc+243a^2+\sqrt{(-2c^3-72bc+243a^2)^2-4(c^2-12b)^3}\right)\wedge(1/3)\right)+\left(\sqrt[3]{2}\,(c^2-12b)\right)\Big/\left(3\left(-2c^3-72bc+243a^2+\sqrt{(-2c^3-72bc+243a^2)^2-4(c^2-12b)^3}\right)\wedge(1/3)\right)\right)}\right)+\frac{4c}{3}-\frac{1}{3\sqrt[3]{2}}\left(\left(-2c^3-72bc+243a^2+\sqrt{(-2c^3-72bc+243a^2)^2-4(c^2-12b)^3}\right)\wedge(1/3)\right)-\left(\sqrt[3]{2}\,(c^2-12b)\right)\Big/\left(3\left(-2c^3-72bc+243a^2+\sqrt{(-2c^3-72bc+243a^2)^2-4(c^2-12b)^3}\right)\wedge(1/3)\right)\right)}\Big\}$$

■ Ferrari's approach to solving the quartic

As before, the term of cubic order can be eliminated by a change of variables, so it suffices to consider quartics with no cubic term

```
quartic = x^4 + p x^2 + q x + r;
```

We can split this object into two pieces:

```
lhs = x^4 + p x^2;
rhs = -q x - r;
```

The original quartic equation is just

```
lhs - rhs == 0
```

$$x^4 + p\,x^2 + q\,x + r = 0$$

Alternatively, we rearrange it as follows:

```
lhs == rhs
```

$$x^4 + p\,x^2 = -r - q\,x$$

Let us add $p^2 + px^2$ to both sides:

```
lhsa = lhs + p x^2 + p^2;
rhsa = rhs + p x^2 + p^2;
```

This makes the left side a perfect square, as is evidenced by:

```
Factor[lhsa]
```

$$(x^2 + p)^2$$

But the right-hand side is still nothing useful:

```
rhsa
```

$$p^2 + x^2\,p - r - q\,x$$

We now add to both sides something that preserves the fact that the left side is a perfect square, but will give us the flexibility to arrange that the right side is also:

```
lhsb = lhsa + 2 (x^2 + p) z + z^2;
rhsb = rhsa + 2 (x^2 + p) z + z^2;
```

The left side is still a perfect square:

```
Factor[lhsb]
```

$$(x^2 + p + z)^2$$

```
rhsb
```

$$p^2 + x^2 p + z^2 - r - q x + 2 (x^2 + p) z$$

If we can choose z so that this new right-hand side is a perfect square, we can take square roots of both sides, leading to two quadratic equations. Now, recall from Chapter 1, that if we take a quadratic in the form

$$\text{quad} = ax^2 + bx + c \tag{3.4}$$

This is already in the form of a perfect square if

$$b^2 = 4ac \tag{3.5}$$

So we need to apply this constraint to **rhsb** to get an equation for z. Let's pull out the various pieces:

```
a = Coefficient[rhsb, x^2]
```

$$p + 2 z$$

```
b = Coefficient[rhsb, x]
```

$$-q$$

```
c = Expand[rhsb - a x^2 - b x]
```

$$p^2 + 2 z p + z^2 - r$$

Now we organize the resulting equation for z:

```
Collect[b^2 - 4 a c, z]
```

$$-4 p^3 - 20 z^2 p + 4 r p - 8 z^3 + q^2 + (8 r - 16 p^2) z$$

Thus we arrive at a cubic equation, the *resolvent cubic*, which can be solved by Tartaglia's method. In Chapter 39 of the *Ars Magna*, Cardano considered an example that is worth investigating. Here we work through it with *Mathematica*.

```
lhs = x^4 - 10 x^2;
rhs = -4 x - 8;

lhsa = lhs - 10 x^2 + 100;
rhsa = rhs - 10 x^2 + 100;
{lhsa, rhsa}
```

$$\{x^4 - 20 x^2 + 100, -10 x^2 - 4 x + 92\}$$

```
Factor[lhsa]
```

$$(x^2 - 10)^2$$

```
lhsb = lhsa + 2 (x^2 - 10) z + z^2;
rhsb = rhsa + 2 (x^2 - 10) z + z^2;
```

```
Factor[lhsb]
```

$$(x^2 + z - 10)^2$$

```
rhsb
```

$$-10\,x^2 - 4\,x + z^2 + 2\,(x^2 - 10)\,z + 92$$

As before, we want this to be a perfect square, so we construct the $b^2 - 4\,ac$ term:

```
a = Coefficient[rhsb, x^2]
```

$$2\,z - 10$$

```
b = Coefficient[rhsb, x]
```

$$-4$$

```
c = Expand[rhsb - a x^2 - b x]
```

$$z^2 - 20\,z + 92$$

Now we organize the resulting equation for z:

```
Collect[b^2 - 4 a c, z]
```

$$-8\,z^3 + 200\,z^2 - 1536\,z + 3696$$

```
cubic = Simplify[% / 8]
```

$$-z^3 + 25\,z^2 - 192\,z + 462$$

Now we get rid of the quadratic piece:

```
reducedcubic = Expand[cubic /. z -> y + 25 / 3]
```

$$-y^3 + \frac{49\,y}{3} + \frac{524}{27}$$

We can solve this using Tartaglia's method, but let's jump to the answer, as we are more interested in quartics right now:

```
y /. Solve[reducedcubic == 0, y]
```

$$\left\{-\frac{4}{3}, \frac{1}{3}\left(2-3\sqrt{15}\right), \frac{1}{3}\left(2+3\sqrt{15}\right)\right\}$$

It looks like algebra will be minimized if we take the first root, so let

```
z = %[[1]] + 25 / 3
```

$$7$$

So the resolvent cubic has a solution at $z = 7$.

```
lhsb /. z -> 7
```

$$x^4 - 20x^2 + 14(x^2 - 10) + 149$$

```
rhsb /. z -> 7
```

$$-10x^2 - 4x + 14(x^2 - 10) + 141$$

Let us define two quantities, which are each side represented as a perfect square:

```
perflhs = Factor[lhsb]
```

$$(x^2 - 3)^2$$

```
perfrhs = Factor[rhsb]
```

$$(2x - 1)^2$$

We now define two quadratic equations by taking the square roots of both sides. It is helpful to use **PowerExpand** here to force *Mathematica* to do the job, and note that we use the double equals to denote equality:

```
quada = PowerExpand[ Sqrt[perflhs] == Sqrt[perfrhs]]
```

$$x^2 - 3 == 2x - 1$$

```
quadb = PowerExpand[ Sqrt[perflhs] == -Sqrt[perfrhs]]
```

$$x^2 - 3 == 1 - 2x$$

```
x /. Solve[quada, x]
```

$$\left\{1-\sqrt{3}, 1+\sqrt{3}\right\}$$

```
x /. Solve[quadb, x]
```

$$\left\{-1-\sqrt{5}, -1+\sqrt{5}\right\}$$

As a check on our answer, we can of course just ask *Mathematica* to do it!

```
x /. Solve[lhs == rhs, x]
```

$$\{1-\sqrt{3},\ 1+\sqrt{3},\ -1-\sqrt{5},\ -1+\sqrt{5}\}$$

3.3 The quintic

Any attempt to discuss this properly would either divert us into a course in Group Theory or be a poor replacement of some work already done using *Mathematica* to illuminate the problem. In the first case it is necessary to study the work of Evariste Galois to understand what specific equations of a given degree admit an algebraic solution, involving just arithmetic operations and *n*th roots. In the second case, one should consult the excellent poster and internet resources associated with the project 'Solving the Quintic' (Adamchik and Trott, 1994). This develops the solution for a quintic in terms of special functions, and contains an excellent bibliography.

3.4 Root movies and root locus plots

It is often useful to have a grip on the location of the roots of a polynomial, within the Complex Plane. *Mathematica* offers several ways of visualizing the locations of roots. If the polynomial under investigation has one or more variable parameters, insight can be gained by allowing one such parameter to vary and looking at the location of the roots. For example, we can check for what values of a parameter some of the roots are real or essentially complex. In more complicated problems, the stability of a system can be influenced by the location of the roots associated with its Laplace transform, and when the system is parametrized by a feedback or other variable, the effect of the parameter on the roots is critical. In general we shall need to explore polynomials of degree greater than four so we shall also employ *Mathematica*'s **NSolve** function.

There are two simple ways of looking at the roots of a parametrized polynomial as the parameter varies. In the first case we can just make a movie using *Mathematica*'s animation functions. In the second case we overlay all the frames of the movie to get a root locus plot. Both of these rely on some common functions that we introduce first. We begin by generalizing our Argand plane plotting routine to allow for a plotting range and a point size:

```
CRPlot[z_List, range_List, size_] :=
Module[{r},
r = Map[{Re[#], Im[#]} &, z];
ListPlot[r, PlotStyle -> PointSize[size],
AspectRatio -> 1, PlotRange -> {range, range},
PlotRegion -> {{0.05, 0.95}, {0.05, 0.95}}]]
```

Next we define a function that extracts the complex roots of a polynomial equation in list form. To keep matters simple, we always work with z as our variable.

```
PolySolver[poly_] := z /. NSolve[poly == 0, z]
```

Here is a working example in the form of a simple quintic:

```
mypoly[z_, λ_] := z^5 + z^3 + z^2 + z + λ;
```

When we apply the solver the the example, we get the five roots in numerical form. If you have done so calculations earlier where a value of z has been assigned you will need to clear it as follows first:

```
Clear[z];
PolySolver[mypoly[z, 3]]
```

$\{-1.11682, -0.30271 - 1.18858\, i,$
$-0.30271 + 1.18858\, i, 0.861121 - 1.0218\, i, 0.861121 + 1.0218\, i\}$

■ Movies

This is done in versions 4 and 5 of *Mathematica* using a simple loop to generate succesive frames of an animation:

```
Do[CRPlot[PolySolver[mypoly[z, λ]], {-2, 2}, 0.05],
  {λ, 0, 4, 0.2}]
```

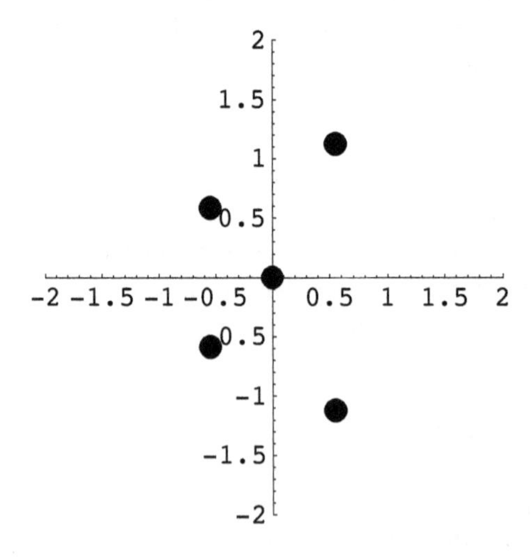

You can execute the animation by selecting a group of frames to animate, then applying 'Animate Selected Graphics' from the Cell Menu of *Mathematica*. If you are using *Mathematica* technology beyond version 5.2, be sure to investigate the options for your current version of *Mathematica*. Also see the on-line supplement and CD.

■ Root locus plots

In this case a table of values of the complex roots is created. The resulting two-dimensional list is flattened into a one-dimensional list of many complex numbers, then fed to the plotting routine.

```
CRPlot[Flatten[
  Table[PolySolver[mypoly[z, λ]], {λ, 0, 4, 0.2}]],
    {-2, 2}, 0.015]
```

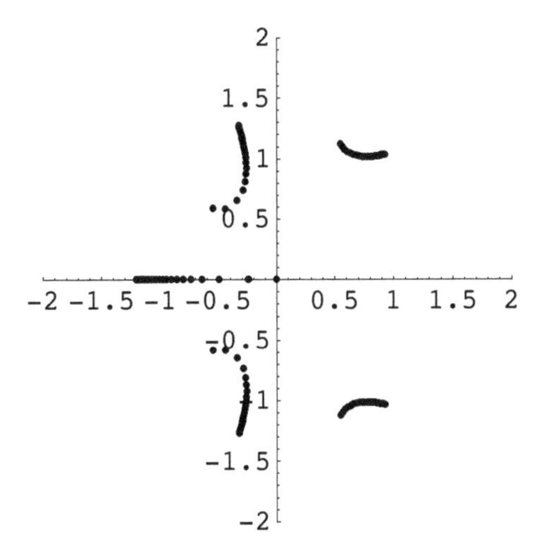

Note that we get a useful measure of the 'speed' of the roots as λ changes, expressed by the separation of the dots. Root locus plots are more traditionally given by joining up the dots to give a smooth curve. This of course discards the velocity information. It is also sometimes difficult to decide what dots to join up – our approach avoids this problem as there is no requirement to join anything up.

Exercises

3.1 Let α and β be the roots of the quadratic equation

$$ax^2 + bx + c = 0$$

By writing the quadratic $ax^2 + bx + c$ in the form $a(x - \alpha)(x - \beta)$ deduce that

$$\alpha + \beta = -\frac{b}{a}, \quad \alpha\beta = \frac{c}{a}$$

3.2 Let α, β, γ be the roots of the cubic equation

$$ax^3 + bx^2 + cx + d = 0$$

By writing the cubic polynomial $ax^3 + bx^2 + cx + d$ in the form $a(x-\alpha)(x-\beta)(x-\gamma)$ deduce that

$$\alpha + \beta + \gamma = -\frac{b}{a}, \quad \alpha\beta + \alpha\gamma + \beta\gamma = \frac{c}{a}, \quad \alpha\beta\gamma = -\frac{d}{a}$$

3.3 Let α, β, γ, δ be the roots of the quartic equation

$$ax^4 + bx^3 + cx^2 + dx + e = 0$$

By writing $ax^4 + bx^3 + cx^2 + dx + e$ in the form $a(x-\alpha)(x-\beta)(x-\gamma)(x-\delta)$ deduce that

$$\alpha + \beta + \gamma + \delta = -\frac{b}{a}$$

$$\alpha\beta + \alpha\gamma + \alpha\delta + \beta\gamma + \beta\delta + \gamma\delta = \frac{c}{a}$$

$$\alpha\beta\gamma + \alpha\beta\delta + \alpha\delta\gamma + \beta\gamma\delta = -\frac{d}{a}$$

$$\alpha\beta\gamma\delta = \frac{e}{a}$$

3.4 Using the results of Exercises 3.1 and 3.2, write down

(i) a quadratic equation with roots $2 + i$ and $2 - i$;

(ii) a cubic equation with roots $3,\ 4 - i,\ 4 + i$.

3.5 Show that $x = -1$ is a root of both the following cubic equations

$$x^3 - 2x - 1 = 0$$
$$x^3 - 3x - 2 = 0$$

and hence solve each equation for all its roots.

3.6 Use Tartaglia's method to solve the cubic equations:

$$x^3 - 3x + 1 = 0$$

$$z^3 + (6 - 9i)z^2 - (18 + 36i)z - 51 = 0$$

3.7 Solve the quartic equation

$$x^4 - 10x^2 + 9 = 0$$

3.8 Show that $x = 1$ is one solution of the quartic equation

$$x^4 - 9x^2 - x + 9 = 0$$

Hence find all roots of the equation.

3.9 Show that $x^2 + x + 1$ is a factor of the quntic expression $x^5 + x + 1$, and find the other (cubic) factor. Hence find all roots of the equation

$$x^5 + x + 1 = 0$$

3.10 ✵ Use the *Mathematica* functions **Expand** and **Collect** to re-work Exercises 3.2 and 3.3 on your computer.

3.11 ✵ Using *Mathematica*'s **Expand** function, and the approach of Exercises 3.1, 3.2 and 3.3, find

(i) a quadratic equation with roots 3 and $2 + i$;

(ii) a cubic equation with roots $3,\ 2 + i,\ 2 - i$;

(iii) a quartic equation with roots $2,\ -2,\ 3 + i,\ 3 - i$.

3.12 ✵ Use *Mathematica*'s **Solve** function to solve directly the cubic equations given in Exercise 3.5.

3.13 ✵ Use *Mathematica*'s **Solve** function to find the solutions of the cubic equations given in Exercise 3.6. What do you notice about the way *Mathematica* presents the solutions?

3.14 ✵ Use *Mathematica*'s **Solve** function to find the solutions of the quartic equation given in Exercise 3.8.

3.15 ✵ Use *Mathematica*'s **Solve** function to find, in exact form, all roots of the quintic equation

$$x^5 + x + 1 = 0$$

3.16 ✵ Using **Solve** and **NSolve**, establish two ways of finding approximate numerical values to all of the complex roots of the quintic equation

$$x^5 + x + 1 = 0$$

3.17 ✵ Using the function **FindRoot**, how many roots can you find of

$$x^5 + x + 1 = 0$$

Hint: Try experimenting with different starting values of x_0 in the solution method:

```
FindRoot[x^5 + x + 1 == 0, {x, x0}]
```

3.18 ✵ Build root movie and root locus plots of the following polynomial, with λ varying:

$$x^4 - 4x^3 + (\lambda^2 + 5)x^2 - 4x\lambda^2 + 5\lambda^2$$

4 Newton–Raphson iteration and complex fractals

Introduction

You may already have seen how to solve polynomial equations numerically in Chapter 3, using the **NSolve** function, or **FindRoot**. How in general can we solve an equation, polynomial or otherwise, numerically? There are many schemes for doing this, with one or perhaps many variables. Given that we cannot solve most polynomials, or indeed other equations, in an exact analytical form, we need to consider a numerical treatment.

Let's look now at the most important such scheme. It leads naturally to the consideration of the solution of polynomial equations by iteration of rational functions, and this chapter is a brief introduction to this theory. Entire books can and have been written about both the art and mathematics of this. In the view of the author there is none better than that by Beardon (1991), which should be consulted by anyone serious about exploring the matter thoroughly. This chapter contains only introductory *analytical* comments on the matter, and for the most part we shall focus on exploring the art with *Mathematica*!

So here we take a novel route, looking at how the business of equation solving, which was the motivation for introducing complex numbers in the 16th century, becomes a whole new area of interest when we combine complex numbers with a computer system.

Note that in this chapter we shall be producing moderately complicated graphics. A machine running at 1 GHz or better is recommended for interactive use of the more complicated examples presented. If you are using an old machine just lower the plot resolution.

4.1 Newton–Raphson methods

In Newton–Raphson iteration you consider the solution of an equation given in the form

$$f(x) = 0 \tag{4.1}$$

The idea is to take a starting value, say x_0, and compute $f(x_0)$. Now consider the tangent to the curve $y = f(x)$ at the point $(x_0, f(x_0))$, which has slope $f'(x_0)$, and consider where this tangent intercepts the x axis. We call the value of x at which this happens x_1. In travelling from x_0 to x_1, the increase in x is $(x_1 - x_0)$, and the increase in y is $(0 - f(x_0))$. The slope of the tangent line is therefore given by

$$\frac{0 - f(x_0)}{x_1 - x_0} \tag{4.2}$$

and this must be equal to $f'(x_0)$. This gives us an equation governing x_1:

$$\frac{0 - f(x_0)}{x_1 - x_0} = f'(x_0) \tag{4.3}$$

which we can solve for x_1, obtaining

$$x_1 = x_0 - \frac{f(x_0)}{f'(x_0)} \tag{4.4}$$

This now defines a new iteration scheme, since we can just repeat this process, defining

$$x_{n+1} = x_n - \frac{f(x_n)}{f'(x_n)} \tag{4.5}$$

This defines the famous 'Newton–Raphson' iteration scheme.

4.2 *Mathematica* visualization of real Newton–Raphson

You can use *Mathematica* to gain a better understanding of the geometry of the Newton–Raphson scheme. Here is a short program, in the form of a Module, that does the iteration and draws the geometrical picture of the iteration scheme:

```
NRIter[func_, xzero_, n_:5] :=
Module[{pointlist = {}, x, xold = xzero, xnew, f, df,
xl, xr,k},
f[x_] = func[x];
df[x_] = D[func[x], x];
Do[
(pointlist = Join[pointlist, {{xold, 0}}, {{xold,
f[xold]}}];
xnew = xold - f[xold]/df[xold];
xold = xnew),
{k,1,n}];
xl = Min[First[Transpose[pointlist]]]-0.5;
xr = Max[First[Transpose[pointlist]]] + 0.5;
Plot[f[x], {x, xl, xr}, PlotRange -> All,
PlotStyle -> {{Thickness[0.001], Dashing[{0.005,
0.005}]}},
Epilog -> {Thickness[0.001], Line[pointlist]}]]
```

The program works by constructing a list, called **pointlist**, to which is added new values of the pairs $(x_n, 0)$, $(x_n, f(x_n))$. We then plot the function together with a line through the final value of pointlist. Here is **NRIter** applied to a straightforward polynomial, $f(x) = x^3 + x^2 - 3$, expressed as a *Mathematica* pure function, where # is the argument:

```
NRIter[(#^3+#^2-3)&, 3.5]
```

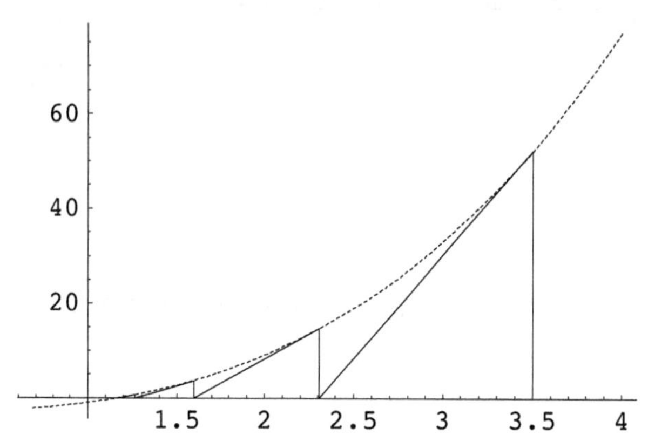

As you can see, you do not have to change the function very much to obtain more complicated behaviour:

```
NRIter[Sin, 1.404091, 10]
```

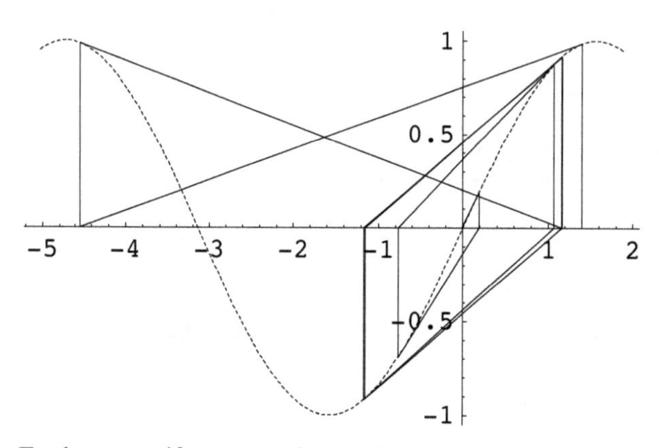

Furthermore, if you vary the starting point, you can wander off to a rather distant root:

```
NRIter[Sin, 1.55, 8]
```

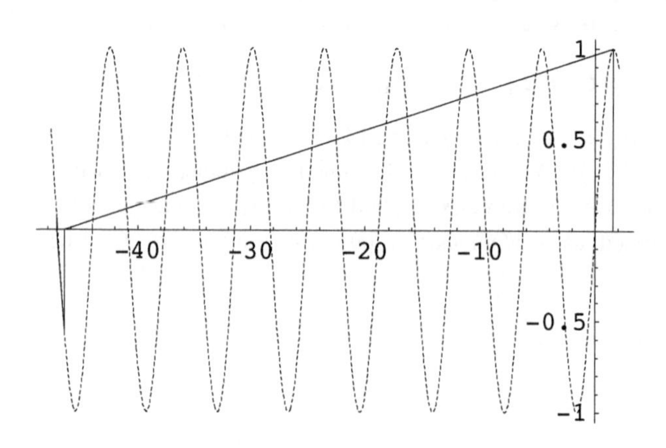

■ The **FindRoot** function

There is a standard kernel function **FindRoot** that, in its default operation on differentiable functions, does Newton–Raphson automatically:

```
? FindRoot
```

```
FindRoot[lhs==rhs, {x, x0}] searches for a numerical
  solution to the equation lhs==rhs, starting with x=x0. FindRoot[
  {eqn1, eqn2, ... }, {{x, x0}, {y, y0}, ... }] searches for
  a numerical solution to the simultaneous equations eqni.
```

This is a very useful function, but one of the lessons of this chapter is that it must be used carefully. The following examples make the point. Let's make a list of six starting values, and apply **FindRoot** for each one:

```
xlist = {1.4, 1.413, 1.414, 1.5, 1.57, 1.5708}
```

{1.4, 1.413, 1.414, 1.5, 1.57, 1.5708}

```
Map[FindRoot[Sin[x] == 0, {x, #}] &, xlist]
```

$\{\{x \to -3.14159\}, \{x \to -6.28319\}, \{x \to -9.42478\}, \{x \to -12.5664\}, \{x \to -25.1327\}, \{x \to 28.2743\}\}$

If we choose starting values in the neighbourhood of a stationary value, we converge to six different multiples of π for the answer!

4.3 Cayley's problem: complex global basins of attraction

Our initial investigations raise various questions. As we vary the starting value, how does the root eventually found vary? Alternatively, for a given root, which starting values will lead eventually to that root? From the point of view of numerical analysis, we aim to find starting values that are sufficiently close to a root. In 1879 A. Cayley formulated this problem in a rather more interesting global context, set in the complex plane. For a given function f, define the Newton–Raphson map NR_f by

$$\mathrm{NR}_f(x) = x - \frac{f(x)}{f'(x)} \tag{4.6}$$

and set $\mathrm{NR}_{f,k}$ to be NR_f applied k times.

$$\mathrm{NR}_{f,1}[x] = \mathrm{NR}_f[x] \tag{4.7}$$

$$\mathrm{NR}_{f,k+1}[x] = \mathrm{NR}_f[\mathrm{NR}_{f,k}[x]] \tag{4.8}$$

Let x^* be a root of the equation $f(x) = 0$. The *global basin of attraction* of x^* is the set

$$B(x^*) = \{x \in \mathbb{C} \mid \mathrm{NR}_k[x] \to x^* \text{ as } k \to \infty\} \tag{4.9}$$

that is, the set of starting values that will yield the given root after many iterations. Cayley was interested in finding the geometrical form of $B(x^*)$ when f is a simple polynomial of low degree, and solved the problem completely for quadratic polynomials. As he said (Cayley, 1879): 'Throwing aside the restrictions as to reality, we have what I call the Newton–Fourier Imaginary Problem ... The solution is easy and elegant in the case of a quadratic equation, but the next succeeding case of a cubic equation appears to present considerable difficulty.' This was something of an understatement. Indeed, it would take about a hundred more years, fast computers, and an appreciation of fractal geometry, before Cayley's difficulty could properly be appreciated. Our main task in this chapter (apart from having some fun with the computer graphics) is to explore these issues.

■ Cayley's solution for the quadratic

Let's consider the equation arising from solving the quadratic equation

$$f(z) = z^2 - 1 = 0 \tag{4.10}$$

The Newton–Raphson mapping for this case is easily seen to be

$$\mathrm{NR}_f(z) = \frac{1}{2}\left(z + \frac{1}{z}\right) \tag{4.11}$$

The solution for the basins of attraction of this problem will represent an early encounter for you with the type of mapping known as a Möbius map. These maps will reappear many times in different applications. Here its role is to make a transformation that makes it obvious what the basins of attraction are for the Newton-Raphson solution of a quadratic equation. The form of the mapping can be appreciated by noting that what we want to do is to transform the point $z = 1$ to the origin, and the point $z = -1$ to infinity. This can be achieved by setting

$$w = M(z) = \frac{z-1}{z+1} \tag{4.12}$$

You can easily check that the corresponding inverse mapping is obtained by setting

$$z = M^{-1}(w) = \frac{1+w}{1-w} \tag{4.13}$$

What mapping does $\mathrm{NR}(z)$ induce in w-coordinates? This is the mapping

$$w \to \tilde{\mathrm{NR}}(w) = M(\mathrm{NR}(M^{-1}(w))) \tag{4.14}$$

Some algebra (it is left for you as an exercise) leads to the observation that

$$\tilde{\mathrm{NR}}(w) = w^2 \tag{4.15}$$

If you work in w-coordinates, and iterate this mapping, it is obvious that $w_n \to 0$ if $|w_0| < 1$, and that $w_n \to \infty$ if $|w_0| > 1$. So in these coordinates, the basins of attraction are the interior and exterior of the unit circle. Points on the unit circle just get mapped around on the unit circle, and the unit circle is the boundary between the two basins. Now you just have to interpret this in z-coordinates. Clearly, if $w_n \to 0$, then $z_n \to 1$, and if $w_n \to \infty$, then $z_n \to -1$, and what is required is an understanding of what the unit circle, in w-coordinates, is in z-coordinates. If you set

$$w = e^{i\phi} \tag{4.16}$$

then it is a simple exercise to show that

$$z = i \cot\left(\frac{\phi}{2}\right) \tag{4.17}$$

so that as ϕ varies the whole of the imaginary axis is obtained. So in z-coordinates the boundary between the two basins is the imaginary axis. In other words, the basins are given by two (left and right) half-planes:

$$B(1) = \{z \in \mathbb{C} \mid \text{Re}(z) > 0\} \tag{4.18}$$

$$B(-1) = \{z \in \mathbb{C} \mid \text{Re}(z) < 0\} \tag{4.19}$$

In the case of a general quadratic mapping similar results apply – the plane is divided into two basins by the perpendicular bisector of the line joining the two roots.

■ *Mathematica* checks on Cayley's algebra

You can use *Mathematica* to help check the algebra developed in Cayley's result, as follows. First, define the Möbius mapping:

```
M[z_] := (z - 1) / (z + 1)
```

Now compute and check the inverse:

```
InverseM[w_] = z /. Solve[w == M[z], z][[1]]
```

$$\frac{-w-1}{w-1}$$

```
Simplify[M[InverseM[x]]]
```

$$x$$

Now define the Newton–Raphson mapping:

```
NR[z_] = 1 / 2 (z + 1 / z)
```

$$\frac{1}{2}\left(z+\frac{1}{z}\right)$$

Now compute the mapping induced on w:

```
M[NR[InverseM[w]]]
```

$$\frac{\frac{1}{2}\left(\frac{-w-1}{w-1}+\frac{w-1}{-w-1}\right)-1}{\frac{1}{2}\left(\frac{-w-1}{w-1}+\frac{w-1}{-w-1}\right)+1}$$

Clearly, some simplification is required!

```
Simplify[%]
```

$$w^2$$

Finally, note that the parametrization of the imaginary axis in z-coordinates can be seen to be the unit circle in w-coordinates, as follows:

```
Simplify[M[I Cot[ϕ / 2]]]
```

$$\cos(\phi) + i \sin(\phi)$$

4.4 Basins of attraction for a simple cubic

Cayley's attempts to establish a similar result for cubics failed. A full appreciation of the reasons why is beyond the scope of this book, but you can get a good grip on why matters are more complicated by simulation using *Mathematica.* If you want to investigate the matter properly, see the text by Beardon (1991). First, let's construct the Newton–Raphson mapping:

```
g[z_] := z^3 - 1
```

```
z - g[z] / g'[z]
```

$$z - \frac{z^3 - 1}{3 z^2}$$

```
Simplify[%]
```

$$\frac{2z}{3} + \frac{1}{3 z^2}$$

What we want to do is to evaluate the progress of iterates of this mapping to each of the three roots of the cubic equation. These three roots of unity were discussed in Chapter 2, and can just as easily be found using *Mathematica*:

```
z /. Solve[g[z] == 0, z]
```

$$\left\{1, -\sqrt[3]{-1}, (-1)^{2/3}\right\}$$

```
N[%]
```

{1., −0.5 − 0.866025 *i*, −0.5 + 0.866025 *i*}

There are various ways of developing an algorithm to manage this. One approach makes use of the standard kernel function **FixedPoint**:

```
? FixedPoint
```

```
FixedPoint[f, expr] starts with expr, then
  applies f repeatedly until the result no longer changes.
```

This will tell us which root has been found, but not how rapid the convergence is. (Look on *MathSource* for fractal generators that implement this approach, and see, for example, Dickau, 1997). Although it requires some more computation, we are interested in developing a *convergence-time* algorithm analogous to the escape-time algorithm we used for the Mandelbrot set. This can also be sorted out by applying the *Mathematica* function **Length** to the function **FixedPointList**, which generates a list of the iterates – the length of this list with a suitable termination criteria gives a convergence time. However, the author's own experiments with this on the Mandelbrot set, as in Chapter 6, suggests that it is less efficient than the compiled *Mathematica* approach. It is left for you to pursue these ideas as programming examples for the Newton–Raphson case, and draw your own conclusions on efficiency, in the first two exercises. The results of this may change as the kernel develops!

■ A convergence-time algorithm

We shall define a function that counts how rapidly progress is made towards a given root, and that separates the roots by mapping the counter into three distinct intervals within the set 0 to 255. In order to evaluate this rapidly, the function to evaluate it is compiled:

```
NewtonCounter = Compile[{{z,_Complex}},
Module[{counter=0, zold=N[z]+1.0, znew=N[z]},
If[Abs[znew] < 10^(-9), znew = 10^(-9)+0.0*I,
znew=znew];
For[counter = 0,
(Abs[zold-znew] > 10^(-6)) && (counter < 85),
counter++,
(zold = znew;znew = 2*zold/3 + 1/(3*zold^2))];
Which[Abs[znew-1] < 10^(-4), counter,
Abs[znew+0.5-0.866025I] < 10^(-4), 85+counter,
Abs[znew+0.5+0.866025I] < 10^(-4), 170+counter,
True,255]]];
```

Let's make a quick check to see that it is working:

```
{NewtonCounter[1.001], NewtonCounter[-10 + I]}
```

{2, 105}

Using **NewtonCounter**, an array of values can be built up for a region of the complex plane:

```
NewtonArray[{{remin_, remax_}, {immin_, immax_}},
steps_] :=
{{{remin, remax}, {immin, immax}},
Table[Re[NewtonCounter[x + y I]],
{y, immin, immax, (immax-immin)/steps},
{x, remin, remax, (remax-remin)/steps}]}
```

■ Colouring schemes

The next part of our plan is to colour points according to how many iterations is takes to get to each root. To get nice pictures it is helpful to introduce a capping function that allows us to amplify colours without breaking *Mathematica*'s rules about the arguments to functions such as **RGBColor[x,y,z]**, which require that their arguments lie between 0 and 1.

```
tr[x_] = Which[x<0,0,x>1,1,True,x];
```

The first colouring scheme will colour all points in the basins of attraction of complex roots black, and colour those approaching $z = 1$ according to the number of iterations.

```
NewtonColorOne[x_] :=
If[x < 0.333, Hue[6x,1,1], Hue[0,0,0]]
```

Now we define two other color schemes, based on the use of three different colours, one for each root:

```
NewtonColorAll[x_] :=
Which[
x <0.333, RGBColor[tr[5(3x)],tr[5(3x)],tr[5(3x)]],
0.334<x<0.666, RGBColor[tr[5(3x-1)],tr[5(3x-1)],0],
0.667<x<1, RGBColor[0,tr[5(3x-2)],tr[5(3x-2)]],
True, RGBColor[0,0,0]]

NewtonColorRGB[x_] :=
Which[
x <0.333, RGBColor[tr[5*3x],0,0],
0.334<x<0.666, RGBColor[0,tr[5*(3x-1)],0],
0.667<x<1, RGBColor[0,0,tr[5*(3x-2)]],
True, RGBColor[0,0,0]]
```

The plotting routine now follows:

```
NewtonPlot[
{{{remin_, remax_}, {immin_, immax_}}, data_},
colorfunc_] :=
ListDensityPlot[data,
AspectRatio -> (immax-immin)/(remax - remin),
Mesh -> False, Frame -> False,
PlotRange -> {0, 255},
ColorFunction -> colorfunc]
```

■ Computing the fractals!

In what follows, you will be doing quite large amounts of computation. You are advised to try out the following routines to get an idea of how long calculations take, and to start with a 200 by 200 array. The following timing is for a Power Macintosh G4 at 1.4 GHz, and is for much larger array.

```
Timing[region = NewtonArray[{{-2, 2}, {-2, 2}}, 1000];]
```

{80.49 Second, Null}

```
NewtonPlot[region, NewtonColorOne]
```

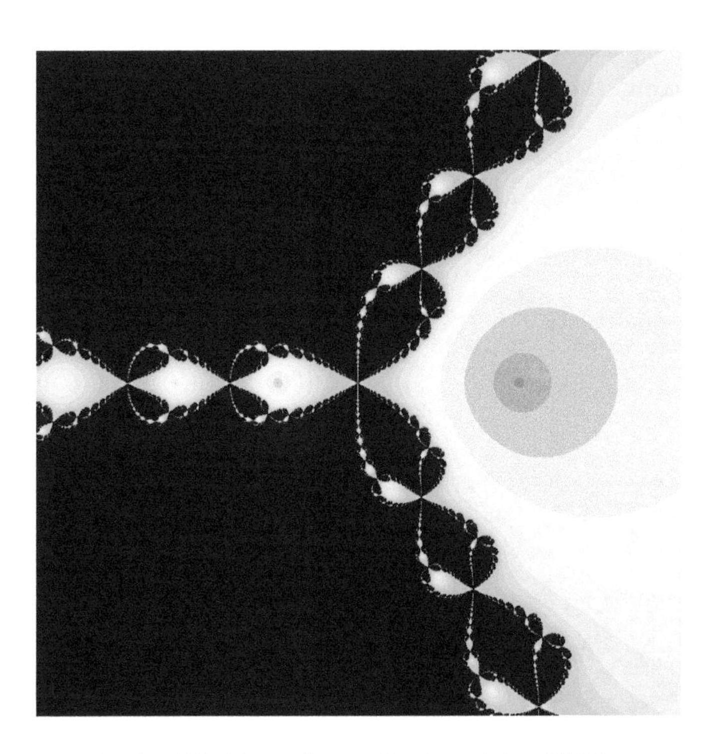

```
NewtonPlot[region, NewtonColorAll]
```

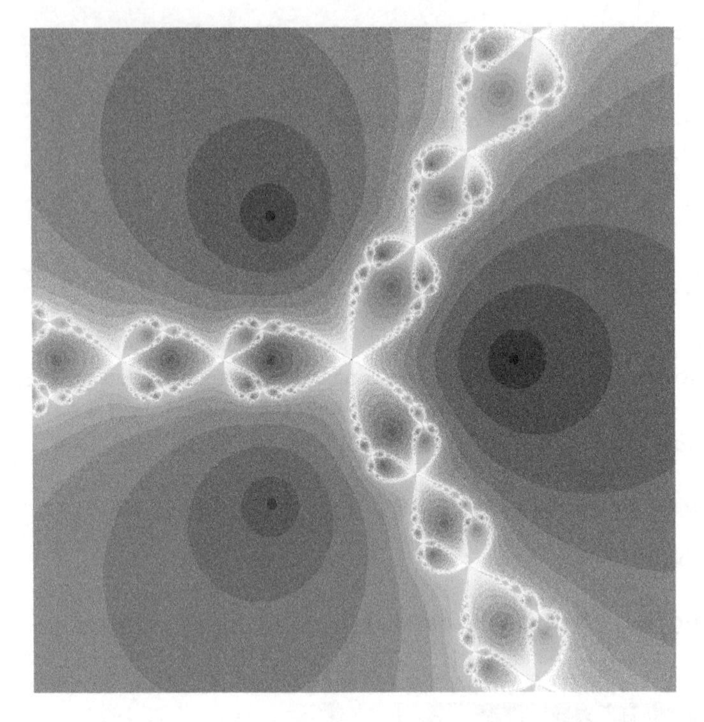

```
NewtonPlot[region, NewtonColorRGB]
```

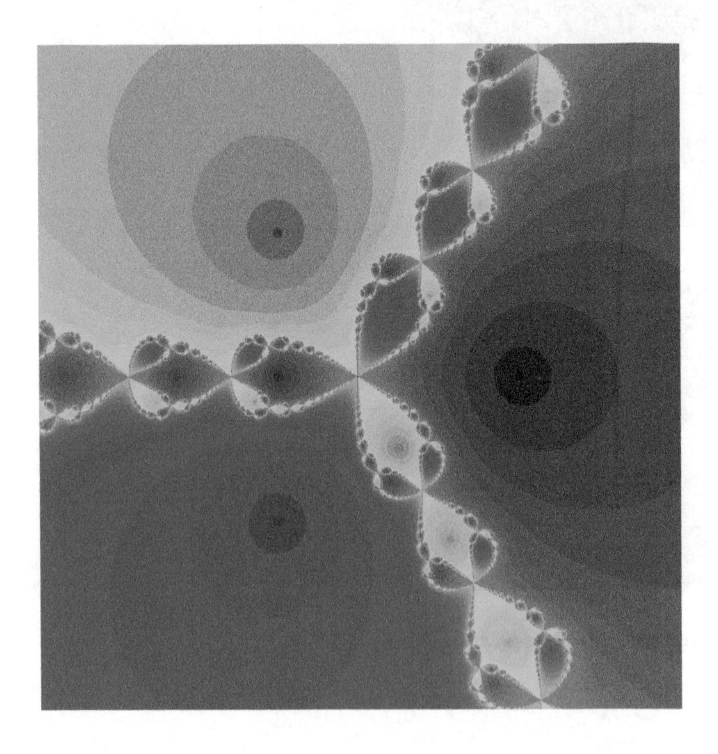

4.5 More general cubics

The plots above suggest that given almost any starting value, convergence to a root will be obtained – there are no regions, except the boundary points, that do not head off to one of the roots. This is absolutely not the case in general. Consider a family of cubics still with one zero at $z = 1$, parametrized by a complex variable r:

```
p[z_, r_] := z^3 + (r-1) z - r
```

```
Factor[p[z,r]]
```

$$(z-1)(z^2+z+r)$$

```
roots[r_] := z /. Solve[p[z,r]==0,z]
```

```
roots[r]
```

$$\left\{1, \frac{1}{2}\left(-\sqrt{1-4r}-1\right), \frac{1}{2}\left(\sqrt{1-4r}-1\right)\right\}$$

Although this might look like a special type of cubic, in fact any cubic is equivalent to one of this family by a linear transformation – so we are dealing with a general cubic (see the exercises). This time the Newton–Raphson mapping is given by:

```
NR[z_, r_] = Together[z - p[z,r]/(D[p[z,r], z])]
```

$$\frac{2z^3+r}{3z^2+r-1}$$

A simple way of treating this map is to supply the parameter r as an additional argument, and the value of just one of the two complex roots. In the following we shall treat r as real.

```
NewNewtonCounter = Compile[{{z,_Complex}, {r,_Real},
{otherroot, _Complex}},
Module[{counter=0, zold=N[z]+1, znew=N[z]},
If[Abs[znew] < 10^(-9), znew = 10^(-9)+0.0*I,
znew=znew];
For[counter = 0,
(Abs[zold-znew] > 10^(-6)) && (counter < 85),
counter++,
(zold = znew;znew = (r + 2*zold^3)/(-1+r+3*zold^2))];
Which[Abs[znew-1] < 10^(-4), counter,
Abs[znew-otherroot] < 10^(-4), 85+counter,
Abs[znew-Conjugate[otherroot]] < 10^(-4), 170+counter,
True,255]]];
```

```
NewNewtonArray[r_, {{remin_, remax_}, {immin_,
immax_}}, steps_] :=Module[{croot = -N[(1+Sqrt[1-4
r])/2]},
{{{remin, remax}, {immin, immax}},
Table[
Re[NewNewtonCounter[x + y I,r,croot]],
{y, immin, immax, (immax-immin)/steps},
{x, remin, remax, (remax-remin)/steps}]}]
```

Let's look first to see what happens when $r = 2$ (try this yourself with array size 200):

```
region = NewNewtonArray[2,{{-2, 2}, {-2, 2}}, 1000];

NewtonPlot[region, NewtonColorRGB]
```

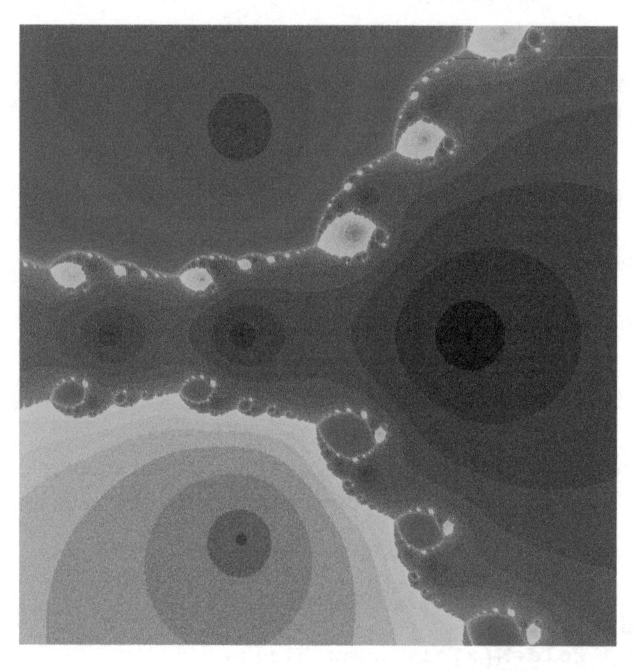

We obtain strikingly different behaviours for different values of r. In the following case there is a large central region that does not converge to any of the roots:

```
region = NewNewtonArray[0.05,{{-2, 2}, {-2, 2}},
1000];
NewtonPlot[region, NewtonColorRGB]
```

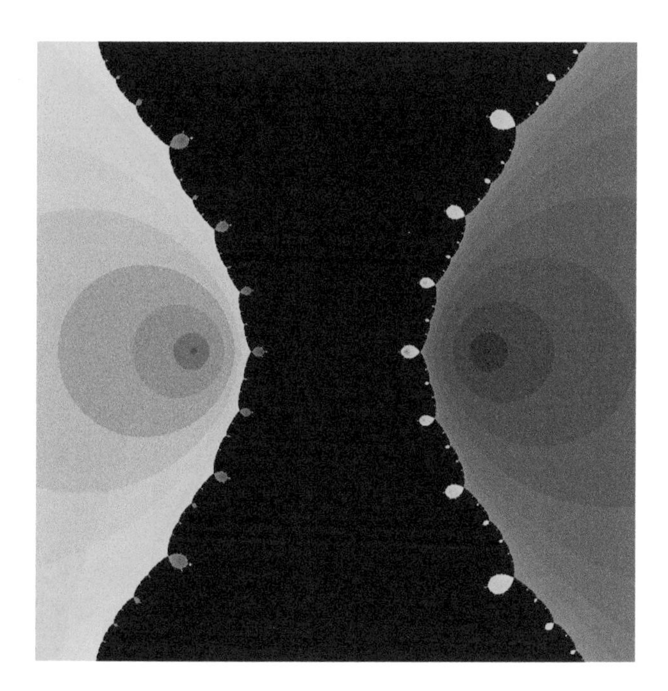

■ An interesting place to visit!

There is a particularly amusing region of the complex plane that emerges in the set for $r = 0.5$. The following speaks for itself. You might like to try other colour schemes!

```
SillyFaceColor[x_] :=
If[EvenQ[Floor[255*x]], RGBColor[0,0,0],
RGBColor[1,1,1]]

SillyFaceColorB[x_] :=
Which[
x <0.333, If[EvenQ[Floor[255*x]], RGBColor[0,0,0],
RGBColor[tr[5(3x)],0,0]],
0.334<x<0.666, If[EvenQ[Floor[255*x]],
RGBColor[0,0,0], RGBColor[0,0,tr[5(3x-1)]]],
0.667<x<1, If[OddQ[Floor[255*x]], RGBColor[0,0,0],
RGBColor[0,0,tr[5(3x-2)]]],
True, RGBColor[0,0,0]]

FaceNewtonPlot[{{{remin_, remax_}, {immin_, immax_}},
data_},colorfn_] :=
ListDensityPlot[Reverse[Transpose[data]],
Mesh -> False, Frame -> False,
PlotRange -> {1, 255}, AspectRatio -> Automatic,
ColorFunction -> colorfn]
```

We shall plot this in black and white at high resolution (this will take a few minutes on a 1.4 GHz machine).

```
region = NewNewtonArray[0.5,{{-2.4, -1.85}, {-0.24,
0.24}}, 2000];
```

```
FaceNewtonPlot[region, SillyFaceColor]
```

You might want to experiment with other colour schemes for this region. Try for example

```
FaceNewtonPlot[region, SillyFaceColorB]
```

and also make up some of your own!

4.6 Higher-order simple polynomials

We now consider polynomials of higher degree, but restrict attention to the simple case:

$$f(z) = z^n - 1 = 0 \tag{4.20}$$

for which the Newton-Raphson mapping is just

$$\left(1 - \frac{1}{n}\right)z + \frac{1}{n\,z^{n-1}} \tag{4.21}$$

We generalize the maps for the simple cubic to treat these mappings:

```
GenNewtonCounter = Compile[{{z,_Complex},
{n,_Integer}}, Module[{counter=0, zold=N[z]+1,
znew=N[z], k, m},
If[Abs[znew] < 10^(-9), znew = 10^(-9)+ 0.0*I,
znew=znew]; For[counter = 0,
(Abs[zold-znew] > 10^(-6)) && (counter < 85),
counter++,
(zold = znew;znew = zold*(1-1/n) + 1/(n*zold^(n-1)))];
For[k = 0, (Abs[znew - Exp[2*Pi*I*k/n]] > 0.1) && (k <
n), k++, counter = counter + 85]; counter]];

GenNewtonArray[{{remin_, remax_}, {immin_, immax_}},
steps_, n_] :=
{{{remin, remax}, {immin, immax}},
Table[GenNewtonCounter[x + y I, n],
{y, immin, immax, (immax-immin)/(steps-1)},
{x, remin, remax, (remax-remin)/(steps-1)}]}

GenNewtonPlot[
{{{remin_, remax_}, {immin_, immax_}}, data_}, n_,
colorfunc_] :=
ListDensityPlot[data,
AspectRatio -> (immax-immin)/(remax - remin),
Mesh -> False, Frame -> False,
PlotRange -> {0, 85*n},
ColorFunction -> (colorfunc[n, #]&)]

tr[x_] = Which[x<0,0,x>1,1,True,x];
```

The colour functions are similar to those defined previously, but we also make one adapted specially to the degree seven case:

```
GenNewtonColorOne[n_, x_] :=
If[x < 1/n, Hue[2*n*x], RGBColor[0,0,0]];

GenNewtonColorSeven[n_, x_] :=
Which[x <0.14285,
RGBColor[tr[(3)7x],tr[(3)7x],tr[(3)7x]],
0.14286<x<0.28571,
RGBColor[tr[(3)(7x-1)],tr[(3)(7x-1)],0],
0.28572<x<0.42857,
RGBColor[0,tr[(3)(7x-2)],tr[(3)(7x-2)]],
0.42858<x<0.57142,
RGBColor[tr[(3)(7x-3)],0,tr[(3)(7x-3)]],
0.57143<x<0.71428, RGBColor[tr[(3)(7x-4)],0,0],
0.71429<x<0.85714, RGBColor[0,tr[(3)(7x-5)],0],
0.85715<x<1,        RGBColor[0,0,tr[(3)(7x-6)]],
True, RGBColor[0,0,0]]

genregion = GenNewtonArray[{{-2, 2}, {-2, 2}}, 1000,7];
```

We can use this program in various ways. First, let's look at the basin of attraction for the real and positive root:

```
GenNewtonPlot[genregion, 7, GenNewtonColorOne]
```

We can identify the basins of attraction for *all* the roots with a different colour scheme:

```
GenNewtonPlot[genregion, 7, GenNewtonColorSeven]
```

4.7 Fractal planets: Riemann sphere constructions

This can consume a substantial amount of memory and time. You build 'fractal planets' by mapping the complex plane onto the unit sphere. In terms of standard polar coordinates, you set $z = e^{i\phi}\cot(\theta/2)$. Here our convention is that θ measures the angle in radians from the North pole. Here is one mapping that gives pleasing results. Try much smaller values for **PlotPoints** first, and increase the resolution to the limit allowed by your computer's memory. This is particularly important when using older versions of *Mathematica* or on systems without virtual memory. The following code is for Versions 4 and 5 of *Mathematica*.

```
ParametricPlot3D[{Sin[θ] Cos[φ], Sin[θ]
Sin[φ],Cos[θ],{EdgeForm[],
NewtonColorRGB[NewtonCounter[Exp[I φ] Cot[θ/2]]/255]}},
{θ, Pi/2, Pi-0.001}, {φ, 0, 2 Pi}, Lighting -> False,
Boxed -> False,
Axes -> False, RenderAll -> False, PlotPoints ->
{400,800},
ViewPoint -> {0,0,-3}, Background -> RGBColor[0,0,0]]
```

This is a picture of the cubic 'Cayley planet' (let's call it 'Cayley 3') viewed from below its South pole. You can explore many fractal planets of this class by suitable modifications for the code. Here, for example, is a view of a corresponding 'Cayley 7' viewed from above the equator:

```
ParametricPlot3D[
{Sin[θ] Cos[ϕ], Sin[θ] Sin[ϕ],Cos[θ],
{EdgeForm[],
GenNewtonColorSeven[7, GenNewtonCounter[Exp[I ϕ]
Cot[θ/2],7]/595]}},
{θ, 0.001, Pi-0.001}, {ϕ, 0, Pi},
Lighting -> False, Boxed -> False,
Axes -> False, RenderAll -> False,
PlotPoints -> {400,800},
ViewPoint -> {0,3,0},Background -> RGBColor[0,0,0]]
```

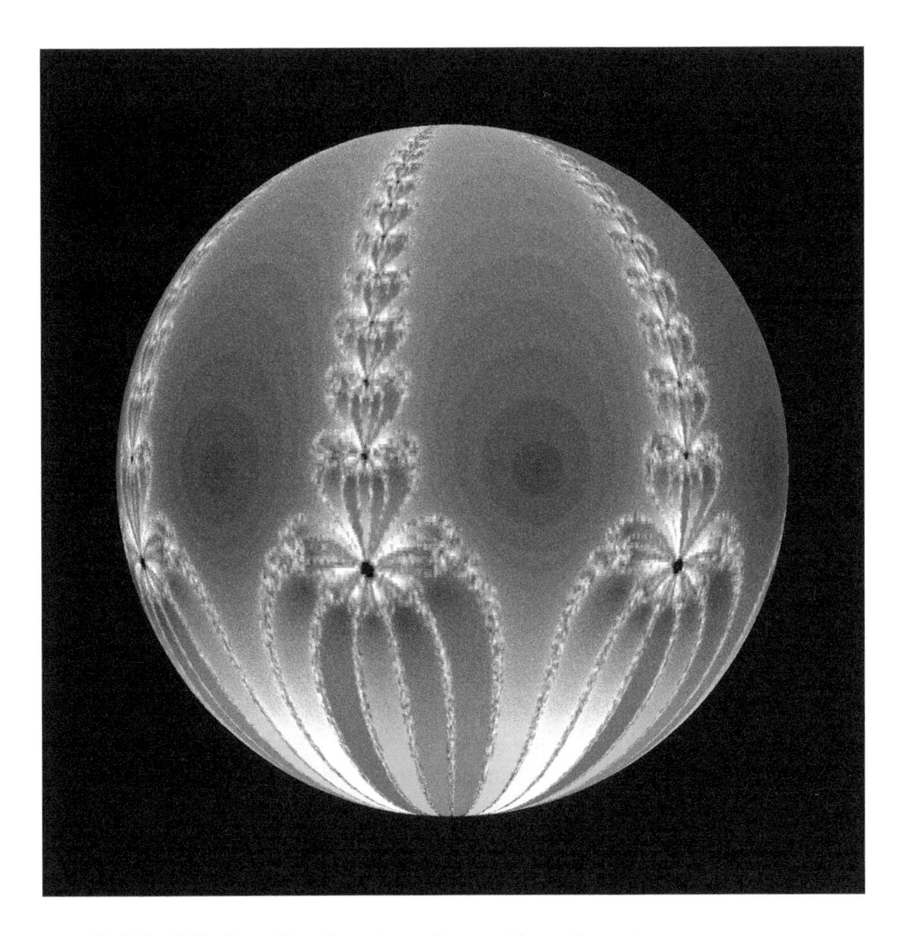

■ Modified code showing the entire planet

In you have problems getting the code giving above to work in your current version of *Mathematica*, you might like to try the following alternative form:

```
ParametricPlot3D[{Sin[θ] Cos[ϕ], Sin[θ]
Sin[ϕ],Cos[θ]},
{θ, 0.001, Pi-0.001}, {ϕ, 0, 2 Pi},
ColorFunction -> (NewtonColorRGB[NewtonCounter[Exp[I
#5] Cot[#4/2]]/255]&),
ColorFunctionScaling -> False,
Lighting -> True, Boxed -> False, Mesh -> False,
Axes -> False, RenderAll -> False,
PlotPoints -> {50,100},
ViewPoint -> {0,0,-3}, Background -> RGBColor[0,0,0]]
```

```
ParametricPlot3D[
{Sin[θ] Cos[ϕ], Sin[θ] Sin[ϕ],Cos[θ]},
{θ, 0.001, Pi-0.001}, {ϕ, 0, 2 Pi},
ColorFunction -> (GenNewtonColorSeven[7,
GenNewtonCounter[Exp[I #5] Cot[#4/2],7]/595]&),
ColorFunctionScaling -> False,
Lighting -> True, Boxed -> False, Mesh -> False,
Axes -> False, RenderAll -> False,
PlotPoints -> {50,100},
ViewPoint -> {0,3,0},Background -> RGBColor[0,0,0]]
```

In these examples a lower value of **PlotPoints** has also been chosen initially in order to get things working on a slower machine. Also, the polar coordinates θ and ϕ run over their full ranges $[0, \pi]$ and $[0, 2\pi]$ respectively, so that you have the entire fractal planet to view, for example by changing the viewpoint. The value of θ is truncated at the North and South pole to avoid kernel complaints about singular behaviour. Also see the on-line supplement and CD.

Exercises

4.1 ❄ Develop a variant of the Newton–Raphson scheme for cubics that uses **FixedPoint** and a simple coloring scheme (i.e. ignore the convergence time issue, and colour only according to which root is found.) Note that **FixedPoint** takes an optional second argument that gives the maximum number of iterations to be used.

4.2. ❄ Extend the analysis of Exercise 4.1 to use the length of **FixedPointList** to generate a colouring scheme for the cubic Newton–Raphson map based on the **Length** function, and that colours the plot also according to convergence time Note that **FixedPointList** takes an optional further argument that gives the maximum number of iterations to be used. Compare the efficiency with that of the compiled routines used in this chapter.

4.3. Show that by the use of a suitable linear transformation a cubic polynomial can be reduced to one of the form

$$z^3 + (r - 1)\,z - r$$

with a complex parameter r.

4.4. ❄ Show how the definition of **NewNewtonCounter** can be generalized to cope with complex r, and explore the resulting pictures for sample complex values of r.

4.5. ❄ For the case of real r, construct a routine to produce an animation of **NewtonPlot** with r playing the role of a frame counter or time coordinate. Try this out first at low resolution, and when you are happy with the results, increase the number of plot points. The interval between $r = 0$ and $r = 2$ is particularly interesting. Explore this first and then zoom in on $0.2 < r < 0.3$. By examining the factorization of the cubic, explain what is special about the point $r = 0.25$. What happens to the iteration scheme at this point, and why?

4.6. (To be considered after reading Chapter 5). Show that if z_0 is a zero of the function f: $f(z_0) = 0$, and

$$F(z) = z - \frac{f(z)}{f'(z)}$$

then z_0 is a *stable* fixed point of F, in the sense of Chapter 5. You may assume that $f(z)$ may be written as

$$f(z) = (z - z_0)^k g(z)$$

for some differentiable $g(z)$ with $g(z_0)$ non-zero. (If you know about the properties of holomorphic functions, you should note that this assumption is guaranteed by the Taylor series, if f is a holomorphic function.)

5 A complex view of the real logistic map

Introduction

You have already read about how to motivate the introduction of complex numbers by the need to solve quadratic equations, and have seen how to solve higher order polynomial equations both through 'pen and paper' analysis and with the help of *Mathematica*. In the previous chapter you looked at Newton–Raphson iteration. This is not the only way of defining an iterative solution method, and there is another approach called 'cobwebbing' which is the subject of this chapter.

You are now in a position to perform a basic investigation of some of the most fascinating topics in modern mathematics: period doubling and transitions to chaos. This topic can be introduced by considering simple quadratic or cubic functions. However, rather than solving a simple quadratic or cubic equation, you are now going to be concerned with applying a function over and over again, given a starting value. Under certain circumstances, this has the effect of finding the solutions to the original equation, but in other situations you will be led to the solutions of other polynomial equations. Hence the need for a complex view.

There are many good reasons for you to investigate these topics. First, you should appreciate the emergence of complexity and beauty from the iteration (repeated application) of a simple quadratic or cubic map. Second, you should appreciate that there is some value in doing 'experimental mathematics'. However, here and elsewhere in this book we shall be concerned with appreciating the special role that complex numbers play. In particular, you should appreciate that on its own 'experimental mathematics' is not particularly useful – understanding *why* things happen is considerably more useful than just seeing what happens. It is not enough to just report the results of feeding non-linear equations to a computer and plotting the (admittedly fascinating) results. This is not, at least in itself, mathematics, at least in the view of the author (unless it is in the context of reporting an entirely new phenomenon), but does raise interesting questions. It is also, however, sometimes fun, and we shall indulge in this later in this chapter and elsewhere.

You will be encouraged to use the power of *Mathematica* to understand, given the theory of complex numbers, why it is likely that certain real iterated maps should behave in a certain way. When this has been accomplished, you will be encouraged to do some numerical experiments to check the theoretical expectations. Later in this book (Chapter 7) you will see how to generalize these real iterated maps to complex iterated maps. We begin this chapter by exploring the theory of 'cobwebbing' as a method for finding the roots of equations. Then the logistic and odd logistic maps are introduced, and the relation to cobwebbing explored. You may wonder why the term 'cobwebbing' is used – this will emerge from one visualization of the theory, later on.

■ The two tracks through this chapter

Some of the material in this chapter is very detailed and may be difficult to take in on a first read. You have a couple of choices about how to read this chapter. The first, and rather uncompromising approach, is to work through in the given order of the sections. The idea of this is that it is possible to predict a lot of what happens with iterated mappings by doing some careful analysis. However, this analysis could easily put you off. A second approach is to read Sections 5.1 and 5.2, and then to skip to Section 5.6 and do the experiments first without having a prediction as to what should happen. Either way, you should work through the exercises at the end to fully appreciate what is going on.

5.1 Cobwebbing theory

We are interested in solving equations that can be written in the form

$$x = f(x) \tag{5.1}$$

The idea is to try to solve such an equation by iteration. We make a starting guess, say x_0, and keep applying the function over and over again. So we set

$$\begin{aligned} x_1 &= f(x_0) \\ x_2 &= f(x_1) \\ &\dots \\ x_{n+1} &= f(x_n) \end{aligned} \tag{5.2}$$

Does this work? Under certain circumstances, it does. Suppose the solution is x, and that our current iterate is x_n. The current error is

$$\epsilon_n = x_n - x \tag{5.3}$$

The error at the next step is

$$\epsilon_{n+1} = x_{n+1} - x = f(x_n) - f(x) \tag{5.4}$$

Suppose that x_n is close enough to x that we can approximate this last relation, assuming differentiability, by a Taylor series expansion:

$$\epsilon_{n+1} = f(x_n) - f(x) = (x_n - x) f'(x) + O(x_n - x)^2 \tag{5.5}$$

So to first order, taking absolute values:

$$|\epsilon_{n+1}| = |\epsilon_n| \, |f'(x)| + O(|\epsilon_n|^2) \tag{5.6}$$

So the errors can only diminish if the absolute value of the derivative is less than unity at the solution. We need

$$|f'(x)| < 1 \tag{5.7}$$

We refer to a solution of Eq. (5.1) satisfying this property as a *stable solution*; otherwise it is an unstable solution. You will see some examples of stable and unstable solutions in the next section. Note that the condition of stability does not guarantee the convergence for any given starting value, it just says that if we are close enough to neglect higher-order terms, we will get closer to the solution.

5.2 ✵ Definition of the quadratic and cubic logistic maps

The logistic maps are mappings parametrized by a variable, λ. There is a great deal of analysis, including work involving *Mathematica*, on the ordinary quadratic logistic map, defined by the equation:

```
OrdLogistic[λ_, x_] := λ x (1 - x)
```

This is a quadratic mapping of great interest, because it is the simplest system exhibiting the phenomena of bifurcations, period doubling and transitions to chaos. It also arises from some fundamental considerations regarding population dynamics. If you wish to explore where this map comes from, see Field and Golubitsky (1992) and the original paper by May (1976). You will be encouraged to investigate this map in the exercises, using methods identical to those given below for a slightly different map. For several reasons, this chapter will develop the ideas using an interesting cubic variant of the usual logistic map, which is the cubic or *odd logistic map*, given by:

```
Logistic[λ_, x_] := λ x (1 - x^2)
```

We shall just use the term **Logistic** to refer to this odd mapping, throughout the text of this chapter. So when we apply this to a variable, x, we obtain:

```
Logistic[λ, x]
```

$$x(1 - x^2)\lambda$$

This has the property that it is an odd function of x, so that it possesses a symmetry. This leads to an additional feature, that of a symmetry-generating bifurcation, not possessed by the ordinary mapping. This is an effect that is particularly important when you come to see how to generalize the logistic maps to complex mappings. Furthermore, the ordinary logistic map has been analysed in detail by several authors working with *Mathematica*, including Gray and Glynn (1991), Wagon (1991) and Maeder (1995). You might like to consult these references if you get stuck on the exercises, but the exercises should be straightforward if you follow the corresponding arguments for the odd map. We shall iterate this map on starting values for x. Our goal is to apply cobwebbing theory to try to predict what will happen. First you will find the fixed points of the logistic map, by solving a cubic equation. This is easy, but you can also get *Mathematica* to sort it out.

5.3 ✡ The logistic map: an analytical approach

Note: this section, and 5.4–5.5, contain rather harder material. You may wish to skip to Section 5.6 on numerical experiments, having just seen the definition of the logistic map, and return to this material later. In that case you are strongly encouraged to come back and review these sections, because they establish the point that complex roots of polynomial equations becoming both real and stable is crucial to a proper understanding of the behaviour of the logistic map.

We shall iterate the cubic logistic map on starting values for x. Our goal is to apply cobwebbing theory to try to predict what will happen. First you find the fixed points of the logistic map, by solving a cubic equation. This is easy, but you can also get *Mathematica* to sort it out:

```
logsol = x /. Solve[x == Logistic[λ, x], x]
```

$$\left\{0, -\frac{\sqrt{\lambda-1}}{\sqrt{\lambda}}, \frac{\sqrt{\lambda-1}}{\sqrt{\lambda}}\right\}$$

So there are three easy solutions. Are they stable? We just work out the derivative mapping and apply it to the solutions:

```
Dlog[λ_, x_] = D[Logistic[λ, x], x]
```

$$(1-x^2)\,\lambda - 2x^2\,\lambda$$

```
Expand[Dlog[λ, logsol]]
```

$$\{\lambda, 3-2\lambda, 3-2\lambda\}$$

This tells us what we want to know right away. The first solution, at zero, is stable if $|\lambda| < 1$, while the other solutions, at $\pm\sqrt{1-\lambda^{-1}}$, are stable provided $1 < \lambda < 2$. So if we iterate the logistic map on a given starting value, we would expect convergence to zero if $|\lambda| < 1$ and convergence to $\pm\sqrt{1-\lambda^{-1}}$ if λ is larger than unity but less than two. Given that we shall focus attention on positive λ, the question therefore arises as to what happens when $\lambda > 2$? We no longer expect to obtain a convergent sequence. But there might be *convergent subsequences*. It is important to appreciate that when we iterate a mapping

$$x = f(x) \tag{5.8}$$

we are simultaneously iterating an infinite number of compound mappings. Embedded within the iterates are the iterates of the mappings

$$\begin{aligned} x &= f(f(x)) = f_2(x); \\ x &= f(f(f(x))) = f_3(x); \\ x &= f(f(f(f(x)))) = f_4(x); \end{aligned} \tag{5.9}$$

and so on. It is very instructive to look also at the sets of stable solutions of the fixed point equations

$$x = f_n(x) \tag{5.10}$$

in the same manner as we have done for f_1, which is just f. This is where we bring the machinery of complex variables into play, since in general we expect to find solutions of the resulting high-order polynomial equations as complex numbers. We shall look carefully at f_2, which can be treated analytically. Then we shall inspect f_3, f_4, and so on, using *Mathematica*'s numerical root finding functions. Then, when we have an idea about what we think will happen, it will be time to do some numerical experimentation. We can use *Mathematica*'s **Nest** function to treat the compound mappings, as illustrated by the case $n = 2$, first on some general function:

```
Nest[f, x, 2]
```

$$f(f(x))$$

Our mapping is a function of two variables, and we want to focus the iteration on the second variable. To iterate the function on the x-argument, we convert the logistic map to a pure function involving its second argument, retaining λ as a parameter

```
Logistic₂[λ_, x_] := Nest[Logistic[λ, #] &, x, 2]
```

```
Expand[Logistic₂[λ, x]]
```

$$\lambda^4 x^9 - 3\lambda^4 x^7 + 3\lambda^4 x^5 - \lambda^4 x^3 - \lambda^2 x^3 + \lambda^2 x$$

Let's take a look at the fixed points of this mapping. In spite of the fact that this is a ninth-order polynomial equation, *Mathematica* has no difficulty in finding the roots!

```
logsol₂ =
 FullSimplify[ x /. Solve[x == Logistic₂[λ, x] , x]]
```

$$\left\{0, -\sqrt{\frac{\lambda-1}{\lambda}}, \sqrt{\frac{\lambda-1}{\lambda}}, -\frac{1}{\sqrt{\frac{\lambda}{\lambda+1}}}, \frac{1}{\sqrt{\frac{\lambda}{\lambda+1}}}, -\frac{\sqrt{1-\frac{\sqrt{\lambda^2-4}}{\lambda}}}{\sqrt{2}},\right.$$

$$\left.\frac{\sqrt{1-\frac{\sqrt{\lambda^2-4}}{\lambda}}}{\sqrt{2}}, -\sqrt{\frac{\sqrt{\lambda^2-4}}{2\lambda}+\frac{1}{2}}, \sqrt{\frac{\sqrt{\lambda^2-4}}{2\lambda}+\frac{1}{2}}\right\}$$

This is useful – we can see that the fixed points of this mapping contain those of f, as they should, but we have found several others also. As before, let us investigate the stability of these fixed points, by considering the derivative map:

```
Dlog2[λ_, x_] = Simplify[D[Logistic2[λ, x], x]]
```

$$(3x^2 - 1)\lambda^2 (3\lambda^2 x^6 - 6\lambda^2 x^4 + 3\lambda^2 x^2 - 1)$$

Now we apply this map to the zeroes, and plot the results as functions of λ

```
der2[λ_] = Simplify[Dlog2[λ, logsol2]]
```

$$\{\lambda^2, (3 - 2\lambda)^2, (3 - 2\lambda)^2, (2\lambda + 3)^2, (2\lambda + 3)^2, 9 - 2\lambda^2, 9 - 2\lambda^2, 9 - 2\lambda^2, 9 - 2\lambda^2\}$$

There are only four distinct functions in this list, so we boil the list down to just these four:

```
relevant[λ_] = Union[der2[λ]]
```

$$\{(3 - 2\lambda)^2, \lambda^2, (2\lambda + 3)^2, 9 - 2\lambda^2\}$$

We see that for some of the solutions, we obtain stability criteria identical to those associated with **Logistic**. Let's plot all four functions together:

```
Plot[Evaluate[relevant[λ]],
 {λ, 0, 3}, PlotRange -> {-5, 5},
    PlotStyle ->
  {{Dashing[{0.01, 0.01}]}, {Thickness[0.005]},
           {Thickness[0.01]}, {Thickness[0.01]}}]
```

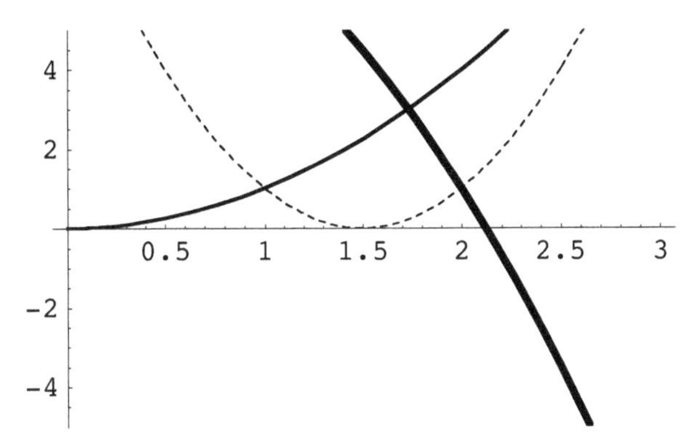

For positive λ, one pair of solutions is ruled out completely – this is the pair of solutions with derivative $(3 + 2\lambda)^2$. Matters are clearer if we force the plot range to be in the interval -1 to 1, in order to see when the various points are stable:

```
Plot[Evaluate[relevant[λ]],
 {λ, 0, 3}, PlotRange -> {-1, 1},
    PlotStyle ->
  {{Dashing[{0.01, 0.01}]}, {Thickness[0.005]},
           {Thickness[0.01]}, {Thickness[0.01]}}]
```

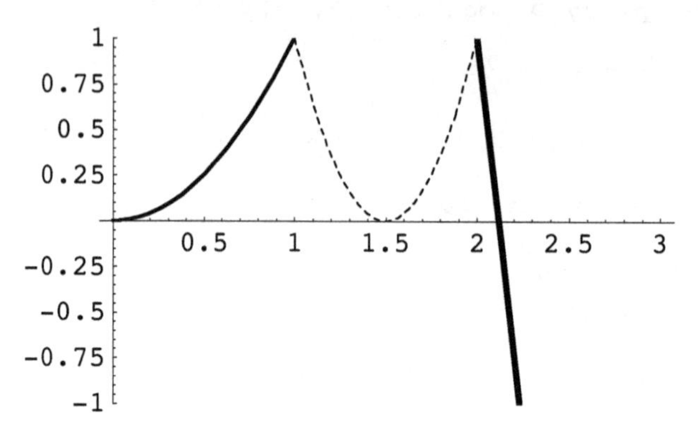

It is then evident that the zeroes of the mapping f_2, which are not fixed points of f, are stable for $\lambda > 2$ up to a slightly larger value of λ, given by solving for the derivative being -1. We identify this from the plot as being the last element of the relevant list, with derivative $9 - 2\lambda^2$:

```
λ /. Solve[9 - 2 λ² == -1]
```

$$\{-\sqrt{5}, \sqrt{5}\}$$

The larger of these two values, and a numerical approximation, is given by:

```
{%[[2]], N[%[[2]], 20]}
```

$$\{\sqrt{5}, 2.2360679774997896964\}$$

We can finally nail down the expected behaviour by observing the action of the logistic map on each of the four fixed points of f_2 that are associated with the derivative map being $9 - 2\lambda^2$. These are the last set of four roots:

```
setoffour = logsol₂[[{6, 7, 8, 9}]]
```

$$\left\{-\frac{\sqrt{1-\frac{\sqrt{\lambda^2-4}}{\lambda}}}{\sqrt{2}}, \frac{\sqrt{1-\frac{\sqrt{\lambda^2-4}}{\lambda}}}{\sqrt{2}}, -\sqrt{\frac{\sqrt{\lambda^2-4}}{2\lambda}+\frac{1}{2}}, \sqrt{\frac{\sqrt{\lambda^2-4}}{2\lambda}+\frac{1}{2}}\right\}$$

Apply the logistic map to these four, to obtain (this may take a little while):

```
mapped = FullSimplify[Logistic[λ, setoffour]]
```

$$\left\{-\frac{\sqrt{2}}{\lambda\sqrt{1-\frac{\sqrt{\lambda^2-4}}{\lambda}}}, \frac{\sqrt{2}}{\lambda\sqrt{1-\frac{\sqrt{\lambda^2-4}}{\lambda}}}, -\frac{\sqrt{2}}{\lambda\sqrt{\frac{\lambda+\sqrt{\lambda^2-4}}{\lambda}}}, \frac{\sqrt{2}}{\lambda\sqrt{\frac{\lambda+\sqrt{\lambda^2-4}}{\lambda}}}\right\}$$

This set is actually just a permutation of the original four. We perform a cyclic permutation of this second set of four values, effectively swapping the first and third elements, and the second and fourth. The division of the original list **setoffour** by the list **mapped**, suitably permuted, should give us a list of ones. To see this you need to force *Mathematica* to use additional simplification rules:

```
setoffour / RotateRight[mapped, 2] /.
  Sqrt[x_] Sqrt[y_] -> Sqrt[x y]
```

$$\left\{\frac{1}{2}\lambda\sqrt{\frac{\left(\lambda+\sqrt{\lambda^2-4}\right)\left(1-\frac{\sqrt{\lambda^2-4}}{\lambda}\right)}{\lambda}},\ \frac{1}{2}\lambda\sqrt{\frac{\left(\lambda+\sqrt{\lambda^2-4}\right)\left(1-\frac{\sqrt{\lambda^2-4}}{\lambda}\right)}{\lambda}},\right.$$

$$\left.\frac{\lambda\sqrt{\left(1-\frac{\sqrt{\lambda^2-4}}{\lambda}\right)\left(\frac{\sqrt{\lambda^2-4}}{2\lambda}+\frac{1}{2}\right)}}{\sqrt{2}},\ \frac{\lambda\sqrt{\left(1-\frac{\sqrt{\lambda^2-4}}{\lambda}\right)\left(\frac{\sqrt{\lambda^2-4}}{2\lambda}+\frac{1}{2}\right)}}{\sqrt{2}}\right\}$$

Now you are almost there:

```
PowerExpand[FullSimplify[%]]
```

{1, 1, 1, 1}

So root number 5 gets mapped to root 7, 6 to 8, 7 to 5, and 8 to 6. Roots number 6 and 8 are the positive roots. So the logistic map just swaps the elements of each of the two new pairs of fixed points. We expect the iterates of the logistic map to settle down to an oscillation between either the positive or negative pair of fixed points of f_2, provided $2 < \lambda < \sqrt{5}$.

Now we take a harder look at f_2. In particular, we are interested in the fixed points, that are the set:

```
setoffour
```

$$\left\{-\frac{\sqrt{1-\frac{\sqrt{\lambda^2-4}}{\lambda}}}{\sqrt{2}},\ \frac{\sqrt{1-\frac{\sqrt{\lambda^2-4}}{\lambda}}}{\sqrt{2}},\ -\sqrt{\frac{\sqrt{\lambda^2-4}}{2\lambda}+\frac{1}{2}},\ \sqrt{\frac{\sqrt{\lambda^2-4}}{2\lambda}+\frac{1}{2}}\right\}$$

We can inspect these values for various values of λ:

```
lambdalist =
  {0.001, 0.9, 1.0, 1.1, 1.9, 2.0, 2.1, 3.0};
```

```
TableForm[
 N[Table[Join[{λ}, setoffour] /. λ -> lambdalist[[i]],
   {i, 8}]], TableSpacing → {1, 1}]
```

0.001	−22.3663 + 22.3551 *i*	22.3663 − 22.3551 *i*	−22.3663 − 22.3551 *i*	22.3663 + 22.3551 *i*
0.9	−0.897527 + 0.552771 *i*	0.897527 − 0.552771 *i*	−0.897527 − 0.552771 *i*	0.897527 + 0.552771 *i*
1.	−0.866025 + 0.5 *i*	0.866025 − 0.5 *i*	−0.866025 − 0.5 *i*	0.866025 + 0.5 *i*
1.1	−0.839372 + 0.452267 *i*	0.839372 − 0.452267 *i*	−0.839372 − 0.452267 *i*	0.839372 + 0.452267 *i*
1.9	−0.71635 + 0.114708 *i*	0.71635 − 0.114708 *i*	−0.71635 − 0.114708 *i*	0.71635 + 0.114708 *i*
2.	−0.707107	0.707107	−0.707107	0.707107
2.1	−0.589529	0.589529	−0.807747	0.807747
3.	−0.356822	0.356822	−0.934172	0.934172

You can look at the fixed points of f_1 in a similar way:

```
TableForm[
 N[Table[Join[{λ}, logsol ] /. λ -> lambdalist[[i]],
   {i, 8}], 3], TableSpacing → {1, 1}]
```

0.001	0	$0.\times 10^{-2} - 31.607\,i$	$0.\times 10^{-2} + 31.607\,i$
0.9	0	$0.\times 10^{-4} - 0.333333\,i$	$0.\times 10^{-4} + 0.333333\,i$
1.	0	0.	0.
1.1	0	−0.301511	0.301511
1.9	0	−0.688247	0.688247
2.	0	−0.707107	0.707107
2.1	0	−0.723747	0.723747
3.	0	−0.816497	0.816497

You see that **setoffour** are *complex* for $\lambda < 2$. Two of the roots in **logsol** are complex for $\lambda < 1$. In the case $\lambda < 1$ the only real solution (it is also stable) is $x = 0$. As lambda passes through 1 the root at zero becomes unstable (check the derivative again). Simultaneously the two complex roots become zero and stable. Then the roots separate and move along the real axis. Each is stable, so one will be chosen by the iteration, depending on the starting value of x. (This is an example of a symmetry-breaking bifurcation, but more of this later.) This continues as λ increases to 2, but the two roots in **logsol** become unstable at this point. At this stage the roots of the second iterate become real and stable, then separate, and move off. We can visualize this very easily using *Mathematica*'s prgramming and graphics tools. First, we make lists containing both the roots and the absolute values of the derivative:

```
values[r_] := {Abs[der2[r]], logsol2 /. λ -> r}
```

Next we make a colouring scheme. Stable roots are large and blue, unstable ones are smaller and red:

```
cfunc[x_] :=
    If[x <= 1,
    {PointSize[0.06], RGBColor[0, 0, 1]},
    {PointSize[0.04], RGBColor[1, 0, 0]}]
```

We can now make a table of points coloured appropriately:

```
pts[r_] :=
 Map[Flatten[{cfunc[#[[1]]], Point[{Re[#[[2]]],
       Im[#[[2]]]}]}] &, Transpose[values[r]]]

plotdata = Table[pts[r], {r, 0.1, 2.5, 0.1}];
```

We can look at the results in various ways. First you can look at the root loci of all points:

```
Show[Graphics[{PointSize[0.02], plotdata}],
 AspectRatio -> 1,
    Frame -> True, FrameTicks -> None]
```

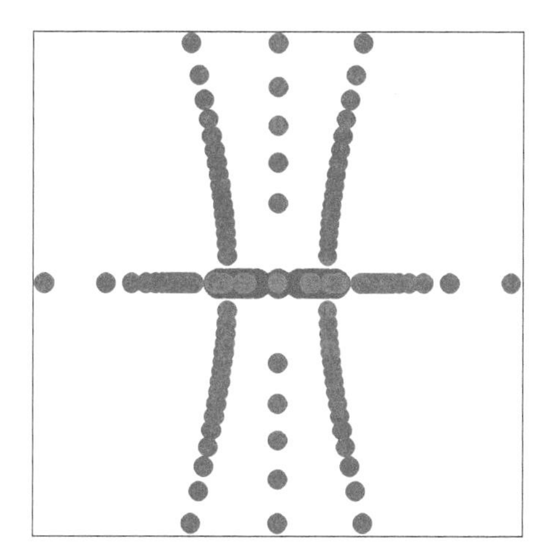

The effect is clearer if you produce a movie, and tidy up a bit (the output is not shown in the printed text). If you have *Mathematica* 6 or later replace the **Do** command by **Manipulate** to get real-time control of the movie in the notebook.

```
Do[frame[k] = Show[Graphics[pts[k / 10]],
        PlotRange -> {{-2, 2}, {-2, 2}},
        Frame -> True, FrameTicks -> None,
        AspectRatio -> 1], {k, 1, 25}]
```

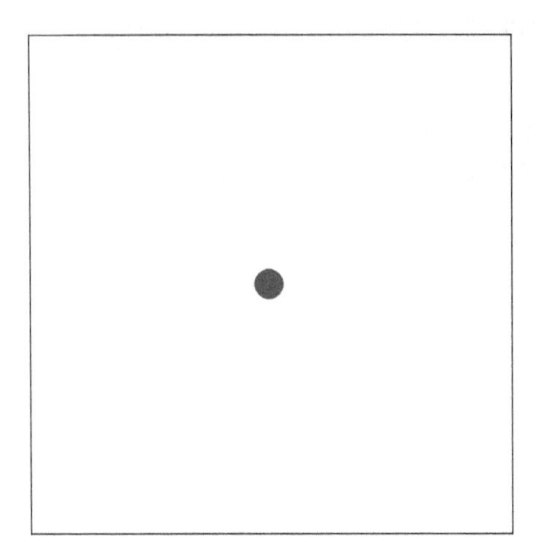

If you are using a computer, the above graphic can be animated. Some of the frames of the film that best communicate what is happening are shown here (this does rely on a **Do** function in the code above):

```
Show[GraphicsArray[
        {{frame[1], frame[4], frame[9]},
            {frame[10], frame[11], frame[16]},
            {frame[19], frame[22], frame[25]}}]]
```

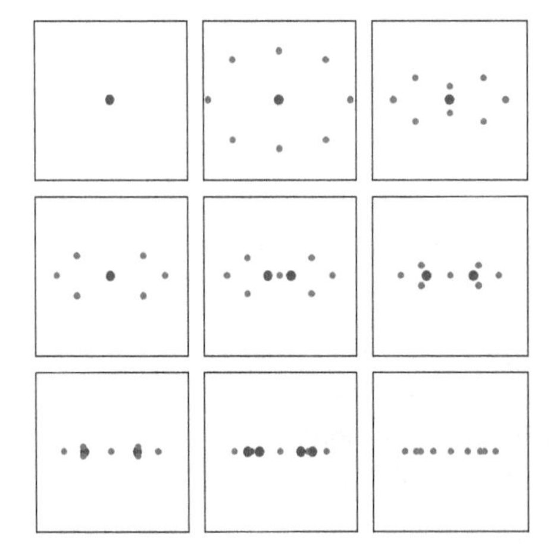

We now have a very clear picture of the process. There are various stable and unstable roots of f_1 and f_2. If we focus attention on just the stable roots, we obtain the phenomenon of bifurcation at the values $\lambda = 1$ and $\lambda = 2$.

We see that a bifurcation is simply the process whereby a previously real and stable root becomes unstable, simultaneously with a pair of previously complex roots becoming both real and stable.

Given that the final quartet of roots becomes unstable at $\lambda = \sqrt{5}$, we might anticipate a further bifurcation there. At $\lambda = 1$ we get a pair of roots, each of which is a stable fixed point of the logistic map. At $\lambda = 2$ we get two pairs of roots of the second iterate of the logistic map. The elements within each positive or negative pair are swapped by a single iteration of the logistic map.

5.4 What about $n = 3, 4, \ldots$?

```
Logistic[λ_, x_, n_] := Nest[Logistic[λ, #] &, x, n];
```

Let's take a look at the fixed points of the first of these mappings The output of the following is suppressed here, try it out for yourself if you have *Mathematica* running!

```
logsol3 = x /. Solve[x == Logistic[λ, x, 3] , x];
```

Unfortunately we no longer have an analytical picture. But instead you can use **NSolve** as follows:

```
Nlogsol[λ_, n_] :=
 N[ x /. Solve[x == Logistic[λ, x, n] , x]]
```

```
Dlog[λ_, x_, n_] := D[Logistic[λ, x, n], x]
```

```
Dlog3[λ_, x_] = Expand[D[Logistic[λ, x, 3], x]]
```

$$-27\,\lambda^{13}\,x^{26} + 225\,\lambda^{13}\,x^{24} - 828\,\lambda^{13}\,x^{22} + 1764\,\lambda^{13}\,x^{20} + 63\,\lambda^{11}\,x^{20} - 2394\,\lambda^{13}\,x^{18} - 399\,\lambda^{11}\,x^{18} + 2142\,\lambda^{13}\,x^{16} + 1071\,\lambda^{11}\,x^{16} - 1260\,\lambda^{13}\,x^{14} - 1575\,\lambda^{11}\,x^{14} - 45\,\lambda^{9}\,x^{14} + 468\,\lambda^{13}\,x^{12} + 1365\,\lambda^{11}\,x^{12} + 195\,\lambda^{9}\,x^{12} - 99\,\lambda^{13}\,x^{10} - 693\,\lambda^{11}\,x^{10} - 330\,\lambda^{9}\,x^{10} + 9\,\lambda^{13}\,x^{8} + 189\,\lambda^{11}\,x^{8} + 270\,\lambda^{9}\,x^{8} + 9\,\lambda^{7}\,x^{8} + 9\,\lambda^{5}\,x^{8} - 21\,\lambda^{11}\,x^{6} - 105\,\lambda^{9}\,x^{6} - 21\,\lambda^{7}\,x^{6} - 21\,\lambda^{5}\,x^{6} + 15\,\lambda^{9}\,x^{4} + 15\,\lambda^{7}\,x^{4} + 15\,\lambda^{5}\,x^{4} - 3\,\lambda^{7}\,x^{2} - 3\,\lambda^{5}\,x^{2} - 3\,\lambda^{3}\,x^{2} + \lambda^{3}$$

Let's take a look at these stable fixed points of f_3, by tabulating them. What we shall do, since we are now dealing with a polynomial equation of degree 27, is to count the number of stable real roots, rather than tabulate them. We start from the value $\lambda = \sqrt{5}$, just tabulating the number of real roots initially:

```
Table[{λ,
  Length[Select[Nlogsol[λ, 3], (Head[#] == Real) &]]},
 {λ, √5, 2.5, 0.02}]
```

$$\begin{pmatrix} 2.23607 & 3 \\ 2.25607 & 3 \\ 2.27607 & 3 \\ 2.29607 & 3 \\ 2.31607 & 3 \\ 2.33607 & 3 \\ 2.35607 & 3 \\ 2.37607 & 3 \\ 2.39607 & 3 \\ 2.41607 & 3 \\ 2.43607 & 3 \\ 2.45607 & 15 \\ 2.47607 & 15 \\ 2.49607 & 15 \end{pmatrix}$$

Clearly the number of real roots jumps between 2.436 and 2.456. Now you can home in on this region. At the same time you can restrict attention to the number of real and stable roots:

```
Table[{λ,
    Length[Select[Nlogsol[λ, 3],
    (Head[#] == Real && Abs[Dlog₃[λ, #]] <= 1) &]]},
  {λ, 2.436, 2.456, 0.002}]
```

$$\begin{pmatrix} 2.436 & 0 \\ 2.438 & 0 \\ 2.44 & 0 \\ 2.442 & 0 \\ 2.444 & 0 \\ 2.446 & 0 \\ 2.448 & 0 \\ 2.45 & 0 \\ 2.452 & 6 \\ 2.454 & 6 \\ 2.456 & 6 \end{pmatrix}$$

By homing in on where these roots come and go, you can establish that the stability interval is [2.45044, 2.46083]. So there is a very small region, centred on about the point $x = 2.456$, where we obtain stable solutions of $x = f_3(x)$. So what happens in the gap between $\lambda = \sqrt{5}$, and 2.45044? This could be investigated along the same lines as was done for f_3, but is a little messy, since it involves seeking real stable roots of a polynomial equation of degree 81! It is time for another approach, but first we summarize our findings so far.

5.5 ✵ Summary of our root-finding investigations

We have analysed the fixed points of the first three multiple iterates of the logistic map, using the criteria of reality and stability. The situation as so far analyzed may be summarized as follows, for λ real and positive:

$\lambda < 1 : x = 0$ is the only stable root of $x = f_1(x)$;
$1 < \lambda < 2 : x = \sqrt{1 - \lambda^{-1}}$ and $-\sqrt{1 - \lambda^{-1}}$ are the only stable roots of $x = f_1(x)$;
$2 < \lambda < \sqrt{5}$: no stable roots of $x = f_1(x)$; two pairs of stable roots of $x = f_2(x)$;
$\sqrt{5} < \lambda$: no stable roots of $x = f_1(x)$; no stable roots of $x = f_2(x)$;
$2.45044 < \lambda < 2.46083$: three stable roots of $x = f_3(x)$.

Just as in the case of the existence of two stable roots, the trio and quartet of stable roots are permuted by f_1. This gives us a picture of what happens when we iterate the odd logistic map. For low positive values of $\lambda < 2$ we expect to converge to one of the two roots of $x = f_1(x)$. When we go above 2 we expect the sequence to settle down to an oscillation between the two stable roots of $x = f_2(x)$, and then when we hit $\sqrt{5}$ we might guess, quite reasonably in fact, that the system will settle down to wander among some set of four roots of $x = f_4(x)$. What happens next? We might hypothesize that this bifurcation process continues, moving to 8, 16, 32 roots in ever smaller intervals. But then something very odd must happen. By the time we get to 2.45044, just three stable roots of $x = f_3(x)$ emerge. What happens in between is part of the subject of our experiments, and is the topic that is genuinely hard to predict.

5.6 ✵ The logistic map : an experimental approach

The experimental approach has been considered by a great many people for the standard logistic map, including Wagon (1991), Gray and Glynn (1991) and, more recently, Maeder (1995). Further investigations are left to the exercises. If you have skipped Sections 5.3 to 5.5 you just need to appreciate that we are now going to explore a very simple question: what happens when we apply the logistic map over and over again? The key function is **NestList**:

```
?NestList
```

```
NestList[f, expr, n] gives a list of
   the results of applying f to expr 0 through n times.
```

In the next few sections you can either just do the experiment or note the analytical hypothesis (given in italics) and then do the experiment. Experiment three in section 5.9 and subsequent plots will also yield the visual justification for the term ‘cobwebbing’.

5.7 Experiment one: $0 < \lambda < 1$

(If you have read Sections 5.3 to 5.5 you will expect that the system will converge to zero, at least if the starting value is close to zero.) We can now explore the behaviour of the iterated logistic map directly. Let's consider a starting value of $x = 0.5$, $\lambda = 0.9$:

```
NestList[Logistic[0.9, #] &, 0.5, 20]
```

{0.5, 0.3375, 0.269151, 0.224688, 0.19201, 0.166438, 0.145645, 0.1283, 0.113569, 0.100894, 0.08988, 0.0802385, 0.0717497, 0.0642423, 0.0575795, 0.0516497, 0.0463607, 0.041635, 0.0374065, 0.0336188, 0.0302227}

Note that there is no need to converge to zero if the initial value is large!

```
NestList[Logistic[0.9, #] &, 10, 10]
```

$\{10, -891., 6.36612\times 10^{8}, -2.32203\times 10^{26}, 1.1268\times 10^{79}, -1.2876\times 10^{237}, 1.921239079075883\times 10^{711}, -6.38244006177965\times 10^{2133}, 2.339929347518819\times 10^{6401}, -1.153056909677763\times 10^{19204}, 1.379732001224673\times 10^{57612}\}$

It is very illuminating to develop a plot that shows the convergence. The form of these plots shows why the term 'cobwebbing' is used. We can regard the iteration as moving between the line $y = x$ and the line $y = f(x)$. We write a function to draw lines between these points.

```
CobWeb[λ_, it_] := Module[{iter = Logistic[λ, it]},
        Line[{{it, it}, {it, iter}, {iter, iter}}]]

Plotter[λ_, xo_, n_] :=
    Plot[Logistic[λ, x], {x, 0, 1},
  PlotRange -> {0, Max[λ / 2 + 0.01, 0.5] },
          Epilog ->
   {Line[{{0, 0}, {1, 1}}], Map[(CobWeb[λ, #] &),
     NestList[(Logistic[λ, #] &), xo, n]]}]

Plotter[0.9, 0.5, 10]
```

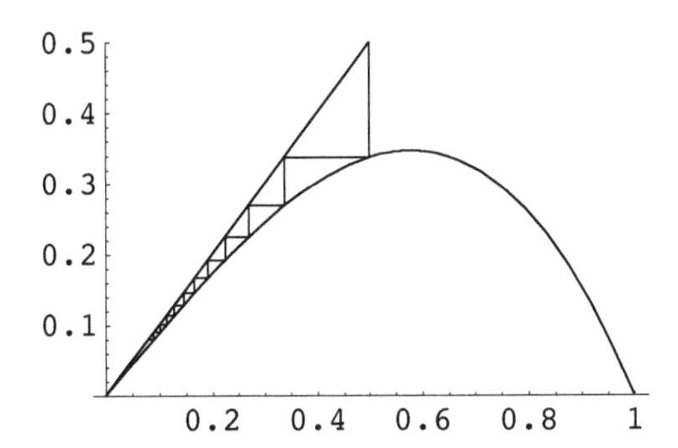

5.8 Experiment two: $1 < \lambda < 2$

(*Based on the analysis of Sections 5.3 to 5.5 we expect that the system will converge to $\sqrt{1-\lambda^{-1}}$, at least if the starting value is close to zero.*) Let's try a starting value of $x = 0.5$, $\lambda = 9/5$:

```
NestList[Logistic[1.8, #] &, 0.4, 20]
```

{0.4, 0.6048, 0.690434, 0.650349, 0.675507, 0.66108, 0.669907, 0.664685, 0.667842, 0.665957, 0.667091, 0.666412, 0.66682, 0.666575, 0.666722, 0.666634, 0.666686, 0.666655, 0.666674, 0.666662, 0.666669}

```
Plotter[1.8, 0.42, 30]
```

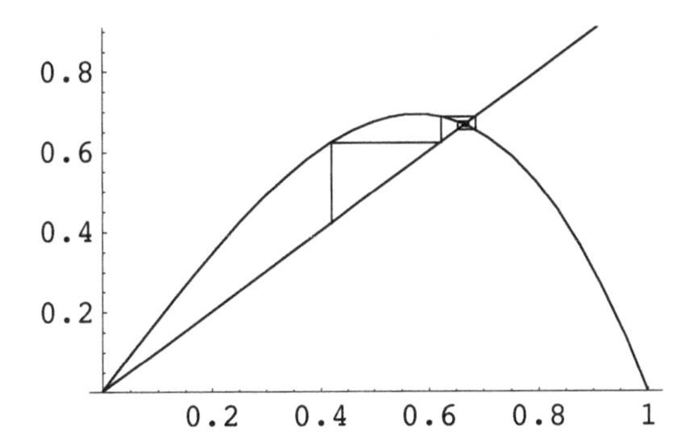

5.9 Experiment three: $2 < \lambda < \sqrt{5}$

(*Sections 5.3. to 5.5 suggests that the system will contain a pair of convergent subsequences, that is, the system will settle down to oscillating between two values. Let's look at a starting value of 0.5, $\lambda = 2.1$, so we expect to converge to the fixed points of f_2 that are not fixed points of f_1. The fixed points of the latter are given by:*)

```
Nlogsol[2.1, 1]
```

{−0.723747, 0., 0.723747}

So we inspect

```
Nlogsol[2.1, 2]
```

{−1.21499, −0.807747, −0.723747,
−0.589529, 0., 0.589529, 0.723747, 0.807747, 1.21499}

```
Map[Dlog2[2.1, #] & , %]
```

{51.84, 0.18, 1.44, 0.18, 4.41, 0.18, 1.44, 0.18, 51.84}

to predict that the system will settle down to an oscillation between 0.589529 and 0.807747.)

Here begins the experiment, where we take a value of $\lambda = 2.1$:

```
NestList[Logistic[2.1, #] &, 0.1, 50]
```

{0.1, 0.2079, 0.41772, 0.724147, 0.723266, 0.724323, 0.723054,
0.724576, 0.722749, 0.724939, 0.722309, 0.725462, 0.721675,
0.726214, 0.720759, 0.727292, 0.719435, 0.728836, 0.717521,
0.731042, 0.71475, 0.734176, 0.710734, 0.738595, 0.704917, 0.74474,
0.696526, 0.753075, 0.684578, 0.763881, 0.668106, 0.776761,
0.647002, 0.789935, 0.623737, 0.800254, 0.604308, 0.805606,
0.593811, 0.807295, 0.590436, 0.807663, 0.589699, 0.807732,
0.58956, 0.807744, 0.589535, 0.807747, 0.58953, 0.807747, 0.589529}

```
Plotter[2.1, 0.1, 100]
```

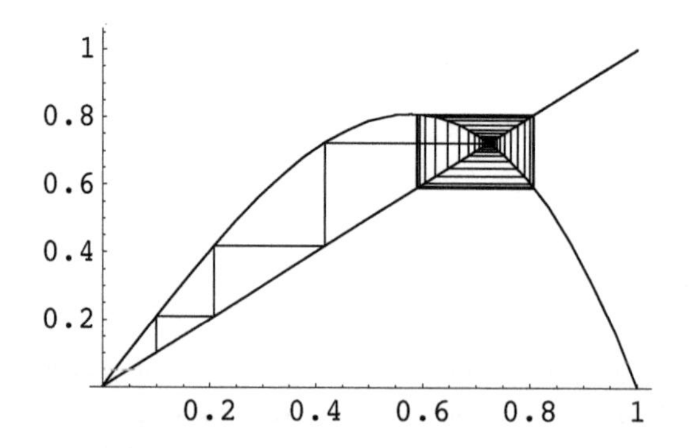

5.10 Experiment four: 2.45044 < λ < 2.46083

(Sections 5.3 to 5.5 suggests that the system will contain three convergent subsequences. Let us test this irst with a starting value o = 0.5, λ = 2.46, so we expect to converge to the fixed points of f_3 that are not fixed points of f_1. The fixed points of the latter are given by:

```
Nlogsol[2.46, 1]
```

{−0.770387, 0., 0.770387}

So we inspect the fixed points of the former that are real and stable:

```
Select[Nlogsol[2.46, 3],
 (Head[#] == Real && Abs[Dlog3[2.46, #]] < 1 &)]
```

{−0.946145, −0.564397, −0.243947, 0.243947, 0.564397, 0.946145}

to predict that the system will settle down to an osccilation between 0.243947, 0.946145, 0.564397.)

Here is our experiment:

```
NestList[Logistic[2.46, #] &, 0.17, 50]
```

{0.17, 0.406114, 0.83427, 0.623887, 0.937379, 0.27976, 0.634346,
0.932558, 0.299003, 0.669787, 0.908504, 0.390262, 0.813826,
0.676055, 0.902977, 0.410127, 0.839209, 0.610516, 0.942078,
0.260695, 0.597725, 0.945065, 0.248417, 0.573394, 0.946788,
0.241277, 0.55899, 0.945433, 0.246894, 0.570336, 0.946646, 0.241868,
0.560189, 0.945612, 0.246154, 0.568849, 0.946548, 0.242274,
0.561011, 0.945728, 0.245675, 0.567885, 0.946475, 0.242578,
0.561627, 0.945811, 0.245332, 0.567192, 0.946417, 0.242817, 0.56211}

It is taking its time. Let's iterate 1000 times and extract the last three:

```
Take[NestList[Logistic[2.46, #] &, 0.17, 1000], -3]
```

{0.564397, 0.946145, 0.243947}

```
Plotter[2.46, 0.17, 100]
```

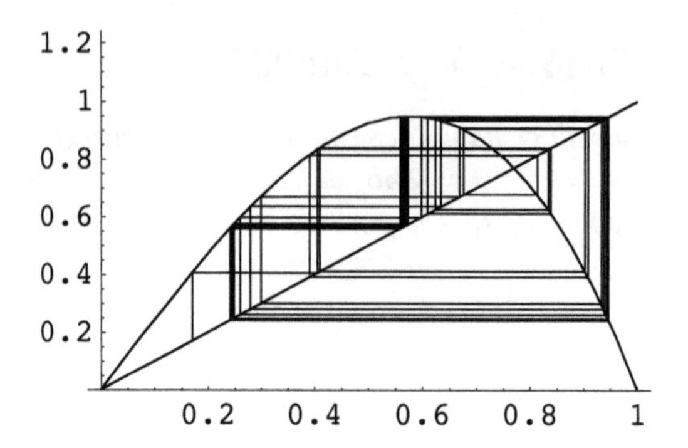

5.11 Experiment five: $\sqrt{5} < \lambda < \sqrt{5} + \epsilon$

(Sections 5.3 to 5.5 suggests that the system will contain four convergent subsequences.) Looking at the last eight elements after 1000 iterations appears to confirm this:

```
Take[NestList[Logistic[√5 + 0.001, #] &, 0.17, 1000], -8]
```

{0.846929, 0.535636, 0.854468, 0.515885,
0.846929, 0.535636, 0.854468, 0.515885}

```
Plotter[√5 + 0.001, 0.17, 100];
```

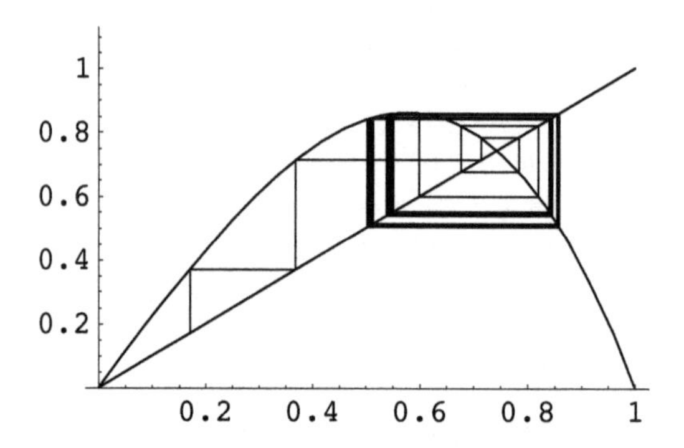

5.12 Experiment six: $\sqrt{5} < \lambda$

We need to go fully numerical now:

```
Sqrt[5] // N
```

2.23607

```
Take[NestList[Logistic[2.287, #] &, 0.17, 1000], -10]
```

{0.82749, 0.59662, 0.878779, 0.457718, 0.82749,
 0.59662, 0.878779, 0.457718, 0.82749, 0.59662}

```
Take[NestList[Logistic[2.288, #] &, 0.17, 1000], -10]
```

{0.827218, 0.597537, 0.879018, 0.457199, 0.82741,
 0.597073, 0.879092, 0.456975, 0.827218, 0.597536}

```
Plotter[2.287, 0.17, 100]
```

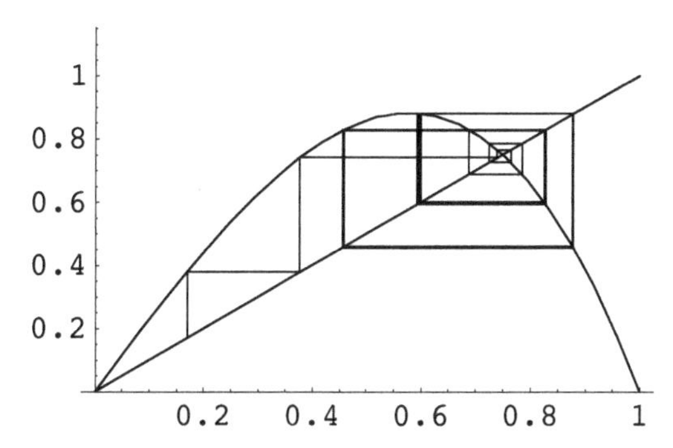

Let's keep increasing λ:

```
Plotter[2.3, 0.17, 100]
```

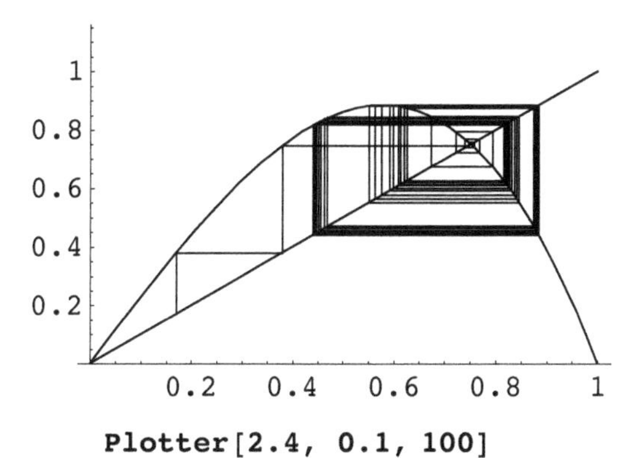


```
Plotter[2.4, 0.1, 100]
```

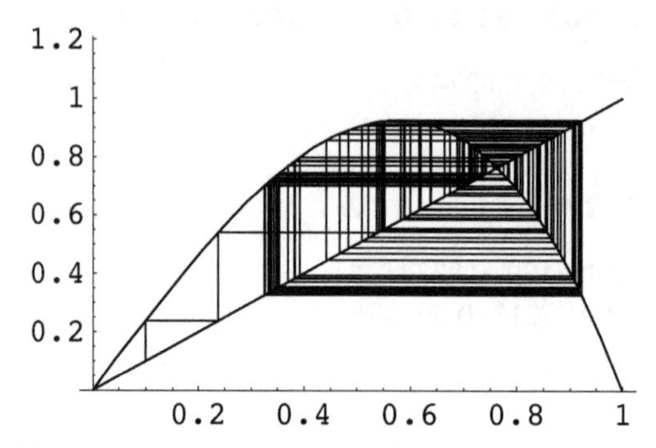

We can no longer see what is happening. A new way of proceeding is just to plot the points that arise after many iterations. We call this the bifurcation diagram.

5.13 Bifurcation diagrams

Now we build some functions that extract the final 128 points after 1000 iterations, and plot them against λ:

```
scatter[λ_] :=
    Map[Point[{λ, #}] &,
  Take[NestList[Logistic[λ, #] &, 0.1, 1000], -128]]

bifurpoints[min_, max_, n_] := {PointSize[0.003],
   Table[scatter[λ], {λ, min, max, (max - min) / n}]};

Show[Graphics[bifurpoints[1.5, 3, 200], Frame -> True]]
```

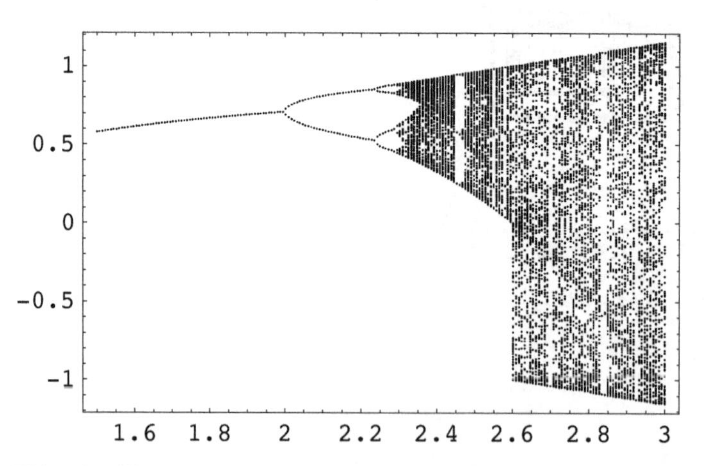

This plot illustrates the behaviour in the region we have been unable to analyse. The doubling continues for a while. This cannot continue indefinitely, as we know a new pattern must emerge to generate the triple of stable fixed points. The doubling continues for a while and then a transition to chaotic behaviour is made. The double and transition to chaos reasserts itself in the triple fixed point region. We see this next.

■ Bifurcation diagram for triple root region

```
Show[Graphics[
  bifurpoints[2.449, 2.47, 200], Frame -> True]]
```

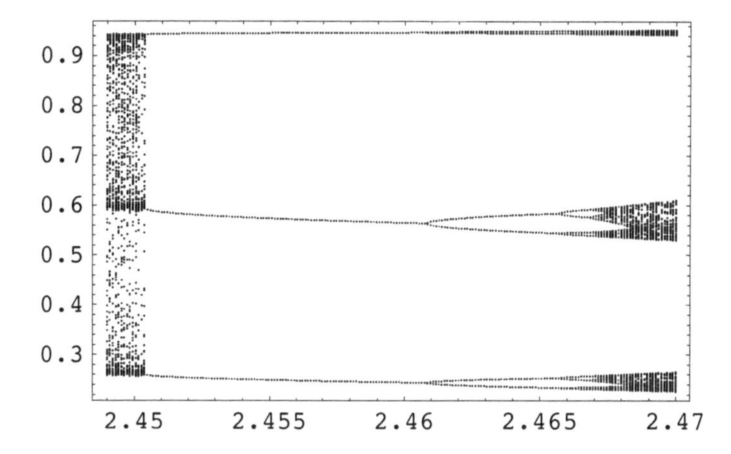

■ Bifurcation diagram for first transition to chaos

You can see the appearance of 2, 4, 8 and (just) 16 points if you zoom in on the relevant region:

```
Show[
 Graphics[bifurpoints[2.23, 2.32, 200], Frame -> True]]
```

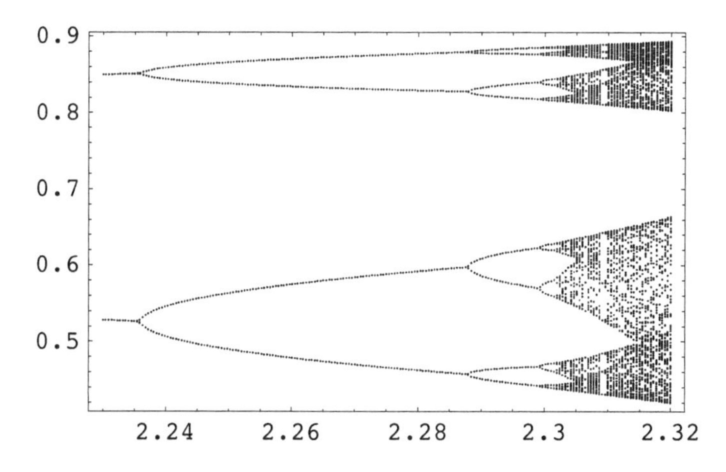

5.14 ✵ Symmetry-related bifurcations

If we zoom out and look at our bifurcation diagams in the large, we can see some other features related to the symmetry inherent in the odd logistic map. We can identify both symmetry-breaking and symmetry-generating bifurcations. Suppose that $0 < \lambda < 1$, then the sequences converge to zero – this is a point of symmetry of the mapping. When λ passes through unity, the sequence converges to a positive value, provided a positive initial value is chosen. This breaks the symmetry. If a negative initial value is chosen (see the final plot below) the sequence converges to a negative value, again breaking the symmetry. There then follow a series of bifurcations with no consequences for the symmetry, until a certain critical value of $\lambda = \lambda_c$, where the plot overflows into negative values, becoming, once again, symmetric. The same happens with the plot with a negative starting value. This is a symmetry-generating bifurcation.

■ Bifurcation diagram for symmetry-generating bifurcations

```
Show[Graphics[bifurpoints[0, 3, 200], Frame -> True]]
```

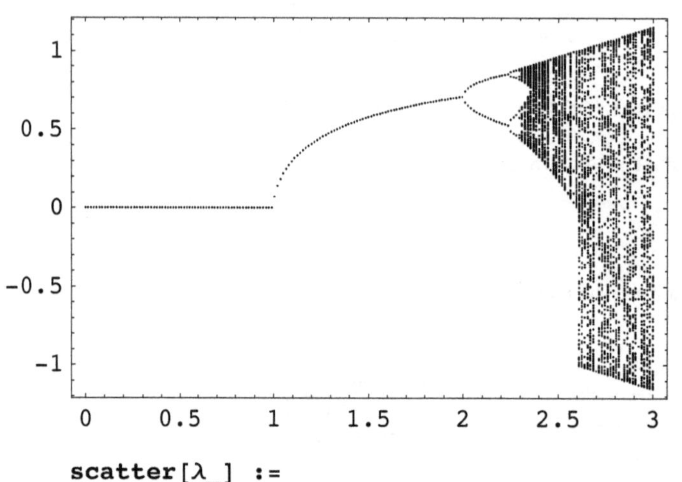


```
scatter[λ_] :=
    Map[Point[{λ, #}] &,
  Take[NestList[Logistic[λ, #] &, -0.1, 1000], -128]]
```

```
Show[Graphics[bifurpoints[0, 3, 200], Frame -> True]]
```

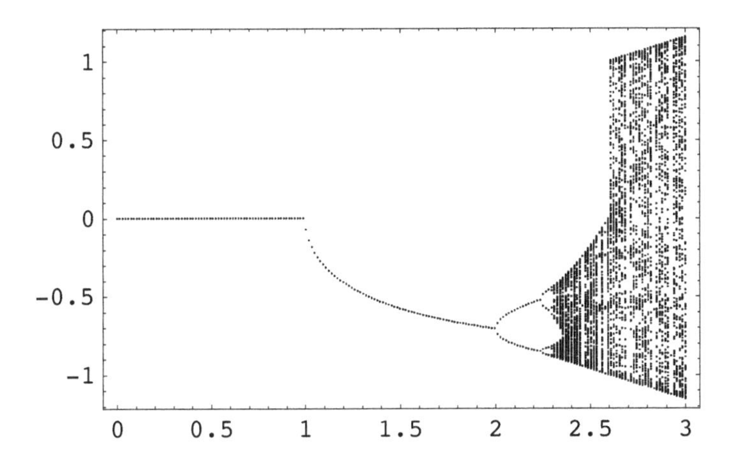

■ Location of the critical point

The value of λ_c is easily deduced. The final values of the iteration spill over zero when zero lies in the image of the map, when applied to positive values of x. Recall now the form of the map:

```
Logistic[λ, x]
```

$$x(1 - x^2)\lambda$$

This can only become zero, for positive x, for $x = 1$. (Note how the bifurcation plot spreads out to zero and one at the same time.) So now we argue that unity has to lie in the image of the logistic map for real positive x. When is this possible? Let's find out:

```
preimages[λ_]  = x /. Solve[Logistic[λ, x] == 1, x]
```

$$\left\{-\frac{\sqrt[3]{\frac{2}{3}}\,\lambda}{\sqrt[3]{9\lambda^2+\sqrt{3}\sqrt{27\lambda^4-4\lambda^6}}}-\frac{\sqrt[3]{9\lambda^2+\sqrt{3}\sqrt{27\lambda^4-4\lambda^6}}}{\sqrt[3]{2}\,3^{2/3}\,\lambda},\right.$$

$$\frac{\left(1+i\sqrt{3}\right)\lambda}{2^{2/3}\sqrt[3]{3}\sqrt[3]{9\lambda^2+\sqrt{3}\sqrt{27\lambda^4-4\lambda^6}}}+\frac{\left(1-i\sqrt{3}\right)\sqrt[3]{9\lambda^2+\sqrt{3}\sqrt{27\lambda^4-4\lambda^6}}}{2\sqrt[3]{2}\,3^{2/3}\,\lambda},$$

$$\left.\frac{\left(1-i\sqrt{3}\right)\lambda}{2^{2/3}\sqrt[3]{3}\sqrt[3]{9\lambda^2+\sqrt{3}\sqrt{27\lambda^4-4\lambda^6}}}+\frac{\left(1+i\sqrt{3}\right)\sqrt[3]{9\lambda^2+\sqrt{3}\sqrt{27\lambda^4-4\lambda^6}}}{2\sqrt[3]{2}\,3^{2/3}\,\lambda}\right\}$$

Evidently the form of **preimages** is influenced by the factor $27\lambda^4 - 4\lambda^6$, which is zero when $\lambda = 3\sqrt{3}/2$. This is indeed the critical value. Let's look at what happens to the preimages in a neighbourhood of this value:

```
N[3 Sqrt[3] / 2]
```

2.59808

```
preimages[2.59]
```

$\{-1.1551, 0.57755 - 0.0263199\, i, 0.57755 + 0.0263199\, i\}$

```
ComplexExpand[preimages[3 Sqrt[3] / 2]]
```

$\{-\frac{2}{\sqrt{3}}, \frac{1}{\sqrt{3}}, \frac{1}{\sqrt{3}}\}$

```
N[%]
```

$\{-1.1547, 0.57735, 0.57735\}$

```
N[ComplexExpand[preimages[26 / 10]]]
```

$\{-1.15461, 0.590126, 0.564479\}$

So for values of λ just less than this critical value, there are no preimages with positive real part that are real, whereas on or above this value there are such points. So the mapping attains the value unity, and hence, on iteration, the value 0, when $\lambda = \lambda_c = 3\sqrt{3}/2$.

5.15 Remarks

Note how much of the behaviour of this system can be extracted merely by looking at the behaviour of certain associated low-order polynomials, and whether their roots are real, complex and stable. Such an analysis allows us to predict the phenomenon of bifurcations, and to appreciate when such a bifurcation is symmetry-generating. We can also make predictions about the outcomes of numerical experiments, confirm our predictions and go on to investigate the transition to chaos.

A detailed examination of the transition to chaos takes us outside the scope of the complex analytical point of view. With enough computer power we could look at several of the higher order polynomials responsible for the continuation of the bifurcation process, but it becomes more efficient to pursue an essentially numerical approach. This can also be done with *Mathematica*, and has been discussed in the *Mathematica* Journal by R. Maeder (1995). You are encouraged to consult this article for further information, particularly regarding the computation of the 'Feigenbaum number' governing the locations of the values of λ at which bifurcations occur.

Exercises

In the following exercises your goal is to explore the behaviour of the standard quadratic logistic function

$$x \to \lambda x(1 - x) \tag{5.11}$$

using methods identical to those used in the text for the corresponding cubic map. Everything, apart from the symmetry-breaking/generating bifurcation should go through in a very similar fashion. To get you started, the first two exercises give very explicit guidance. After that, it is up to you to adapt the material in the text to treat the quadratic case.

5.1 ✵ Define the ordinary (quadratic) logistic map by the *Mathematica* function

```
Logistic[λ_, x_] := λ x (1 - x)
```

(In this and all the other exercises, **Logistic** now refers to this standard quadratic mapping.) Find the set of fixed points of this mapping, and identify them with the *Mathematica* list **logsol**. Construct a suitable derivative mapping **Dlog** and show that the positive fixed point is zero provided $1 < \lambda < 3$.

5.2 ✵ Construct the first compound mapping f_2

```
Logistic2[λ_, x_] := Nest[Logistic[λ, #] &, x, 2]
```

and find the fixed points of this mapping using

```
logsol2 = x /. Solve[x == Logistic2[λ, x] , x]
```

Next, work out

```
Dlog2[λ_, x_] = D[Logistic2[λ, x], x]
```

and

```
der2[λ_] = Simplify[Dlog2[λ, logsol2]]
```

Hence, using plotting techniques similar to that used for the odd logistic map, show that the fixed points of the first compound mapping f_2 (that are not fixed points of the logistic map) are stable provided $3 < \lambda < 1 + \sqrt{6}$. What does the original logistic map do to the pair of stable fixed points that arise in this interval?

5.3 ✵ Construct visualizations of the first bifurcation of the logistic map, in the Argand plane, using methods similar to those employed for the odd logistic map.

5.4 ✵ Show that, for the approximate interval $3.8284 < \lambda < 3.8415$, there are three stable roots of the thrice-iterated logistic map.

5.5 ❄ Show, by some numerical investigations on the fourth-iterated mapping, that there are four real and stable fixed points of this mapping in the approximate interval $1 + \sqrt{6} < \lambda < 3.55409$.

5.6 ❄ Show that there are no stable real roots of any of the first four iterated mappings in the approximate interval $3.55409 < \lambda < 3.8284$.

5.7 ❄ Construct bifurcation plots for the quadratic logistic map.

5.8 ❄ Using the bifurcation plot, zoom in on the doubling region and see how many doublings you can identify.

5.9 ❄ Using the bifurcation plot, zoom in on the triple fixed point region and see how the doubling and transition to chaos reasserts itself.

5.10 'I would therefore urge that people be introduced to, say, equation (5.11) early in their mathematical education. This equation can be studied phenomenologically by iterating it on a calculator, or even by hand. Its study does not involve as much conceptual sophistication as does elementary calculus. Such study would greatly enrich the student's intuition about nonlinear systems.' This is a quote (verbatim apart from the equation reference) from the conclusion of Sir Robert May's original paper (May, 1976) on the logistic map. Discuss whether you agree with May's view.

6 The Mandelbrot set

Introduction

In Chapter 5 we looked at the logistic map given by the equation

$$\text{Logistic}(x, \lambda) = \lambda x(1 - x^k)$$

for values of $k = 1, 2$. Although this is a real mapping, considerable insight was gained into its behaviour by allowing x to be complex and investigating fixed points of the iterated logistic map. We will now consider the first of two approaches to complexifying this map, by allowing both x and λ to be complex. In the first case we shall consider $k = 1$, but consider a general complex quadratic map. In the second case we shall investigate complex generalizations of the $k = 2$ case that possess various degrees of symmetry. We defer this until Chapter 7.

As in Chapter 5, there are two routes you can take through this chapter. If you really want to understand what is happening, you are encourage to read through in order. If you just want to see how to use *Mathematica* to make pictures of the Mandelbrot set and zoom in to interesting regions, then feel free to skip to Section 6.4.

The author is grateful to Professor Sir Roger Penrose, F.R.S., for encouraging me to think about the issues involved in computing *the* Mandelbrot set, within a purist black and white representation. A brief sketch of the issues involved is given in Section 6.7.

6.1 ❁ From the logistic map to the Mandelbrot map

Consider first a general quadratic map, for x complex:

```
quad[x_] := a₂ x^2 + a₁ x + a₀
```

Superficially it looks like we have a three-parameter family of mappings. In fact there is only one parameter, since we can make a change of variables to relate all quadratic mappings to a single simpler family. To see this, let

```
z[x_] := a₂ x + a₁ / 2
```

```
sqd = Expand[z[x]^2]
```

$$\frac{a_1^2}{4} + x\, a_2\, a_1 + x^2\, a_2^2$$

```
c = Simplify[z[quad[x]] - sqd]
```

$$-\frac{a_1^2}{4} + \frac{a_1}{2} + a_0\, a_2$$

So we can write

```
z[quad[x]] = z[x]^2 + c;
```

Working in the z-coordinates, we can use

```
quadc[z_] := z^2 + c
```

It suffices to consider variations only in the parameter c, to get genuinely different mappings. This family of maps are called the *Mandelbrot maps*. All the other quadratic mappings are obtained from these by a linear transformation. Note that c is easily related to the λ parameter of the quadratic logistic map:

```
c /. {a0 -> 0, a1 -> λ, a2 -> -λ}
```

$$\frac{\lambda}{2} - \frac{\lambda^2}{4}$$

We can therefore repeat our investigations of the logistic map in these new coordinates, characterizing the bifurcation behaviour in terms of values of c. We shall not do this in any detail, but will just check that the iterated map bifurcates where it should. First we standardize our function name:

```
Remove[c];
Quad[z_, c_] := z^2 + c
```

```
Quad[z, c]
```

$$z^2 + c$$

Then we find the fixed points of **Quad** itself:

```
fixeda = z /. Solve[Quad[z, c] == z, z]
```

$$\left\{\frac{1}{2}\left(1 - \sqrt{1 - 4c}\right), \frac{1}{2}\left(\sqrt{1 - 4c} + 1\right)\right\}$$

Therefore **Quad** has a pair of real fixed points provided $c < 1/4$. What about stability? The derivative of **Quad** is just $2z$:

```
Map[2 # &, fixeda]
```

$$\left\{1 - \sqrt{1 - 4c}, \sqrt{1 - 4c} + 1\right\}$$

Clearly the second is excluded for real solutions, and the first works provided $c > -3/4$. We expect the fixed points of the first iterated map to become real here, and they do:

```
fixedb = z /. Solve[Nest[Quad[#, c] &, z, 2] == z, z]
```

$$\left\{\frac{1}{2}\left(-\sqrt{-4c-3}-1\right), \frac{1}{2}\left(\sqrt{-4c-3}-1\right), \frac{1}{2}\left(1-\sqrt{1-4c}\right), \frac{1}{2}\left(\sqrt{1-4c}+1\right)\right\}$$

It is an easy matter to check that the first bifurcation appears where it should. Let's take a starting value of zero for z. We iterate 100 times and then inspect the next four values:

```
start = Nest[Quad[#, 0.2] &, 0, 100]
```

0.276393

```
NestList[Quad[#, 0.2] &, start, 4]
```

{0.276393, 0.276393, 0.276393, 0.276393, 0.276393}

```
start = Nest[Quad[#, -0.8] &, 0, 100]
```

−0.276392

```
NestList[Quad[#, -0.8] &, start, 4]
```

{−0.276392, −0.723607, −0.276392, −0.723607, −0.276393}

This would lead us down the same path as already described in Chapter 5, so we shall pursue this particular route no further. But we do ask: what happens if $c > 1/4$?

```
NestList[Quad[#, 0.3] &, 0, 20]
```

{0, 0.3, 0.39, 0.4521, 0.504394, 0.554414, 0.607375, 0.668904, 0.747432, 0.858655, 1.03729, 1.37597, 2.19329, 5.11051, 26.4173, 698.175, 487448., 2.37606×10^{11}, 5.64565×10^{22}, 3.18734×10^{45}, 1.01591×10^{91}}

Clearly rapid progress to infinity is obtained. For large enough $|c|$ we would expect to see this type of behaviour. But we also know that for small c the system can converge to a limit, or bounce around periodically or chaotically.

6.2 ❄ Stable fixed points: complex regions

Recall that the fixed points and their stability are given by looking at the derivative map applied to the fixed points of the basic map. We rewrite this in a form that allows a simple generalization to higher iterates:

```
fixeda = z /. Solve[Nest[Quad[#, c] &, z, 1] == z, z]
```

$$\left\{\frac{1}{2}\left(1-\sqrt{1-4c}\right), \frac{1}{2}\left(\sqrt{1-4c}+1\right)\right\}$$

```
dmap[z_, c_] = D[Nest[Quad[#, c] &, z, 1], z]
```

$2z$

```
derivs = Expand[Map[dmap[#, c] &, fixeda]]
```

$\{1-\sqrt{1-4c}, \sqrt{1-4c}+1\}$

What we want to do is to characterize the regions where the derivative is less than one in magnitude. To do this we parametrize the derivative:

```
c /. Solve[derivs[[1]] == μ, c][[1]]
```

$\frac{1}{4}(2\mu-\mu^2)$

So the stability region is given parametrically by

```
cstable[μ_] := 1/4 (2 μ - μ^2)
```

for $|\mu| < 1$. Let's plot this region:

```
stableregion = ParametricPlot[
    Evaluate[Table[{Re[cstable[r Exp[I θ]]],
      Im[cstable[r Exp[I θ]]]}, {r, 0, 1, 0.2}]],
  {θ, 0, 2 Pi}, AspectRatio -> 1,
  PlotStyle -> Hue[6 / 7], PlotRange → All]
```

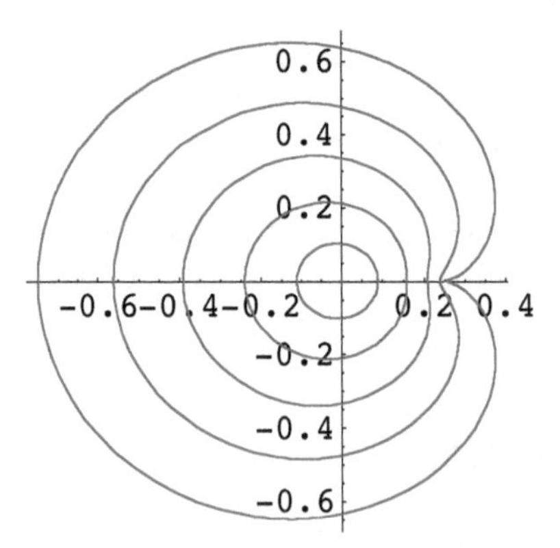

■ Stable regions for the once-iterated map

We can do the analysis for the basic mapping again for the first iterate:

```
fixedb = z /. Solve[Nest[Quad[#, c] &, z, 2] == z, z]
```

$$\left\{\frac{1}{2}\left(-\sqrt{-4c-3}-1\right), \frac{1}{2}\left(\sqrt{-4c-3}-1\right), \frac{1}{2}\left(1-\sqrt{1-4c}\right), \frac{1}{2}\left(\sqrt{1-4c}+1\right)\right\}$$

Let's take out the fixed points of the basic map:

```
fixedbonly = Complement[fixedb, fixeda]
```

$$\left\{\frac{1}{2}\left(-\sqrt{-4c-3}-1\right), \frac{1}{2}\left(\sqrt{-4c-3}-1\right)\right\}$$

Now we construct the derivative of the iterated map and apply it to the fixed points:

```
dmap2[z_, c_] = D[Nest[Quad[#, c] &, z, 2], z]
```

$$4z(z^2+c)$$

```
derivsb = Expand[Map[dmap2[#, c] &, fixedbonly]]
```

$$\{4c+4, 4c+4\}$$

The result is remarkably simple, and it is easy to parametrize the derivative map:

```
c /. Solve[derivsb[[1]] == μ, c][[1]]
```

$$\frac{\mu-4}{4}$$

So the stability region is given parametrically by

```
cstableb[μ_] := (μ - 4) / 4
```

for $|\mu| < 1$. Let's plot this region – it's obviously a circle!

```
stableregionb = ParametricPlot[
    Evaluate[Table[{Re[cstableb[r Exp[I θ]]],
     Im[cstableb[r Exp[I θ]]]}, {r, 0, 1, 0.2}]],
  {θ, 0, 2 Pi}, AspectRatio -> 1, PlotStyle -> Hue[5 / 7]]
```

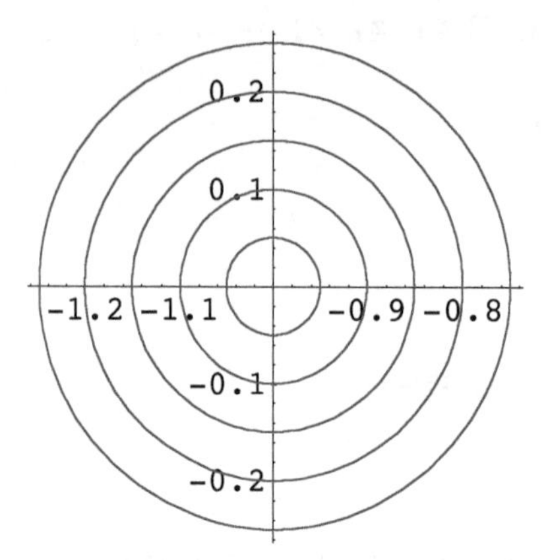

Unfortunately the higher order-iterates do not have an simple analytic solution allowing us to parametrize the stable regions in this way, so we take this route no further. But we can do something else to find out what this map is doing!

6.3 Periodic orbits

We shall now look for values of c that generate periodic orbits. First we seek those containing the origin. What we do is to solve the resulting polynomial equations, using numerical methods when the degree is high.

```
fixed[n_] :=
  c /. Solve[Nest[Quad[#, c] &, 0, n] == 0, c];
Nfixed[n_] :=
  c /. NSolve[Nest[Quad[#, c] &, 0, n] == 0, c];
```

```
one = fixed[1];
none = N[one]
```

{0.}

```
two = fixed[2]
```

{−1, 0}

Now one of these points is the fixed point (zero) for $n = 1$, so we pull this out to leave the fixed point of the iterated map that is not a fixed point of the map itself:

```
twoonly = Complement[two, one];
ntwo = N[twoonly]
```

{−1.}

What about the fixed points of the map iterated twice?

```
three = fixed[3];
```

We remove the fixed point already found:

```
threeonly = Complement[three, one]
```

$$\left\{\frac{1}{3}\left(-2-\sqrt[3]{\frac{2}{25-3\sqrt{69}}}-\sqrt[3]{\frac{1}{2}\left(25-3\sqrt{69}\right)}\right),\right.$$

$$-\frac{2}{3}+\frac{1}{6}\left(1+i\sqrt{3}\right)\sqrt[3]{\frac{1}{2}\left(25-3\sqrt{69}\right)}+\frac{1-i\sqrt{3}}{3\,2^{2/3}\sqrt[3]{25-3\sqrt{69}}},$$

$$\left.-\frac{2}{3}+\frac{1}{6}\left(1-i\sqrt{3}\right)\sqrt[3]{\frac{1}{2}\left(25-3\sqrt{69}\right)}+\frac{1+i\sqrt{3}}{3\,2^{2/3}\sqrt[3]{25-3\sqrt{69}}}\right\}$$

and extract the numerical values:

```
nthree = N[threeonly]
```

$\{-1.75488, -0.122561-0.744862\,i, -0.122561+0.744862\,i\}$

We can now proceed semi-automatically: $n = 4$ gets new fixed points, and we delete those already found. Note how the symbolic power of *Mathematica* allows us to take out the elements already found without a messy numerical comparison:

```
four = fixed[4];
fouronly = Complement[four, one, two];
nfour = N[fouronly]
```

$\{-1.9408, -1.3107, -0.15652-1.03225\,i,$
$-0.15652+1.03225\,i, 0.282271-0.530061\,i, 0.282271+0.530061\,i\}$

```
five = fixed[5];
fiveonly = Complement[five, one];
nfive = N[fiveonly];

six = fixed[6];
sixonly = Complement[six, one, two, three];
nsix = N[sixonly];

seven = fixed[7];
sevenonly = Complement[seven, one];
nseven = N[sevenonly];
```

We shall stop here as higher orders strain the computer somewhat!

```
data =
  {none, ntwo, nthree, nfour, nfive, nsix, nseven};

Map[Length, data]
```

{1, 1, 3, 6, 15, 27, 63}

We shall plot the various periodic cycles coloured according to the period. To this end we attach the periodicity as a list:

```
aux = Range[1, 7]
```

{1, 2, 3, 4, 5, 6, 7}

```
plotinfo = Transpose[{aux, data}];

realplotdata =
  Map[{Hue[1 - #[[1]] / 7], PointSize[0.025],
     Map[ Point[{Re[#], Im[#]}] &, #[[2]]]} &, plotinfo];

plota = Show[Graphics[realplotdata]]
```

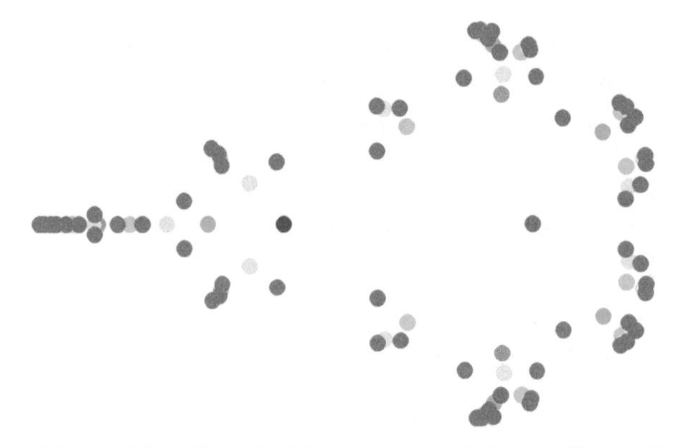

This graphic will probably start to remind you of some familiar pictures.

■ The origin as a pre-periodic point

We can also locate the first view values of c where the origin ends up in a periodic orbit after a few iterations. This is a simple generalization of our map **fixed** to two variables:

```
fixed[n_, k_] := c /. Solve[
            Nest[Quad[#, c] &, 0, n] ==
      Nest[Quad[#, c] &, 0, n - k], c];
Nfixed[n_, k_] := c /. NSolve[
            Nest[Quad[#, c] &, 0, n] ==
      Nest[Quad[#, c] &, 0, n - k], c];
```

Let's look at the first few values of c so obtained – in each case we delete fixed points already found:

```
threeone = fixed[3, 1];
threeoneonly = Complement[threeone, one];
nthreeoneonly = N[threeoneonly]
```

{–2.}

```
fourone = fixed[4, 1];
fouroneonly = Complement[fourone, one, threeoneonly];
nfouroneonly = N[fouroneonly]
```

{–1.54369, –0.228155 + 1.11514 *i*, –0.228155 – 1.11514 *i*}

```
fourtwo = fixed[4, 2];
fourtwoonly =
  Complement[fourtwo, one, two, threeoneonly];
nfourtwoonly = N[fourtwoonly]
```

{0. – 1. *i*, 0. + 1. *i*}

If we plot these, we obtain the first few *Misiurewicz* points – here we colour them black:

```
datab =
  Flatten[{nthreeoneonly, nfouroneonly, nfourtwoonly}];
plotdatab = {PointSize[0.04], Map[
        Point[{Re[#], Im[#]}] &, datab]};
plotb = Show[Graphics[plotdatab],
        PlotRegion -> {{0.05, 0.95}, {0.05, 0.95}}]
```

Now we overlay this with our first plot of periodic orbits containing the origin, AND the plot of the stable regions:

```
Show[plota, plotb, stableregion, stableregionb,
 PlotRegion -> {{0.05, 0.95}, {0.05, 0.95}},
 AspectRatio -> 1,
 PlotRange -> {{-2, 1 / 2}, {-5 / 4, 5 / 4}}]
```

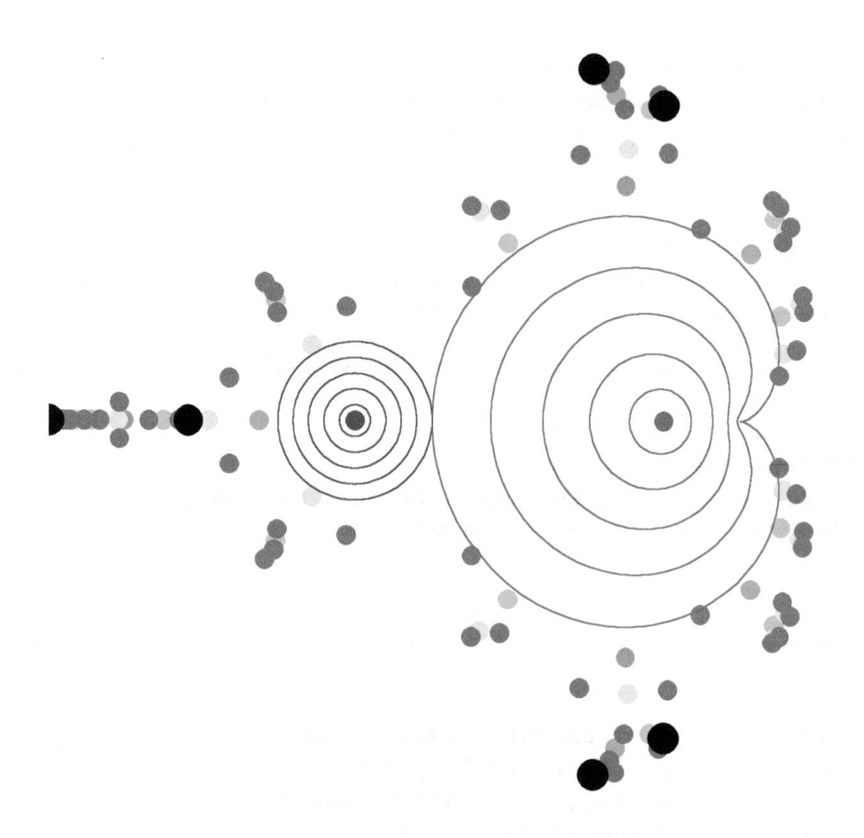

What we have done is a partial characterization of the Mandelbrot set in terms of stable regions and periodic orbits. We have carried out our analysis using only root-finding algorithms and stability analyses based on the derivatives of certain polynomial maps. However, we have stretched the symbolic approach significantly to get this information. We wish to characterize this set in more detail – at the same time it would be interesting to characterize the behaviour of the exterior region. Clearly the point at infinity is a stable attractor for this system – we now turn to a numerical approach based on the 'escape-time algorithm' to investigate further.

6.4 ❄ Escape-time algorithm for the Mandelbrot set

The idea is to compute how rapidly the iterates of the Mandelbrot map move to infinity. What one does is to define a large circle (in the following it will be of radius 100, centred on the origin), and count how many iterations it takes, for a given starting value of z and a value of c, for the iterated map to get outside this circle. That is all there is to it!

Well, not quite! This caveat is related to the point that we have not yet defined what 'the Mandelbrot set' is. It is the set of values of c for which the iterated map *never* reaches infinity. In our initial investigations below we will not dwell on this too much, but you should appreciate that the nicely coloured illustrations you will develop here, and that feature in many of the books on the matter, are representations of the escape time, usually through some colouring scheme. While they give an indication of the shape of the Mandelbrot set itself, they often give a rather distorted picture. It is actually rather difficult to draw the pure Mandelbrot set with a computer, and one should also appreciate that a proper picture of it would really just be black and white (or indeed any two different colours), where black would denote points in the set and white those that are not. This will be discussed in more detail in Section 6.7.

It is actually quite straightforward to implement the *escape-time* algorithm, but it raises some interesting programming issues. We will use the following discussion not only to illuminate the detailed structure of the Mandelbrot set, but also to show how to use *Mathematica* in various different ways. If you follow the coding issues carefully, you will be in a much better position to understand how to use *Mathematica* to do other types of simulation. We shall use *Mathematica* in three different ways to solve the problem, based on:

(a) simple interpreted *Mathematica*;
(b) compiled *Mathematica*;
(c) use of *MathLink* with a C-program add-in.

Most of the time we use *Mathematica* in *interpreted* mode. Here we shall see how to compile a numerical operation for efficient evaluation. We shall also show how a C program can be written to be added in to *Mathematica*, and look at some of the details of how *Mathematica* renders graphics in order to optimize our C further. It should be noted, however, that most of the benefit comes from just using the *Mathematica* compiler – it is virtually no effort to take a well-written *Mathematica* expression and compile it, whereas the creation of a separate C program and its linking to *Mathematica* requires some effort.

The following discussion follows a similar logic (interpreted, compiled, *MathLink)* to that developed by the author and J. Tigg in *Applied Mathematica* (Shaw and Tigg, 1993). However, the following code contains a number of improvements, in that variables are treated as essentially complex, and the final *MathLink* version is much more efficient. In one implementation we have borrowed and extended some interesting ideas from Tom Wickham Jones' excellent book, (Wickham Jomes, 1994) '*Mathematica* Graphics, Techniques and Applications', which is a very useful reference on graphics generally. His algorithm for fast black and white fractals is generalized here to include colour management.

■ Interpreted *Mathematica*

First we define a function to count the number of iterations it takes to leave a circle of radius 100. The number of iterations is capped at 100:

```
IterationsToLeave[z_, c_] :=
Module[{cnt, nz=N[z], nc=N[c]},
For[cnt=0,
(Abs[nz] <100) && (cnt<100), cnt++,
nz = nz*nz+nc];
N[cnt]];
```

To define the graphics of interest, we fix a starting value of z and allow c to vary over a rectangular region in the complex plane:

```
FracM[z_, {{remin_, remax_}, {immin_, immax_}},
steps_]:=
{{{remin, remax}, {immin, immax}},
Table[IterationsToLeave[z, x+I*y],
{y, immin, immax, (immax-immin)/steps},
{x, remin, remax, (remax-remin)/steps}]};
```

Next we define a simple function for coloring our plots. The idea is that points that hit the 100 iteration cap will be coloured black, while all others will be coloured according to the number of iterations:

```
FracColor = (If[#==1, Hue[1,1,0], Hue[5 #/6]]&);

FracPlot[{{{remin_, remax_}, {immin_, immax_}},
matrix_},
colorfn_] :=
ListDensityPlot[matrix,
Mesh -> False, Frame -> False,
ColorFunction -> colorfn,
PlotRange -> {0, 100},
AspectRatio -> (immax-immin)/(remax - remin)]
```

So let's try it out on a small block, with $z = 0$ as our starting value – you should try something like this first to make sure things are working. The way the picture and output statement (including the timing) are combined will depend on the version of *Mathematica* you are using.

```
Timing[
FracPlot[FracM[0,{{-2.1,0.5},{-1.2,1.2}},50],
FracColor]
]
```

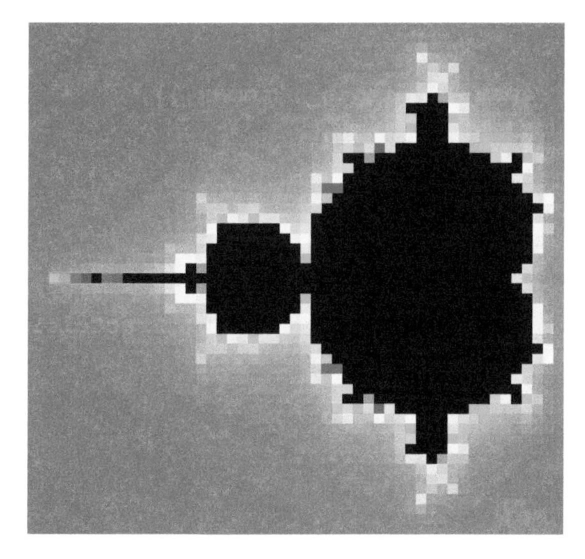

{1.57 Second, - DensityGraphics -}

That is great, and we can already see the shape computed previously analytically, complete with tendrils! However, it is a bit slow. Here and elsewehere the timings have all been done on a Power Macintosh G4 running at 1.4 GHz. With a GHz-class machine it is a little time-consuming to make lots of pictures this way. However, it is easy to make it go much faster, without the need to go out and buy newer hardware, as we shall see.

■ Compiled *Mathematica*

The built-in compiler allows one to produce optimized numerical routines from *Mathematica* code. We used this without much explanation in Chapter 4. Here is what **Compile** does in more detail:

```
? Compile
```

```
Compile[{x1, x2, ... }, expr] creates a compiled function which evaluates
   expr assuming numerical values of the xi. Compile[{{x1, t1}, ... },
   expr] assumes that xi is of a type which matches ti. Compile[{{x1, t1,
   n1}, ... }, expr] assumes that xi is a rank ni array of objects each of
   a type which matches ti. Compile[vars, expr, {{p1, pt1}, ... }] assumes
   that subexpressions in expr which match pi are of types which match pti.
```

So let's make some minor adjustments and implement this operation. We only need to make small changes to two of our functions:

```
FastIter = Compile[
{{z, _Complex}, {c, _Complex}},
Module[{cnt, nz=N[z], nc=N[c]},
For[cnt=0,
(Abs[nz] <100) && (cnt<100), cnt++,
nz = nz*nz+nc];
N[cnt]]];

FracMC[z_, {{remin_, remax_}, {immin_, immax_}},
steps_]:=
{{{remin, remax}, {immin, immax}},
Table[FastIter[z, x+I*y],
{y, immin, immax, (immax-immin)/steps},
{x, remin, remax, (remax-remin)/steps}]};

Timing[
FracPlot[FracMC[0,{{-2.1,0.5},{-1.2,1.2}},50],FracColor
]]
```

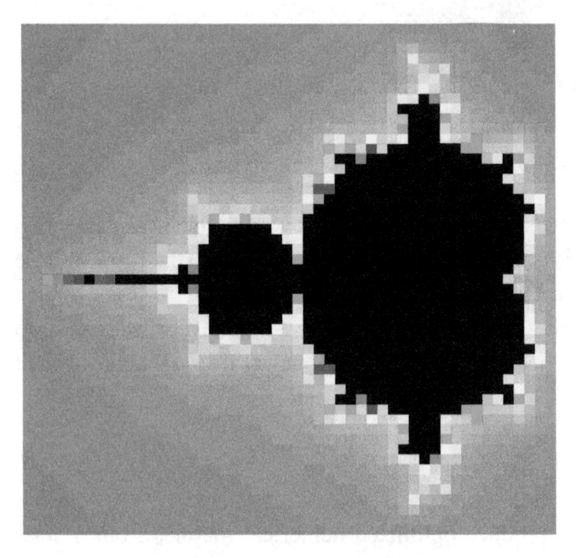

{0.19 Second, - DensityGraphics -}

The calculation is now going nearly ten times faster! Now we can generate some serious pictures.

```
Timing[
FracPlot[FracMC[0,{{-2.1,0.5},{-1.2,1.2}},200],FracColo
r]]
```

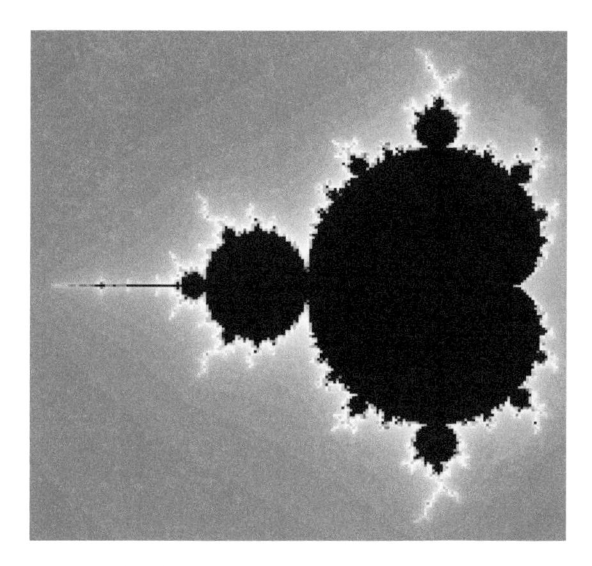

{3.16 Second, - DensityGraphics -}

At the time of final editing we know that computers running at over 3 Ghz can go *at least* twice as fast.

■ Using **FixedPointList**

A tempting and compact alternative to **FastIter** is to use the built-in function **Fixed-PointList** and the length of the list it generates subject to termination conditions that parallel those already used. The following modification to **FastIter** may be tried, but it turns out to take at least half as long again. (You might want to revisit this, as with different versions of *Mathematica* there are different optimizations of the built-in functions.)

```
FastIterFP = Compile[{{z, _Complex}, {c, _Complex}},
   N[Length[FixedPointList[#^2 + c &, z,
      100, SameTest -> (Abs[#2] > 100.0 &)]]]];
```

```
FracMC[z_, {{remin_, remax_}, {immin_, immax_}},
steps_]:=
{{{remin, remax}, {immin, immax}},
Table[FastIterFP[z, x+I*y],
{y, immin, immax, (immax-immin)/steps},
{x, remin, remax, (remax-remin)/steps}]};
```

```
Timing[
FracPlot[FracMC[0,{{-2.1,0.5},{-1.2,1.2}},200],FracColo
r]]
```

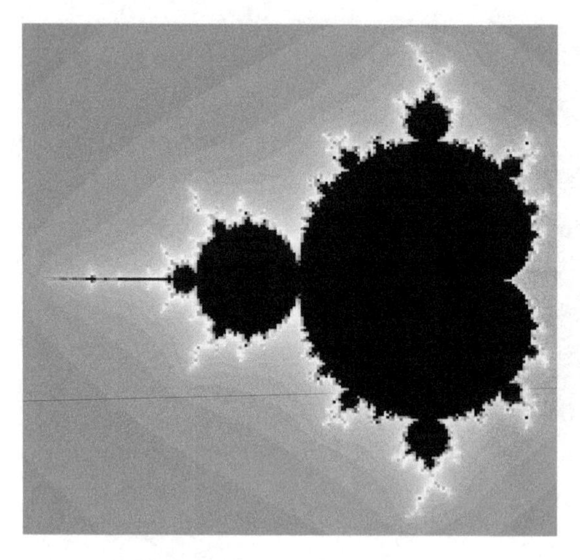

{4.95 Second, - DensityGraphics -}

6.5 ✵ *MathLink* versions of the escape-time algorithm

The following material assumes that you are able to use the *MathLink* system, either by loading the plug-ins supplied (see the enclosed CD for binaries for popular platforms) or by recompiling the code. If you adopt the latter approach, please consult the *MathLink* documentation thoroughly before proceeding. The idea is to create a C program to take over some of the computation, compile it with a special type of *MathLink* template file, and add the resulting program into *Mathematica*. The template file does the job of assocating *Mathematica* variable and function names with associated C variable and function names.

■ *MathLink* version 1

Here is the template file:

```
:Begin:
:Function:    fractalml
:Pattern:
FractalML[a_?NumberQ,b_?NumberQ,{{c_?NumberQ,d_?NumberQ},{e_?NumberQ,
f_NumberQ}},g_Integer]
:Arguments:  { N[a],N[b],N[c],N[d],N[e],N[f],g }
:ArgumentTypes: { Real,Real,Real,Real,Real,Real,Integer}
:ReturnType:   Manual
:End
```

Here is the C implementation of the escape-time algorithm, as given in the file `mandel.c`. It uses the same algorithm as before, just written out in C. Note (a) the include statement (we use *MathLink*, not stdio); (b) the MLPut statements to send results down the link to the *Mathematica* kernel; (c) the form of the Main block. These are the ingredients that characterize a *MathLink* C program.

```
#include "mathlink.h"

void fractalml(double x0,double y0, double xmin, double xmax, double
ymin, double ymax, int divs)
{
  double x, y, dist, xadd, yadd, temp;
  int xco, yco, counts;

  MLPutFunction(stdlink,"List",divs);

  for(yco = 0; yco <divs; yco++){
  yadd = ymin + (ymax-ymin)*yco/(divs-1.0);
  MLPutFunction(stdlink,"List",divs);

  for(xco = 0; xco <divs ; xco++){

  x = x0;
  y = y0;
  xadd = xmin + (xmax-xmin)*xco/(divs-1.0);

  dist = x0*x0+y0*y0;
  for(counts = 0; counts<100 && dist < 10000; counts++){
  temp = x*x-y*y+xadd;
  y = 2*x*y+yadd;
  x = temp;
  dist = x*x+y*y;
  }
  MLPutInteger(stdlink,counts);
  }
  }
}

int main(int argc, char *argv[])
{
return MLMain(argc, argv);
}
```

▪ Loading and using the *MathLink* escape-time program

First we take the executable, called ‘mandel’, and install it into *Mathematica*. The first task is to tell *Mathematica* where the binary *MathLink* executables are:

```
SetDirectory["/Books/ComplexMath2005/MathLink/binOSX"];
```

Note that this path should be set to wherever you have put them – the above works only on the author's computer! *See the CD notes for location and other information of the files on the enclosed CD.*

```
Install["mandel"]
```

LinkObject[./mandel, 2, 2]

We can query the operation of the function as with any built-in function, The query returns information contained in the template file and references to the fact that it is a **LinkObject**:

```
?FractalML
```

```
Global`FractalML

FractalML[a_?NumberQ, b_?NumberQ,
  {{c_?NumberQ, d_?NumberQ}, {e_?NumberQ, f_?NumberQ}}, g_Integer] :=
 ExternalCall[LinkObject[./mandel, 2, 2],
  CallPacket[0, {N[a], N[b], N[c], N[d], N[e], N[f], g}]]
```

We adapt our previous routine to call this function:

```
FracML[z_, {{remin_, remax_},{immin_, immax_}},
steps_] :=
{{{remin, remax},{immin, immax}},
FractalML[Re[z], Im[z],
{{remin, remax},{immin, immax}},
steps]}
```

Finally, for some variety, we define some new colour schemes:

```
FracColorA := (
If[#==1,
RGBColor[0,0,0],
If[#<=0.5, RGBColor[2#,2#,2#],
RGBColor[2-2#,2-2#,1]]]&)

FracColorB := (
If[#==1,
RGBColor[0,0,0],
If[#<=0.5, RGBColor[1,1-2#,0], Hue[1.5-#]]]&)

FracPlot[{{{remin_, remax_}, {immin_, immax_}},
matrix_},
colorfn_] :=
ListDensityPlot[matrix,
Mesh -> False, Frame -> False,
ColorFunction -> colorfn,
PlotRange -> {0, 100},
AspectRatio -> (immax-immin)/(remax - remin)]
```

We can now try it out. We not not use a **Timing** command as this reports kernel time only, but this routine is about four times faster than the compiled *Mathematica*:

```
FracPlot[FracML[0,
{{-2.1,0.5},{-1.2,1.2}},200],FracColorA]
```

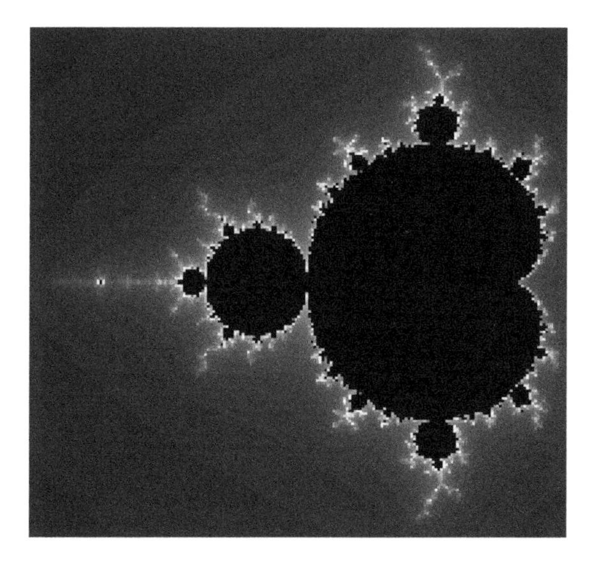

It is now a simple matter to switch colour scheme:

```
FracPlot[FracML[0,
{{-2.1,0.5},{-1.2,1.2}},200],FracColorB]
```

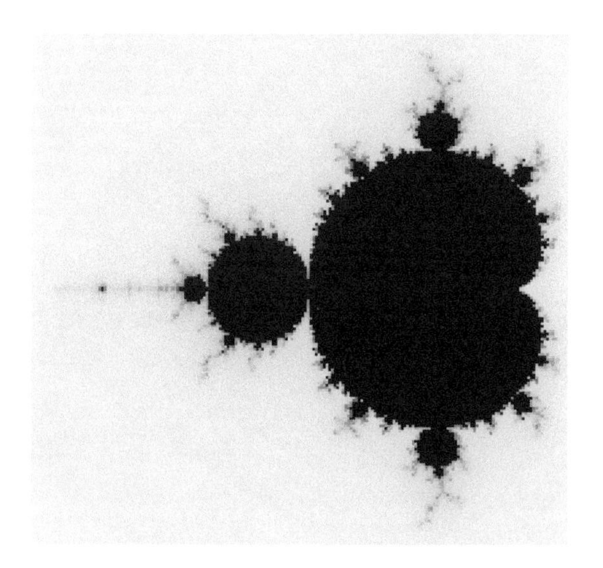

Note that with the `mandel.c` code, only the counting of iterations is farmed out to external code. This means that this code generates fractals that can be coloured as you choose, as above.

■ A Turbo-charged direct PostScript^{TM} *MathLink* version!

Having optimized the computation time we are now still waiting for these images to render. It is time for some dirty tricks to speed things up further, at the price of some lack of obviousness in our code. There are two places to look for clues as to how to do this. One is in the notebook file itself, the other is in Tom Wickham Jones' excellent book, "*Mathematica* Graphics, Techniques and Applications", where a fast program for computing black and white fractals is described in outline. Sadly, Tom provides neither the C code nor a colour algorithm for his fractals, so we are compelled to do a little snooping around *Mathematica*'s use of PostScript in order to see what needs to be done. (This particular implementation is for use with *Mathematica* 4 and 5 – the author canot guarantee such low-level tricks will work with *Mathematica* technology beyond version 5.2. See the on-line supplement and CD for any further information on this.)

Having done so, we find that we can talk 'native' to the front end, producing the image in precisely the format that the front end needs to render. In the following program, we wrap a *MathLink* program that sends a list of six character strings of the form FF00FF, with *Mathematica* code that supplies the necessary PostScript sandwich and an operation to merge all the resulting strings into one large one. (Devotees of TWJ's book will note that he creates one very large string to send down *MathLink*. My own attempts to do this have resulted in code slower than that used here, though this may be due to my ignorance of issues regarding memory management and *MathLink* when dealing with very large strings. Furthermore, we have avoided the use of a temporary local variable to store the image in *Mathematica* – leaving it out further reduces computation time, and we just send the *MathLink* output straight into the middle of the PostScript!)

```
Clear[HexaFracPlot];

Install["hexaman"];

?HexaFrac
```

```
Global`HexaFrac
```

```
HexaFrac[a_?NumberQ, b_?NumberQ,
  {{c_?NumberQ, d_?NumberQ}, {e_?NumberQ, f_?NumberQ}}, g_Integer] :=
 ExternalCall[LinkObject[./hexaman, 3, 3],
  CallPacket[0, {N[a], N[b], N[c], N[d], N[e], N[f], g}]]
```

```
HexaFracPlot[z_, {{remin_, remax_},{immin_, immax_}},
steps_]:=
Module[{nstring, tnstring},
nstring = ToString[steps];
tnstring = ToString[3*(steps)];
Show[Graphics[PostScript[
"0 0 translate",
"1 1 scale",
tnstring <> " string",
StringReplace["nn nn 8 [nn 0 0 nn 0 0]","nn" ->
nstring],
"{ currentfile 1 index readhexstring pop } false 3
Mcolorimage",
StringJoin[
HexaFrac[Re[z], Im[z], {{remin, remax},{immin,
immax}}, steps]
],
"pop"],
PlotRange -> {{remin,remax},{immin, immax}},
AspectRatio -> 1]]]
```

Here is the result of a call to **HexaFrac** of the type used in the Module.

```
StringJoin[HexaFrac[0,0, {{-2, 1},{-1.5, 1.5}}, 6]]
```

```
FF0000FF00000000FF0000FF0000FFFF0000
0000FF0000FFFF0000FF0000FF00000000FF
0000FFFF0000FF0000000000000000FF0000FF
0000FFFF0000FF0000000000000000FF0000FF
0000FF0000FFFF0000FF0000FF00000000FF
FF0000FF00000000FF0000FF0000FFFF0000
```

```
HexaFracPlot[0, {{-2, 1}, {-1.5,1.5}}, 800]
```

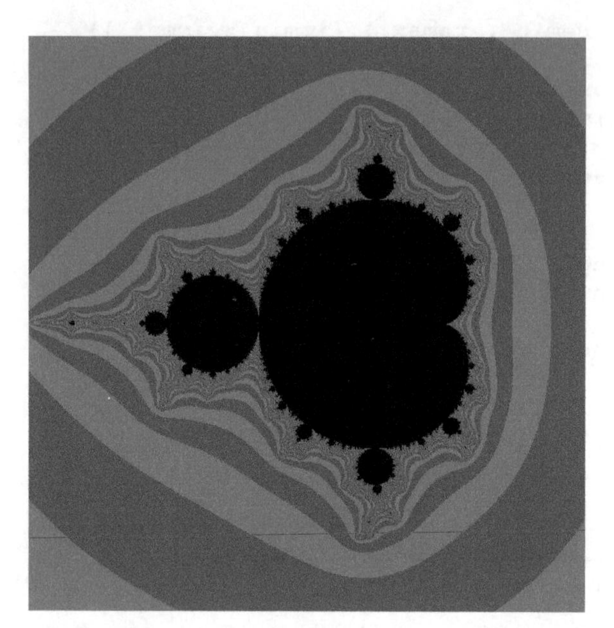

We do not report a kernel timing as it is again misleading, as it does not include the time taken to render the image by the front end. But this method is, overall, significantly faster than our previous method. The kernel is actually doing very little work by now, and not much effort is actually required to join the strings together, as the following computation illustrates:

```
u = Flatten[Table["AA00FF", {k, 1, 200}, {l, 1,
200}]];
Timing[v = StringJoin[u];]
```

{0.01 Second, Null}

6.6 ✵ Diving in to the Mandelbrot set: fractal movies

Now that we have a set of efficient tools for drawing our fractals, we can begin to explore properly. An excellent way of doing so is to 'dive' into the fractal. The idea is to pick a point – a little experimentation at low resolution is useful for locating interesting points first. Then, you zoom in on this point by drawing a rectangle centred on the point and adjusting the length of its sides by a contraction factor for each new frame. This is easily done by the following function:

```
g[pointx_, pointy_, n_, scale_] :=
Module[{width = scale^n},
{{pointx-width,pointx+width},{pointy-width,pointy+width
}}]
```

■ Places to visit

The following place to visit was suggested by some guesswork. It is chosen as it demonstrates the concept of self-similarity, revealing a copy of the Mandelbrot set deep within itself. In *Mathematica* 4 and 5 the resulting movie can be animated by selecting the block of cells and choosing 'Animate Selected Graphics' from the Cell Menu. You should see what options are available in your current version of *Mathematica* for animation. (See the on-line supplement and CD.) If you have not already done so, install the C binary file with the following:

```
Install["mandel"];
```

In the printed version you will see one frame from the movie at full resolution. If you have a computer running at less than 1 GHz you should consider halving the size of the grid in the arguments of **FracML**, given below as 288.

▪ Making the movie

Here a high-resolution version of a frame from near the middle ($n = 10$) of the movie is shown in the output.

```
Do[
    FracPlot[
    FracML[0, g[-1.40835915,0.13627737,n,0.8],288],
    FracColorB],
{n, -3, 30}]
```

■ A visit to 'sea-horse valley'

Here a high-resolution version of a frame from near the middle ($n = 12$) of the movie is shown in the output.

```
Do[FracPlot[FracML[0, g[-0.78, 0.15, n, 0.8], 288],
  FracColorB], {n, -4, 30}]
```

If you a version of *Mathematica* compatible with hexaman, you can similarly try the following:

```
Install["hexaman"];
Do[
    HexaFracPlot[0, g[-0.78,0.15,n,0.8],288],
{n, -4, 30}]
```

■ Other ways of coding the set

There are other ways to farm out the processing. One that gives the job of generating the colours to the C code, but is less painful than producing the Hex output, is embodied in **rasterman**. This produces an array that can be shown with the **RasterArray** function. If you are using *Mathematica* technology beyond version 5.2, see the on-line supplement and CD for further information about the operation of **RasterArray** methods.

```
Install["rasterman"];
```

This returns a matrix of RGB values:

```
RasterFrac[0,0, {{-2, 1},{-1.5, 1.5}}, 3]
```

$$\begin{pmatrix} \text{RGBColor}[1., 0.92, 0.] & \text{RGBColor}[1., 0.9, 0.] & \text{RGBColor}[1., 0.92, 0.] \\ \text{RGBColor}[0., 0., 0.] & \text{RGBColor}[0., 0., 0.] & \text{RGBColor}[1., 0.9, 0.] \\ \text{RGBColor}[1., 0.92, 0.] & \text{RGBColor}[1., 0.9, 0.] & \text{RGBColor}[1., 0.92, 0.] \end{pmatrix}$$

and you can see it with, for example,

```
Show[
 Graphics[RasterArray[RasterFrac[0, 0, {{-0.8, -0.75},
     {0.1, 0.15}}, 600]]], AspectRatio → 1]
```

6.7 ✡ Computing and drawing *the* Mandelbrot set

If you wish to try to get a better grip on the Mandelbrot itself, you need to confront several issues:

(1) You are either in the set or out of it, so pretty and detailed colouring schemes are really rather pointless, and you might as well stick to black and white.
(2) The threshold for the number of iterations is technically infinite, so in practice you need to compute with larger and larger caps on the iteration, and explore what happens.
(3) The set is composed of very fine elements, which can easily be missed in a black and white representation at finite resolution. In fact, the highly coloured schemes make it easier to perceive some aspects of the structure. You should try out some of the movies above to see this.

The major difficulty is the coupling between issues (2) and (3). Finally, if you want to view the results, whether on paper or on a computer screen, you have to deal with the finite resolution of the viewing device.

■ A black and white colouring scheme

This is easy to implement. We just take, for example:

```
FracColorBW = (If[# == 1, GrayLevel[0], GrayLevel[1]] &);
```

■ Parametrizing the cap on iterations

To get a feel for the influence of the cap on iterations we now parametrize it. The radius at which we stop counting is not a critical parameter – we do in fact have room to lower it, and take the value of 10.

```
FastIter = Compile[{{z, _Complex}, {c, _Complex},
    {cap, _Integer}}, Module[{cnt, nz = z, nc = c},
    For[cnt = 0, (Abs[nz] < 10) && (cnt < cap),
     cnt ++, nz = nz * nz + nc];
    N[cnt]]];

FastIter[0, 0.0 + 0.5 I, 500]
```

500.

```
FracMC[cap_, z_, {{remin_, remax_}, {immin_, immax_}},
   steps_] := {{{remin, remax}, {immin, immax}},
   Table[FastIter[z, x + I * y, cap],
    {y, immin, immax, (immax - immin) / steps},
    {x, remin, remax, (remax - remin) / steps}]};

FracPlot[{{{remin_, remax_}, {immin_, immax_}},
matrix_},
colorfn_, newrange_] :=
ListDensityPlot[matrix,
Mesh -> False, Frame -> False,
ColorFunction -> colorfn,
PlotRange -> {0, newrange},
AspectRatio -> (immax-immin)/(remax - remin)]
```

Computing a picture of the set in the large can be done as before (we increase the resolution for this discussion – you should try significantly lower numbers than 2000 to get started!):

```
FracPlot[
 FracMC[300, 0, {{-21 / 10, 1 / 2}, {-12 / 10, 12 / 10}},
  2000], FracColorBW, 300]
```

If you compare this with the figure computed in the sub-section on compiled *Mathematica*, or the figure computed with **FixedPointList**, it looks like we have 'misplaced' some filamental structure joining some of the outlying black regions. This issue is highlighted very dramatically if we dive deep into the fractal to look at a well-known region:

```
g[pointx_, pointy_, n_, scale_] := Module[
  {width = scale^n}, {{pointx - width, pointx + width},
   {pointy - width, pointy + width}}]
```

The region of interest is given by the following range of coordinates.

```
g[-748 / 1000, 107 / 1000, 30, 8 / 10] // N
```

$$\begin{pmatrix} -0.749238 & -0.746762 \\ 0.105762 & 0.108238 \end{pmatrix}$$

Let's take a look a reasonable resolution (2000×2000) computation with caps of 100, 300, 500, 700 and 900:

```
newcap = 100;
FracPlot[
 FracMC[newcap, 0, g[-748 / 1000, 107 / 1000, 30, 8 / 10],
  2000], FracColorBW, newcap]
```

```
newcap = 300;
FracPlot[
 FracMC[newcap, 0, g[-748 / 1000, 107 / 1000, 30, 8 / 10],
  2000], FracColorBW, newcap]
```

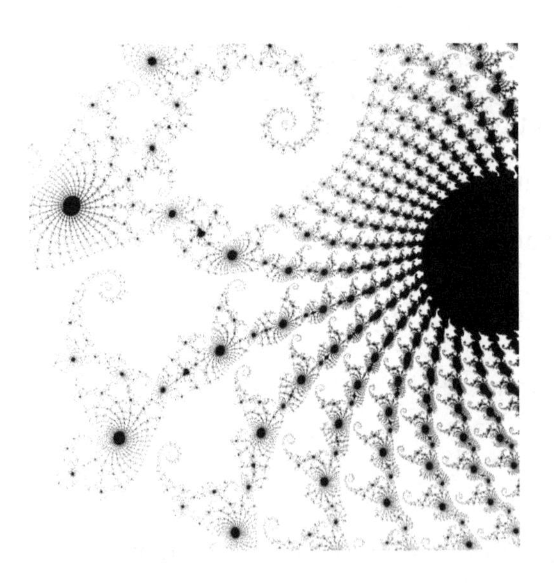

```
newcap = 500;
FracPlot[
 FracMC[newcap, 0, g[-748 / 1000, 107 / 1000, 30, 8 / 10],
  2000], FracColorBW, newcap]
```

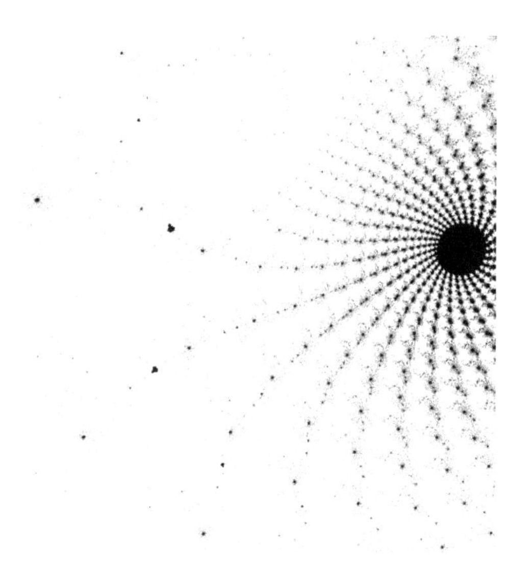

```
newcap = 700;
FracPlot[
 FracMC[newcap, 0, g[-748 / 1000, 107 / 1000, 30, 8 / 10],
  2000], FracColorBW, newcap]
```

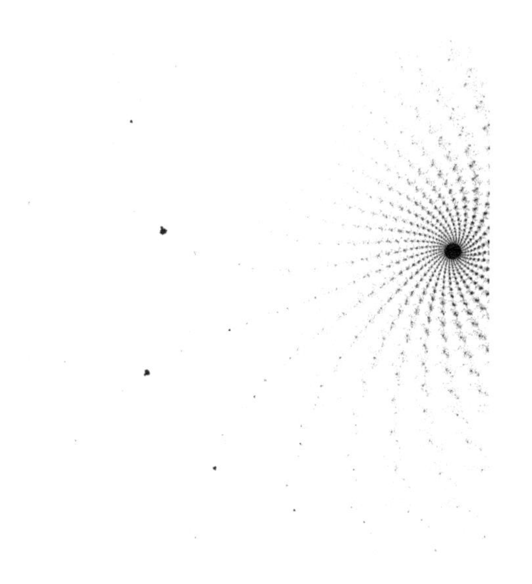

```
newcap = 900;
FracPlot[
 FracMC[newcap, 0, g[-748 / 1000, 107 / 1000, 30, 8 / 10],
  2000], FracColorBW, newcap]
```

If we drop our purist view and go to one of our jollier colouring schemes for this last region we get a rather different outcome. The reader is left to make their own judgement about the relative merits!

```
newcap = 900;
FracPlot[
 FracMC[newcap, 0, g[-748 / 1000, 107 / 1000, 30, 8 / 10],
  2000], FracColorA, newcap]
```

The reader might like to compare these last few illustrations with Plates 37–39 of Peitgen and Sauper (1986), The region illustrated here is centred slightly to the 'left' of their plate 38. It is also, with a cap of 200, the region shown by Penrose (2004), Figure

1.2d. An alternative representation of parts of the set is given in Chapter 3 of Penrose (1999).

■ Some purist projects

You might like to explore more cunning ways of visualizing the 'true' Mandelbrot set. One set of approaches involves fixing a display resolution (start with 500 × 500 and go up as resources allow), and defining the value of a pixel to be black or white according to a sub-sampling search algorithm. The idea would be to sample within each pixel k times, and colour it black if at least one of the k points does not escape. The k sub-samples could be picked by

(1) placing them on a rectangular or square grid;
(2) Monte Carlo sampling;
(3) sampling using a space-filling algorithm, such as Sobol sequences.

Exercises

6.1 ❄ Construct and explore compiled escape-time algorithms for each of the following mappings:

$$z \to z^3 + c$$

$$z \to \sin(z) + c$$

In each case construct suitable coloured visualizations, find some interesting regions, and zoom in on them.

6.2 ❄ Show that, by a linear transformation of coordinates, any cubic polynomial mapping can be reduced to the standard form:

$$z \to z^3 + (\lambda - 1)\, z - \lambda$$

Use *Mathematica*'s **`Solve`** function to find the fixed points of this mapping, and establish where these fixed points are stable. Numerical studies are helpful here!

6.3 ❄ Build a compiled escape-time visualizer for the cubic polynomial of Exercise 6.2, and identify the stable fixed point zones on your plot.

6.4 ❄ Construct a compiled escape-time visualizer for the mapping:

$$z \to \begin{cases} \frac{z-1}{c} & \text{Re}(z) \geq 0 \\ \frac{z+1}{\bar{c}} & \text{Re}(z) < 0 \end{cases}$$

Using the zooming function **`g`**, construct an animation of a dive into a neighbourhood of the point $c = 0.4131 + 0.62155\, i$, based on a starting value of $z = 0$. As a tip, one of your sequence should look roughly like this:

Appendix: C Code listings

■ C code for hexaman

```
#include "mathlink.h"
#include "math.h"
void fractalml(double x0, double y0, double xmin, double xmax, double ymin, double
ymax, int divs)
{double x, y, dist, xadd, yadd, temp;
  int xco, yco, counts;
  MLPutFunction(stdlink,"List",divs*(divs+1)-1);

  for(yco = 0; yco <divs; yco++){
  yadd = ymin + (ymax-ymin)*yco/(divs-1.0);

  for(xco = 0; xco <divs ; xco++){
  x = x0;
  y = y0;
  xadd = xmin + (xmax-xmin)*xco/(divs-1.0);
  dist = x0*x0+y0*y0;
  for(counts = 0; counts<100 && dist < 10000; counts++){
  temp = x*x-y*y+xadd;
  y = 2*x*y+yadd;
  x = temp;
  dist = x*x+y*y;
  }
  if(counts==100)
  {
  MLPutString(stdlink,"000000");
  }
  else
  {
  if(counts%2==0)
  {
  MLPutString(stdlink,"FF0000");
  }
  else
```

```
  {
  MLPutString(stdlink,"0000FF");
  }}}
  if(yco < divs-1)
  {MLPutString(stdlink,"\n");
  }
  else
  {}}
}
int main(int argc, char *argv[])
{return MLMain(argc, argv);}
```

■ C code for rasterman

```
#include "mathlink.h"
#include "math.h"
void fractalml(double x0, double y0, double xmin, double xmax, double ymin, double
ymax, int divs)
{double x, y, dist, xadd, yadd, temp;
  int xco, yco, counts;
  MLPutFunction(stdlink,"List",divs);

  for(yco = 0; yco <divs; yco++){
  yadd = ymin + (ymax-ymin)*yco/(divs-1.0);
  MLPutFunction(stdlink,"List",divs);

  for(xco = 0; xco <divs ; xco++){
  x = x0;
  y = y0;
  xadd = xmin + (xmax-xmin)*xco/(divs-1.0);
  dist = x0*x0+y0*y0;
  for(counts = 0; counts<100 && dist < 10000; counts++){
  temp = x*x-y*y+xadd;
  y = 2*x*y+yadd;
  x = temp;
  dist = x*x+y*y;
  }
  if(counts==100){
  MLPutFunction(stdlink,"RGBColor",3);
  MLPutReal(stdlink,0.0);
  MLPutReal(stdlink,0.0);
  MLPutReal(stdlink,0.0);}
  else
  {
  if(counts<51){
  MLPutFunction(stdlink,"RGBColor",3);
  MLPutReal(stdlink,1.0);
  MLPutReal(stdlink,1.0-counts/50.0);
  MLPutReal(stdlink,0.0);}
  else
  {
  MLPutFunction(stdlink,"Hue",1);
  MLPutReal(stdlink,1.5-counts/100.0);}
  }}}}
int main(int argc, char *argv[])
{return MLMain(argc, argv);}
```

7 Symmetric chaos in the complex plane

Introduction

The existence of functions such as **Nest** and **NestList**, together with extensive visualization tools, ensure that *Mathematica* is a natural system for investigating the iteration of mappings. One favourite is the logistic map that was considered in Chapter 5. Complex numbers play a very natural role, from two quite different points of view. First, we wish to understand why maps such as the logistic map behave the way they do, without relying merely on numerical simulation. Second, we wish to extend the use of numerical simulation to mappings of the complex plane.

You explored the first point in Chapter 5, and the second in Chapters 4 and 6, for simple polynomial maps. Now you will consider some other mappings of the complex plane into itself. These mappings are non-holomorphic (in the sense that will be defined properly in Chapter 10 – for now it suffices to realize that this means the functions involve complex conjugation in an essential way). Normally, when considering the theory of complex numbers, one's interest is quite rightly focused on the analytical properties of holomorphic or meromorphic functions. I hope the constructions described here will suggest that there is much that is both beautiful and interesting in complex structures that are not holomorphic. My approach is based on that of Field and Golubitsky (1992), and the use of *Mathematica* on this topic was first given by the author (Shaw, 1995). Readers are encouraged (a) to get the original book by Field and Golubitsky, (b) to see M. Field's web site at `http://nothung.math.uh.edu/~mike/` and (c) to pay attention to the copyright notices on that web site!

The odd logistic map considered in Chapter 5 can be extended very trivially to the complex plane, complete with the symmetry of a circle, by introducing

$$\text{Logistic}(\lambda, z) = \lambda\, z\, (1 - (|z|)^2) \qquad (7.1)$$

This is symmetric since

$$\text{Logistic}(\lambda, z\, e^{i\phi}) = e^{i\phi}\, \text{Logistic}(\lambda, z) \qquad (7.2)$$

if ϕ is real. However, this is not terribly interesting. We will obtain the same dynamics as the odd logistic map, but along a ray in the complex plane. Matters are significantly more interesting, if, rather than the symmetry of the circle, we require that maps are symmetric under the action of the discrete groups Z_n (the cyclic group consisting of rotations by $2\pi/n$ and D_n (the dihedral group, consisting of such rotations and a flip). In this chapter you will explore the simplest possible non-linear maps possessing the discrete symmetries. For polynomial maps, these take the form

$$g(z, \lambda, \alpha, \beta, \gamma, \omega, n) = \gamma\, \bar{z}^{n-1} + z\,(\lambda + i\,\omega + z\,\alpha\,\bar{z} + \beta\,\text{Re}(z^n)) \qquad (7.3)$$

with λ, α, β, γ and ω real, and n an integer. These possess the cyclical symmetry

$$g\left(e^{\frac{2\pi i}{n}} z, \lambda, \alpha, \beta, \gamma, \omega, n\right) = e^{\frac{2\pi i}{n}} g(z, \lambda, \alpha, \beta, \gamma, \omega, n) \tag{7.4}$$

If, in addition, $\omega = 0$, they also inherit the additional symmetry expressed by

$$g(\bar{z}) = \overline{g(z)} \tag{7.5}$$

which ensures full dihedral symmetry. There is a simple non-polynomial generalization, with extra parameters, and this will be given below. The experiments presented here investigate the structure of the resulting complex attractors, in an attempt to understand the material presented by Field and Golubitsky (1992). You are strongly advised to first carry out the simulations with, say, 1/10 the number of iterations used in the supplied examples, in order to get a feel for the time required. Some of the simulations take some time – the results are usually worth waiting for! They illustrate the sense in which symmetry (implying a certain harmony of form) can be consistent with chaos (implying rather a lack of form).

■ Acknowledgement

I am very grateful to Professors Brian Twomey and Donal Hurley for useful discussions on these and other topics during a visit to University College, Cork.

7.1 Creating and iterating complex non-linear maps

We can make the simulations in a more straightforward way by introducing a function that automatically compiles our nonlinear map. Note that this function assumes default values of zero for the parameters ω, δ and p. These can be omitted unless explicitly required.

```
MakeMap[λ_, α_, β_, γ_, m_, ω_:0, δ_:0, p_:1] =
Compile[{{z, _Complex}},
((λ + α*z*Conjugate[z] + β*Re[z^m] +
δ*Abs[z]*Cos[Arg[z]*m*p] + I ω)*z +
γ*Conjugate[z]^(m-1))];
```

We can also automate the generation of a starting value and subsequent list. The idea is to first iterate the map for a while (2000 times) to eliminate the effect of any transients, then to proceed to further iteration.

```
makedata[func_, n_] :=
Module[{start=Nest[func, 0.5+0.5I,2000]},
NestList[func,start,n]];
```

Finally, we make a function that does it all and plots the resulting attractor:

```
MakeSymChaos[myfunc_, n_, color_, size_]:=
Module[{data},
data = makedata[myfunc,n];
ListPlot[Map[{Im[#],Re[#]}&, data],
PlotStyle -> {PointSize[size],color},
AspectRatio -> 1, Axes -> None]];
```

Now we make the compiled maps for some interesting objects.

- **Dihedral group: polynomial**

```
threegadget = MakeMap[1.56, -1.00, 0.10, -0.82, 3];
flint       = MakeMap[2.50, -2.50, 0.00, 0.90, 3];
sanddollar  = MakeMap[-2.34, 2.00, 0.20, 0.10, 5];
pentagon    = MakeMap[2.60, -2.00, 0.00, -0.50, 5];
churwin     = MakeMap[2.409, -2.50, -0.20, 0.81, 24];

MakeSymChaos[threegadget, 25000, RGBColor[0,0,1],
0.005]
```

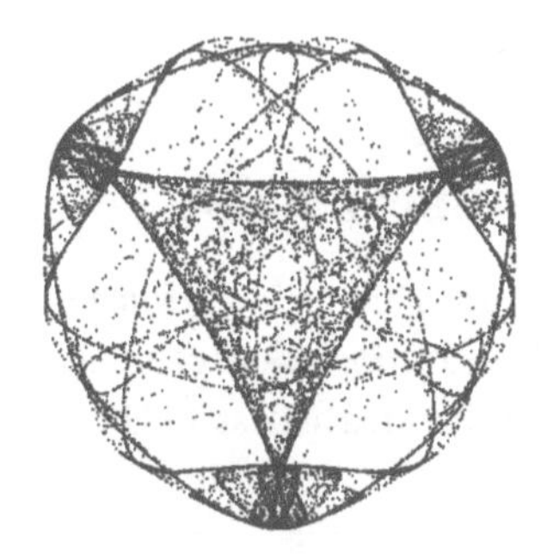

```
MakeSymChaos[flint, 25000, RGBColor[0,0.5,0.5], 0.005]
```

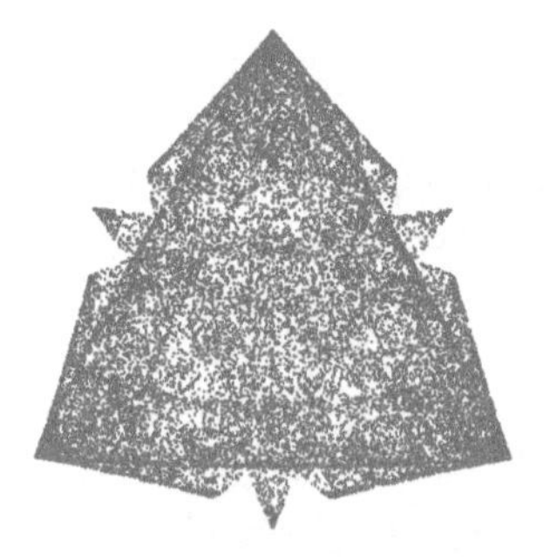

```
MakeSymChaos[sanddollar, 25000, RGBColor[1,0,0], 0.005]
```

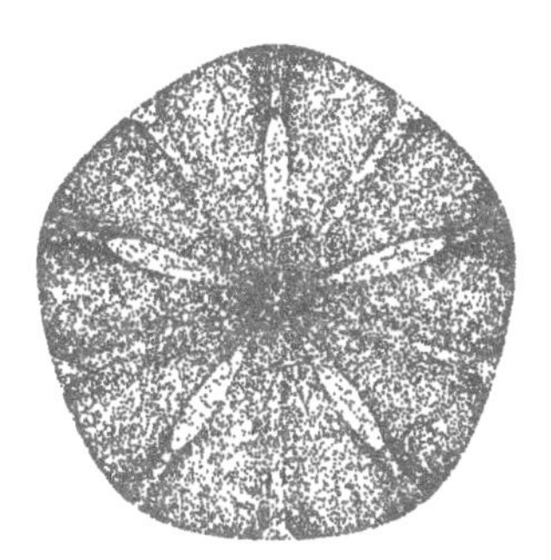

```
MakeSymChaos[pentagon, 25000, RGBColor[1,0,1], 0.005]
```

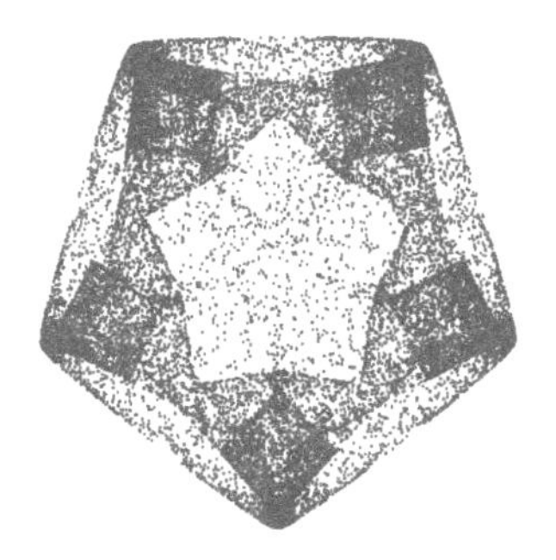

```
MakeSymChaos[churwin, 50000, RGBColor[0.5,0,0.5],
0.005]
```

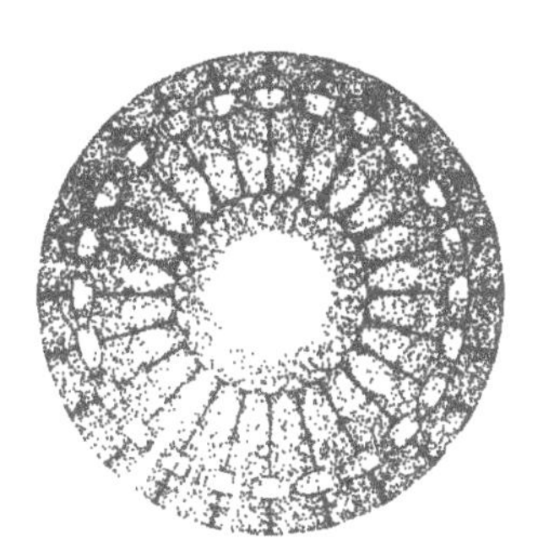

What you should appreciate from these pictures is that extensive iteration of the mappings, in common with the odd logistic map for some values of its parameters, reinstates the symmetry in the original mapping – this is shown very dramtically in the pictures. Later on we shall see more explicitly how the symmetry can be initially broken and then reinstated.

■ Cyclic group: polynomial

Now we break the full dihedral symmetry down to cyclical symmetry only, by introducing a complex term into the function. This produces some rotation or 'swirl' in the resulting plots. To induce this we supply an extra parameter to our **MakeMap** function – previously this assumed its default value of zero.

```
swirlygig = MakeMap[-1.86, 2.00, 0.00, 1.00, 4, 0.10];
flower = MakeMap[-2.50, 5.00, -1.90, 1.00, 5, 0.188];

MakeSymChaos[swirlygig, 25000, RGBColor[1,0,0], 0.005]
```

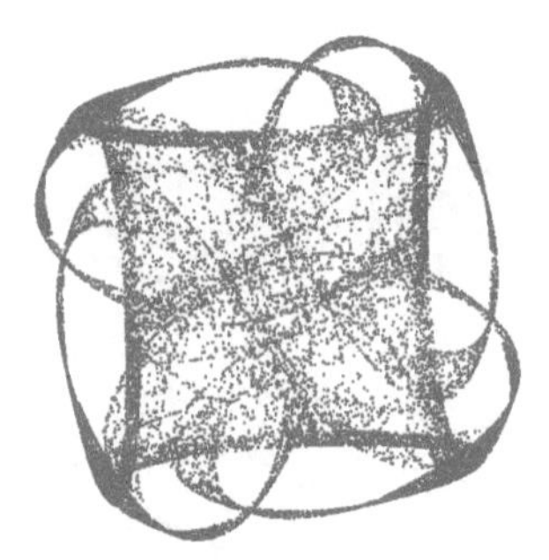

```
MakeSymChaos[flower, 25000, RGBColor[1,0.5,0], 0.005]
```

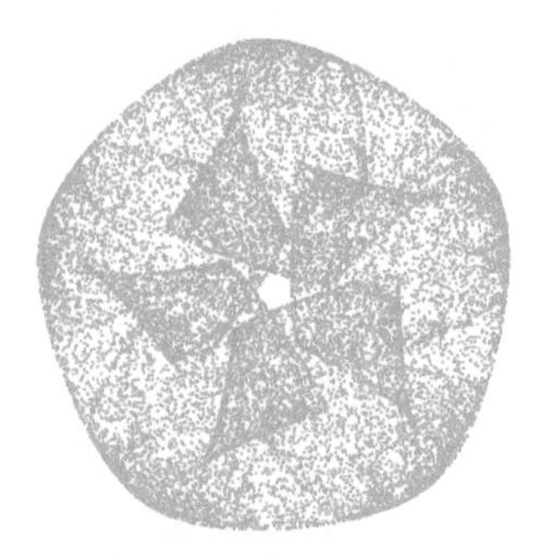

■ Dihedral group: non-polynomial

As a third variant, we put back the dihedral symmetry but introduce a non-polynomial term into the mapping:

```
familiar =
MakeMap[1.00, -2.10, 0.00, 1.00, 3, 0.00, 1.00, 1];
star =
MakeMap[-2.42, 1.00, -0.04, 0.14, 6, 0.00, 0.088, 0];

MakeSymChaos[familiar, 25000, RGBColor[0,0,1], 0.005]
```

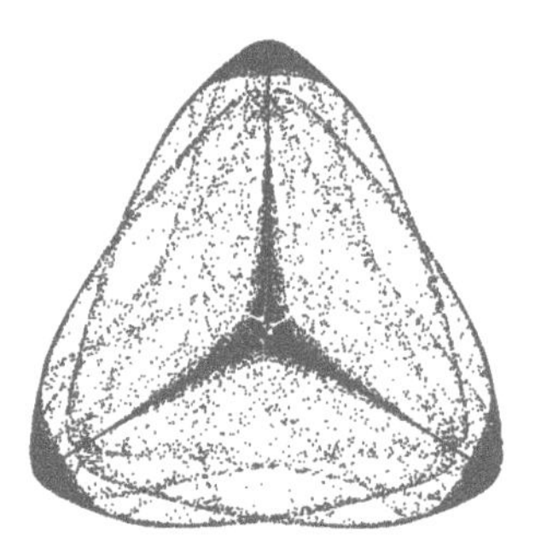

```
MakeSymChaos[star, 25000, RGBColor[1,0,0], 0.005]
```

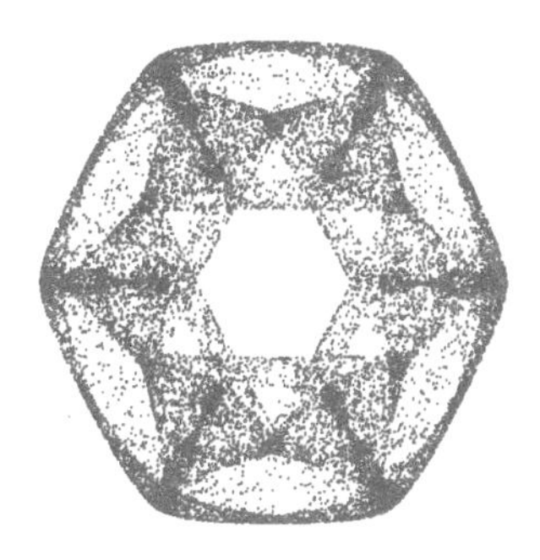

7.2 A movie of a symmetry-increasing bifurcation

All of the plots given above are fully symmetric. Just as in the case of the odd logistic map discussed in Chapter 5, the symmetric state is only achieved for certain values of the map parameters. You can explore the creation of the symmetry by creating a movie of the scatter plots, based on varying a relevant parameter. It is worth exploring this for any of the maps described above. The code for just one is given here. The **Do**-loop creates a movie, while the subsequent graphics array shows several frames of the movie. If you are using a *Mathematica* notebook, you can animate the movie using the 'Animate Selected Graphics' command on the Cell menu. If you are using *Mathematica* technology beyond version 5.2, you should explore the options provided in your current version of *Mathematica* for animation – see also the enclosed CD and on-line supplement.

```
ScaledSymChaos[myfunc_, n_, color_, size_, range_:1]:=
Module[{data},
data = makedata[myfunc,n];
ListPlot[Map[{Im[#],Re[#]}&, data],
PlotStyle -> {PointSize[size],color},
AspectRatio -> 1, Axes -> None,
PlotRange -> {{-range, range},{-range, range}}]];
```

```
Do[
(tri = MakeMap[2.16+0.005*k,-1.0, 0.0,-0.5, 3];
frame[k] = ScaledSymChaos[tri, 2000, RGBColor[0,0,1],
0.01, 2.0];),{k, 0, 35}]
```

For printing purposes, one can just show selected frames that show the overall increase in symmetry. Note how the movie shows that this process is not in any sense monotonic. Having gone chaotic once, it collapses to a periodic orbit and then goes chaotic again – this mirrors the behaviour of the logistic map (islands of periodicity within chaos) very nicely.

```
Show[GraphicsArray[{{frame[0], frame[7], frame[14]},
{frame[21], frame[28], frame[35]}}]]
```

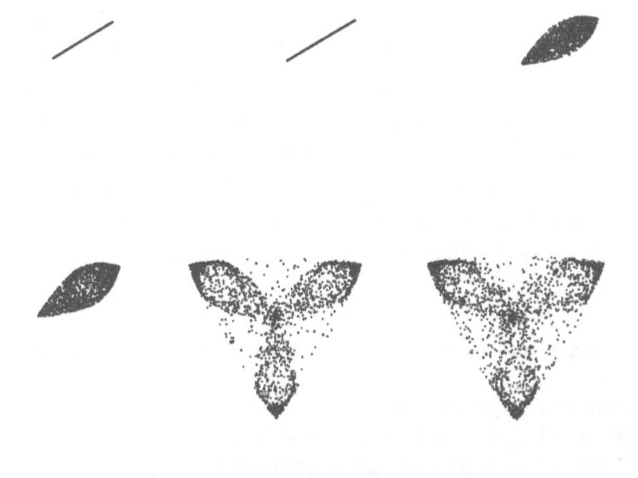

7.3 ❀ Visitation density plots

Some regions in the complex plane are visited significantly more frequently than others. There are more dots in such regions. How can we quantify this effect? One approach is to split the complex plane into a large array of neighbourhoods, and count the number of times the iterates of the map visit each neighbourhood. To do this effectively we need a significantly larger number of iterations (millions) and an efficient means of identifying the neighbourhood with which each iterate should be associated.

In the following function, each iterate is computed, used to calculate which neighbourhood is visited, and then discarded. Thus the only significant memory requirement is that needed to store a two-dimensional array used to build up the visitation density. The value of the iteration is used directly to compute which array element is to be incremented.

```
MakeVisitDensity[myfunc_, itmax_, nop_] :=
Module[{d, zold, znew,i,j,m},
d = Table[0, {nop}, {nop}];
znew = Nest[myfunc, -0.1+0.6I,2000];
Do[(zold = znew;
 znew = Evaluate[myfunc[zold]];
 i =
 Min[Max[Floor[nop*(Re[znew]+1)/2],1],nop];
 j = Min[Max[Floor[nop*(Im[znew]+1)/2],1],nop];
d[[j,i]] += 1),{m, 1, itmax}];d]
```

We can do a quick one to make sure it is working:

```
pdata = MakeVisitDensity[familiar, 20000, 80];
```

The conjugation symmetry is more easily perceived if we transpose the data so that the reflection symmetry is about the vertical axis – I therefore transpose before plotting:

```
ListDensityPlot[Transpose[pdata],
Mesh -> False, Frame -> False]
```

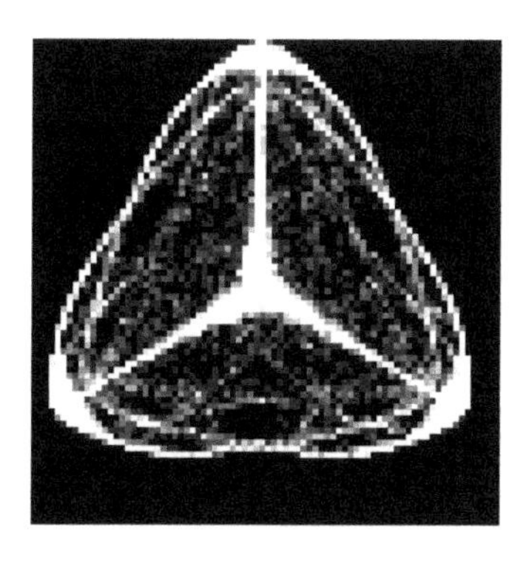

7.4 ❁ High-resolution plots

Once we have a working algorithm, we can try increasing both the number of iterations and the resolution of our plots. This takes a longer time, but gives tantalizing results! If you try this on your own computer bear in mind the results and time taken in the previous example. The timing below is on a 1.4 GHz machine.

```
Timing[pdata = MakeVisitDensity[familiar,100000,250];]
```

{17.24 Second, Null}

```
ListDensityPlot[Transpose[pdata],
Mesh -> False, Frame -> False]
```

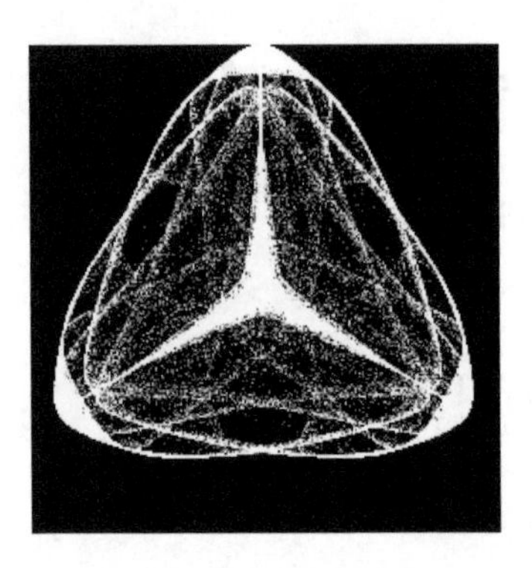

7.5 ❁ Some colour functions to try

While you have **pdata** stored in the kernel, you may wish to try using some of the following colour functions to bring out features in the data. This first group adopts a single background colour, usually black, for neighbourhoods that are not visited, and then we take a colour directive that is a single continuous function of the number of visits.

```
cfunca =
(If[#==0, RGBColor[0,0,0], RGBColor[1-#,1-#,1]]&);
cfunch =
(If[#==0, Hue[0.5,0.5,0], Hue[2#/3]]&);
cfuncb =
(If[#==0, RGBColor[0,0,0], RGBColor[0,0,#]]&);
cfuncr =
(If[#==0, RGBColor[0,0,0], RGBColor[#,0,0]]&);
cfuncrb =
(If[#==0, RGBColor[0,0,1], RGBColor[#,0,0]]&);
gold =
(If[#==0, RGBColor[0,0,0], RGBColor[1,1-#,0]]&);
```

The following two functions split the range into two and assigns a separate function to each range. This is a useful way of arranging that sufficient contrast is applied to ensure that the detail in infrequently visited regions is not washed out.

```
metallica =
(Which[#==0,          RGBColor[0,0,0],
       0<#<0.5,       RGBColor[2#,2#,2#],
       0.5<=#<=1.0,   RGBColor[2-2#,2-2#,1]
       ]&);

stellartipped =
(Which[#==0,          RGBColor[0,0,0],
       0<#<=0.5,      RGBColor[1,1-2#,0],
       0.5<#<=1.0,    Hue[1.5-#]
       ]&);

ListDensityPlot[Transpose[pdata],
Mesh -> False, Frame -> False,
ColorFunction -> stellartipped]
```

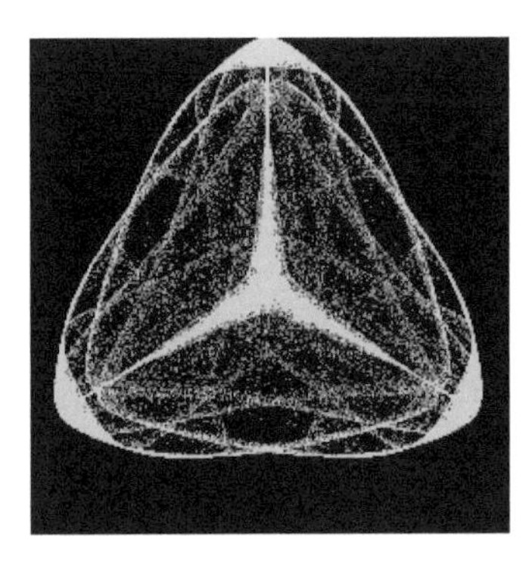

Is it left for you (in Exercise 7.1) to explore what this looks like with other colour functions.

7.6 Hit the turbos with *MathLink*!

Our plots so far take some time to compute and the results are somewhat grainy. Clearly we need to be more efficient, or just faster. An approach that gives very satisfying results very quickly is to use some heavily customized C code with *MathLink* to speed up the computation. *Mathematica* then does the rendering job. Note that our *Mathematica* code is now a useful prototype for the C code – we just have to implement the same algorithm using C. We have also made some changes in the details, working with real and imaginary parts, and writing C explicitly for the cases of symmetries of order 3, 4, 5 and 6. The C program and associated template file are given in Appendix 7.

Once this program is compiled and linked, it can be installed in the usual way. Note that here the author sets a path to the *MathLink* executables that needs to be adapted to your own system:

```
SetDirectory[
  "/Books/ComplexMath2005/AAMathLink/binMacOSX"];

Install["symchaos"]
```

LinkObject[./symchaos, 2, 2]

```
?SymChaos
```

```
Global`SymChaos

SymChaos[lambda_?NumberQ, alpha_?NumberQ, beta_?NumberQ,
  gamma_?NumberQ, sym_Integer, omega_?NumberQ, delta_?NumberQ,
  p_Integer, radius_?NumberQ, itmax_Integer, nopoints_Integer] :=
 ExternalCall[LinkObject[./symchaos, 2, 2],
  CallPacket[0, {N[lambda], N[alpha], N[beta], N[gamma],
    sym, N[omega], N[delta], p, N[radius], itmax, nopoints}]]
```

The parameters play exactly the role they do in *Mathematica*. For now we consider omega = 0 and radius = 1, and now **itmax** is to be measured in thousands of iterations. We also supply the extent of the symmetry through a parameter **sym**. This can take any integer value, and we have carried out additional C optimization for low values of the integer parameter **sym** that gives the degree of symmetry. We can also automate the plot routines as follows:

```
SymChaosPlot[object_,colorfunc_] :=
ListDensityPlot[Transpose[object],
Mesh -> False, Frame -> False,
ColorFunction -> colorfunc]
```

As a test, try the following (on a 1.4 GHz machine, this takes about 1/4 second to calculate using *one million* iterations):

```
Timing[sanddollar=SymChaos[-2.34,2,0.2,0.1,5,0,0,0,1.0,
100,200];]
```

{0.27 Second, Null}

```
SymChaosPlot[sanddollar, metallica]
```

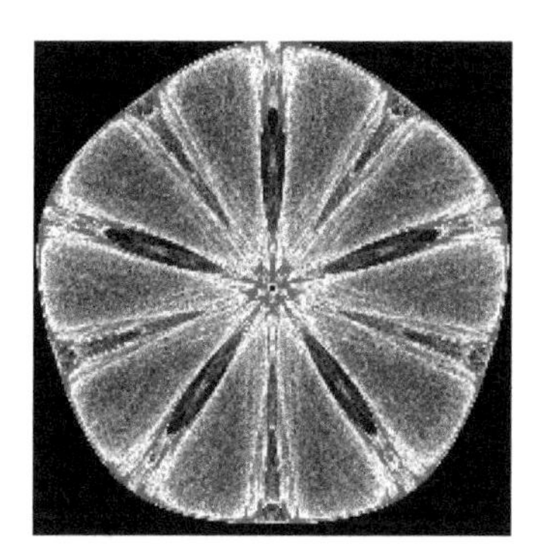

You should experiment with the settings to determine what is practical for your own system. I suggest computing a few medium resolution samples, based on 60 million iterations, and a visitation density based on a 400×400 grid. This involves the last two parameters of **SymChaos** being set to 6000, 400. Below I have shown, for printing purposes, plots based on a billion iterations and a 1000×1000 array. The number of iterations governs the execution time for the C code, while the grid size influences the time taken by *Mathematica* to render the image. On a 1.4 GHz G4 machine, the real elapsed time is less than 1 minute for the calculation of **sanddollar** with the parameters I suggest you try. Note that the binary files supplied are not optimized – if you are an expert with compiler settings you may be able to achieve significant speed improvements!

7.7 Billion iterations picture gallery

```
sanddollar =
SymChaos[-2.34,2,0.2,0.1,5,0,0,0,1.0,100000,1000];

SymChaosPlot[sanddollar,metallica]
```

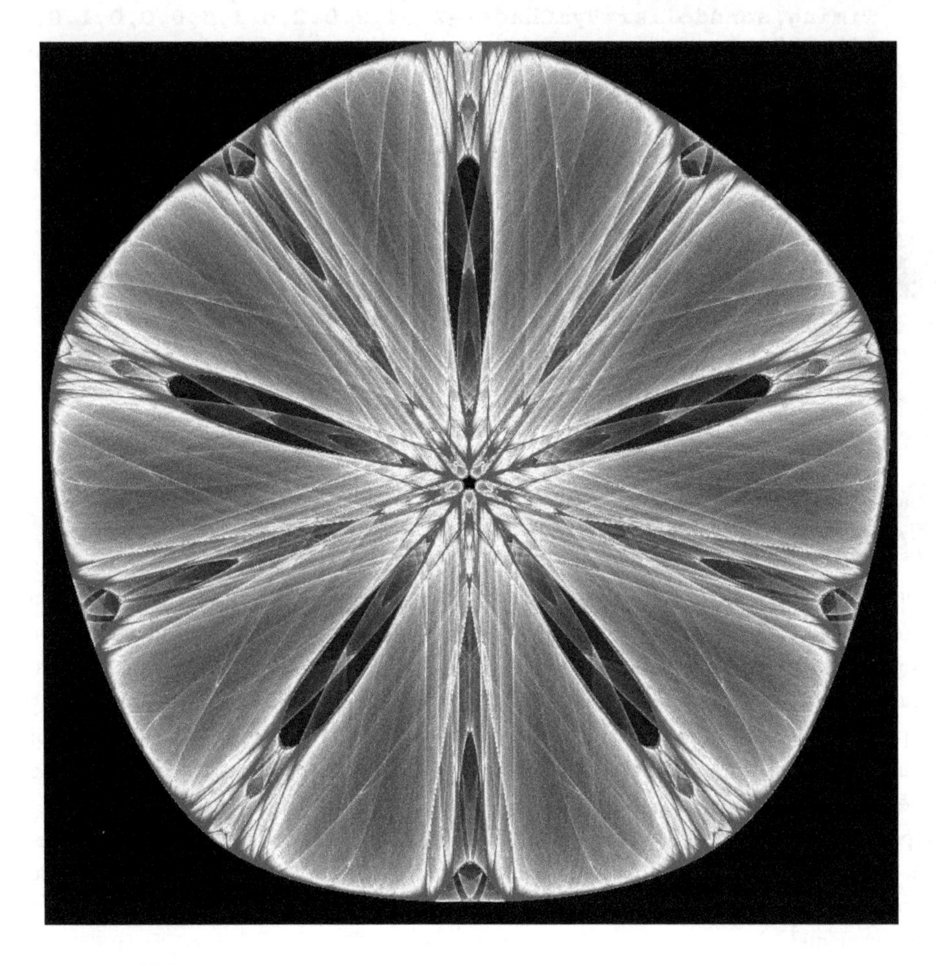

```
quadgig =
SymChaos[-1.86,2,0.0,1.0,4,0,0,0,1.0,100000,1000];
SymChaosPlot[quadgig,stellartipped]
```

```
pentibug =
SymChaos[-2.32,2.32,0,0.75,5,0,0,0,1.0,100000,1000];
SymChaosPlot[pentibug,gold]
```

■ Using the radius parameter

Depending on the values of the supplied parameters, the iteration may wander out of the standard square $-1 < \mathrm{Re}[z], \mathrm{Im}[z] < 1$ used by default. It is good practice to do some sample iterations with the scatter plot technique to see how far out the iterates go in the complex plane. In the example below we have reset the radius parameter to 1.5 in order to accommodate a larger attractor.

```
flintstone =
SymChaos[2.5,-2.5,0,0.9,3,0,0,0,1.5,100000,1000];
SymChaosPlot[flintstone,gold]
```

```
familiar =
SymChaos[1,-2.1,0,1,3,0,1,1,1.1,100000,1000];
SymChaosPlot[familiar,gold]
```

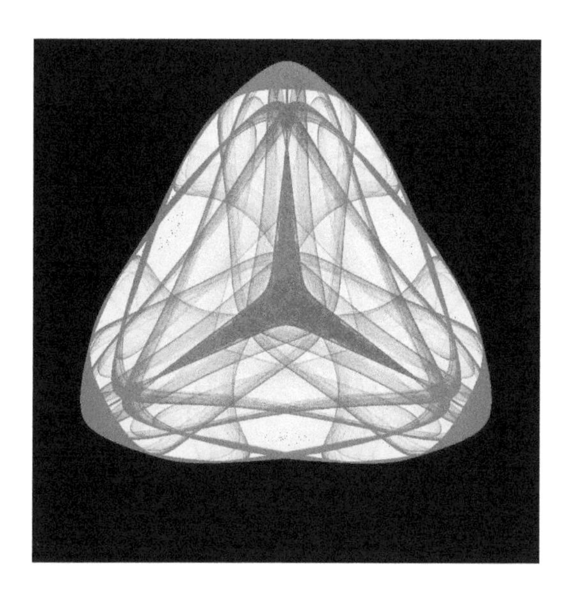

■ Breaking the dihedral group to cyclic group

All our examples so far have had full dihedral symmetry. We can break the dihedral symmetry by introducing a non-zero value to the parameter **omega**. This gives attractors with Z_n symmetry only. Our visualization below of **swirlygig** is not reflection symmetric about the vertical axis.

```
swirlygig =
SymChaos[-1.86,2.0,0,1,4,0.1,0,0,1.0,100000,1000];
SymChaosPlot[swirlygig,stellartipped]
```

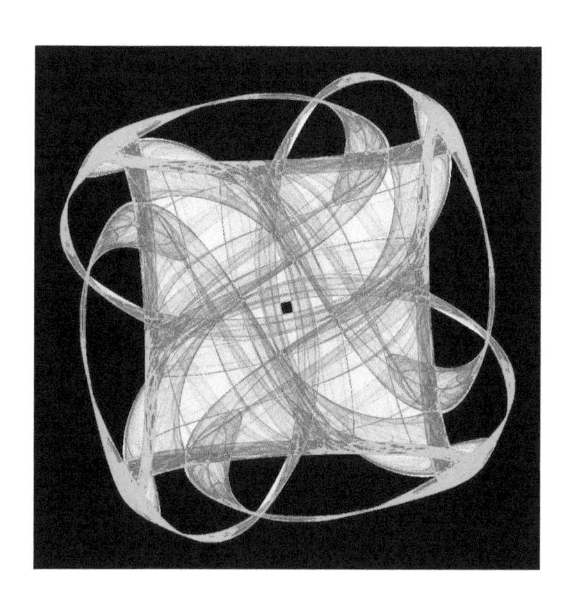

```
flowerv =
SymChaos[-2.5,5.0,-1.9,1.0,5,0.188,0,0,0.75,100000,1000
];SymChaosPlot[flowerv,gold]
```

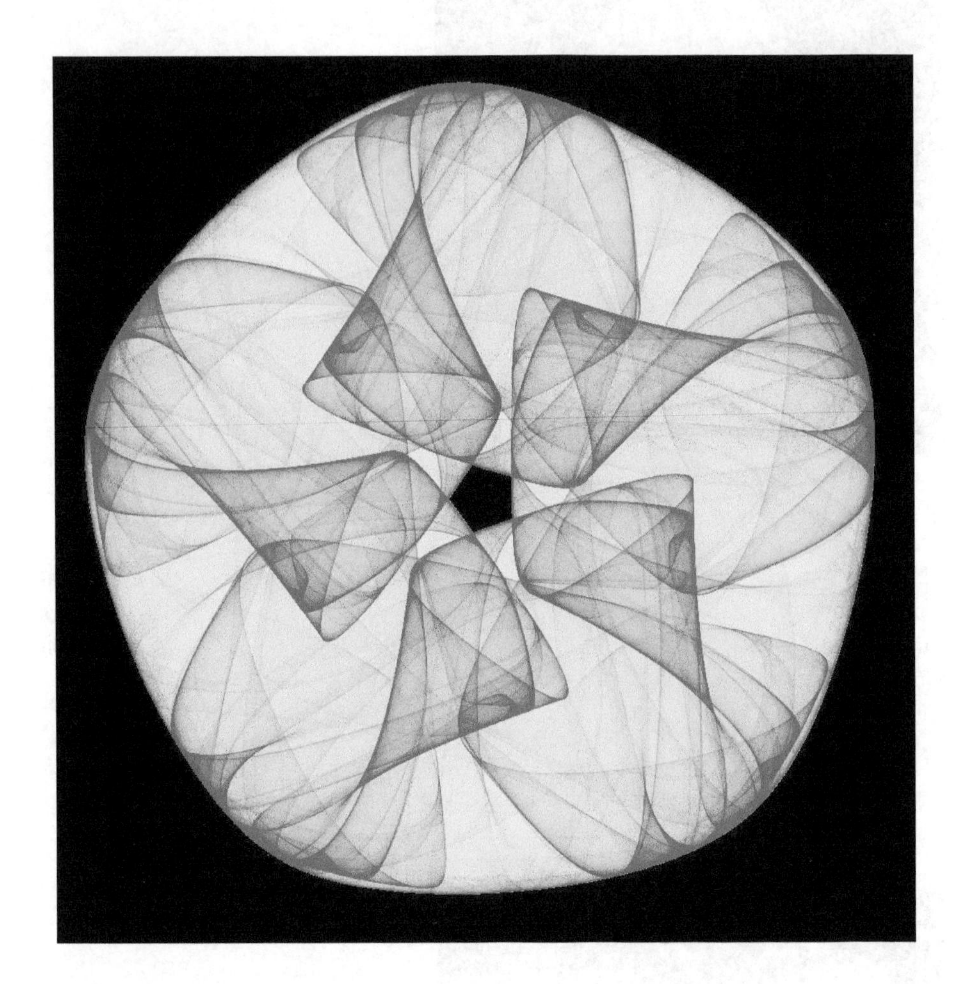

■ Cleaning up

Remove the installed routine with:

```
Uninstall["symchaos"];
Remove[SymChaos]
```

Exercises

7.1 Recreate and explore the data set **pdata**, and view it with various choices of colour function, such as **cfuncx**, **metallica**, **stellartipped** and **gold**.

7.2 ⚙ Load the **symchaos** C *MathLink* program, and the definitions of **SymChaos-Plot** and the colour functions of Exercise 7.1. Explore the visitation densities of the following systems, and of those arising from small variations in parameters from those given here:

```
washington =
SymChaos[2.6,-2,0,-0.5,5,0,0,0,1.4,6000,400];
SymChaosPlot[washington,metallica];

halloween =
SymChaos[-2.7,5,1.5,1,6,0,0,0,1.0,6000,400];
SymChaosPlot[halloween,metallica];

bracelet =
SymChaos[-2.08,1,-0.1,0.167,7,0,0,0,1.3,6000,400];
SymChaosPlot[bracelet,stellartipped];

glass =
SymChaos[-2.05,3,-16.79,1,9,0,0,0,0.8,6000,400];
SymChaosPlot[glass,gold];
```

7.3 ⚙ Create the analogs of the symmetry-generating movie of Section 7.2 for some of the other examples in Section 7.1, using the scatterplot technique.

7.4 ⚙ If you have a fast computer, try to create a movie of the symmetry-generating process viewed by a visitation density. Use of the **symchaos** add-in is recommended!

Appendix: C code listings

The following *MathLink* template and C source code were used to produce the installable executables. A few comments are pertient. In general, we recommend the use of the vector and matrix memory allocation/deallocation routines described in *Numerical Recipes* (Press *et al.* 1992). Indeed, I strongly recommend this book for its discussion of setting up arrays in C if you have it and use of the `nrutil.c` utility package for creating arrays. The authors of *Numerical Recipes* have kindly put this in the public domain. Here, however, I wish to use a square array of long integers, in order to ensure that I don't get integer overflow in any cell when using many millions of iterations. I have therefore modified and simplified the `nrutil.c` function imatrix to this end. You should investigate `nrutil.c` to see how to treat other cases of vectors and matrices. It is a fairly standard template file, but note the use of the `Manual` return type. To return a list, or, as here, a list of lists down *MathLink,* it is simpler to use this approach. In the C code note how this is expressed by sending the `List` header down the link, and then the value of each integer in the array describing the visitation density. This is done with `MLPutLongInteger`. Note that, for portability, I have used explicit real variables, and done some optimization for each of the cases 3, 4, 5 and 6. As noted in the code, the function `MLPutLongIntegerArray` could be used to clean up a little.

```
#include "mathlink.h"
#include <stdlib.h>
#include <math.h>

/* P() is a simple macro defined in mathlink.h that allows function pro-
toypes to be written in the same way for ANSI and K&R C */

long **isqmatrix P((int nsize));
void freeisqmatrix P((long **m, int nsize));
void symchaos P((double lambda, double alpha, double beta, double gamma,
int sym, double omega, double delta, int pee, double radius, int itmax,
int nopoints));
int main P((int argc, char** argv));

long **isqmatrix(nsize)
int nsize;
{int i;
 long **mat;
 mat = (long **) malloc((unsigned) (nsize + 1)*sizeof(long*));
 for(i=0;i<nsize;i++)
 {mat[i] = (long *) malloc((unsigned) (nsize + 1)*sizeof(long));
 };
 return mat;
 }

void freeisqmatrix(m, nsize)
long **m;
int nsize;
 {int i;
 for (i=0;i<nsize;i++) free(m[i]);
 free(m);
 }

void symchaos(lambda,alpha,beta,gamma,sym,omega,delta,pee,radius,itmax,
nopoints)
double lambda,alpha,beta,gamma,omega,delta,radius;
int itmax,nopoints,sym,pee;
{ double l,a,b,g,w,de,r;
  double xold, yold, xnew, ynew, xsq, ysq, xfo, yfo;
  double xsqysq, temp, argum, absz, abssq;
  int i, n, j, k, p, q, nsym, pe;
  long **d;

  d = isqmatrix(nopoints);
  l = lambda;
  a = alpha;
  b = beta;
  g = gamma;
  w = omega;
  de = delta;
  i = itmax;
  n = nopoints;
  r = radius;
```

```
  nsym = sym;
  pe = pee;

  for(j = 0; j < n; j++)
        {
        for(k = 0; k < n; k++)
                {
                d[j][k] = 0;
                }
        }
        xold = -0.1;
        yold = 0.6;

/* Do the iterations, with 10000 to start off to remove transients */
       for(q=0; q < 10000; q++)
        {
    for(j=0; j < i+1; j++)
  {  xsq = xold*xold;
  ysq = yold*yold;
  argum = atan2(yold,xold);
  abssq = xsq + ysq;
  absz = sqrt(abssq);

switch(nsym)
{ case 3:
  temp = 1+a*abssq+b*xold*(xsq - 3.0*ysq)+de*absz*cos(3.0*pe*argum);
  xnew = xold*temp + g*(xsq - ysq) - w*yold;
  ynew = yold*temp - 2.0*g*xold*yold + w*xold;
  break;

  case 4:
  temp=1+a*abssq+b*(abssq*abssq - 8.0*ysq*xsq)+de*absz*cos(4.0*pe*argum);
  xnew = xold*temp + g*xold*(xsq - 3.0*ysq) - w*yold;
  ynew = yold*temp - g*yold*(3.0*xsq - ysq) + w*xold;
  break;

  case 5:
  xfo = xsq*xsq;
  yfo = ysq*ysq;
  xsqysq = xsq*ysq;
  temp = 1+a*abssq+b*xold*(xfo-10*xsqysq+5*yfo)+de*absz*cos(5.0*pe*argum);
  xnew = xold*temp + g*(xfo-6*xsqysq+yfo) - w*yold;
  ynew = 4*g*xold*yold*(ysq - xsq) + yold*temp + w*xold;
  break;

  case 6:
  xfo = xsq*xsq;
  yfo = ysq*ysq;
  xsqysq = xsq*ysq;
temp=1+a*abssq+b*(xfo*xsq-15*xfo*ysq+15*xsq*yfo-yfo*ysq)+de*absz*cos(6.0*p
e*argum);
  xnew = xold*temp + g*xold*(xfo-10*xsqysq+5*yfo) - w*yold;
  ynew = g*yold*(10*xsqysq - 5*xfo -yfo) + yold*temp + w*xold;
```

```
  break;

  default:
 temp=
  1+a*abssq+b*pow(absz,nsym)*cos(nsym*argum)+de*absz*cos(nsym*pe*argum);
  xnew = xold*temp + g*pow(absz,nsym-1.)*cos((nsym-1.)*argum) - w*yold;
  ynew = yold*temp - g*pow(absz,nsym-1.)*sin((nsym-1.)*argum) + w*xold;
}

  xold = xnew;
  yold = ynew;
  if (j > 0)
  {
  k = floor(0.5*n*(xold+radius)/radius);
  p = floor(0.5*n*(yold+radius)/radius);
  if ((p < n) && (k < n) && (p > -1) && (k > -1))
  {
    d[p][k] += 1;
  }}}}

/* The following could be simplified by use of MLPutLongIntegerArray*/
  MLPutFunction(stdlink,"List", n);
    for(j = 0; j < n; j++)
        {
         MLPutFunction(stdlink,"List", n);
        for(k = 0; k < n ; k++)
                {
                MLPutLongInteger(stdlink,d[j][k]);
                }
        }
   freeisqmatrix(d,nopoints);
}
int main(argc,argv)
int argc; char* argv[];
{return MLMain(argc, argv);}
```

Here is the template file:

```
:Begin:
:Function: symchaos
:Pattern: SymChaos[lambda_?NumberQ,alpha_?NumberQ,beta_?NumberQ,gamma_?Num-
berQ, sym_Integer, omega_?NumberQ, delta_?NumberQ, p_Integer, radius_?Num-
berQ,itmax_Integer,nopoints_Integer]
:Arguments: { N[lambda], N[alpha], N[beta], N[gamma], sym, N[omega],
N[delta], p, N[radius],itmax, nopoints}
:ArgumentTypes:{Real,Real,Real,Real,Integer,Real,Real,Integer,Real,
Integer,Integer}
:ReturnType:   Manual
:End:
```

8 Complex functions

Introduction

In this section we give a more precise characterization of complex functions and review their basic properties. We also introduce some formal concepts, such as neighbourhoods and open sets, in order to lay the foundations for a discussion of continuity and differentiability. We shall then make a first definition of basic functions such as the exponential and trigonometric functions, and their inverses, by referring back to real definitions. This will be revisited in Chapter 9 from a power series perspective. We shall also look at the concept of branch points, and the extended complex plane or 'Riemann sphere'.

We shall also explore various ways of visualizing complex functions using *Mathematica*. We can build various routines for looking at functions. The first one we will consider takes a two-dimensional point of view, where functions are regarded as mappings taking one region of the complex plane to another. The second regards the function as a pair of functions of two real variables, and we show how to use *Mathematica*'s three-dimensional plotting routines to view simultaneously both the modulus and argument of complex functions. Then we shall develop some plot routines tailored to bring out the folded structure of certain complex functions. Note that, in this chapter, the output of all *Mathematica* computations is set to appear in **`TraditionalForm`**. If you are using *Mathematica* technology beyond version 5.2, you should explore the options provided in your current version for managing graphics. See also the on-line supplement and enclosed CD.

8.1 Complex functions: definitions and terminology

Let $\mathbb{C}$ be the set of all complex numbers and let U be a subset of $\mathbb{C}$. A *function* f defined on U is a rule that assigns a complex number w to each $z \in U$. We write

$$w = f(z) \tag{8.1}$$

The set U is called the *domain of definition* of f, and the set of points

$$\{w \in \mathbb{C} \mid w = f(z),\, z \in U\} \tag{8.2}$$

is called the *range* of f, often simply denoted by $f(U)$. Such a function will normally be characterized by supplying a definition of U, such as the upper half-plane $\text{Im}[z] \geq 0$, and a formula, such as

$$f(z) = e^z \tag{8.3}$$

Often, however, one leaves out a detailed specification of the set U, and U is to be somehow, but reasonably, inferred from the formula, as 'wherever this formula makes sense'. A good example of this would be the function

$$f(z) = \frac{1}{z^3} \tag{8.4}$$

where U would be inferred to be the set of *all* complex numbers *except* zero.

It has come to be a standard to describe complex variables using certain variable names. There is no compulsion to use these lettering schemes, but many books use them. We set

$$z = x + i y \tag{8.5}$$

and a function $f(z)$ is often written as a new complex variable w, that is also decomposed into real and imaginary parts as

$$w = f(z) = u(x, y) + i v(x, y) \tag{8.6}$$

or, alternatively, as

$$w = f(z) = \phi(x, y) + i \psi(x, y) \tag{8.7}$$

The use of one of the choices u and v, or ϕ and ψ, is almost universal. As an example, consider

$$f(z) = z^n \tag{8.8}$$

for $n = 2, 3$. In the case $n = 2$, we have

$$u = \phi = x^2 - y^2; \ v = \psi = 2 x y \tag{8.9}$$

Then for $n = 3$ we have

$$u = \phi = x^3 - 3 x y^2; \ v = \psi = 3 x^2 y - y^3 \tag{8.10}$$

■ Using *Mathematica* to extract real and imaginary parts

This section is quite important, because *Mathematica* can be rather obstinate about extracting real and imaginary parts, because of its default assumption that everything is complex unless informed otherwise. Consider the following:

```
expr = Expand[(x + I y)^2]
```

$$x^2 + 2 i y x - y^2$$

```
Re[expr]
```

$$\mathrm{Re}(x^2 - y^2) - 2\,\mathrm{Im}(x y)$$

This is true, but not what we want! In general, this operation can be made to do what we want by doing one of several things, the approach depending on the complexity of the expression.

■ ComplexExpand revisited

We already visited this function in Chapter 2, where we looked at it applied to simple arithmetical functions. It works very well in the extraction of real and imaginary parts:

```
? ComplexExpand
```

```
ComplexExpand[expr] expands expr assuming that all variables
   are real. ComplexExpand[expr, {x1, x2, ... }] expands expr
   assuming that variables matching any of the xi are complex.
```

```
ComplexExpand[Re[expr]]
```

$x^2 - y^2$

In general, we define a pair of new functions, **re** and **im**, that apply **ComplexExpand** appropriately, but also reserving z and w as complex expressions.

```
re[expression_] :=
  ComplexExpand[Re[expression], {z, w}];
im[expression_] := ComplexExpand[
   Im[expression], {z, w}];
```

```
{re[expr], im[expr]}
```

$\{x^2 - y^2, 2\, x\, y\}$

```
re[(x + I y)^4]
```

$x^4 - 6\, y^2\, x^2 + y^4$

```
re[(x + I y)^6 + z^3]
```

$x^6 - 15\, y^2\, x^4 + 15\, y^4\, x^2 - y^6 + \text{Re}(z)^3 - 3\, \text{Im}(z)^2\, \text{Re}(z)$

However, this approach is not perfect. In some earlier versions of *Mathematica* you might obtain, for example,

```
re[1 / (x + I y)]
```

$$\frac{x}{|x + i\, y|^2}$$

In version 5 or later this works just fine:

```
re[1 / (x + I y)]
```

$$\frac{x}{x^2 + y^2}$$

Before we explain how to fix this, note the following:

```
Timing[re[(x + I y)^10]]
```

$$\{0.02\,\text{Second},\ x^{10} - 45\,y^2\,x^8 + 210\,y^4\,x^6 - 210\,y^6\,x^4 + 45\,y^8\,x^2 - y^{10}\}$$

Our functions operate reasonably quickly! There is a package that boosts the functionality of **Re** and **Im** that forces decomposition into the required pieces, but at a price. This is certainly necssary in older versions of *Mathematica*, but is less of a necessity in versions 5 or later.

```
Needs["Algebra`ReIm`"]
```

This package operates by adding a list of rules about real and imaginary parts. The price that is to be paid is that searching the list and trying the rules takes some time, and the search may be carried out even when quite simple expressions are supplied.

```
Timing[re[(x + I y)^10]]
```

$$\{0.27\,\text{Second},\ x^{10} - 45\,y^2\,x^8 + 210\,y^4\,x^6 - 210\,y^6\,x^4 + 45\,y^8\,x^2 - y^{10}\}$$

Whether you are using an older version with this package, or a recent version, we can manage quite tricky expressions:

```
exprb = (x + I y)^(-3) Exp[x + I y] * Sin[2 (x + I y)]
```

$$\frac{e^{x+i\,y}\,\sin(2\,(x+i\,y))}{(x+i\,y)^3}$$

```
parts = {re[exprb], im[exprb]};
```

```
Together[parts]
```

$$\Big\{\frac{1}{(x^2+y^2)^3}\,(e^x\,(\cos(y)\cosh(2\,y)\sin(2\,x)\,x^3 - \cos(2\,x)\sin(y)\sinh(2\,y)\,x^3 + 3\,y\cosh(2\,y)\sin(2\,x)\sin(y)\,x^2 + 3\,y\cos(2\,x)\cos(y)\sinh(2\,y)\,x^2 - 3\,y^2\cos(y)\cosh(2\,y)\sin(2\,x)\,x + 3\,y^2\cos(2\,x)\sin(y)\sinh(2\,y)\,x - y^3\cosh(2\,y)\sin(2\,x)\sin(y) - y^3\cos(2\,x)\cos(y)\sinh(2\,y))),$$
$$\frac{1}{(x^2+y^2)^3}\,(e^x\,(\cosh(2\,y)\sin(2\,x)\sin(y)\,x^3 + \cos(2\,x)\cos(y)\sinh(2\,y)\,x^3 - 3\,y\cos(y)\cosh(2\,y)\sin(2\,x)\,x^2 + 3\,y\cos(2\,x)\sin(y)\sinh(2\,y)\,x^2 - 3\,y^2\cosh(2\,y)\sin(2\,x)\sin(y)\,x - 3\,y^2\cos(2\,x)\cos(y)\sinh(2\,y)\,x + y^3\cos(y)\cosh(2\,y)\sin(2\,x) - y^3\cos(2\,x)\sin(y)\sinh(2\,y)))\Big\}$$

A good working practice is to use our definitions **re** and **im** as is unless this fails. Only then should you load the **ReIm** package. Other tricks for getting things to work include liberal applications of **PowerExpand**, and possibly introducing your own supplementary rules – use of a subset of the rules given in the **ReIm** package can be quite efficient. T. Bahder (1995), in his book *Mathematica for Scientists and Engineers*, has suggested that **ReIm** be avoided altogether on efficiency grounds, and that your own customized rules sets be used in preference. I recommend that you consult Section 5.11 of that text for further information if you find yourself confronted with a problem that the approach described here cannot treat.

8.2 Neighbourhoods, open sets and continuity

Now we need to add a little extra structure to the complex plane, and define various important types of sets. The first important type of set is a *neighbourhood*, also called an *open disc*. Given a point a, the ε-neighbourhood of a is the set:

$$N_\varepsilon(a) = \{z \in \mathbb{C} : |z - a| < \varepsilon\} \tag{8.11}$$

It may also be denoted by $D(a;\ \varepsilon)$. Geometrically, it consists of the points that are a distance strictly less than ε from a.

Sometimes, in dealing with functions that are singular at a point, but defined elsewhere, it will be convenient to deal with a neighbourhood with a hole in it, usually called a *punctured disc* or *deleted neighbourhood*. These are the sets:

$$\{z \in \mathbb{C} : 0 < |z - a| < \varepsilon\} \tag{8.12}$$

There are also structures called closed discs (the term neighbourhood tends to be reserved for the open objects), where the boundary points are added. These are sets of the form:

$$\overline{D}(a;\ \varepsilon) = \overline{N}_\varepsilon(a) = \{z \in \mathbb{C} : |z - a| \leqslant \varepsilon\} \tag{8.13}$$

■ Open sets

Armed with the concept of an open neighbourhood, we can now define open sets in general. A set $U \subset \mathbb{C}$ is said to be *open*, if given any $z \in U$, there is a real positive ε (which can and usually will depend on z), such that $N_\varepsilon(z) \subset U$. That is, there is room to wander a small distance in any direction and stay within U. An example is the set of points whose distance from a given point is strictly less than a given radius – this is just a standard neighbourhood.

■ Closed sets

A set $U \subset \mathbb{C}$ is said to be *closed*, if its complement, $\mathbb{C} - U = \{z \in \mathbb{C} : z \notin U\}$, is open. An example is the set of points whose distance from a given point is less than *or equal to* a given radius – this is just a standard closed disk.

■ Limits and continuity

Informally, a complex function is continuous at z, if its limiting value approaching z from any direction agrees with its value at z. Formally, we phrase this in terms of neighbourhoods and limits. We introduce the formal characterization in two stages:

■ Limits

Suppose that f is a function defined on a set that contains a punctured disk centred on z_0. We say that the limit of f as z tends to z_0 is w_0, or,

$$\lim_{z \to z_0} f(z) = w_0 \tag{8.14}$$

if, given any $\epsilon > 0$, there is a positive number δ such that $0 < |z - z_0| < \delta$ implies that

$$|f(z) - w_0| < \epsilon \tag{8.15}$$

We can rephrase this as saying that the image of a deleted δ-neighbourhood lies within an ϵ-neighbourhood about w_0.

■ Continuity

Suppose that f is a function defined on a set that contains a disk centred on z_0. So now $f(z_0)$ is defined. We say f is continuous at z_0 if

$$\lim_{z \to z_0} f(z) = f(z_0) \tag{8.16}$$

We need these definitions for future reference. There are various basic theorems that can be proved about limits of sums, products, quotients, and compositions of functions, and the relationship between the complex forms and corresponding real forms. We shall not labor the proofs of these points, as they belong in a basic analysis course, but shall just remark that if $f(z)$ and $g(z)$ are continuous at z_0, then

(1) $\lambda f(z) + \mu g(z)$ is continuous at z_0 for any complex numbers λ, μ;
(2) $f(z) \times g(z)$ is continuous at z_0;
(3) $f(z) / g(z)$ is continuous at z_0 provided $g(z_0) \neq 0$.

Furthermore, if f is continuous at $w_0 = g(z_0)$, then $f(g(z))$ is continuous at z_0.

Note that when we come to define complex differentiability in Chapter 10, we shall give proper proofs of the corresponding results for sums, products, quotients and compositions. Note that it is also true that a complex function $f(z)$ is continuous at z_0 if and only if its real and imaginary parts $u(x, y)$, $v(x, y)$ are continuous at (x_0, y_0), where $z_0 = x_0 + i\, y_0$. *This is most emphatically not the case for differentiation*! We shall see why in Chapter 10.

8.3 Elementary vs. series approach to simple functions

How do we define functions of complex variables? Giving a formula and a domain of definition is how we have set it up. For functions of most interest (the differentiable ones), the most powerful and general prescription is to write down a power series for the function. We can then throw results on power series and differentiation (stated or derived in Chapters 9 and 10) at our series definition. Furthermore, most of the interesting 'special functions' of applied mathematics are in fact defined through a series description. Such series descriptions arise typically from the series solution of some differential equation associated with a physical problem of practical interest. A good example would be Bessel functions, defined by a series solution of Bessel's equation, which arises in connection with various wave and potential problems with cylindrical symmetry.

In spite of the power and generality of such a series approach to functions, it is nevertheless useful to first go down a rather more friendly route that builds on our existing understanding of a subset of corresponding real functions, and some simple identities, to construct a subset of interesting simple functions of complex variables. When we finally get a grip on series methods, we can redefine this subset again using such series, and also extend our supply of functions to include Bessel and Hypergeometric functions and all sorts of other exotic objects.

It may seem a little inefficient to define things twice, but since it is not actually necessary to introduce series in order to get a grip on simple exponential, trigonometric and hyperbolic functions and their inverses, we will take a good look at these first.

■ The exponential function

We have already done the work to deal with this, in Chapter 2. Recall that we have the identity

$$e^{iy} = \cos(y) + i\sin(y) \tag{8.17}$$

We therefore define

$$e^{z} = e^{x+iy} = e^{x}\,(\cos(y) + i\sin(y)) \tag{8.18}$$

This definition has an important consequence related to the fact that the function is not 1:1 as a mapping. Suppose that

$$e^{z_1} = w \text{ and } e^{z_2} = w \tag{8.19}$$

Then

$$e^{z_1 - z_2} = 1 \tag{8.20}$$

Writing

$$z_1 - z_2 = x + iy \tag{8.21}$$

it follows that

$$e^x \cos(y) = 1 \text{ and } e^x \sin(y) = 0 \tag{8.22}$$

which has solutions given by

$$x = 0, \;\; y = 2k\pi \tag{8.23}$$

for k an integer. That is,

$$z_1 - z_2 = 2k\pi i \tag{8.24}$$

The exponential function is not only not 1:1, it is actually *periodic* in the imaginary direction.

■ Trigonometric and hyperbolic functions

The starting point for a definition of trigonometric functions is the following equation, considered in Chapter 2, for real θ:

$$e^{i\theta} = \cos(\theta) + i\sin(\theta) \tag{8.25}$$

from which it follows that

$$\cos(\theta) = \frac{1}{2}\left(e^{i\theta} + e^{-i\theta}\right) \tag{8.26}$$

$$\sin(\theta) = \frac{1}{2i}\left(e^{i\theta} - e^{-i\theta}\right) \tag{8.27}$$

We therefore define the trigonometric functions, for complex z, by the formulae:

$$\cos(z) = \frac{1}{2}\left(e^{iz} + e^{-iz}\right) \tag{8.28}$$

$$\sin(z) = \frac{1}{2i}\left(e^{iz} - e^{-iz}\right) \tag{8.29}$$

It follows that these functions have a periodicity property inherited from that of the exponential function, given by

$$\cos(z + 2\pi) = \cos(z); \quad \sin(z + 2\pi) = \sin(z) \tag{8.30}$$

This extends the usual real periodicity property into the complex plane. We know that in the real case, the real zeroes of $\sin(x)$ and $\cos(x)$ are located at $n\pi$ and $(n + \frac{1}{2})\pi$ respectively. We can ask where the zeroes are in the complex plane. For example, if

$$\sin(z) = 0 \tag{8.31}$$

then

$$e^{iz} = e^{-iz} \tag{8.32}$$

$$e^{2iz} = 1 \tag{8.33}$$

which leads us to deduce that, for some integer n,

$$2iz = 2n\pi i \tag{8.34}$$

$$z = n\pi \tag{8.35}$$

So there are no additional complex zeroes in the complex plane. Similarly, the zeros of $\cos(z)$ are located at $z = (n + \frac{1}{2})\pi$. The proof of this is left for you as an exercise. In the case of hyperbolic functions we also make the obvious extension of the real definition to the complex plane:

$$\cosh(z) = \frac{1}{2}(e^z + e^{-z}) \tag{8.36}$$

$$\sinh(z) = \frac{1}{2}(e^z - e^{-z}) \tag{8.37}$$

These are periodic in the imaginary direction:

$$\cosh(z + 2\pi i) = \cosh(z); \quad \sinh(z + 2\pi i) = \sinh(z) \tag{8.38}$$

The hyperbolic cosine now has zeroes at

$$\left(n + \frac{1}{2}\right) i\pi \tag{8.39}$$

and the hyperbolic sine has zeroes at

$$ni\pi \tag{8.40}$$

With any of these functions, some solutions for the zeroes may be found by direct use of *Mathematica*, but others may need to be inferred from periodicity properties. You do get a warning about what solutions are being found:

```
z /. Solve[Cosh[z] == 0, z]
```

— *Solve::ifun : Inverse functions are being used by Solve, so some solutions may not be found.*

$$\left\{-\frac{i\pi}{2}, \frac{i\pi}{2}\right\}$$

We will consider inverse functions shortly.

■ Simple identities: `ComplexExpand` and `TrigExpand`

`ComplexExpand` works with the simpler transcendental functions as it does with polynomial functions, and can be used to generate all the standard complex trigonometric identities for complex arguments. Various standard addition formulas, but in complex form, can also be generated using **`TrigExpand`**. The following lists can all be generated by referring back to the definitions in terms of exponential functions, but we get *Mathematica* to tabulate them. You are strongly recommended to work out some of these using pen and paper.

```
ComplexExpand[Exp[x + I y]]
```

$e^x \cos(y) + i\, e^x \sin(y)$

```
ComplexExpand[Sin[x + I y]]
```

$\cosh(y) \sin(x) + i \cos(x) \sinh(y)$

```
ComplexExpand[Cos[x + I y]]
```

$\cos(x) \cosh(y) - i \sin(x) \sinh(y)$

```
ComplexExpand[Sinh[x + I y]]
```

$i \cosh(x) \sin(y) + \cos(y) \sinh(x)$

```
ComplexExpand[Cosh[x + I y]]
```

$\cos(y) \cosh(x) + i \sin(y) \sinh(x)$

```
TrigExpand[Sin[z + w]]
```

$\cos(z) \sin(w) + \cos(w) \sin(z)$

```
TrigExpand[Cos[z + w]]
```

$\cos(w) \cos(z) - \sin(w) \sin(z)$

```
TrigExpand[Sinh[z + w]]
```

$\cosh(z) \sinh(w) + \cosh(w) \sinh(z)$

```
TrigExpand[Cosh[z + w]]
```

$\cosh(w) \cosh(z) + \sinh(w) \sinh(z)$

Finally, the reverse identities can be generated by the use of **`TrigReduce`** - here we just invert the last example:

```
TrigReduce[%]
```

cosh($w + z$)

8.4 Simple inverse functions

For functions that are simple nth powers, the inverse 'function' is the nth root function that we discussed in Chapter 2. We have put the term 'function' in quotes as the operation of taking nth roots yields many answers. This mapping is not strictly a function at all, but a 'many-valued function' or 'multifunction'. For functions that are simple polynomials, the corresponding inverse function corresponds to finding the roots of the equation arising from equating the polynomial with a given complex number. Again this is many-valued.

In the case of the simple exponential, trigonometric and hyperbolic functions defined above, we face similar issues related to the many-valued behaviour of the logarithm function. As discussed in Chapter 2, this is defined by

$$\log(z) = \log(|z|) + i \operatorname{Arg}(z) \tag{8.41}$$

and we must make a choice about what value of the argument is to be taken. This is of course related to the fact that the exponential function is periodic, i.e. it is many-to-one as a function. A picture of the log function showing its many-valued character was given in Section 2.11.

The inverse trigonometric and hyperbolic functions also have this many-valued property, again arising from the fact that the trigonometric and hyperbolic functions are periodic. The problem can also be seen by expressing the inverse trigonometric and hyperbolic functions in terms of logarithms. Let's do an example of this. Suppose first that

$$\sinh(w) = z \tag{8.42}$$

Writing this in terms of the exponential function gives

$$e^{w} - e^{-w} = 2z \tag{8.43}$$

Now make the substitution

$$q = e^{w} \tag{8.44}$$

so that

$$q - \frac{1}{q} = 2z \tag{8.45}$$

Rearranging, we obtain the quadratic equation

$$q^2 - 2zq - 1 = 0 \tag{8.46}$$

which has the following pair of solutions:

$$\left\{z - \sqrt{z^2+1},\, z + \sqrt{z^2+1}\right\} \tag{8.47}$$

Now we demand that $z = 0$ implies $w = 0$, and hence that $q = 1$. This involves picking the second root. Finally taking logs suggests that we set

$$\operatorname{arcsinh}(z) = \log\left(z + \sqrt{z^2+1}\right) \tag{8.48}$$

Similar arguments for the other functions lead to

$$\operatorname{arccosh}(z) = \log\left(z + \sqrt{z^2-1}\right) \tag{8.49}$$

$$\operatorname{arctanh}(z) = \frac{1}{2}\log\left(\frac{1+z}{1-z}\right) \tag{8.50}$$

Similarly, for the trigonometric functions

$$\arcsin(z) = -i\log\left(i\,z + \sqrt{1-z^2}\right) \tag{8.51}$$

$$\arccos(z) = -i\log\left(z + i\sqrt{1-z^2}\right) \tag{8.52}$$

$$\arctan(z) = \frac{1}{2}\,i\log\left(\frac{i+z}{i-z}\right) \tag{8.53}$$

■ *Mathematica* description of simple inverse functions

All of the trigonometric, hyperbolic functions and their inverses are available within *Mathematica*. Their names are **Sin**, **Cos**, **Tan**, **Sinh**, **Cosh**, **Tanh**, **ArcSin**, **ArcCos**, **ArcTan**, **ArcSinh**, **ArcCosh** and **ArcTanh**. The corresponding names for the reciprocal functions and their inverses are **Csc**, **Sec**, **Cot**, **Csch**, **Sech**, **Coth**, **ArcCsc**, **ArcSec**, **ArcCot**, **ArcCsch**, **ArcSech** and **ArcCoth**. We can enquire about any of them in the usual way:

```
?ArcSin
```

```
ArcSin[z] gives the arc sine of the complex number z.
```

We will explore some of the interesting properties of this function in Section 8.6.

8.5 Branch points and cuts

For those functions that are many-valued there is a standard approach to forcing them to be single-valued. The idea is to draw a line or curve in the complex plane, and state that one cannot cross from one side of this curve to the other. Such a curve is called a *branch cut*. It is in part defined by the locations of certain points, called *branch points*. Such points have the property that a loop traversed around a branch point leads to a change in the value of the function as one moves from the beginning to the end of the loop. Branch cuts are drawn between branch points. There is often an implicit branch point at infinity, with cuts being drawn from key finite branch points out to infinity.

We have already encountered such phenomena, without actually describing them in terms of branch structures, in Chapter 2, and again in this chapter. Both the nth root functions and the log function are many-valued. Traversing any loop around the origin produces a change in the value of the function. So the origin is a branch point. We draw a line emanating from the origin for the branch cut. In fact for different purposes we may consider putting this line at any angle we please, but it has become conventional to place it along the negative real axis. Let's take a look at the impact of this on the functions of interest – we draw the square root, cube root and log functions. These also make the point that *Mathematica* conforms to the convention of locating the branch cut along the negative real axis.

```
Plot3D[Im[Sqrt[(x + I y)]], {x, -3, 3}, {y, -3, 3},
        PlotPoints -> 40]
```

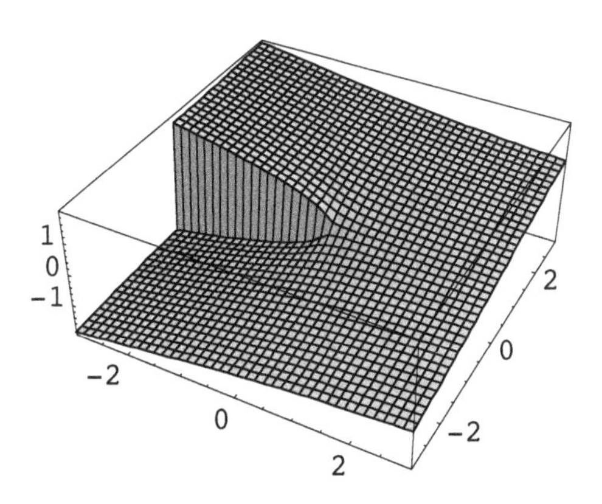


```
Plot3D[Im[(x + I y)^(1/3)], {x, -3, 3}, {y, -3, 3},
        PlotPoints -> 40]
```

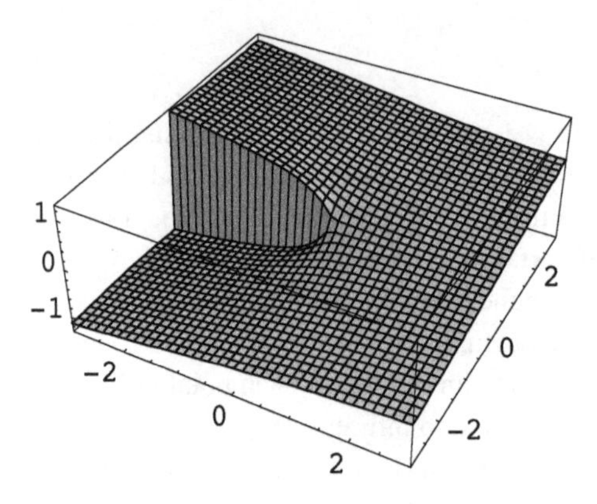


```
Plot3D[Im[Log[(x + I y)]], {x, -3, 3}, {y, -3, 3},
        PlotPoints -> 40]
```

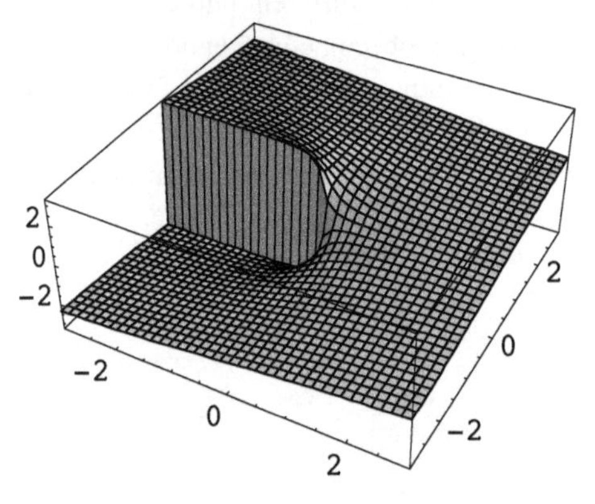

Other representations of branch cuts can also be given – we defer this until necessary, when we consider the evaluation of integrals involving multi-valued functions.

■ Other branch cut locations and *Mathematica* conventions

In general the family of branch points and cuts associated with a given function may be very complicated. A good working rule is to locate the branch points associated with any fractional powers and logs in an expression, and to consider how they (and possibly infinity) can be joined up. As an example, consider the expression for the inverse of the sine function:

$$\arcsin(z) = -i \log\left(iz + \sqrt{1 - z^2}\right) \tag{8.54}$$

Examination of this function reveals that we do not have to worry here about the log function, as its argument does not vanish, but we take the square roots of both $1 - z$ and $1 + z$. So there are branch points at 1 and -1. How should these, and possibly infinity, be linked by branch cuts? One might start by guessing that we take the line joining -1 to 1 along the real axis. But this would be foolish, as we would be cutting away precisely that region on which arcsin(x) makes sense as a real function! Remember that as a real function sin(x) attains values between -1 and 1, so ideally we would like to keep a complex neighbourhood of this region within the domain of the inverse function. An alternative is to place branch cuts along the intervals $(-\infty, -1)$ and $(1, \infty)$ - this is what is normally done and is also *Mathematica*'s convention. Let's take a look at the imaginary part of the inverse sine function - the branch cuts are clearly visible.

```
Plot3D[Im[ArcSin[(x + I y)]], {x, -3, 3}, {y, -3, 3},
       PlotPoints -> 40]
```

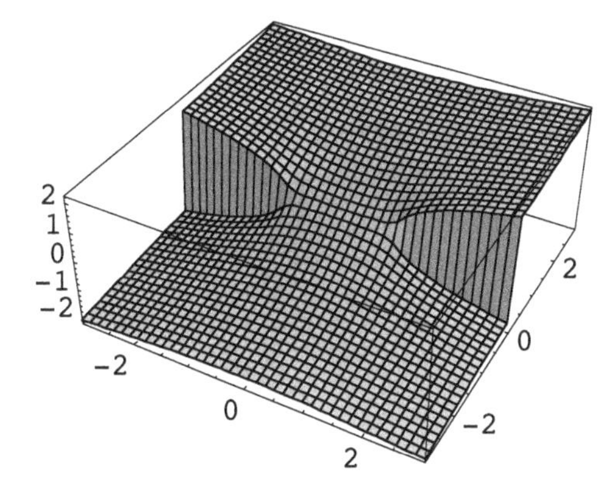

This is quite a good example as it illustrates the fact that there is no hard and fast rule for drawing branch cuts, and that we may be influenced by the real form of the function of interest. Sometimes we may wish to make other choices that at first sight seem perverse. For example, there are sometimes good reasons to take the branch cut of the log function to be along the positive real axis.

Mathematica's conventions for branch cuts for various multi-valued functions are described in detail in relevant sections of *The Mathematica Book* (Wolfram, 2003). In the text for versions 3, 4 and 5, the conventions for the inverse trigonometric and hyperbolic, as well as nth root and log functions, are given in Section 3.2.7.

It is well worth exploring what *Mathematica* does when confronted with, for example, trigonometric functions and their many-valued inverses, as this raises two interesting issues. In the real setting, we might have got used to the inverse sine function being relevant only to the interval $(-1, 1)$, but in the full complex setting the inverse sine works for general complex arguments, for example,

```
{ArcSin[12], ArcSin[12.0]}
```

$\{\sin^{-1}(12), 1.5708 - 3.17631\, i\}$

If we take the sine of these two complex numbers, we get

```
{Sin[ArcSin[12]], Sin[ArcSin[12.0]]}
```

$\{12, 12. - 7.32232 \times 10^{-16}\, i\}$

For the second result, we can simply **Chop** it:

```
Chop[%[[2]]]
```

12.

or indeed use more precision:

```
Sin[ArcSin[12.0000000000000000000]]
```

$12.0000000000000000000 + 0. \times 10^{-19}\, i$

But aside from these numerical complications, **Sin[ArcSin[z]]** returns z for all complex z. On the other hand, if we apply the single-valued forward function first and then the multi-valued inverse, we may return to a different branch. This is illustrated by the following sample calculations:

```
ArcSin[Sin[12]]
```

$12 - 4\pi$

```
StandardForm[Table[{n, ArcSin[Sin[n]]}, {n, -6, 6}]]
```

```
{{-6, -6 + 2 π}, {-5, -5 + 2 π}, {-4, 4 - π}, {-3, 3 - π},
 {-2, 2 - π}, {-1, -1}, {0, 0}, {1, 1}, {2, -2 + π},
 {3, -3 + π}, {4, -4 + π}, {5, 5 - 2 π}, {6, 6 - 2 π}}
```

```
Plot[ArcSin[Sin[x]], {x, -12, 12}]
```

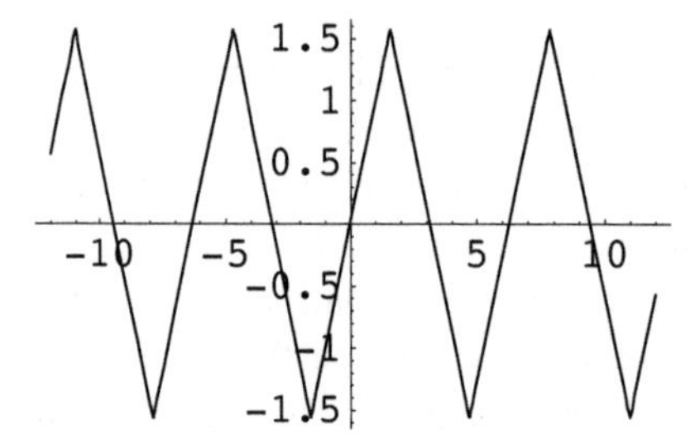

8.6 The Riemann sphere and infinity

There is one other type of special point in complex variable theory, called the point at infinity. This is a point that we add to the complex plane to make it topologically equivalent to a sphere, called the Riemann sphere. The complex plane so extended is often called the extended complex plane. Note that there are various different versions based on different forms of the stereographic projection that is involved – our approach is tailored to later applications of these ideas.

The simplest route to understanding the geometry is to let (θ, ϕ) be polar coordinates on a sphere. With our conventions, $\theta = 0$ at the north pole, $\theta = \pi$ at the south pole, and is the co-latitude, ϕ is the longitude. You may be used to conventions where ϕ and θ are swapped, in which case they should be interchanged in the formulas in this section. The precise location and radius of the sphere are intially irrelevant, and we set

$$\zeta = e^{i\phi} \cot\left(\frac{\theta}{2}\right) \tag{8.55}$$

We use ζ as our complex variable to avoid confusion with a Cartesian coordinate z introduced later. The south pole, where $\theta = \pi$, gives the origin. The equator, where $\theta = \pi/2$, gives

$$\zeta = e^{i\phi} \tag{8.56}$$

which is the unit circle. As we decrease θ we obtain values of ζ of larger modulus, and with ϕ varying from 0 to 2π we obtain all possible arguments. We can see that points of the sphere, apart from the north pole, are in 1:1 correspondence with finite points in the complex plane. The Riemann sphere, or extended complex plane, is obtained by just adding back the north pole.

We can make the relationship between this hypothetical sphere and the complex plane more explicit by introducing a sphere of radius R centred at the origin, and to letting

$$\{x, y, z\} = \{R\sin(\theta)\cos(\phi),\ R\sin(\theta)\sin(\phi),\ R\cos(\theta)\} \tag{8.57}$$

be the standard mapping between the spherical polar and Cartesian coordinates for the sphere

$$x^2 + y^2 + z^2 = R^2 \tag{8.58}$$

Then some trigonometric algebra leads to the explicit relation

$$\{x, y, z\} = \left\{\frac{R(\zeta + \bar{\zeta})}{|\zeta|^2 + 1},\ -\frac{iR(\zeta - \bar{\zeta})}{|\zeta|^2 + 1},\ \frac{R(|\zeta|^2 - 1)}{|\zeta|^2 + 1}\right\} \tag{8.59}$$

This relation can be interpreted as a stereographic projection from the north pole of the sphere to a point on the Argand plane viewed as a horizontal plane in three-dimensional space. Details of this interpretation for the case $R = 1$ are in the exercises.

The point at infinity can be treated in a systematic way by introducing reciprocal coordinates

$$\tilde{\zeta} = \frac{1}{\zeta} \tag{8.60}$$

Given any function f, we can ask questions about its behaviour at infinity by asking corresponding questions about the function

$$\tilde{f}(\tilde{\zeta}) = f(\zeta) \tag{8.61}$$

in a neighbourhood of zero. For example, the question of continuity at infinity is now just posed as a question of continuity at the origin in $\tilde{\zeta}$ coordinates. This notion of a Riemann sphere and a point at infinity are more than a geometrical and analytical convenience. This sphere, together with its family of tangent planes, define a complex space of considerable interest in its own right, with implications for both functions and surfaces in real three-dimensional space and relativity. See, for example, Chapters 23 and 24.

8.7 Visualization of complex functions

There are a host of ways of visualizing a complex function of a complex variable. We have to bear in mind that we are dealing with two real functions of two real variables – an essentially four-dimensional problem. We consider the most useful techniques that do not discard any of the information about the function – clearly other methods become feasible if one plots only the absolute value, dropping the phase, or if one plots only the real part, and ignores the imaginary part. We divide our considerations into two-dimensional views and three-dimensional views.

■ Two-dimensional views of a complex function

In this approach we consider a standard coordinate system for the plane, either Cartesian or Polar, and either:

(1) plot the images of the coordinate lines or curves under the mapping;
(2) plot pre-images of the coordinate lines or curves under the mapping.

The former can be done with a package supplied with *Mathematica.* The latter is essentially a pair of overlaid contour plots of either the real and imaginary parts, or the modulus and phase, of the function. The construction of such a composite plot can be done directly with the supplied package, if a suitable inverse function can be written down, or alternatively constructed by hand from *Mathematica*'s built-in contour plotting routines. This type of plot is often used in control theory, where it is called a Nichols chart.

■ ComplexMap package

This package is supplied with *Mathematica* and is documented in the packages documentation. It was written by Roman Maeder and he has documented the programming issues involved extensively in his book 'Programming in *Mathematica*', (Maeder, 1997) where it is used as an example to illustrate how one builds a *Mathematica* package. We will not therefore go into the construction of this package and the reader is referred to Maeder's text for programming details. In what follows we shall illustrate the use of the plot routines with two different functions:

(1) the simple built-in trigonometric function **Sin**;
(2) a simple user-defined function.

We begin by loading the **ComplexMap** package:

```
Needs["Graphics`ComplexMap`"]
```

We inquire what has been loaded:

```
?Graphics`ComplexMap`*
```

Graphics`ComplexMap`

```
CartesianMap Lines PolarMap $Lines
```

Let's investigate the first function:

```
?CartesianMap
```

```
CartesianMap[f, {x0, x1, (dx)}, {y0, y1, (dy)}] plots the
  image of the cartesian coordinate lines under the function
  f. The default values of dx and dy are chosen so that the
  number of lines is equal to the value of the option Lines.
```

When you use **CartesianMap** with a built-in function with a single argument, you can just supply the name of the function.

```
SetOptions[ParametricPlot,
  PlotStyle → AbsoluteThickness[0.1]];

CartesianMap[Sin, {-2, 2, 0.1},
 {-2, 2, 0.1}, PlotRange -> All, PlotPoints → 60]
```

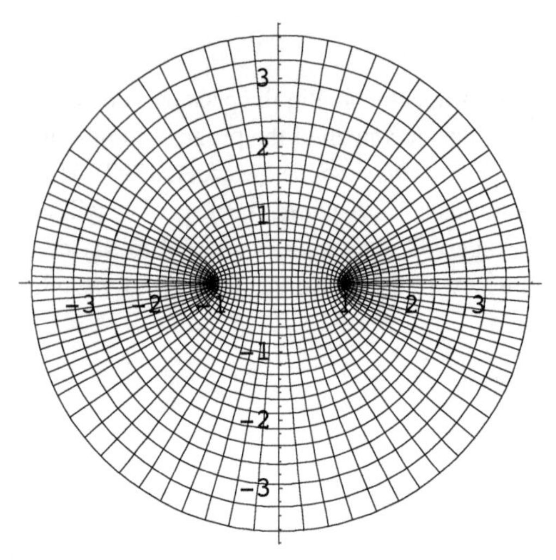

If you wish to use your own custom functions with this package it is easiest if you supply them as pure functions, where the variable being plotted is labelled # and the function is completed by an ampersand (&). If you wish to use a built-in function that takes several arguments, the same idea applies: use the # for the variable, and set any other parameters explicitly (e.g. **BesselJ[0, #]&**). Here are some examples. Be warned – some of these take a while to generate!

```
CartesianMap[Exp[-#^2] * Erfc[-I * #] &,
 {0, 4, 0.1}, {0, 4, 0.1}, PlotPoints → 60]
```

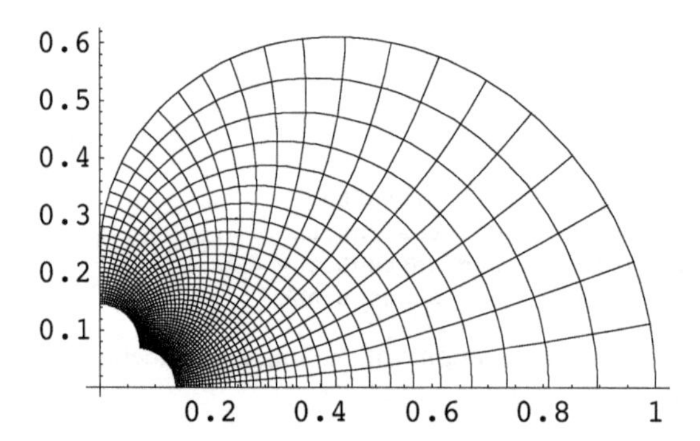

■ Polar version

The package is also set up to treat polar coordinates:

```
? PolarMap
```

```
PolarMap[f, {r0:0, r1, (dr)}, {phi0, phi1, (dphi)}] plots the image of the
   polar coordinate lines under the function f. The default for the phi
   range is {0, 2Pi}. The default values of dr and dphi are chosen so
   that the number of lines is equal to the value of the option Lines.
```

```
PolarMap[Sin, {0, 2, 0.1}, {0, 2 Pi, Pi / 10},
 PlotRange -> All, PlotPoints → 60]
```

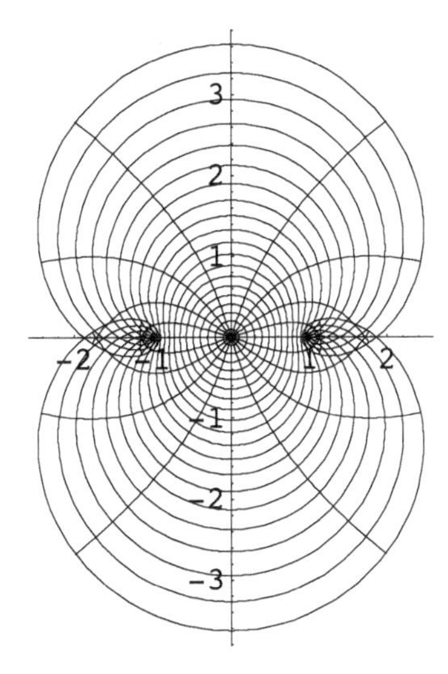


```
PolarMap[Exp[-#^2] * Erfc[-I * #] &,
 {0.1, 2, 0.1}, {0, 2 Pi, Pi / 10}, PlotPoints → 60]
```

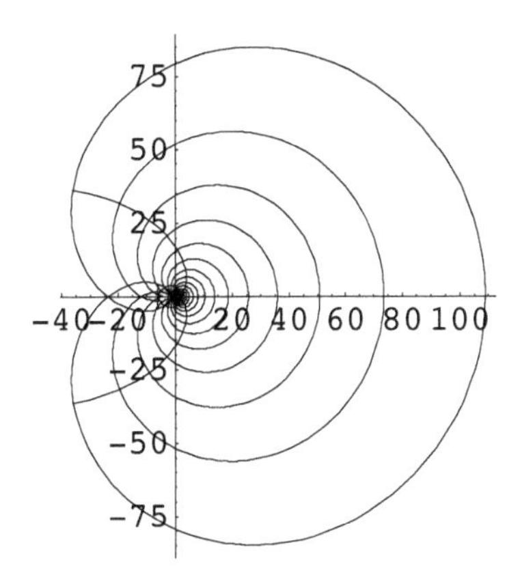

Watch out for singularities in the function – sometimes you may have to remove the origin from the range of polar coordinates, if you do not want *Mathematica* to complain! You may be unfamiliar with the definitions of some of the functions used here, and might consider asking *Mathematica* for the definition of the **Erfc** and **Zeta** functions.

■ Pre-image graphics

In this case we wish to see the function the other way round. That is, we wish to locate the pre-images of a given set of coordinate curves or lines under the mapping. If the function has an explicit inverse you can feed it to **CartesianMap** or **PolarMap** as before:

```
CartesianMap[ArcSin, {-2, 2, 0.1},
 {-2, 2, 0.1}, PlotRange -> All, PlotPoints → 60]
```

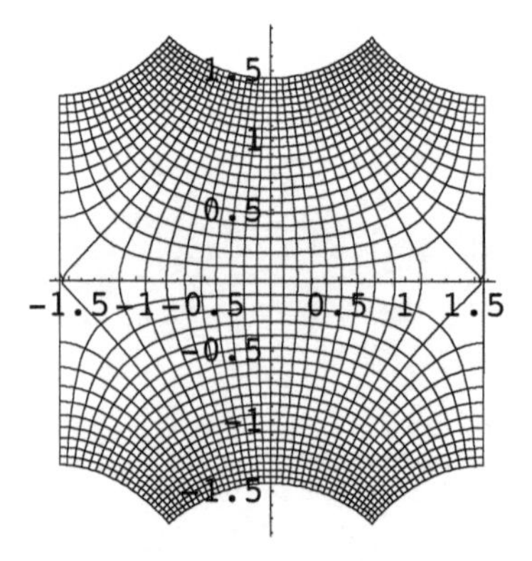

More frequently, an explicit inverse will not be available and the plot will have to be constructed differently. Note that now the x and y ranges refer to the plot domain, and not to the extent of the axes range under consideration:

```
SetOptions[ContourPlot,
  ContourStyle → AbsoluteThickness[0.1]];

CartesianPreImage[func_,
  xrange_, yrange_, options___] :=
    Module[{tempa, tempb, xlist, ylist},
    tempa = ContourPlot[
    Re[func[x + I y]], Evaluate[Prepend[xrange, x]],
    Evaluate[Prepend[yrange, y]],
    ContourShading -> False,
    DisplayFunction -> Identity, options];
    tempb = ContourPlot[Im[func[x + I y]],
    Evaluate[Prepend[xrange, x]],
    Evaluate[Prepend[yrange, y]],
    ContourShading -> False,
    DisplayFunction -> Identity, options];
    Show[tempa, tempb,
   DisplayFunction -> $DisplayFunction]]
```

```
CartesianPreImage[Sin, {-1.5, 1.5},
 {-1.5, 1.5}, Contours -> 60, PlotPoints -> 60]
```

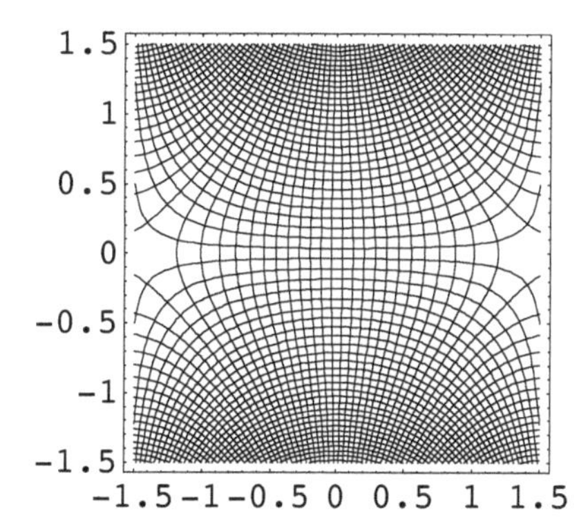

Of course, this can also be used with the inverse function:

```
CartesianPreImage[ArcSin, {-2, 2},
 {-2, 2}, Contours -> 40, PlotPoints -> 60]
```

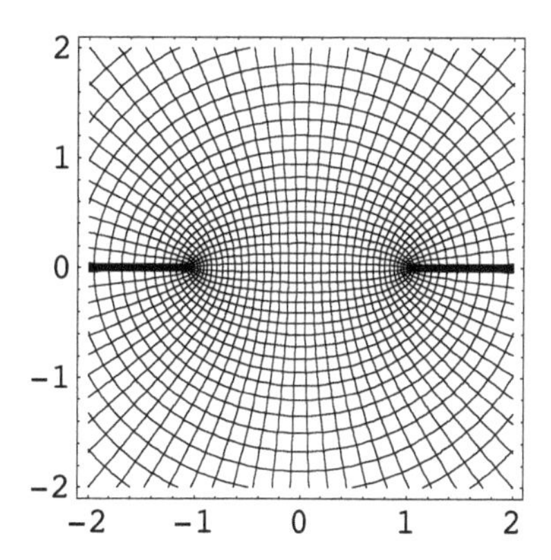

The following three examples are some of the nicer plots that can be extracted this way. The first one shows the complementary error function and the second explores the Riemann zeta function. The third looks at the inverse sine function using polar coordinates.

```
CartesianPreImage[Exp[-#^2] * Erfc[-I * #] &,
 {-1.5, 1.5}, {-1.5, 1.5},
 Contours -> 30, PlotPoints -> 60]
```

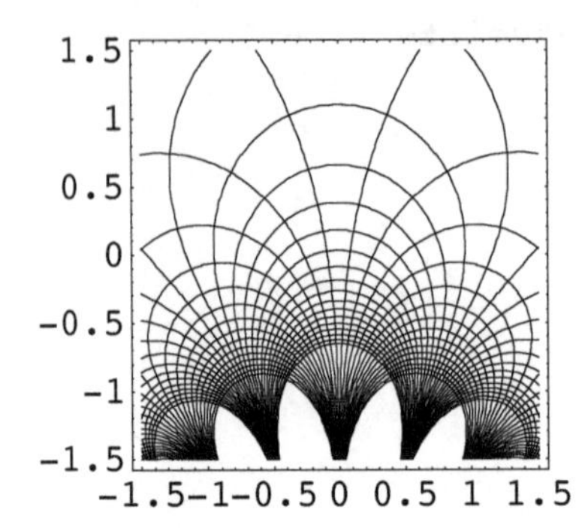

Let's also look at the Riemann zeta function in this way:

```
CartesianPreImage[Zeta, {-1.5, 1.5},
 {-1.5, 1.5}, Contours -> 40, PlotPoints -> 60]
```

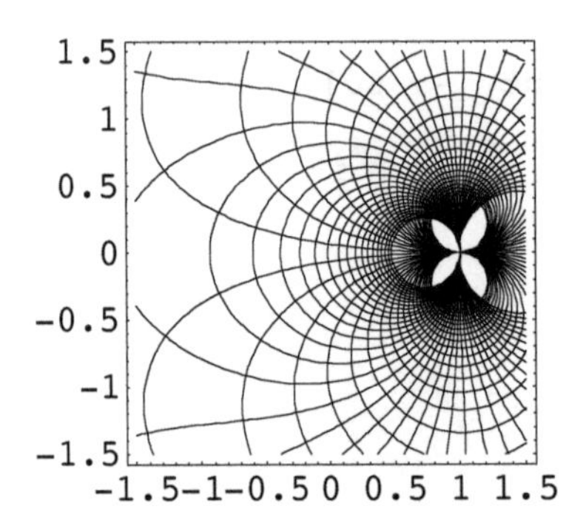


```
PolarPreImage[func_, xrange_, yrange_, options___] :=
    Module[{tempa, tempb, xlist, ylist},
        tempa = ContourPlot[
    Abs[func[x + I y]], Evaluate[Prepend[xrange, x]],
    Evaluate[Prepend[yrange, y]],
    ContourShading -> False,
        DisplayFunction -> Identity, options];
        tempb = ContourPlot[Arg[func[x + I y]],
    Evaluate[Prepend[xrange, x]], Evaluate[
     Prepend[yrange, y]], ContourShading -> False,
    DisplayFunction -> Identity, options];
        Show[tempa, tempb,
   DisplayFunction -> $DisplayFunction]]
```

```
PolarPreImage[ArcSin, {-2, 2},
 {-2, 2}, Contours -> 60, PlotPoints -> 60]
```

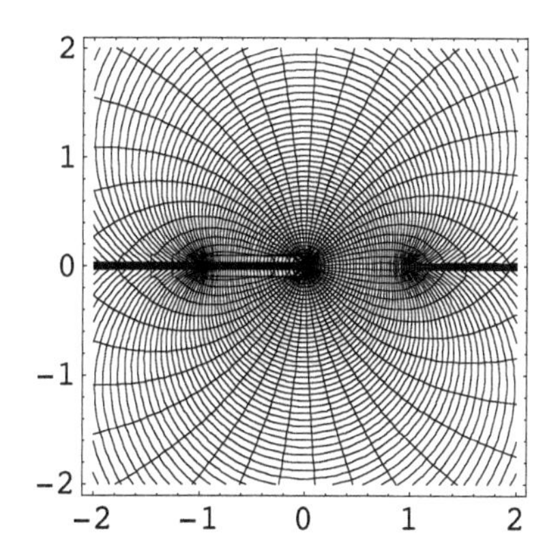

It is left to the reader to try other **PolarPreImage** examples. These types of pre-image plots, particularly the Cartesian version, are very useful, not just for seeing what a complex function looks like, but also for understanding the geometry to which they relate in applications. We will use them extensively later in this book in applications such as fluid dynamics. As a foretaste of this, we have superimposed a unit circle on a plot that will turn out to represent flow around a cylinder (this is discussed in detail in Chapter 19).

```
CartesianPreImage[(# + 1 / #) &, {-3, 3},
 {-3, 3}, Contours -> 40, PlotPoints -> 60,
   Epilog -> {Thickness[0.01], Circle[{0, 0}, 1]}]
```

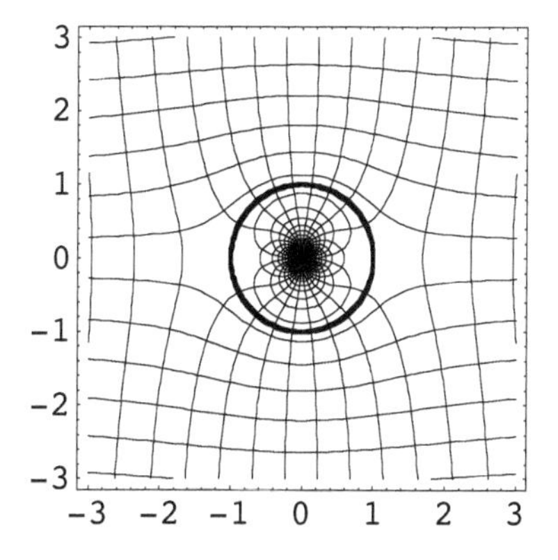

8.8 Three-dimensional views of a complex function

A simple way of using three dimensions to treat a four-dimensional structure is to colour the function according to the argument, with the height representing the modulus. Before doing so, in order to avoid an error in the plot routine, we define the argument of zero:

```
Unprotect[Arg]; Arg[0] = 0;
ComplexPlot3D[func_, xrange_, yrange_, options___] :=
    Plot3D[{Abs[func[x + I y]],
   Hue[N[(Pi + Arg[func[x + I y]]) / (2 Pi)]]},
        xrange, yrange, options]
```

```
ComplexPlot3D[Exp[-#^2] * Erfc[-I * #] &,
 {x, -2, 2}, {y, -2, 2},
    PlotPoints -> 30, Mesh -> False, FaceGrids -> All,
    ViewPoint -> {2, 2, 1}, PlotRange -> All]
```

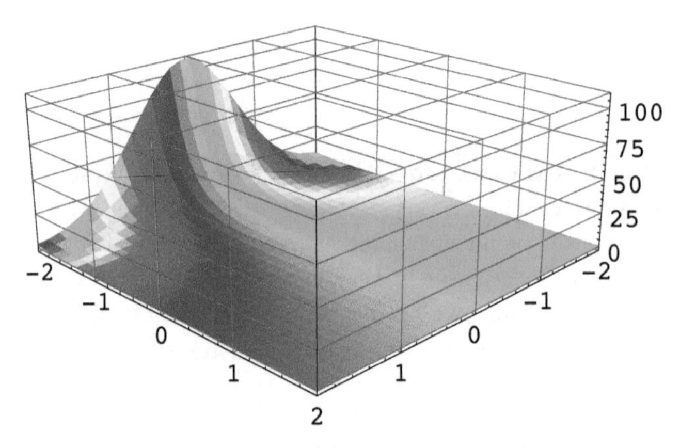


```
ComplexPlot3D[Sin, {x, -2, 2}, {y, -2, 2},
    PlotPoints -> 30, Mesh -> False, FaceGrids -> All]
```

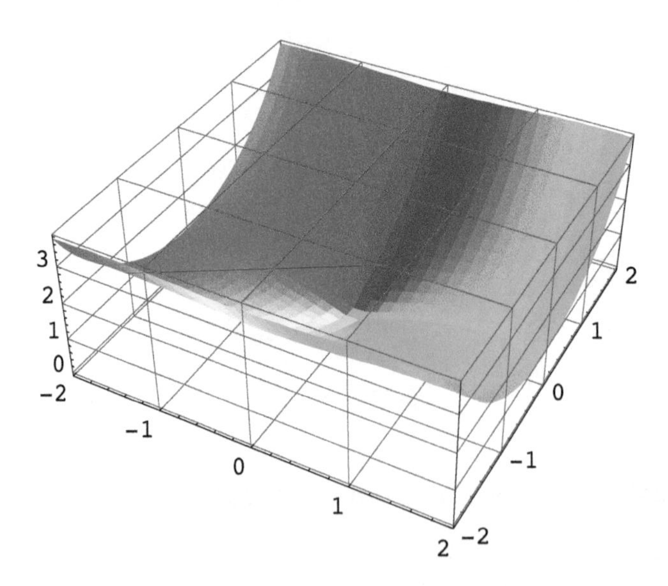

```
ComplexPlot3D[Gamma, {x, -4, 1}, {y, -1, 1},
    PlotPoints -> 100,
 Mesh -> False, PlotRange -> {0, 20}]
```

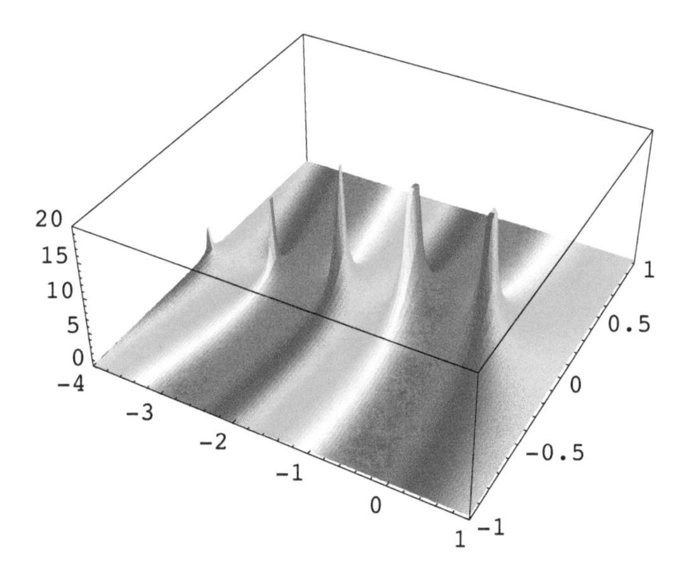

■ Alternative routines

There is nothing to stop you building your own plotting routines, using some custom colouring functions. The following is particularly dramatic, and is especially effective when you are limited to a black and white display or printer. The idea is to colour the function according to its phase, just using two alternating colours – here black and white. Make up some more of your own!

```
Whacky[x_] := If[EvenQ[Floor[x]],
  RGBColor[0, 0, 0], RGBColor[1, 1, 1]]

WhackyComplexPlot3D[{func_, levels_},
  xrange_, yrange_, options___] :=
    Plot3D[{Abs[func[x + I y]],
   Whacky[N[levels * (Pi + Arg[func[x + I y]]) / (2 Pi)]]},
        xrange, yrange, options]

WhackyComplexPlot3D[{Sin, 20}, {x, -2, 2},
 {y, -2, 2},     PlotPoints -> 100, Mesh -> False,
 PlotRange -> All, ViewPoint -> {2, 2, 4}]
```

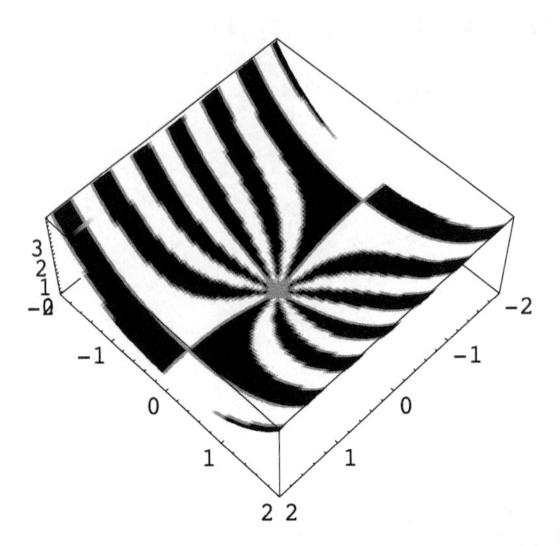

Sometimes the limited resolution of the plot routines can be used to generate some interesting special effects, but this is more for fun than for understanding the function!

```
WhackyComplexPlot3D[{#^10 &, 20},
 {x, -1, 1}, {y, -1, 1},
 PlotPoints -> 150, Mesh -> False, PlotRange -> All]
```

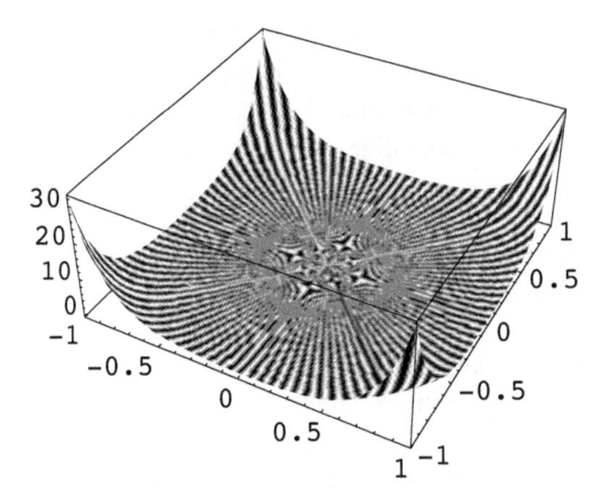

8.9 Holey and checkerboard plots

Frequently a complex function of interest may fold over itself several times, and it may be useful to see the folding. *Mathematica*'s standard plot routines do not allow for transparency so some special treatment is required. One approach is to use a visualization tool, such as 'Dynamic Visualizer' (*Mathematica* Applications Package – Dynamic Visualizer, 1999), that allows various levels of transparency to be set, in addition to a host of other real-time manipulation features being enabled. Another approach is to plot surfaces with many holes in them, so that you can see though the surface. One approach to this was given by the author (Shaw, 1995b). Another more sophisticated approach has been developed by M. Trott and V. Adamchik for the 'Solving the Quintic' project (Adamchik and Trott, 1994). The simple approach given below has the advantage that there is an obvious modification to allow high-contrast checkerboard plots. One might have imagined that the 'Whacky' code given above would give sharp black and white polygons in the plot. However, *Mathematica* interpolates the **ColorFunction** values to produce intermediate shades of colour - greys in this case. Although this makes nicer plots in general, it thoroughly undermines attempts to produce sharp colour divisions. Finally, if you are using *Mathematica* technology beyond version 5.2, you should see the CD and on-line supplement.

First, we give a simple routine for plotting surfaces with holes in them:

```
HoleyPlot3D[{f_,g_,h_},{u_,v_},{c_,d_},{n_,m_},
{cola_}, opts___] :=
Module[{data, plotdata},
data = Table[N[Through[{f, g, h}
[u+x*(v-u)/(n-1),c+y*(d-c)/(m-1)]]],
{x,0,n-1},{y,0,m-1}];
plotdata = Table[{cola,
Polygon[{data[[x,y]],data[[x+1,y]],
data[[x+1,y+1]],data[[x,y+1]]}]},
{x,1,n-1},{y,If[EvenQ[x], 2, 1],m-1, 2}];
Show[Graphics3D[plotdata], opts,
Lighting -> False, Boxed -> False]]
```

Now we apply it to redraw one of the surfaces given in Chapter 2, showing the real part of the fourth root of z:

```
HoleyPlot3D[{((#1)*Cos[#2]&),((#1)*Sin[#2]&),
((#1)^(1/4)*Cos[(#2)/4]&)},
{0.01,1},{0, 8 Pi},{15,120}, {RGBColor[0,0,1]},
ViewPoint -> {-3, -3, 0}]
```

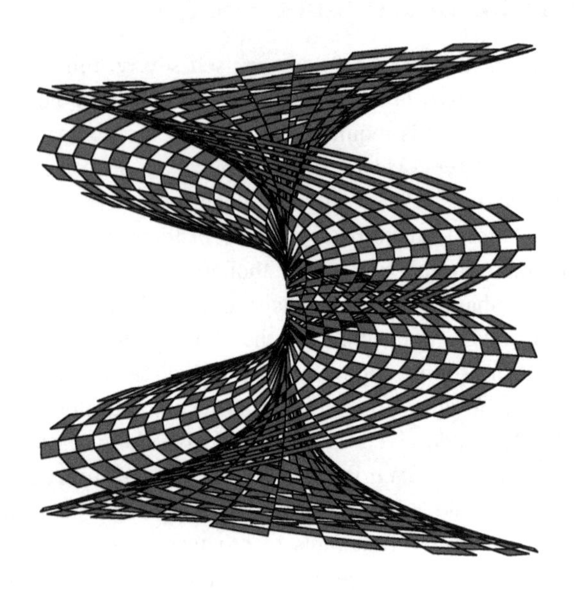

By filling in a second colour rather than a hole we obtain the following:

```
ChekerPlot3D[{f_,g_,h_},{u_,v_},{c_,d_},{n_,m_},
{cola_, colb_}, opts___] :=
Module[{data, plotdata},
data = Table[N[Through[{f, g, h}
[u+x*(v-u)/(n-1),c+y*(d-c)/(m-1)]]],
{x,0,n-1},{y,0,m-1}];
plotdata = Table[{If[EvenQ[x+y], cola, colb],
Polygon[{data[[x,y]],data[[x+1,y]],
data[[x+1,y+1]],data[[x,y+1]]}]},
{x,1,n-1},{y,1, m-1}];
Show[Graphics3D[plotdata], opts,
Lighting -> False, Boxed -> False]]
```

Here is a filled-in checkerboard plot of the (real part of the) cube root function:

```
ChekerPlot3D[{((#1)*Cos[#2]&),((#1)*Sin[#2]&),
((#1)^(1/3)*Cos[(#2)/3]&)},
{0.01,1},{0, 6 Pi},{15,120}, {RGBColor[1,1,1],
RGBColor[0, 0, 0]},
ViewPoint -> {-3, -3, 0}]
```

Such routines can often make the folds and intersections of a complex surface much clearer.

8.10 Fractals everywhere?

This section makes the point that iteration of some of our elementary transcendental functions can also produce interesting fractal structures. This example is inspired by Chapter 8 of the wonderful book by Gray and Glynn (1991), which I thoroughly encourage you to consult for details. You can try out examples such as the following by combining **ComplexPlot3D** with generalizations of the escape-time algorithm to special functions. In the following we just iterate **Sin** first twice and then three times. Note the fluted structure that is suggestive of early fractal development/

```
ComplexPlot3D[(Sin[Sin[#]] &),
 {x, -Pi, Pi}, {y, -Pi, Pi},
   PlotPoints -> 50, PlotRange -> {0, 100}]
```

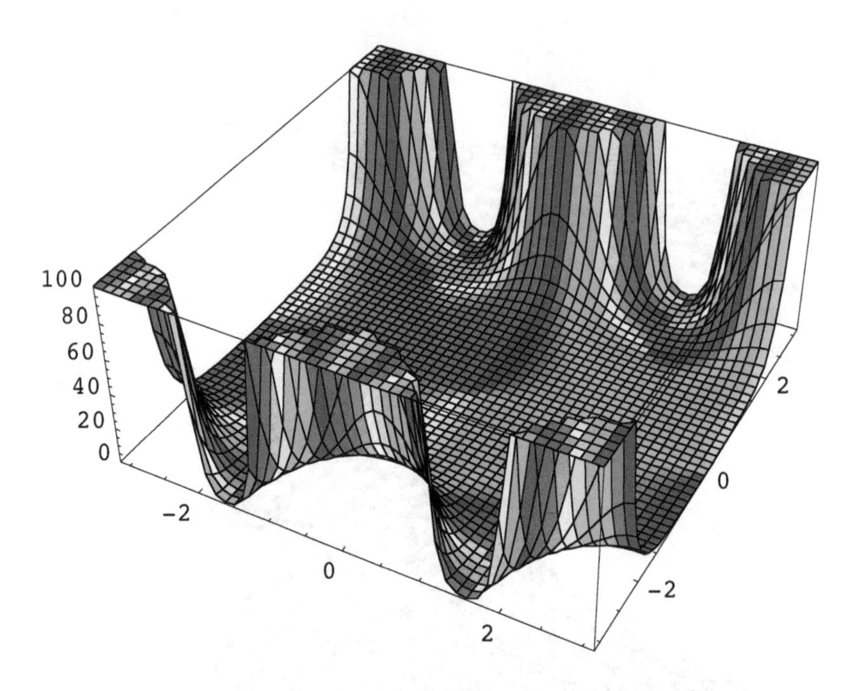


```
ComplexPlot3D[(Sin[Sin[Sin[#]]] &),
 {x, -Pi, Pi}, {y, -Pi, Pi},
    PlotPoints -> 60, PlotRange -> {0, 10^6}]
```

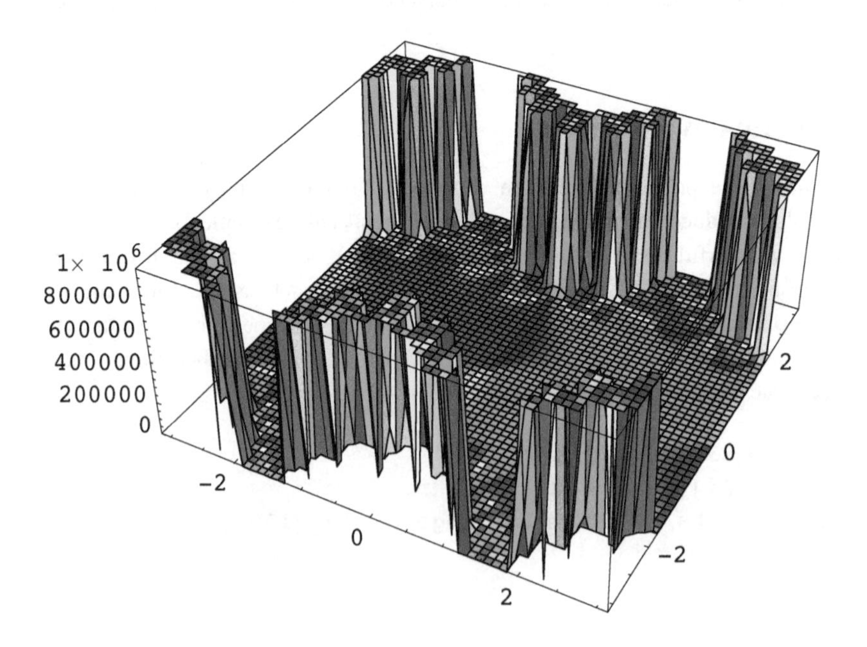

It is a simple matter to modify the escape-time algorithm to treat the case of the sine function, or any other that you choose:

```
SinIter = Compile[{{z, _Complex}},
Module[{cnt, nz=N[z]},
For[cnt=0,
(Abs[nz] <100) && (cnt<100), cnt++,
nz = Sin[nz]];
N[cnt]]];

SinFrac[z_, {{remin_, remax_}, {immin_, immax_}},
steps_]:=
{{{remin, remax}, {immin, immax}},
Table[SinIter[x+I*y],
{y, immin, immax, (immax-immin)/steps},
{x, remin, remax, (remax-remin)/steps}]};

FracColor = (If[#==1, Hue[1,1,0], Hue[2 #]]&);
FracPlot[{{{remin_, remax_}, {immin_, immax_}},
matrix_},
colorfn_] :=
ListDensityPlot[matrix,
Mesh -> False, Frame -> False,
ColorFunction -> colorfn,
PlotRange -> {0, 100},
AspectRatio -> (immax-immin)/(remax - remin)]
```

Now we can make the development of a *periodic fractal* manifest:

```
FracPlot[SinFrac[0,
{{-1.5 Pi,1.5 Pi},{-1.5 Pi,1.5 Pi}},600],FracColor]
```

Exercises

8.1 Show that the zeroes of cos(z) are located at $(n + \frac{1}{2})\pi$.

8.2 Using the definition of the sine and cosine functions in terms of the exponential functions, show that for all complex z,

$$\cos^2(z) + \sin^2(z) = 1$$

8.3 Using the definition of the sinh and cosh functions in terms of the exponential functions, show that for all complex z,

$$\cosh^2(z) - \sinh^2(z) = 1$$

8.4 Using the definition of the sine and cosine functions in terms of the exponential functions, show that for all complex z and w,

$$\sin(w + z) = \cos(z)\sin(w) + \cos(w)\sin(z)$$

$$\cos(w + z) = \cos(w)\cos(z) - \sin(w)\sin(z)$$

8.5 Using the definition of the sinh and cosh functions in terms of the exponential functions, show that for all complex z and w,

$$\sinh(w + z) = \cosh(z)\sinh(w) + \cosh(w)\sinh(z)$$

$$\cosh(w + z) = \cosh(w)\cosh(z) + \sinh(w)\sinh(z)$$

8.6 Work out the real and imaginary parts of the following functions, assuming that x and y are real:

$$(x + i y)^5$$

$$\cos(x + i y)$$

$$\sin(x + i y)$$

$$\cosh(x + i y)$$

$$\sinh(x + i y)$$

8.7 (Stereographic Projection) Let S denote the unit sphere $x^2 + y^2 + z^2 = 1$. Let P be a point on S with coordinates (x, y, z), and let P' denote the point on the plane $z = 0$ obtained by drawing a line from the north pole of S through P and extending it to the plane $z = 0$. (P' is its intersection) Let P' have coordinates $(x', y', 0)$ and let

$$\zeta = x' + i y'$$

Show that

$$\zeta = \frac{x + i y}{1 - z}$$

Deduce that

$$\{x,\ y,\ z\} = \left\{\frac{(\zeta + \overline{\zeta})}{|\zeta|^2 + 1},\ -\frac{i\,(\zeta - \overline{\zeta})}{|\zeta|^2 + 1},\ \frac{(|\zeta|^2 - 1)}{|\zeta|^2 + 1}\right\}$$

If polar coordinates are introduced according to the rule

$$\{x,\ y,\ z\} = \{\sin(\theta)\cos(\phi),\ \sin(\theta)\sin(\phi),\ \cos(\theta)\}$$

show also that

$$\zeta = e^{i\phi}\cot\left(\frac{\theta}{2}\right)$$

8.8 Where are the branch points of the following functions:

$$\cos^{-1}(z)$$

$$\cosh^{-1}(z)$$

$$\tanh^{-1}(z)$$

Hint: use the logarithmic form of the inverse function. What branch cuts are possible for these functions? Which choices make sense given the behaviour of these functions when z is real?

Before trying out the following *Mathematica* examples, enter these definitions into the kernel:

```
re[expression_] :=
  ComplexExpand[Re[expression], {z, w}];
im[expression_] := ComplexExpand[
   Im[expression], {z, w}];
```

8.9 ✵ Use *Mathematica* to work out the real and imaginary parts of

$$(x + iy)^5$$

$$\cos(x + iy)$$

$$\sin(x + iy)$$

$$\cosh(x + iy)$$

$$\sinh(x + iy)$$

8.10 ✵ Use *Mathematica* to plot the inverse functions $\cos^{-1}(z)$, $\cosh^{-1}(z)$ and $\tanh^{-1}(z)$. Where do these plots suggest the branch cuts are? Do the plots reflect the answers you have given for Exercise 8.8?

9 Sequences, series and power series

Introduction

In this chapter we introduce the concepts of sequences and series, and explore their convergence properties. We also look briefly at how elementary functions can be defined by series. We begin by reviewing basic concepts of convergence. This is intended to be a review only, and to supply statements of a number of key definitions and theorems without the proofs that would be traditionally the domain of a rigorous course in real analysis, where the extension of results to complex numbers is a largely trivial matter. We focus our attention, when it comes to supplying proofs, on matters related to power series, since it is such series that are of critical interest to complex analysis. Students of pure mathematics should ensure that they are aware of the background material, whereas more applications-oriented people can just take note of the results. The main result of this section, on the radius of convergence of a power series, is of critical importance to all.

9.1 Sequences, series and uniform convergence

■ Sequences of complex numbers

A *sequence* of complex numbers is respresented by prescribing a mapping from the positive integers into $\mathbb{C}$. That is, we have a set of complex numbers

$$z_1, z_2, \ldots, z_n, \ldots \tag{9.1}$$

We need to describe various properties of sequences. We say that such a sequence is *bounded* if there is a real number M such that

$$|z_n| < M \tag{9.2}$$

for all n. We say that such a sequence has a limit a, or is convergent to a, if, given any $\epsilon > 0$, we can find an integer N such that, for all $n > N$,

$$|z_n - a| < \varepsilon \tag{9.3}$$

■ Sequences of functions and uniform convergence

A sequence of complex functions is represented by prescribing a mapping from the positive integers into the set of mappings from $\mathbb{C}$ to $\mathbb{C}$. That is, we have a set of complex functions

$$f_1(z), f_2(z), \ldots, f_n(z), \ldots \tag{9.4}$$

We say that such a sequence is *convergent* to $f(z)$, if, for each z, given any $\epsilon > 0$, we can find an integer N such that, for all $n > N$,

$$|f_n(z) - f(z)| < \epsilon \tag{9.5}$$

In general, we may need to choose a different ϵ for each z. If we can find one that will do for all z, we say that the sequence is *uniformly convergent*. More formally, we say that such a sequence is *uniformly convergent* to $f(z)$, if, given any $\epsilon > 0$, we can find an integer N such that, for all z, and all $n > N$,

$$|f_n(z) - f(z)| < \varepsilon \tag{9.6}$$

■ Series of functions and uniform convergence

We obtain a *series* of functions by considering the partial sums

$$S_1(z) = f_1(z) \tag{9.7}$$

$$S_2(z) = f_1(z) + f_2(z) \tag{9.8}$$

$$S_n(z) = f_1(z) + f_2(z) + \ldots + f_n(z) \tag{9.9}$$

Such a series is said to be *convergent* if the sequence $S_n(z)$ is convergent. We write this limit as

$$S(z) = \sum_{n=1}^{\infty} f_n(z) \tag{9.10}$$

Similarly, we say that the series is uniformly convergent if the partial sums are uniformly convergent as a sequence.

■ Absolute convergence

The series

$$S(z) = \sum_{n=1}^{\infty} f_n(z) \tag{9.11}$$

is said to be *absolutely convergent* if the series

$$A(z) = \sum_{n=1}^{\infty} |f_n(z)| \tag{9.12}$$

is convergent. Note that the convergence of the absolute values implies the convergence of the original series. If the series itself converges, but the corresponding series of absolute values does not, we say that the series is *conditionally* convergent.

9.2 Theorems about series and tests for convergence

These results are really the province of a course on basic analysis. We list them here for completeness, but refer the reader to a text such as Rudin (1976) for proofs.

■ Theorem 9.1: The terms of a convergent series tend to zero

If the series

$$S(z) = \sum_{n=1}^{\infty} f_n(z) \tag{9.13}$$

is convergent then $f_n(z) \to 0$ *as* $n \to \infty$.

■ Theorem 9.2: A bounded sequence has a convergent subsequence

A *subsequence* is defined by a set of integers

$$n_1 < n_2 < n_3 < \ldots < n_k < \ldots \tag{9.14}$$

Then the theorem takes the following form. *If the sequence*

$$z_1, z_2, \ldots, z_n, \ldots \tag{9.15}$$

is bounded then there is a subsequence

$$z_{n_1}, z_{n_2}, \ldots, z_{n_k}, \ldots \tag{9.16}$$

that converges to a limit.

■ Theorem 9.3: Monotone bounded real sequences converge

Let a_n be a real sequence that is bounded, i.e. $|a_n| < M$. *Then if the bounded sequence* a_n *satisfies either the condition that it is monotone increasing:*

$$a_n \leqslant a_{n+1} \tag{9.17}$$

or, that it is monotone decreasing:

$$a_n \geqslant a_{n+1} \tag{9.18}$$

then the sequence converges.

■ Theorem 9.4: Cauchy's convergence criterion

Suppose that given $\epsilon > 0$, *we can find an* N *such that* $p > N \cap q > N$ *implies*

$$|a_p - a_q| < \varepsilon \tag{9.19}$$

then the sequence a_n is convergent. Note the use of $\bigcap$ to indicate that *both* inequalities have to be satisfied.

Operations with absolutely convergent series

The terms of an absolutely convergent series may be reorganized without affecting the sum of the terms. Furthermore, two absolutely convergent series may be added, subtracted and multiplied to obtain new absolutely convergent series whose limits are the sum, difference and product of the two. Multiplication is not necessarily possible with conditionally convergent series.

Basic tests for convergence

Theorem 9.5: The comparison test

If

$$\sum_{n=1}^{\infty} |a_n| \tag{9.20}$$

converges and $|b_n| \leqslant |a_n|$, then

$$\sum_{n=1}^{\infty} b_n \tag{9.21}$$

is absolutely convergent.

Theorem 9.6: The ratio test

If

$$\lim_{n\to\infty} \left| \frac{a_{n+1}}{a_n} \right| = L \tag{9.22}$$

then

$$\sum_{n=1}^{\infty} a_n \tag{9.23}$$

converges absolutely if $L < 1$ and diverges if $L > 1$. No information is available (the test fails) if $L = 1$.

Theorem 9.7: The nth root test

If

$$\lim_{n\to\infty} |a_n|^{1/n} = L \tag{9.24}$$

then

$$\sum_{n=1}^{\infty} a_n \tag{9.25}$$

converges absolutely if $L < 1$ and diverges if $L > 1$. No information is available (the test fails) if $L = 1$.

▪ Theorem 9.8: Dirichlet's test

If b_r is a real, monotone-decreasing sequence, with $b_r \to 0$ as $r \to \infty$, and a_n is a complex sequence with the property that

$$\sum_{r=1}^{n} a_r \tag{9.26}$$

is bounded for all n, then

$$\sum_{r=1}^{\infty} a_r\, b_r \tag{9.27}$$

is convergent.

▪ Theorem 9.9: The integral test for convergence

Suppose that $f(x)$ is non-negative and decreases monotonically for

$$x \geqslant x_0 \tag{9.28}$$

Then

$$\sum_{n=1}^{\infty} f(n) \tag{9.29}$$

is convergent (divergent) if

$$\lim_{M \to \infty} \int_{x_0}^{M} f(x)\, dx \tag{9.30}$$

is convergent (divergent).

▪ Theorem 9.10: The alternating series test

If $a_n \geqslant 0$ and $a_n \geqslant a_{n+1}$ and $a_n \to 0$, then

$$\sum_{n=1}^{\infty} (-1)^n\, a_n \tag{9.31}$$

converges. Note that this is a special case of Dirichlet's test. The alternating series test can often be used to check the behaviour of a power series at $z = -1$, whereas the Dirichlet test can be used to test behaviour elsewhere on the unit circle – we shall look at this later in Section 9.3.

■ Review of results on uniform convergence

The property of a sequence or series of functions that they are uniformly convergent is of particular importance. First a reminder:

A sequence of functions $f_n(z)$ is *uniformly convergent* on a subset $U \subset \mathbb{C}$, to $f(z)$, if, given any $\epsilon > 0$, we can find an integer N such that, *for all* z, and all $n > N$, $|f_n(z) - f(z)| < \varepsilon$.

A number of results follow from this. A full statement and proof of some of them must wait until we have defined complex integration and differentiation properly, but it is also convenient to group them here.

▪ Theorem 9.11: Uniform convergence and continuity

Suppose $f_n(z) \to f(z)$ *uniformly on* U. *If* $f_n(z)$ *is continuous for all* n, *then* $f(z)$ *is continuous*. Proof: Given $\epsilon > 0$, we can find an N such that $n > N$ implies

$$|f_n(z) - f(z)| < \epsilon \tag{9.32}$$

for all $z \in U$. Let $w \in U$. Then, since f_n is continuous at w, we can find a δ such that $|z - w| < \delta \Rightarrow |f_n(z) - f_n(w)| < \epsilon/3$. Then, using the triangle inequality, given that

$$f(z) - f(w) = (f(z) - f_n(z)) + (f_n(z) - f_n(w)) + (f_n(w) - f(w)) \tag{9.33}$$

we can see that

$$\begin{aligned} |f(z) - f(w)| &\leqslant |f(z) - f_n(z)| + |f_n(z) - f_n(w)| + |f_n(w) - f(w)| \\ &\leqslant \epsilon/3 + \epsilon/3 + \epsilon/3 = \epsilon \end{aligned} \tag{9.34}$$

Hence f is continuous on U.

This is a good place to remind you that in constructing any sort of '$\epsilon - \delta$' proof of this type, it is not necessary to pull 'out of a hat', in advance, convenient factors such as $\epsilon/3$. The point is to show that $|f(z) - f(w)|$ can be made as small as you like. This often involves many applications of the triangle inequality.

▪ Theorem 9.12: Uniform convergence and integration or differentiation

We have not officially defined complex integration and differentiation yet, so we cannot yet state this properly. It will be revisited in Chapters 10 and 11. However, the result is easy to state informally: *suppose that* $f_n(z)$ *is continuous and* $f_n(z) \to f(z)$ *uniformly. Then*

$$\int f_n(z)\, dz \to \int f(z)\, dz \tag{9.35}$$

This can be applied to a series just as easily as to a sequence. This then says that we can integrate a uniformly convergent series term by term. The corresponding differentiation theorem for series is slightly weaker. *Suppose that each $f_n(z)$ is differentiable and that the series of derived terms*

$$\sum_{n=1}^{\infty} f_n{}'(z) \tag{9.36}$$

is uniformly convergent, and that

$$\sum_{n=1}^{\infty} f_n(z) \tag{9.37}$$

is convergent on a set U. Then

$$\frac{\partial}{\partial z} \sum_{n=1}^{\infty} f_n(z) \tag{9.38}$$

exists and equals $\sum_{n=1}^{\infty} f_n{}'(z)$. This follows by applying the theorem on integration to the derived series. Note that the form of the theorem as stated here is the weak form of the differentiation theorem that also applies in the case of real functions. There is a stronger form of the theorem, known as Weierstrass's Theorem, which asserts that the limit of a uniformly convergent sequence of holomorphic functions is holomorphic, and that the derivative sequence converges to the derivative of the limit. A discussion of this is beyond the scope of this text. For a discussion, see for example Ahlfors (1979), Section 5.1.

■ Theorem 9.13: The Weierstrass M-test

This result is best stated as the following theorem. *Suppose that $|f_n(z)| \leqslant M_n$ on a set U where M_n is independent of z within U, and that*

$$\sum_{n=1}^{\infty} M_n \tag{9.39}$$

is convergent. Then

$$\sum_{n=1}^{\infty} f_n(z) \tag{9.40}$$

is uniformly convergent in U. The proof is straightforward, and proceeds by considering the remainder

$$R_N = \sum_{n=N+1}^{\infty} f_n(z) \tag{9.41}$$

$$|R_N| = \left| \sum_{n=N+1}^{\infty} f_n(z) \right| \leqslant \sum_{n=N+1}^{\infty} |f_n(z)| \leqslant \sum_{n=N+1}^{\infty} M_n \tag{9.42}$$

Now since the series of M_n is convergent, given $\epsilon > 0$, we can find an N_0 such that $N > N_0$ implies

$$\sum_{n=N+1}^{\infty} M_n < \epsilon \tag{9.43}$$

Since N_0 is clearly independent of z, the series of f_n is manifestly uniformly convergent. Note that by the comparison test the series is also absolutely convergent.

■ Uniform convergence counter examples

It is important to understand that the conditions of the results just stated are necessary. This is made clear by the following examples, which use real variables.

■ Continuity failure example

Investigate the uniform convergence of

$$f_n(x) = \frac{1}{n\,x + 1} \tag{9.44}$$

(i) on a region defined by $x > \alpha > 0$; (ii) on the region $x \geqslant 0$.

■ Integration failure example

Let $f(x)$ be defined by:

$$f_n(x) = \begin{cases} 0 & \text{if } x = 0 \\ n^2\,x & \text{if } 0 < x \leqslant 1/n \\ n - n^2(x - 1/n) & \text{if } 1/n < x \leqslant 2/n \\ 0 & \text{if } x > 2/n \end{cases} \tag{9.45}$$

To what function of x does $f_n(x)$ converge pointwise? Is the convergence uniform? What is the integral of $f_n(x)$ over positive x? What is the integral of the limit function over positive x? Try the following *Mathematica* input to investigate matters visually:

```
plotf[x_, n_]  := Which[x == 0, 0,
        0 < x <= 1 / n,  n^2 * x,
  1 / n < x <= 2 / n,  n - n^2 (x - 1 / n),  True,  0]
```

```
plotn[n_] :=
  Plot[plotf[x, n], {x, 0, 2}, PlotRange -> {0, 20}];
```

Have a look at **plotn[3]**, **plotn[7]** to get yourself going.

9.3 Convergence of power series

The basic property of a power series that is of interest is the concept of a radius of convergence. This is so important we make it into a Theorem:

■ Theorem 9.14: Existence of a radius of convergence

Given a series

$$\sum_{n=0}^{\infty} a_n z^n \tag{9.46}$$

one of three things happens:

(1) *the series converges for $z = 0$ only;*
(2) *the series converges for all (finite) z;*
(3) *there is a real number R such that the series is convergent for all z with $|z| < R$ and divergent for all z with $|z| > R$.*

Note that no information is available about what happens when $|z| = R$. In fact, anything can happen, as discussed below. The key to understanding the existence of a disk inside which the series converges is the following result. Suppose that the series

$$\sum_{n=0}^{\infty} a_n z^n \tag{9.47}$$

converges for $z = z_1 \neq 0$. Then the series is absolutely convergent at each point in the open set defined by

$$|z| < |z_1| \tag{9.48}$$

To prove this, note that the terms in the series with z_1 are bounded :

$$|a_n z_1^n| \leq M \tag{9.49}$$

for some positive real number M. Now suppose that

$$|z| < |z_1| \tag{9.50}$$

and let

$$r = \left|\frac{z}{z_1}\right| \tag{9.51}$$

then

$$|a_n z^n| = |a_n z_1^n| \left|\frac{z}{z_1}\right|^n \leqslant M r^n \tag{9.52}$$

with $r < 1$. This allows us to use the comparison test with a convergent series to infer that

$$\sum_{n=0}^{\infty} |a_n z^n| \tag{9.53}$$

is convergent. Hence the series is absolutely convergent when $|z| < |z_1|$. What we want for R is the largest value of $|z_1|$ for which this works. We define it by the following:

$$R = \sup\left\{ |z| \ : \ \sum_{n=0}^{\infty} |a_n z^n| \text{ converges}\right\} \tag{9.54}$$

We find the radius by using simple tests such as the ratio test.

▪ Examples of differing radii of convergence

We can give explicit examples of the various types of behaviour. You should check, using the ratio test, that

$$\sum_{n=1}^{\infty} z^n\, n! \tag{9.55}$$

has radius of convergence zero, that

$$\sum_{n=1}^{\infty} \frac{z^n}{n!} \tag{9.56}$$

has radius of convergence infinity, and that

$$\sum_{n=1}^{\infty} \left(\frac{z}{4}\right)^n \tag{9.57}$$

has radius of convergence 4.

▪ What happens on the boundary circle?

The family of power series:

$$s(k, z) = \sum_{n=1}^{\infty} \frac{z^n}{n^k} \tag{9.58}$$

is instructive in this respect.We can use the ratio test to demonstrate that the radius of convergence is 1 for all k. If $k = 0$ then the terms do not go to zero anywhere on the boundary, so that the series is not convergent anywhere on the boundary. This occurs in spite of the fact that

$$s(0, z) = \frac{1}{1 - z} \tag{9.59}$$

for

$$|z| < 1 \tag{9.60}$$

and this function may be evaluated everywhere in the complex plane except $z = 1$. When $k = 1$ matters are more complicated. If $z = 1$ the series is divergent (to see this use the 'old trick' of grouping the terms in pieces, each of which is greater than $1/2$). For other z on the unit circle, we can use the Dirichlet test to show that the series is convergent. When $k = 2$, we use the fact that

$$\sum_{n=1}^{\infty} \frac{1}{n^2} \tag{9.61}$$

is convergent to infer that $s(2, z)$ is absolutely convergent on $|z| = 1$. On the boundary, the series may converge everywhere, nowhere, or on a strict subset of the boundary. The convergence of these types of series can be visualized with *Mathematica* (see below).

▪ Theorem 9.15: Uniform convergence of power series

Suppose that a power series

$$\sum_{n=1}^{\infty} a_n z^n \tag{9.62}$$

converges for $z = z_0$. *Then it converges uniformly for* $|z| \leq |z_1|$, *where* $|z_1| < |z_0|$. The proof uses the Weierstrass M-test in a very simple way. Set

$$M_n = \frac{|z_1|^n}{|z_0|^n} \tag{9.63}$$

and note that $\sum_{n=1}^{\infty} M_n$ converges by the geometric series property. Now, since

$$\sum_{n=1}^{\infty} a_n z_0{}^n \tag{9.64}$$

converges, then

$$\lim_{n\to\infty} a_n z_0{}^n = 0 \tag{9.65}$$

and so, for large enough n, we must have

$$|a_n| < \frac{1}{|z_0|^n} \tag{9.66}$$

It follows that

$$|a_n\, z^n| < M_n \tag{9.67}$$

for $|z| \leqslant |z_1|$, so by the M-test, $\sum_{n=1}^{\infty} a_n\, z^n$ is uniformly convergent. Therefore a power series is uniformly convergent in any disc entirely inside the circle of convergence.

9.4 Functions defined by power series

We can write down series definitions of many common functions. Here are a few:

$$\exp(z) = \sum_{n=0}^{\infty} \frac{z^n}{n!} = 1 + z + \frac{z^2}{2!} + \frac{z^3}{3!} + \ldots \tag{9.68}$$

$$\sin(z) = \sum_{n=1}^{\infty} \frac{(-1)^{n-1}\, z^{2n-1}}{(2n-1)!} = z - \frac{z^3}{3!} + \frac{z^5}{5!} - \ldots \tag{9.69}$$

$$\log(1+z) = \sum_{n=1}^{\infty} \frac{(-1)^{n-1}\, z^n}{n} = z - \frac{z^2}{2} + \frac{z^3}{3} - \ldots \tag{9.70}$$

Finding the radii of convergence of these and related series is left to you as an exercise.

9.5 ✵ Visualization of series and functions

The behaviour of the convergence of power series, suitably truncated, can be explored with *Mathematica*. First, load a package to help with plotting:

```
Needs["Graphics`ParametricPlot3D`"]
```

Next, define a family of truncated power series and a plot routine:

```
UnitCoefPowerSeries[z_, a_, N_Integer] :=
  Sum[z^k/k^a, {k, 1, N}]

  plot[n_, ptsr_, ptsang_, trunc_] :=
ParametricPlot3D[{r*Cos[theta], r*Sin[theta],
Abs[UnitCoefPowerSeries[r*Exp[I*theta], n, trunc]],
Hue[Arg[UnitCoefPowerSeries[r*Exp[I*theta], n,
trunc]]/Pi]},
{r, 0.001, 0.99}, {theta, 0, 2 Pi},
PlotRange -> All, ViewPoint -> {-2, -2, 2},
Lighting -> False, Axes -> False,
PlotPoints -> {ptsr, ptsang}];
```

Finally, try it out for $n = 2$ and 1, and explore other parameters.

```
plot[2, 30, 60, 20]
```

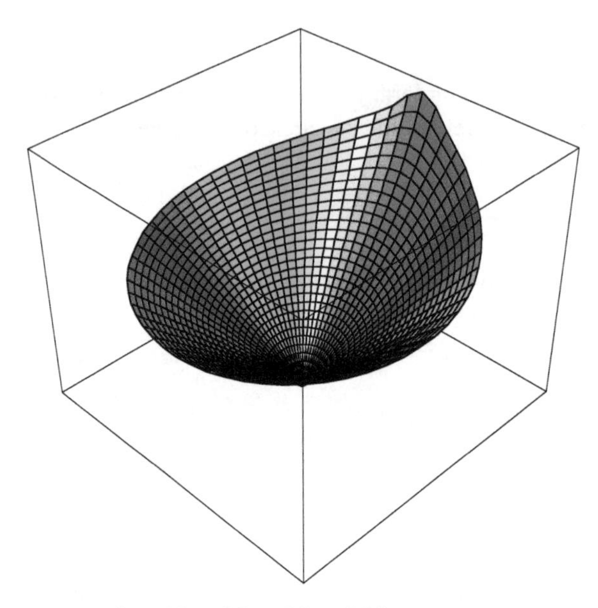

```
plot[1, 30, 60, 30]
```

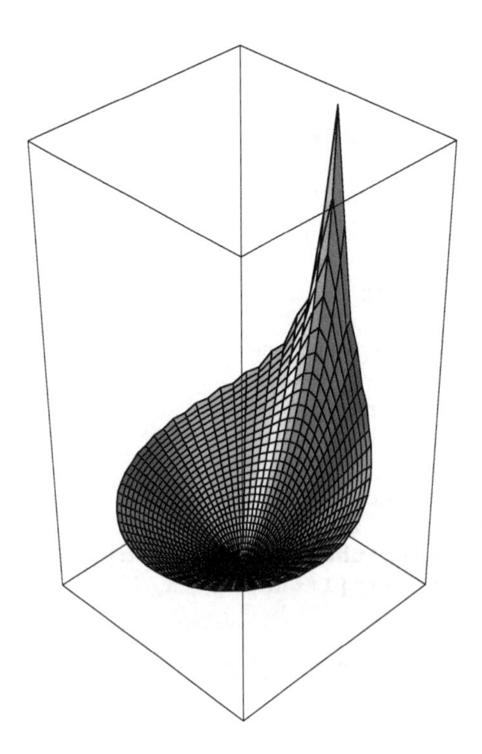

Exercises

9.1 Use the ratio test to show that the radii of convergence of each of the series (in powers of z) for $\exp(z)$, $\sin(z)$, $\cos(z)$, $\log(1+z)$, are, respectively, ∞, ∞, ∞, 1.

9.2 Let m be a posititve integer. Expand the function

$$\frac{1}{(1-z)^m}$$

in a power series about the point $z = 0$. What is the radius of convergence of this series? Repeat this exercise by expanding about the point $z = -1$. What is the radius of convergence of this new series. (Note that to expand about $z = -1$ you write the function in terms of $w = z + 1$ and expand in powers of w.)

9.3 What is the radius of convergence of the series

$$\sum_{n=1}^{\infty} n!\, z^n \ ?$$

9.4 What is the radius of convergence of the series

$$\sum_{n=1}^{\infty} n^4\, z^n \ ?$$

9.5 Bessel functions $J_\nu(z)$ are defined for ν real and non-negative by the power series

$$\sum_{n=1}^{\infty} \frac{(-1)^n \left(\frac{z}{2}\right)^{2n+\nu}}{n!\,(n+\nu)!}$$

What is the radius of convergence of this series?

9.6 ✵ Use the definition of the function **UnitCoefPowerSeries** to explore the behaviour of this function with $a = 0, 1, 2$, using **ParametricPlot3D**.

9.7 ✵ *Mathematica* can generate power series using the **Series** function. Try, e.g.,

```
Series[Exp[z], {z, 0, 7}]
```

Explore the power series for $\sin(z)$, $\cosh(z)$ and $J_0(z)$ in terms of powers of z. Use **?Series** to find out how Series works, and find series of these functions about $z = 1$.

9.8 ✵ Sometimes you can generate series about infinity also. Explore the action of

```
Series[Exp[z], {z, Infinity, 2}]
```

with **Exp[z]** replaced in turn by **Sin[z]**, **Cosh[z]** and **BesselJ[0,z]**.

10 Complex differentiation

Introduction

In this chapter we introduce the concept of differentiation of a complex function of a complex variable. There are several ways of approaching this topic, and we shall consider at least two. Given that a complex number may be regarded as a pair of real numbers, we need to make very clear the distinction between differentiability of a function that has two *real* arguments and differentiability of a function with a single *complex* argument, and shall take as our starting point a review of the differentiation of functions of two real variables. This approach has the merit of generalizing in a straightforward way to functions of many real or complex variables. We shall also consider another simple approach to differentiation based on the limit of a ratio. This latter approach is perhaps more familiar if you have taken a course in one-variable real calculus, but does not generalize to functions of several real or complex variables. A key result that we will establish is that a complex function is differentiable in the complex sense if (a) it is differentiable when considered as two real functions of two real variables and also (b) the Cauchy–Riemann equations (partial differential equations) apply. These equations link the real and imaginary parts of the function. After proving some basic results about complex differentiability, e.g., the product, quotient and chain rules, we then derive one of the principal results of basic complex analysis – that power series are differentiable within their radius of convergence. This establishes the differentiability of a large class of functions and explains why, as in Chapter 9, we like to think about functions in terms of power series.

In this chapter you will also see traditional and novel ways of recovering a holomorphic function from either its real or its imaginary part. This is accomplished traditionally by solving the Cauchy–Riemann equations. You will see here, in Section 10.10, how this can be accomplished algebraically by considering holomorphic functions of *two* complex variables. This is another good reason for approaching differentiation in a way that can be applied to two or more variables. This result has a rather intriguing history. The author uncovered some of the details of the history as a result of an appeal for information in an article published in SIAM Review (Shaw, 2004) where a *Mathematica* implementation of the main result was first given. The author wishes to expresses his thanks to Prof. Harold Boas and others for responding and submitting information leading to the discussion in Section 10.10. The title of that section is 'The Ahlfors–Struble(?) Theorem', the ? reflecting the nagging feeling that the result might be older than the original text by Ahlfors (1953). Further information on the result, as given, predating Ahlfors' 1953 text, would be appreciated.

Our first task, to which we now turn, is to define what it means for a complex function to be differentiable at a point.

10.1 Complex differentiability at a point

■ Differentiation of functions of two real variables

Since a complex number can be viewed as a pair of real numbers, a complex function of a complex variable is at once a pair of functions of two real variables:

$$f(z) = f(x + iy) = u(x, y) + iv(x, y) \tag{10.1}$$

Let's define differentiability of a real function of two real variables. We say that a function $u(x, y)$ is *differentiable* at (x, y) if we can write, for some pair of real numbers A, B

$$u(x + h, y + k) = u(x, y) + Ah + Bk + r(h, k, x, y) \tag{10.2}$$

for all (h, k) with $\sqrt{h^2 + k^2} < \epsilon$, subject to the condition that

$$\left| \frac{r}{\sqrt{h^2 + k^2}} \right| \to 0 \tag{10.3}$$

as $(h, k) \to (0, 0)$. This says that we can find a linear approximation to the function in the neighbourhood of the point of interest. This is the proper definition of differentiability, and it says, geometrically, that the surface defined by f has a tangent plane at the point in question. It is related in quite a subtle way to the existence of ordinary partial derivatives, as the following two observations will point out.

■ Differentiability implies the existence of partial derivatives

We can ask the question: does the limit: $\lim_{h\to 0} \frac{u(h+x, y)-u(x, y)}{h}$ exist? Using our formula, we see that in a neighbourhood of the point (x, y)

$$\frac{u(x + h, y) - u(x, y)}{h} = A + \frac{r}{h} \tag{10.4}$$

so that the limit can be taken, revealing that

$$A = \frac{\partial u(x, y)}{\partial x} \tag{10.5}$$

Similarly, it follows that the partial derivative with respect to y exists and equals B.

■ Existence of partial derivatives does not imply differentiability

In fact, matters are rather worse than the title of this section might imply, for it is possible for partial derivatives to exist at a point (or even everywhere) without a function even being continuous at that point. The following exercise makes this clear:

Exercise: Let $u(0, 0) = 0$, and for other values of x and y, let

$$u(x, y) = \frac{xy}{x^2 + y^2} \tag{10.6}$$

Show that both the partial derivatives

$$\frac{\partial u(x, y)}{\partial x}, \frac{\partial u(x, y)}{\partial y} \tag{10.7}$$

exist everywhere. [Hint: At the origin you need to apply the basic definition.] By considering $u(r\cos(\theta), r\sin(\theta))$, show that u is *not continuous* at the origin. [Hint: Look at what happens to the function as you approach the origin from various directions.] At the end of this chapter, Exercise 10.16 gives another nicely perverse example, illustrating the fact that having one-dimensional differentiability in *all* directions (not just the x- and y-axis directions) is still not sufficient to guarantee differentiability in the two-dimensional sense!

■ Visualization of discontinuous function with partial derivatives

The following plots help to illuminate the somewhat hazardous relationship between the existence of partial derivatives and differentiability. The function considered in the previous exercise is plotted, first in Cartesian coordinates, then, more smoothly, in its polar form. In each case it is clear that there is a discontinuity at the origin. But the function is zero along both axes – so the partial derivatives exist at the origin.

```
Plot3D[x y / (x^2 + y^2),
 {x, -1, 1}, {y, -1, 1}, PlotPoints -> 50]
```

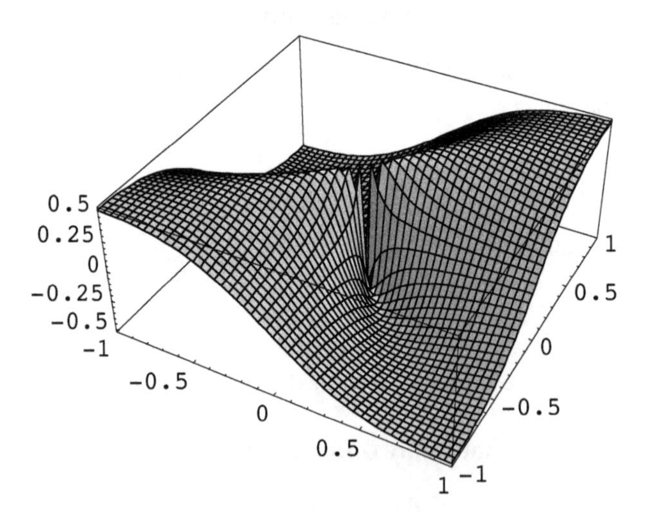


```
ParametricPlot3D[{r Cos[θ], r Sin[θ], Sin[2 θ] / 2},
    {r, 0, 1}, {θ, 0, 2 π}, PlotPoints -> {20, 40}]
```

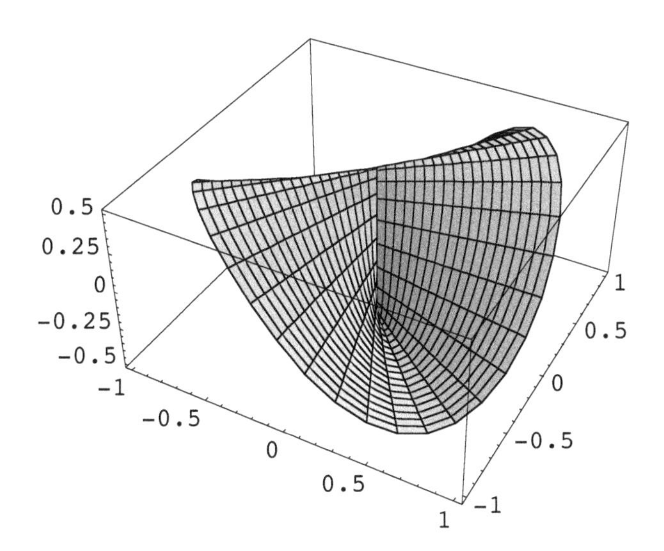

10.2 Real differentiability of complex functions

Now suppose that we have a complex function related to a pair of real functions as follows:

$$f(z) = f(x + iy) = u(x, y) + iv(x, y) \tag{10.8}$$

We say that the *pair* of *real* functions u v are *differentiable* at (x, y) if there are real numbers A, B, C, D such that we can write, with u and v defined in a neighbourhood of (x, y)

$$u(x + h, y + k) = u(x, y) + Ah + Bk + r(h, k, x, y) \tag{10.9}$$

$$v(x + h, y + k) = v(x, y) + Ch + Dk + s(h, k, x, y) \tag{10.10}$$

for all (h, k) in an ε-neighbourhood of $(0, 0)$, subject to the conditions that

$$\left|\frac{r}{\sqrt{h^2 + k^2}}\right| \to 0, \quad \left|\frac{s}{\sqrt{h^2 + k^2}}\right| \to 0 \tag{10.11}$$

as $(h, k) \to (0,0)$. This again says that we can find a linear approximation to each real function in the neighbourhood of the point of interest. This is the proper definition of differentiability of mappings $\mathbb{R}^2 \to \mathbb{R}^2$. So far so good. We see that the differentiability of such a map relies in part on the existence of numbers A, B, C and D, which may be, and usually will be, independent. This approach allows us to define differentiation of vector-valued functions of vectors using a matrix formalism, since we can write

$$\begin{pmatrix} u(h + x, k + y) \\ v(h + x, k + y) \end{pmatrix} = \begin{pmatrix} u(x, y) \\ v(x, y) \end{pmatrix} + \begin{pmatrix} A & B \\ C & D \end{pmatrix} \cdot \begin{pmatrix} h \\ k \end{pmatrix} + \begin{pmatrix} r \\ s \end{pmatrix} \tag{10.12}$$

The matrix with components A...D is a linear map that we call the derivative of the mapping $\mathbb{R}^2 \to \mathbb{R}^2$. It should now be obvious what the derivative of a mapping $\mathbb{R}^m \to \mathbb{R}^n$ is – it is a matrix, as here in the case $m = n = 2$.

10.3 Complex differentiability of complex functions

Now suppose, precisely as before, that we have a complex function and a pair of real functions defined through the relation

$$f(z) = f(x + iy) = u(x, y) + iv(x, y) \tag{10.13}$$

We write $z = x + iy$, $w = h + ik$, and suppose that f is defined in a neighbourhood of z. We say that f is differentiable at z, considered as a *complex* function, if there is a *complex* number E such that we can write

$$f(z + w) = f(z) + Ew + \sigma(w, z) \tag{10.14}$$

for all w in an ε-neighbourhood of zero, subject to the condition that

$$\left|\frac{\sigma}{w}\right| \to 0 \tag{10.15}$$

as $w \to 0$. This again says that we can find a linear approximation to the function in the neighbourhood of the point of interest. This is the proper definition of differentiability of mappings $\mathbb{C} \to \mathbb{C}$. So just how does it differ from that we have written down for two real functions of two real variables? It's easy to find out. We just write

$$E = \alpha + i\beta \tag{10.16}$$

$$\sigma = r + is \tag{10.17}$$

and expand our definition:

$$\begin{aligned} &u(h + x, k + y) + iv(h + x, k + y) \\ &= u(x, y) + iv(x, y) + (\alpha + \beta i)(h + ik) + r + is \end{aligned} \tag{10.18}$$

When we take the real and imaginary parts, we obtain

$$u(x + h, y + k) = u(x, y) + \alpha h - \beta k + r \tag{10.19}$$

$$v(x + h, y + k) = v(x, y) + \beta h + \alpha k + s \tag{10.20}$$

This is just like the pair of real equations, except that the four independent numbers A, B, C, D, have been replaced, respectively, by α, $-\beta$, β, α. So there is a constraint between the partial derivatives of u and v. If we write this in conventional form, we see that

$$\frac{\partial u}{\partial x} = \frac{\partial v}{\partial y} \tag{10.21}$$

$$\frac{\partial u}{\partial y} = -\frac{\partial v}{\partial x} \tag{10.22}$$

These relations are very important and are given a special name: 'The Cauchy–Riemann equations'. This analysis has established the following important result:

■ Theorem 10.1: differentiability of complex functions

A complex function is differentiable as a mapping $\mathbb{C} \to \mathbb{C}$ if and only if:

(1) *the Cauchy–Riemann equations hold, AND,*
(2) *the function is differentiable in the real sense, as a mapping* $\mathbb{R}^2 \to \mathbb{R}^2$.

The number $E = \alpha + i\beta$, is called the complex derivative of the map, and is usually referred to simply as $f'(z)$.

10.4 Definition via quotient formula

The approach we have taken so far makes clear the relationship between real and complex differentiation, and has the advantage of generalizing nicely to functions of several real or complex variables – we just assume that the function has a linear approximation in the neighbourhood of a point. However, within the confined scope of just considering one complex function of one complex variable, a simpler approach can be used. We can say that $f(z)$ is differentiable at z, if f is defined in a neighbourhood of z, and the limit

$$\lim_{w \to 0} \frac{f(w + z) - f(z)}{w} \tag{10.23}$$

exists, and is independent of the manner in which $w \to 0$. If this limit exists, it is denoted $f'(z)$, as before.

You should carry out the following two-part exercise:

(1) Show that this definition is equivalent to the one we have given.
(2) By considering w real, and then imaginary, derive the Cauchy–Riemann equations from this definition.

Note that the definition by the quotient method is very popular in elementary texts, but does not, for obvious reasons (how do you divide by a vector?) generalize to many variables. For this reason this test will develop the theory within the approach of Section 10.3.

10.5 Holomorphic, analytic and regular functions

We use the word 'differentiable' to denote behaviour at a single point. If a function is differentiable at every point in an open set $U \subset \mathbb{C}$, we say that it is 'holomorphic' on U, or 'analytic' (or, less frequently, 'regular') on U. These terms will be regarded as synonymous - different books and papers may use different terminology. We shall use the terms holomorphic and analytic essentially interchangeably. The *important* thing is to remember that *differentiability* refers to the behaviour of a function at a *point*, whereas *holomorphic/analytic/regular* refers to the behaviour of a function on a specified *open set.*

Some books, with which I have no serious argument, prefer use the word 'analytic' to refer to functions having a convergent power series expression, with 'holomorphic' being reserved for differentiability on an open set. These terms become synonymous once one has the theorems expressing the equivalence of (a) being differentiable on an open set and (b) having a power series. The fact that power series are differentiable where they converge is proved in Section 10.8, and the converse is discussed in Section 13.2. In order to make sure that one has terminology and other matters under control, you are recommended to try Exercises 10.2 and 10.3 at the end of this chapter.

10.6 Simple consequences of the Cauchy–Riemann equations

There are a few basic consequences of the Cauchy–Riemann equations that are easily derived, and that have very important consequences, especially in the development of applications in applied mathematics.

■ Orthogonality of curves of constant u and v

Consider the curves $u(x, y) = \text{constant}$, $v(x, y) = \text{constant}$. The normal to the curve $u(x, y) = \text{constant}$ is the vector grad u with components:

$$\text{grad}\, u = \begin{pmatrix} \frac{\partial u}{\partial x} \\ \frac{\partial u}{\partial y} \end{pmatrix} \tag{10.24}$$

and similarly for v. The Cauchy–Riemann equations imply that

$$\left(\frac{\partial u}{\partial x}\right)^2 + \left(\frac{\partial u}{\partial y}\right)^2 = \left(\frac{\partial v}{\partial x}\right)^2 + \left(\frac{\partial v}{\partial y}\right)^2 \tag{10.25}$$

and that

$$\frac{\partial u}{\partial x}\frac{\partial v}{\partial x} + \frac{\partial u}{\partial y}\frac{\partial v}{\partial y} = 0 \tag{10.26}$$

so that the normals to the two curves have the same length, and are, by the second relation, orthogonal.

■ Laplace's equation and harmonic functions

If we can differentiate again, and also swap the order of differentiation (and we shall show later that this is the case), we can use the Cauchy–Riemann equations to derive an important result.

$$\frac{\partial^2 u}{\partial x^2} = \frac{\partial\left(\frac{\partial u}{\partial x}\right)}{\partial x} = \frac{\partial\left(\frac{\partial v}{\partial y}\right)}{\partial x} = \frac{\partial\left(\frac{\partial v}{\partial x}\right)}{\partial y} = -\frac{\partial\left(\frac{\partial u}{\partial y}\right)}{\partial y} = -\frac{\partial^2 u}{\partial y^2} \tag{10.27}$$

Hence we see that

$$\frac{\partial^2 u}{\partial x^2} + \frac{\partial^2 u}{\partial y^2} = 0 \tag{10.28}$$

so that u satisfies Laplace's equation. Such a function is said to be 'harmonic'. We leave it to you to show that the same holds for v. This leads to some very practical applications of analytic functions to any application of potential theory, including special types of fluid flow, electro- and magnetostatics, and two-dimensional gravity models.

10.7 Standard differentiation rules

All the basic rules that apply to differentiation of real functions extend to complex functions. Suppose that we have two functions f and g, each satisfying, in a neighbourhood of z,

$$f(z + w) = f(z) + Ew + \sigma(w) \tag{10.29}$$

$$g(z + w) = g(z) + Fw + \tilde{\sigma}(w) \tag{10.30}$$

We shall prove all the basic rules, using the 'approximate linear mapping' definition of differentiability. Within this framework the proofs are all easy. Note the order of the proofs of the rules: sum, product, chain, $1/z$, $1/f$, quotient. As long as one can write down the rule for $1/z$, it is not necessary to do much else to establish the quotient rule.

■ Sum rule

Let $h = f + g$, then

$$\begin{aligned} h(z+w) &= f(z+w) + g(z+w) = f(z) + Ew + \sigma(w) + g(z) + Fw + \tilde{\sigma}(w) \\ &= f(z) + g(z) + (E + F)w + (\sigma + \tilde{\sigma}) \\ &= h(z) + h'(z)\,w + \Sigma \end{aligned} \tag{10.31}$$

where $\Sigma = \sigma + \tilde{\sigma}$ and $h'(z) = E + F = f'(z) + g'(z)$. Hence h is differentiable and the derivative is the sum of those of f and g.

■ Product rule

To make the operation of multiplication rather more explicit, we temporarily use the 'times' symbol × rather than an implicit multiplicative space.

Let $h = f \times g$, then

$$\begin{aligned} h(z+w) &= f(z+w) \times g(z+w) \\ &= (f(z) + Ew + \sigma(w)) \times (g(z) + Fw + \tilde{\sigma}(w)) \\ &= f(z) \times g(z) + (f(z) \times F + g(z) \times E)w + R \end{aligned} \tag{10.32}$$

where the remainder R is given by:

$$R = (f(z) + Ew) \times \tilde{\sigma}(w) + (g(z) + Fw) \times \sigma(w) + EFw^2 + \sigma(w)\,\tilde{\sigma}(w) \tag{10.33}$$

This clearly satisfies $|R/w| \to 0$ as $w \to 0$, so h is differentiable and

$$h'(z) = F \times f(z) + E \times g(z) = g(z) \times f'(z) + f(z) \times g'(z) \tag{10.34}$$

■ Chain rule

Let $p = f(z)$ and suppose that f is differentiable at z and that g is differentiable at p, i.e.

$$f(z + w) = f(z) + Ew + \sigma(w) \tag{10.35}$$

$$g(p + q) = g(p) + Fq + \tilde{\sigma}(q) \tag{10.36}$$

Let $h(z) = g(f(z))$. Then

$$h(z+w) = g(f(z+w)) = g(f(z) + Ew + \sigma(w)) = g(p+q) \tag{10.37}$$

where $p = f(z)$ and $q = Ew + \sigma(w)$.

$$\begin{aligned} h(z+w) &= g(p) + F(Ew + \sigma(w)) + \tilde{\sigma}(Ew + \sigma(w)) \\ &= g(f(z)) + FEw + S \end{aligned} \tag{10.38}$$

where the remainder

$$S = F\sigma(w) + \tilde{\sigma}(Ew + \sigma(w)) \tag{10.39}$$

It is easily checked that $|S/w| \to 0$ as $w \to 0$, so h is differentiable and

$$h'(z) = EF = f'(z)\, g'(f(z)) \tag{10.40}$$

■ $1/z$ rule

The quotient rule is best derived by working with the function $f = 1/z$ and applying the rules already developed. First let's do some basic algebra:

$$\begin{aligned} 1/(z+w) &= 1/z + (1/(z+w) - 1/z) \\ &= 1/z + (z - (z+w))/(z*(z+w)) \\ &= 1/z - w/(z*(z+w)) \end{aligned} \tag{10.41}$$

$$= 1/z - w/z^2 + (w/z^2 - w/(z*(z+w)))$$
$$= 1/z - w/z^2 + (w/z)*(1/z - 1/(z+w))$$
$$= 1/z - w/z^2 + w^2/(z^2(z+w))$$

That is, with $\sigma = w^2/(z^2(z+w))$, we have

$$1/(z+w) = 1/z + (-1/z^2)w + \sigma \tag{10.42}$$

so, provided $z \neq 0$, $1/z$ is differentiable with derivative $-1/z^2$, which is a familiar result in other contexts.

■ $1/f$ rule

The rule for differentiating $1/f(z)$ now follows by the result for $1/z$ and the chain rule, applied as above with f given and $g(p) = 1/p$. So the derivative of $1/f(z)$ exists provided $f(z) \neq 0$ and equals

$$-\frac{f'(z)}{f(z)^2} \tag{10.43}$$

■ f/g rule

The quotient rule follows from the product rule and the result for $1/g$. The derivative is

$$-\frac{g'(z)f(z)}{g(z)^2} + \frac{f'(z)}{g(z)} \tag{10.44}$$

which is usually written in the more familiar form

$$(f(z)/g(z))' = \frac{f'(z)g(z) - g'(z)f(z)}{g(z)^2} \tag{10.45}$$

10.8 Polynomials and power series

■ Powers and polynomials

Note that given what we know already, it suffices to check that z^n is differentiable for all positive integers n in order to establish differentiability of any polynomial. This is because (a) the product rule applied to $\alpha * z^n$, for any constant α ensures $\alpha * z^n$ is differentiable, (b) the sum rule applied to any linear combination of such functions then shows that any polynomial is differentiable. So let's deal with simple powers. By the binomial theorem, we can make the expansion

$$(w+z)^n = z^n + nwz^{n-1} + w^2 g(z, w) \tag{10.46}$$

where g is a polynomial in z and w. Hence z^n is differentiable with derivative

$$nz^{n-1} \tag{10.47}$$

You probably expected this from the corresponding real result.

■ Theorem 10.2: differentiability of power series

The corresponding results for power series are a major step up in both the form of the proof and significance of the result. Here is a statement of the result. Informally it says that power series are differentiable provided they converge, and the result is the sum of the derivatives of each term differentiated as though it is a simple one-term polynomial. Formally, we have

Let $f(z) = \sum_{n=0}^{\infty} a_n z^n$ *have radius of convergence* $R > 0$*. Then within the circle of convergence* $f'(z)$ *exists and is given by* $\sum_{n=1}^{\infty} n a_n z^{n-1}$.

It is a good idea to prove this in two parts, and to first consider the origin.

▪ Proof at the origin

Consider f in a neighbourhood of the origin. We can clearly write

$$f(z) = a_0 + a_1 z + S(z) \tag{10.48}$$

where

$$S(z) = \sum_{n=2}^{\infty} a_n z^n \tag{10.49}$$

Consider

$$\frac{S(z)}{z^2} = \sum_{n=2}^{\infty} a_n z^{n-2} \tag{10.50}$$

Now, within the circle of convergence of $f(z)$, the series for $f(z)$ is absolutely convergent, so by setting $|z| = R/2$, we can see that

$$\sum_{n=0}^{\infty} |a_n| \left(\frac{R}{2}\right)^n \tag{10.51}$$

is convergent, and so is

$$M = \sum_{n=2}^{\infty} |a_n| \left(\frac{R}{2}\right)^{n-2} \tag{10.52}$$

If we call this last sum M, then, for $|z| < R/2$

$$\left|\frac{S(z)}{z}\right| \leqslant |z| \left|\sum_{n=2}^{\infty} a_n z^{n-2}\right| \leqslant |z| \sum_{n=2}^{\infty} \left|a_n z^{n-2}\right| \leqslant |z| \sum_{n=2}^{\infty} |a_n| \left(\frac{R}{2}\right)^{n-2} \leqslant M|z| \tag{10.53}$$

Hence $|S(z)/z| \to 0$ as $z \to 0$. Hence f is differentiable at the origin, and its derivative is given by

$$f'(0) = a_1 \tag{10.54}$$

■ Proof at other points

To prove the result at a general point within the circle of convergence, we do some reorganization with the binomial theorem to reduce the general case to the one we have just proved. So now suppose that $|w| < R$. We consider w close enough to z such that

$$|z - w| < R - |w| \tag{10.55}$$

It follows that

$$|z - w| + |w| < R \tag{10.56}$$

and hence that

$$\sum_{n=0}^{\infty} |a_n| \left(|w| + |z - w|\right)^n \tag{10.57}$$

is convergent. Now we use the binomial theorem to deduce that

$$\sum_{n=0}^{\infty} |a_n| \left(\sum_{r=0}^{n} \binom{n}{r} |w|^{n-r} |z - w|^r \right) \tag{10.58}$$

is convergent. Now consider our original series with $z = w + (z - w)$. The binomial theorem gives us

$$\sum_{n=0}^{\infty} a_n z^n = \sum_{n=0}^{\infty} a_n \left(\sum_{r=0}^{n} \binom{n}{r} w^{n-r} (z - w)^r \right) \tag{10.59}$$

We now know that this series is absolutely convergent, so we can rearrange it to give

$$\sum_{n=0}^{\infty} a_n z^n = \sum_{r=0}^{\infty} b_r (z - w)^r \tag{10.60}$$

where

$$b_r(w) = \sum_{n=r}^{\infty} \binom{n}{r} w^{n-r} a_n \tag{10.61}$$

Now we can use our previous result to infer differentiability at w, with

$$f'(w) = b_1 = \sum_{n=1}^{\infty} n w^{n-1} a_n \tag{10.62}$$

■ Obvious special functions

It follows from Theorem 10.2 that a great many well-known functions are differentiable, including the exponential, trigonometric and hyperbolic functions discussed previously.

10.9 A point of notation and spotting non-analytic functions

It is often convenient to use another notation for complex differentiation, based on the chain rule. We have, using the bar notation for complex conjugate,

$$z = x + iy \tag{10.63}$$

$$\bar{z} = x - iy \tag{10.64}$$

$$x = \frac{1}{2}(z + \bar{z}) \tag{10.65}$$

$$y = \frac{-i}{2}(z - \bar{z}) \tag{10.66}$$

$$\frac{\partial f}{\partial z} = \frac{\partial f}{\partial x}\frac{\partial x}{\partial z} + \frac{\partial f}{\partial y}\frac{\partial y}{\partial z} = \frac{1}{2}\left(\frac{\partial f}{\partial x} - i\frac{\partial f}{\partial y}\right) \tag{10.67}$$

$$\frac{\partial f}{\partial \bar{z}} = \frac{\partial f}{\partial x}\frac{\partial x}{\partial \bar{z}} + \frac{\partial f}{\partial y}\frac{\partial y}{\partial \bar{z}} = \frac{1}{2}\left(\frac{\partial f}{\partial x} + i\frac{\partial f}{\partial y}\right) \tag{10.68}$$

Hence we can use the alternative forms:

$$\frac{\partial}{\partial z} = \frac{1}{2}\left(\frac{\partial}{\partial x} - i\frac{\partial}{\partial y}\right) \tag{10.69}$$

$$\frac{\partial}{\partial \bar{z}} = \frac{1}{2}\left(\frac{\partial}{\partial x} + i\frac{\partial}{\partial y}\right) \tag{10.70}$$

Exercise 10.3 at the end of this chapter confirms that the Cauchy–Riemann equations are equivalent to

$$\frac{\partial f}{\partial \bar{z}} = 0 \tag{10.71}$$

This is more than just a notational convenience – it also gives a handy way of spotting when a simple function is or is not holomorphic. If the formula involves $\bar{z}$ the function is not holomorphic, since then $\frac{\partial f}{\partial \bar{z}} \neq 0$. Now since

$$|z|^2 = z\bar{z} \tag{10.72}$$

$$\text{Re}(z) = \frac{1}{2}(z + \bar{z}) \tag{10.73}$$

$$\text{Im}(z) = \frac{1}{2}(-i)(z - \bar{z}) \tag{10.74}$$

it is easy to see why functions involving the modulus, real or imaginary parts in some way cannot be holomorphic, as they all involve $\bar{z}$ in some way.

10.10 The Ahlfors–Struble(?) theorem

The ? in the title of this section will be discussed in due course! First we will explain the situation that this result applies to. Suppose that you are given a function $u(x, y)$ and are required to find out what holomorpic function it came from. How do you do this? A good first step is to check that u satisfies Laplace's equation as discussed in Section 10.6. The next step is to solve the Cauchy–Riemann equations. A simple example will make it clear how this works. Let $u = x^2 - y^2$. The Cauchy–Riemann equations give us

$$\frac{\partial u}{\partial x} = \frac{\partial v}{\partial y} = 2x \tag{10.75}$$

$$\frac{\partial u}{\partial y} = -\frac{\partial v}{\partial x} = -2y \tag{10.76}$$

You can integrate each of these equations for v to obtain

$$v = 2xy + g(x) \tag{10.77}$$

$$v = 2xy + h(y) \tag{10.78}$$

where g and h are arbitrary functions of their arguments. Consistency demands that both of these functions must be the same constant, say c. So $v = 2xy + c$ for some real constant c, and furthermore

$$f(z) = x^2 - y^2 + 2ixy + ci = (x + iy)^2 + ci = z^2 + ci \tag{10.79}$$

This procedure can be summarized as follows: (i) differentiate u; (ii) integrate the Cauchy–Riemann equations for v, picking arbitrary functions for consistency; (iii) identify $f(z)$.

You should try this out yourself for the function $u(x, y) = x^3 - 3xy^2$, and some further examples are given in Exercise 10.17. If you persevere with Exercise 10.17 you will rapidly come to the conclusion that an easier route would be helpful. The following theorem, believed to be due to Ahlfors and Struble (see later) is very useful.

■ Theorem 10.3: the holomorphic extraction formula

Let $f(z)$ be holomorphic in a neighbourhood of the point a, and let us suppose that $f(z)$ has a real part $u(x, y)$ and imaginary part $v(x, y)$, where $z = x + iy$. Then the Ahlfors–Struble (holomorphic extraction) theorem states that

$$f(z) = 2u\left(\frac{z + \bar{a}}{2}, \frac{z - \bar{a}}{2i}\right) - \overline{f(a)} = 2i\,v\left(\frac{z + \bar{a}}{2}, \frac{z - \bar{a}}{2i}\right) + \overline{f(a)} \tag{10.80}$$

This result for the case $a = 0$ was given by Ahlfors (1953) and given as an exercise by Spiegel (1981). A full discussion and a simple proof of the general case has been given by this author (see Shaw, 2004 and 'a simple view', below). In the proof of this we will make use of the holomorphic reflection of f, given by

$$\hat{f}(z) = \overline{f(\overline{z})} \tag{10.81}$$

You might like to satisfy yourself that if f is holomorpic on $U \subset \mathbb{C}$ then $\hat{f}$ is holomorphic on $\overline{U} \subset \mathbb{C}$. This is the reflection of U in the real axis (not its closure).

■ A simple view of the proof

Let's consider the relationship between u and f. Consider the function U of two *complex* variables w_1 and w_2 given by

$$U(w_1, w_2) = \frac{1}{2}\left(f(w_1 + i w_2) + \hat{f}(w_1 - i w_2)\right) \tag{10.82}$$

This function is a holomorphic function of the two complex variables w_1, w_2, in the sense of equation (2) with x, y, h, k now all complex. Observe that if $w_1 = x$ and $w_2 = y$ with x, y real, then

$$\begin{aligned} U(x, y) &= \frac{1}{2}\left(f(x + i y) + \hat{f}(x - i y)\right) \\ &= \frac{1}{2}(f(x + i y) + \overline{f(x + i y)}) = u(x, y) \end{aligned} \tag{10.83}$$

so that U is the same as u, but considered now as a function of two complex variables. We now identify U as u and we set

$$w_1 = \frac{z + \overline{a}}{2} \tag{10.84}$$

$$w_2 = \frac{z - \overline{a}}{2i} \tag{10.85}$$

and observe that Eqs. (10.82), (10.84) and (10.85) combine to tell us that

$$u\left(\frac{z + \overline{a}}{2}, \frac{z - \overline{a}}{2i}\right) = \frac{1}{2}\left(f(z) + \hat{f}(\overline{a})\right) \tag{10.86}$$

from which the relation

$$f(z) = 2\,u\left(\frac{z + \overline{a}}{2}, \frac{z - \overline{a}}{2\,i}\right) - \overline{f(a)} \tag{10.87}$$

follows. The link between v and f follows by considering a similar route, with

$$V(w_1, w_2) = \frac{1}{2i}\left(f(w_1 + i w_2) - \hat{f}(w_1 - i w_2)\right) \tag{10.88}$$

▪ An example

A good illustration of the method is to consider the function

$$u(x, y) = \frac{x}{x^2 + y^2} \tag{10.89}$$

The application of Theorem 10.3 says that we double u, and replace x by

$$\frac{z + \overline{a}}{2} \tag{10.90}$$

and y by

$$\frac{z - \overline{a}}{2\,i} \tag{10.91}$$

This implies that we replace $x^2 + y^2$ by $z\overline{a}$. So our function f is given by

$$f(z) = 2\left(\frac{z + \overline{a}}{2}\right)\left(\frac{1}{z\,\overline{a}}\right) - \overline{f(a)} \;=\; \frac{1}{z} + \frac{1}{\overline{a}} - \overline{f(a)} \tag{10.92}$$

If we inspect this relation at $z = a$, we deduce that the real part of $f(a)$ is the real part of $1/\overline{a}$, so we deduce that, for some real constant β,

$$f(z) = \frac{1}{z} + i\beta \tag{10.93}$$

▪ Some further examples for you to try

For each of the following functions, considered as the real part u of some holomorphic f, find f. Consider carefully what base point to use – a good initial guess is to try $a = 0$, but consider carefully what you need to do when the origin is not a point at which the functions make sense.

$$\begin{gathered} x^2 - y^2 \\ 4\,x\,y\,(y^2 - x^2) \\ e^x \cos(y) \\ e^{x/(x^2+y^2)} \cos\left(\frac{y}{x^2 + y^2}\right) \\ \frac{1}{2}\log(x^2 + y^2) \\ \sqrt[4]{x^2 + y^2}\,\cos\left(\frac{1}{2}\tan^{-1}(x, y)\right) \end{gathered} \tag{10.94}$$

You can check the answers against an automatic calculation with *Mathematica*, the code for which is given below.

▪ Uniqueness of the complexification

When the argument given above was published by the author (Shaw, 2004) I received numerous comments, including one from B. Margolis (BM below) that I did not spell out the precise conditions under which the complexification U of u is uniquely defined. Ahlfors was careful to restrict attention to functions that are rational in x and y, probably because of this issue. The first thing to point out is that *given* f, the natural complexifications, U, V of u, v are indeed uniquely defined, having being constructed explicitly as in the simple proof above. The issue is whether, for example, given only 'some formula' for $u(x, y)$, whether we can uniquely specify its complexification. I did not address this in the paper (Shaw, 2004) or indeed above, but my view on this is that the answer is 'yes' once u has been specified in real analytic form, i.e. expressed as a real power series in x and y about the relevant base point, or, more usefully, in terms of real functions that are expressible as such power series. Since it turns out that *any* complex function that is holomorphic on an open set containing the base point is then analytic (i.e. it has a power series) in a neighbourhood of that base point (by Taylor's Theorem as discussed in Chapter 13, Section 13.2) then any u can be so expressed, so this is not a restriction at all. However, you do have to write it in real analytic form. The *Mathematica* code given below now forces this, in contrast to what was originally published by the author (Shaw, 2004). If one does not work in this representation one can end up with different complexifications. Of course, if one restricts attention to power series then one can give an alternative direct and constructive proof (see below), but this does not make clear the elegant role of the reflection principle embodied in the 'simple view'.

To see why this matters, consider the following example (kindly supplied by BM). Let's suppose we start with the formula $u(x, y) = x$. We can write u in various ways. Let's consider

$$u(x, y) = x \tag{10.95}$$

$$u(x, y) = \frac{1}{2}((x + i\, y) + (x - i\, y)) \tag{10.96}$$

$$u(x, y) = \frac{1}{2}((x + i\, y) + (x + i\, y)^*) \tag{10.97}$$

$$u(x, y) = \frac{1}{2}((x - i\, y) + (x - i\, y)^*) \tag{10.98}$$

The first two work out fine, while Eqs. (10.97) and (10.98) give x and 0 respectively if you just blindly apply the extraction formula for f! So it is important that you *first* simplify to the real analytic form of u. Another way of looking at this problem is that one can contrive a zero u to have a non-zero complexification, but not if you limit attention to the real analytic case. This is related to the reason why *Mathematica*'s **`ComplexExpand`** function has to have its target functions set to **`Re`** and **`Im`**, (see below), in order to provide a purely real characterization of u.

▪ The power series version of the proof

Bearing in mind that the 'simple view' tells us what we should we aiming for, it is possible to give a now very simple argument based on power series, which shows that we can specify U uniquely, up to an imaginary constant. (This also makes it clear why Eqs. (10.95–96) above are right for complexification and Eqs. (10.97–98) are wrong.) This does involve bringing forward the results of Chapter 13.

So suppose that $u(x, y)$ is the real part of a holomorphic function $f(z)$ with the property that f is holomorphic in a neighbourhood of $z = a = a_1 + i\, a_2$. Then, by Taylor's Theorem (see Section 13.2), there is a unique power series

$$f(z) = \sum_{n=0}^{\infty} c_n\,(z-a)^n \;. \tag{10.99}$$

It follows that we can write:

$$u(x, y) = \frac{1}{2}\sum_{n=0}^{\infty} c_n\,(x + i\,y - a)^n + \frac{1}{2}\sum_{n=0}^{\infty} \overline{c}_n\,(x - i\,y - \overline{a})^n \;. \tag{10.100}$$

We also note that *given a knowledge of* $u(x, y)$ *alone*, on the assumption that it is real anaytic about (a_1, a_2) *and* a solution of Laplace's equation, such a representation must exist and be unique (apart from $\mathrm{Im}[c_0]$). This is because the Laplace condition is just

$$\frac{\partial^2 u}{\partial z\,\partial \overline{z}} = 0 \;\Leftrightarrow\; u \;=\; g(z) \;+\; h(\overline{z}). \tag{10.101}$$

and reality constrains h to be expressed in terms of the conjugate of g. Then the existence of a series supplies Eq. (10.100) uniquely, apart from the imaginary part of c_0. Now define $U(w_1, w_2)$ *uniquely, up to the imaginary part of* c_0, by

$$U(w_1, w_2) = \frac{1}{2}\sum_{n=0}^{\infty} c_n\,(w_1 + i\,w_2 - a)^n + \frac{1}{2}\sum_{n=0}^{\infty} \overline{c}_n\,(w_1 - i\,w_2 - \overline{a})^n \;, \tag{10.102}$$

and observe that, first

$$U(x, y) = \frac{1}{2}\sum_{n=0}^{\infty} c_n\,(x + i\,y - a)^n + \frac{1}{2}\sum_{n=0}^{\infty} \overline{c}_n\,(x - iy - \overline{a})^n \;=\; u(x, y) \;, \tag{10.103}$$

and second,

$$U\!\left(\frac{z+a}{2},\; \frac{z-\overline{a}}{2\,i}\right) = \frac{1}{2}\sum_{n=0}^{\infty} c_n\,(z-a)^n + \frac{1}{2}\sum_{n=0}^{\infty} \overline{c}_n\,(\overline{a}-\overline{a})^n \;= \\ \frac{1}{2}\,f(z) + \frac{1}{2}\,\overline{c}_0 \;=\; \frac{1}{2}\,f(z) + \frac{1}{2}\,\overline{f(a)} \tag{10.104}$$

The result then follows. This is essentially the same argument as that given by Cartan (1961) for the case $a = 0$.

■ *Mathematica* implementation of the extraction formula

This is now a straightforward purely algebraic operation so is easily implemented. Let's do the automation for the extraction of f from its real part. We use **ComplexExpand** with the option **TargetFunctions → {Re, Im}** to get the right form of the function for complexification.

```
RealToHolo[expr_, anum_, {xsym_, ysym_, zsym_}] :=
 Module[{abar = Conjugate[anum], exprf},
  exprf =
   ComplexExpand[expr, TargetFunctions → {Re, Im}];
  func =
   2 * exprf /. {xsym → (zsym + abar) / 2,
     ysym → (zsym - abar) / (2 * I)}; basecorr =
   - exprf /. {xsym → Re[anum], ysym → Im[anum]};
  FullSimplify[func + basecorr + I * β]]
```

Here are all the functions given for you to try in the previous sub-section, expressed as a *Mathematica* list:

```
TestUSet = {x^2 - y^2, 4 x y (y^2 - x^2), Exp[x] Cos[y],
   Exp[x / (x^2 + y^2)] Cos[y / (x^2 + y^2)],
   1 / 2 Log[x^2 + y^2],
   (x^2 + y^2) ^ (1 / 4) Cos[1 / 2 ArcTan[x, y]]};
```

Let's try this first with the origin playing the role of the base point a (you will get some warnings that are not shown here).

```
Map[RealToHolo[#, 0, {x, y, z}] &, TestUSet]
```

$\{z^2 + i\,\beta,\ i\,(z^4 + \beta),\ i\,\beta + e^z,\ \text{Indeterminate},\ \text{Indeterminate},\ i\,\beta + \text{Interval}[\{0, 0\}]\}$

The extraction is failing when the origin is not a point at which the functions are defined. If we shift the base point to, for example, $a = 1$, then all is well:

```
Map[RealToHolo[#, 1, {x, y, z}] &, TestUSet]
```

$\left\{z^2 + i\,\beta,\ i\,(z^4 + \beta - 1),\ i\,\beta + e^z,\ i\,\beta + e^{\frac{1}{z}},\ i\,\beta + \log(z),\ i\,\beta + \sqrt{z}\right\}$

Let's see if we can induce any non-uniqueness by writing down u in various different ways.

```
PerverseUSet = {x, 1/2 ((x+I y) + (x - I y)),
  1/2 ((x+I y) + Conjugate[(x + I y)]),
  1/2 ((x-I y) + Conjugate[(x - I y)])}
```

$$\left\{x,\ x,\ \frac{1}{2}\,(x+i\,y+(x+i\,y)^*),\ \frac{1}{2}\,(x-i\,y+(x-i\,y)^*)\right\}$$

```
Map[RealToHolo[#, 0, {x, y, z}] &, PerverseUSet]
```

$$\{z+i\,\beta,\ z+i\,\beta,\ z+i\,\beta,\ z+i\,\beta\}$$

The corresponding formula for obtaining the holomorphic function from its imaginary part is:

```
ImToHolo[expr_, anum_, {xsym_, ysym_, zsym_}] :=
 Module[{abar = Conjugate[anum], exprf},
  exprf =
   ComplexExpand[expr, TargetFunctions → {Re, Im}];
  func =
   2*I*exprf /. {xsym → (zsym+abar)/2,
     ysym → (zsym-abar)/(2*I)}; basecorr =
   -I*exprf /. {xsym → Re[anum], ysym → Im[anum]};
  FullSimplify[func+basecorr+β]]
```

We test it with the following set of test problems:

```
TestVSet = {2*x*y, x^4 + y^4 - 6*x^2*y^2,
   Exp[x]*Sin[y], (-E^(x/(x^2 + y^2)))*
    Sin[y/(x^2 + y^2)], ArcTan[x,y], (x^2+y^2)^(1/4)
Sin[1/2 ArcTan[x,y]]};

Map[ImToHolo[#, 1, {x, y, z}] &, TestVSet]
```

$$\left\{z^2+\beta-1,\ i\,z^4+\beta,\ \beta+e^z-e,\ \beta+e^{\frac{1}{z}}-e,\ \beta+\log(z),\ \beta+\sqrt{z}-1\right\}$$

■ The inverse of `ComplexExpand[Re[]]`

The extraction of the holomorphic function from just its real part can be seen as an inversion of the composite operation given by **`ComplexExpand[Re[]]`**. The following examples make this clear – you might like to try some more of your own. The important thing is to make sure **`ComplexExpand`** generates an expression just involving x and y. We do this with:

```
SetOptions[ComplexExpand, TargetFunctions → {Re, Im}];
```

Of course, this is also done within **`RealToHolo`**, but we also wish to exhibit the expressions in the right form here. Now consider the following:

```
ComplexExpand[Re[1/(x+I y)]]
```

$$\frac{x}{x^2 + y^2}$$

```
RealToHolo[%, 1, {x, y, z}]
```

$$i\beta + \frac{1}{z}$$

▪ Some more exotic examples

```
ComplexExpand[Im[Exp[1 / (x + I y) ^2]]]
```

$$-e^{\frac{x^2-y^2}{(x^2+y^2)^2}} \sin\left(\frac{2xy}{(x^2+y^2)^2}\right)$$

```
ImToHolo[%, 1, {x, y, z}]
```

$$\beta + e^{\frac{1}{z^2}} - e$$

```
ComplexExpand[Re[(x + I y) ^ (1 / n)],
 TargetFunctions → {Re, Im}]
```

$$(x^2 + y^2)^{\frac{1}{2n}} \cos\left(\frac{\tan^{-1}(x, y)}{n}\right)$$

```
RealToHolo[%, 1, {x, y, z}]
```

$$i\beta + 2\sqrt{z^{\frac{1}{n}}} \cosh\left(\frac{\log(z)}{2n}\right) - 1$$

▪ Some history

This approach to reconstructing f from its real or imaginary part has an interesting history, and for much of the following information I wish to thank Prof. Harold Boas ('HB' in further comments below). First, it is interesting to note that this result has been rediscovered several times over the years! Indeed, the author's interest in this method stemmed from an observation by John Ockendon that the result for the case $a = 0$ holds, and he issued a challenge to explain it and figure out what happens if f is singular at the origin. This and the desire to find a *Mathematica* implementation resulted in the paper (Shaw, 2004).

I am grateful to Bill Margolis for e-mailing me to let me know that Theorem 10.3 with $a = 0$, was given as an exercise (Exercise 3 of Chapter III) by J. D'Angelo in his monograph (D'Angelo 2002). HB asked J. D'Angelo about his source and D'Angelo recalls discovering the formula for himself in 1988.

Moving a step further back into history, the book by R.P. Boas (Boas, 1987), has a statement and proof of the result and some other short-cut methods (see pages 158-164). One of these short-cut methods is particularly interesting from a historical point of view as its development is curiously interwined with the extraction theorem. The result is that, if one already knows $u(x, y)$ and $v(x, y)$ then

$$f(z) = u(z, 0) + i v(z, 0) \tag{10.105}$$

From my point of view this result is rather less interesting, as you do need to know *both* u and v, whereas the extraction result allows the construction of f from just *one* of u and v. However, the historical link is very interesting. Let's call Eq. (10.105) the 'alternative method' for getting $f(z)$. So far as I am aware, and according to E.V. Laitone (Laitone, 1977), this alternative method was first given by L.M. Milne-Thomson (Milne-Thomson, 1937) and it also appeared in the first edition of his 'Theoretical Hydrodynamics' (Milne-Thomson, 1938). It also appeared in the second (1949) and third (1955) editions. In the fourth (1962) and later editions Milne-Thomson dropped the alternative method and replaced it with the algebraic extraction theorem in the style of Ahlfors, and did so far as I can tell, without comment or attribution (see note 5.32 on pp. 130-131 of the fourth edition). The story of Milne-Thomson's contribution was discussed up by Laitone (1977). Laitone's emphasis was on the alternative method, and he describes the extraction theorem as 'a very dubious procedure that utilized the complex conjugate'!

All of these discussions refer to the theorem with $a = 0$, where the base point is the origin. Indeed, the note by Laitone (1977) goes on to say that 'Milne-Thomson's method (i.e. in his 4th ed) cannot be justified in general since it results in $(x^2 + y^2) = 0$ so that any $f(z)$ containing z^{-n} cannot be obtained from any combination of u and v'. So the issue, at least in this particular thread of papers, of how to cope with a general base-point, was not properly understood in 1977.

My best understanding, based on papers to which I have had access, is that the issue of the base point was resolved by R.A. Struble (Struble, 1979). It should be noted however that an outline discussion of the result with a general base point has also been given by Volkovyskii *et al* (1960). Struble gave a proof of what I called Theorem 10.3 based on Taylor series, and made the point that the identities 'should be better known than they appear to be'. He also made the explicit comment that (in the context of fluid dynamics, for which further discussion is given in Chapter 19) 'each stream function can be expressed algebraically in terms of the potential function, and vice versa.' On the basis of Ahlfors (1953) being the first to shed light on the matter, and Struble (1979) sorting out the general base point, I have called this result the Ahlfors–Struble theorem. However, this is only on the basis of information received by the author thus far! Some may view the contribution of Cartan (1961) as being more critical. It has also been suggested that the ideas are implicit in the work of I.N. Vekua from the 1940s. Work on polar representations has also been done (Huilgol, 1981; Shaw, 2004).

This, for now, is the end of the historical discussion. We now turn back to other things you can do with this result and *Mathematica*.

■ Checking the Laplacian

Note that as defined, neither of the functions described above care about whether the u or v you supply to them are the real or imaginary parts of a holomorphic function. They return results whatever you give them. For example,

```
RealToHolo[x^2 + y^2, 0, {x, y, z}]
```

$i\beta$

```
RealToHolo[x^2 + y^2, 1, {x, y, z}]
```

$2z + i\beta - 1$

Note the dependence on the base point. It is far from clear whether such transformations have any useful interpretation, and they certainly do not serve as an inverse to **ComplexExpand**. For example

```
ComplexExpand[Re[% /. {z → x + I y, β → 0}]]
```

$2x - 1$

So it is a good idea to include a check to exclude this case. In the following extended code we include an explicit check that the Laplacian operator gives zero. Note that this check involves (a) invoking **FullSimplify** to boil down the Laplacian as much as possible, having used **Together** to put things over a common denominator, (b) the use of the 'identically equals' within *Mathematica*, given by ===.

```
RealToHoloCheck[expr_,
  anum_, {xsym_, ysym_, zsym_ }] :=
 Module[{abar = Conjugate[anum], exprf,
   (*This may need improving to stop false negatives *)
   laplacian = FullSimplify[Together[
      D[expr, {xsym, 2}] + D[expr, {ysym, 2}]]]]},
  exprf = ComplexExpand[expr,
    TargetFunctions → {Re, Im}];
  If[laplacian === 0,
   func =
    2 * exprf /. {xsym → (zsym + abar) / 2,
      ysym → (zsym - abar) / (2 * I)}; basecorr =
    - exprf /. {xsym → Re[anum], ysym → Im[anum]};
   FullSimplify[func + basecorr + I * β],
   Print["This expression is not the
      real part of a holomorphic function -
      its Laplacian is "]; laplacian]]
```

```
ImToHoloCheck[expr_, anum_, {xsym_, ysym_, zsym_ }] :=
 Module[{abar = Conjugate[anum], exprf,
   (*This may need improving to stop false negatives *)
   laplacian = FullSimplify[Together[
      D[expr, {xsym, 2}] + D[expr, {ysym, 2}]]]]},
  exprf = ComplexExpand[expr,
    TargetFunctions → {Re, Im}];
  If[laplacian === 0,
   func =
    2 * I * exprf /. {xsym → (zsym + abar) / 2,
      ysym → (zsym - abar) / (2 * I)}; basecorr =
    -I * exprf /. {xsym → Re[anum], ysym → Im[anum]};
   FullSimplify[func + basecorr + β],
   Print["This expression is not the
      imaginary part of a holomorphic
      function - its Laplacian is "]; laplacian]]
```

Note that the output is a message together with the Laplacian if the check fails, so you do at least have the chance to do further manual simplifications of the Laplacian if you really think the input is harmonic, and then use the non-checking code to finalize matters.

```
ImToHoloCheck[(x^2 - y^2)^4, 1, {x, y, z}]
```

This expression is not the imaginary part of a holomorphic function – its Laplacian is

$$48\,(x^2-y^2)^2\,(x^2+y^2)$$

```
Map[RealToHoloCheck[#, 1, {x, y, z}] &, TestUSet]
```

$$\left\{z^2+i\,\beta,\, i\,(z^4+\beta-1),\, i\,\beta+e^z,\, i\,\beta+e^{\frac{1}{z}},\, i\,\beta+\log(z),\, i\,\beta+\sqrt{z}\right\}$$

```
Map[ImToHoloCheck[#, 1, {x, y, z}] &, TestVSet]
```

$$\left\{z^2+\beta-1,\, i\,z^4+\beta,\, \beta+e^z-e,\, \beta+e^{\frac{1}{z}}-e,\, \beta+\log(z),\, \beta+\sqrt{z}-1\right\}$$

Comments on how to improve these functions are very welcome, especially if you, the reader, find cases where the Laplacian check fails.

■ Finding harmonic conjugates

This is now a matter of sticking together our new function (now with the check) with **ComplexExpand.** We do not need to introduce a complex variable here in the arguments, so it is just used internally.

```
HarmonicConjugate[expr_, anum_, {xsym_, ysym_}] :=
 Module[{abar = Conjugate[anum], zsym,
   exprf, laplacian = FullSimplify[Together[
      D[expr, {xsym, 2}] + D[expr, {ysym, 2}]]]]},
  exprf = ComplexExpand[expr,
    TargetFunctions → {Re, Im}];
  If[laplacian === 0,
   func =
    2 * exprf /. {xsym → (zsym + abar) / 2,
      ysym → (zsym - abar) / (2 * I)}; basecorr =
    - exprf /. {xsym → Re[anum], ysym → Im[anum]};
   ComplexExpand[Im[FullSimplify[
      (func + basecorr + I * β) /. zsym → xsym + I * ysym]]],
   Print["This expression is not the real
      part of a holomorphic function -
      its Laplacian is "]; laplacian]]
```

```
ComplexExpand[Re[(x + I y)^4]]
```

$x^4 - 6\,y^2\,x^2 + y^4$

```
ComplexExpand[Im[(x + I y)^4]]
```

$4\,x^3\,y - 4\,x\,y^3$

```
HarmonicConjugate[p * (x^4 - 6 x^2 y^2 + y^4), 0, {x, y}]
```

$4\,p\,y\,x^3 - 4\,p\,y^3\,x + \beta$

```
Map[HarmonicConjugate[#, 1, {x, y}] &, TestUSet]
```

$$\Big\{2\,x\,y + \beta,\ x^4 - 6\,y^2\,x^2 + y^4 + \beta - 1,\ \beta + e^x \sin(y),$$
$$\beta - e^{\frac{x}{x^2+y^2}} \sin\!\left(\frac{y}{x^2+y^2}\right),\ \beta + \tan^{-1}(x, y),\ \beta + \sqrt[4]{x^2+y^2}\,\sin\!\left(\frac{1}{2}\tan^{-1}(x, y)\right)\Big\}$$

```
TableForm[Transpose[{TestUSet, %}]]
```

$x^2 - y^2$	$2\,x\,y + \beta$
$4\,x\,y\,(y^2 - x^2)$	$x^4 - 6\,y^2\,x^2 + y^4 + \beta - 1$
$e^x \cos(y)$	$\beta + e^x \sin(y)$
$e^{\frac{x}{x^2+y^2}} \cos\!\left(\frac{y}{x^2+y^2}\right)$	$\beta - e^{\frac{x}{x^2+y^2}} \sin\!\left(\frac{y}{x^2+y^2}\right)$
$\frac{1}{2}\log(x^2 + y^2)$	$\beta + \tan^{-1}(x, y)$
$\sqrt[4]{x^2+y^2}\,\cos(\frac{1}{2}\tan^{-1}(x, y))$	$\beta + \sqrt[4]{x^2+y^2}\,\sin(\frac{1}{2}\tan^{-1}(x, y))$

You can use other variables if you want:

```
HarmonicConjugate[a^4 + b^4 - 6 a^2 b^2, 0, {a, b}]
```

$4ba^3 - 4b^3 a + \beta$

You will not get anywhere if the function is not harmonic:

```
HarmonicConjugate[a^4 + b^4, 0, {a, b}]
```

This expression is not the real part of a holomorphic function – its Laplacian is

$12(a^2 + b^2)$

Exercises

10.1 Show that

$$f(z) = |z|^2$$

is differentiable as a mapping $\mathbb{R}^2 \to \mathbb{R}^2$ everywhere, and differentiable as a mapping $\mathbb{C} \to \mathbb{C}$ at the origin. Show further that the Cauchy–Riemann equations are satisfied at the origin, but nowhere else. Deduce that there is no open set $U \subset \mathbb{C}$ on which f is holomorphic.

10.2 (Harder!) Show that the function given by $f(0) = 0$, and otherwise

$$f(z) = \frac{z^5}{|z|^4}$$

satisfies the Cauchy–Riemann equations at $z = 0$ but is not holomorphic anywhere.

10.3 By taking real and imaginary parts, show that the Cauchy–Riemann equations are equivalent to

$$\frac{\partial f}{\partial \bar{z}} = 0$$

10.4 An 'anti-holomorphic' function f is defined by the condition

$$\frac{\partial f}{\partial z} = 0$$

Derive the analogue of the Cauchy–Riemann equations for such a function, by taking real and imaginary parts of this condition.

10.5 A 'double number' is defined as a formal pair

$$p = x + jy$$

subject to the multiplication rule obtained by requring that the symbol j satisfies

$$j^2 = +1$$

Derive the analogue of the Cauchy–Riemann equations for functions $f(p) = u(x, y) + j\, v(x, y)$ that are 'holomorphic' functions of such double numbers, and show, assuming as much differentiability as you require, that the 'real' and 'imaginary' parts u and v both satisfy the wave equation, e.g., for u,

$$\frac{\partial^2 u}{\partial x^2} = \frac{\partial^2 u}{\partial y^2}$$

10.6 Calculate the complex derivatives of the following functions, explaining which properties of complex differentiation you are using:

$$f(z) = 4z^4 + (3 + 2i)\,z^2 + (1 - i)z - 2i + 3$$

$$g(z) = \frac{1}{(z^2 + 4)^5}$$

$$h(z) = \frac{z^3 + 4}{z^2 - 7}$$

10.7 Where, if anywhere, are the functions f, g, h defined in Exercise 10.6 not differentiable?

10.8 By making direct use of the definition of differentiability, prove L'Hospital's rule: if $f(z)$ and $g(z)$ are differentiable at z_0 and $f(z_0) = g(z_0) = 0$, but $g'(z_0) \neq 0$, show that

$$\lim_{z \to z_0} \frac{f(z)}{g(z)} = \frac{f'(z_0)}{g'(z_0)}$$

10.9 By using the result of Exercise 10.8, evaluate

$$\lim_{z \to i} \frac{z^2 + 1}{z^6 + 1}$$

Note: in questions 10–12 you may assume that U is an open disc or neighbourhood. If you know the term *connected*, this will also suffice – it is defined in this book in Section 12.1. You might like to consider what happens, for example, if U is the disjoint union of two neighbourhoods.

10.10 If $f(z)$ is holomorphic in an open set U and is real-valued in U, show that f is constant.

10.11 If $f(z)$ and $\overline{f(z)}$ are both holomorphic in an open set U, show that f is constant.

10.12 If $f(z)$ is holomorphic in an open set U and $|f(z)|$ is constant, show that f is constant. (Hint: differentiate $|f(z)|^2$.)

10.13 If $f = u + iv$ is written as a function of polar coordinates (r, θ), show that the Cauchy–Riemann equations become:

$$\frac{\partial u}{\partial r} = \frac{1}{r}\frac{\partial v}{\partial \theta}$$
$$\frac{\partial v}{\partial r} = -\frac{1}{r}\frac{\partial u}{\partial \theta}$$

and hence derive the Laplace equation in polar form:

$$\frac{\partial^2 u}{\partial r^2} + \frac{1}{r}\frac{\partial u}{\partial r} + \frac{1}{r^2}\frac{\partial^2 u}{\partial \theta^2} = 0$$

and similarly for v.

10.14 What are the real and imaginary parts of the function

$$i \log(z)$$

expressed in terms of Cartesian coordinates? Check that these satisfy Laplace's equation. Repeat the exercise with the real and imaginary parts expressed in polar coordinates.

10.15 Show *in two different ways* that the functions

$$r^n \cos(n\,\theta) \text{ and } r^n \sin(n\,\theta)$$

satisfy Laplace's equation.

10.16 (Harder) Show that the function f given by $f(0, 0) = 0$, and otherwise

$$f(x, y) = \frac{x^3}{x^2 + y^2}$$

is continuous at $(0, 0)$, and, by considering the function

$$g(s) = f(s\cos(t), s\sin(t))$$

for fixed t, show that g is differentiable at $s = 0$ for all t. By establishing this you are showing that f is differentiable in a 'one-dimensional' sense along any line through the origin. Show, nevertheless, that f is not differentiable. (Hint: try writing f in approximate linear form, using the partial derivatives as coefficients, and see what goes wrong.)

10.17 Given the following functions u, find the corresponding imaginary functions v and hence find the holomorphic $f(z)$ whose real part is u:

$$u(x, y) = 4\,x\,y\,(y^2 - x^2)$$

$$u(x, y) = e^{x/x^2+y^2} \cos\left(\frac{y}{x^2 + y^2}\right)$$

$$u(x, y) = \sqrt[4]{x^2 + y^2} \cos\left(\frac{1}{2} \tan^{-1}\left(\frac{y}{x}\right)\right)$$

10.18 ❋ Produce a surface plot (i.e. use **`Plot3D`**) of the function discussed in Exercise 10.1.

10.19 ❁ Produce a surface plot (i.e. use **Plot3D**) of the function discussed in Exercise 10.3.

10.20 ❁ Use *Mathematica*'s **D** and **Simplify** functions to calculate the derivatives of the functions defined in Exercise 10.7, and compare with your pen and paper calculations.

10.21 ❁ Use the *Mathematica* function **ContourPlot**, with the option **ContourShading -> False**, to visualize and overlay the contours of the following pairs of functions:

$$u = x^2 - y^2;\ v = 2xy$$

$$u = \frac{1}{2}\log(x^2 + y^2);\ v = \tan^{-1}(x, y)$$

To what holomorphic function do each of these function pairs correspond? (Hint: You may need to increase the value of **PlotPoints** to obtain a good plot, and you should note for the last example that *Mathematica* has a special **ArcTan** function that accounts for the precise location of x and y in the complex plane.)

10.22 ❁ Use *Mathematica*'s **D** and **Simplify** functions to show that each of the four functions defined in Exercise 10.19 satisfy Laplace's equation.

10.23 ❁ Use *Mathematica*'s **Limit** function to evaluate

$$\lim_{z \to i} \frac{z^2 + 1}{z^6 + 1}$$

(Use **?Limit** to find out how this function works if you need to.)

10.24 ❁ Use *Mathematica* to check your results for Exercise 10.17, including a check that the given examples of u are harmonic.

10.25 ❁ Use *Mathematica* to find the harmonic conjugates of the examples of u given in Exercise 10.17.

11 Paths and complex integration

Introduction

In this chapter we introduce the concept of integration of a complex function of a complex variable. This relies on understanding what is meant by a 'path' in the complex plane, because integrals are defined with respect to a path. This chapter is mostly concerned with defining integration by referring back to the real case. You will also see how to generalize the fundamental theorem of calculus to the complex setting. Throughout this chapter we will refer to complex functions of a real variable, and refer to them as being, respectively, continuous or differentiable. All this means is that the real and imaginary parts are, respectively, continuous or differentiable in the ordinary real sense.

11.1 Paths

A *path* ϕ is a continuous mapping from a segment of the real axis into the complex numbers:

$$\phi : [a, b] \subset \mathbb{R} \to \mathbb{C} \tag{11.1}$$

The square brackets indicate that the domain of definition of ϕ contains end-points, i.e. $\phi(x)$ is defined for

$$a \leqslant x \leqslant b \tag{11.2}$$

A path is said to be *simple* if it is 1:1 as a mapping. That is, $\phi(x) = \phi(y) \Rightarrow x = y$. It is said to be *closed* if $\phi(a) = \phi(b)$. If ϕ is closed, and is simple except at a and b, ϕ is a *simple closed* path. As a point of notation, the image of the path is denoted ϕ^*. Thus

$$\phi^* = \{\phi(t) : a \leqslant t \leqslant b\} \tag{11.3}$$

A path ϕ is a *differentiable path* if its real and imaginary parts are differentiable in the usual sense for $t \in (a, b)$, i.e. $a < t < b$ and one-sided derivatives exist at the end-points a and b.

▪ Theorem 11.1: Mean value theorem for the modulus

Let $\phi : [a, b] \to \mathbb{C}$ be differentiable, then

$$|\phi(b) - \phi(a)| \leqslant (b - a) \sup_{t \in (a,b)} |\phi'(t)| \tag{11.4}$$

Remark 1: If you have not taken a formal course in real analysis, you can think of sup, short for *supremum*, as being the maximum value along the path. More formally, the supremum S of a set of reals A is the smallest number such that $x \in A \Rightarrow x \leqslant S$.

Remark 2: Note that the 'obvious' generalization, to complex numbers, of the real mean value theorem

$$\phi(b) - \phi(a) = (b - a)\ \phi'(t) \tag{11.5}$$

for some $t \in (a, b)$ is simply untrue. You might like to check this by considering the path $\phi(t) = e^{it}$ for $0 \leqslant t \leqslant 2\pi$.

Proof of theorem: Suppose that $\phi(b) \neq \phi(a)$, for otherwise there is nothing to prove, and let $\mu = \phi(b) - \phi(a)$, $\psi(t) = \overline{\mu}\,\phi(t)$. Clearly, $\psi'(t) = \overline{\mu}\,\phi'(t)$, and

$$\psi(b) - \psi(a) = \mu\,\overline{\mu} = |\phi(b) - \phi(a)|^2 \tag{11.6}$$

which is real. So if we write $\psi(t) = \psi_1(t) + i\,\psi_2(t)$,

$$|\phi(b) - \phi(a)|^2 = \psi_1(b) - \psi_1(a) = (b - a)\,\psi_1'(c) \tag{11.7}$$

for some $c \in (a, b)$, by the real mean value theorem. But

$$(b - a)\,\psi_1'(c) \leqslant (b - a)\,|\,\overline{\mu}\,||\,\phi'(c)\,| \tag{11.8}$$

and so

$$|\phi(b) - \phi(a)|^2 \leqslant (b - a)\,|\,\overline{\mu}\,||\,\phi'(c)\,| \tag{11.9}$$

Now divide the result by $|\,\overline{\mu}\,|$ to obtain

$$|\phi(b) - \phi(a)| \leqslant (b - a)\,|\,\phi'(c)\,| \leqslant (b - a) \sup\, |\phi'(t)| \tag{11.10}$$

■ Smooth and piecewise smooth paths

A path ϕ is said to be *smooth* if it is a differentiable path, and furthermore, the derivative map: $\phi' : [a, b] \to \mathbb{C}$ is continuous. So in this case the tangent to the path is a continuous function. It is frequently useful to consider paths that consist of several smooth pieces, but where there may be finitely many points where the tangent is discontinuous, that is, the path may have corners. Formally, a path is piecewise smooth if there is a finite set of real numbers, $a_0, a_1, \ldots, a_n$, with $a = a_0 < a_1 < a_2 < \ldots < a_n = b$, such that

$$\phi : [a_{i-1}, a_i] \to \mathbb{C} \tag{11.11}$$

is smooth, for $i = 1 \ldots n$. A simple example of a piecewise smooth path would be a polygonal path, consisting of just a collection of straight lines. Such a path is piecewise linear. Another word for a piecewise smooth path is a *contour*.

■ Drawing piecewise linear paths in *Mathematica*

It is useful to define a standard form representing a piecewise smooth path consisting of a bunch of straight lines, which you can use to both draw the path and carry out integration:

```
PlotPath[path_ , options___] :=
 Module[{argand = Map[{Re[#], Im[#]} &, path]},
  grdata = {Line[argand],
    {PointSize[0.03], Map[Point, argand]}};
Show[Graphics[grdata, AspectRatio -> 1,
    Axes -> True, options]]]
```

```
psp = {1, I, -1, -I}; PlotPath[psp]
```

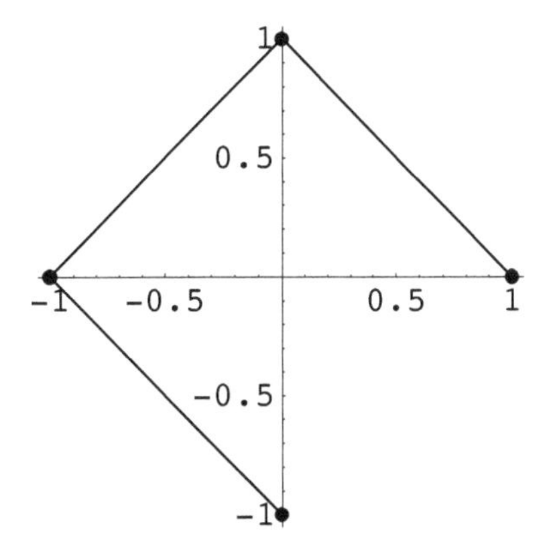


```
cpsp = {1, I, -1, -I, 1}; PlotPath[cpsp]
```

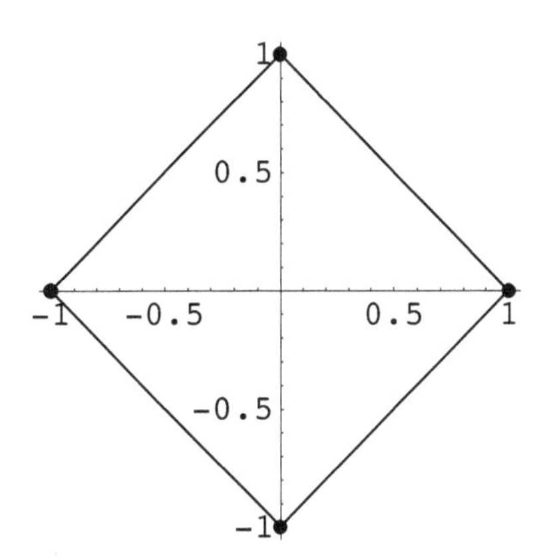

Here is a piecewise linear path that gives a polygon that tends to a circle as its argument n gets large:

```
path[n_] := Table[Exp[2 * I * Pi * k / n], {k, 0, n}]
```

```
PlotPath[path[30]]
```

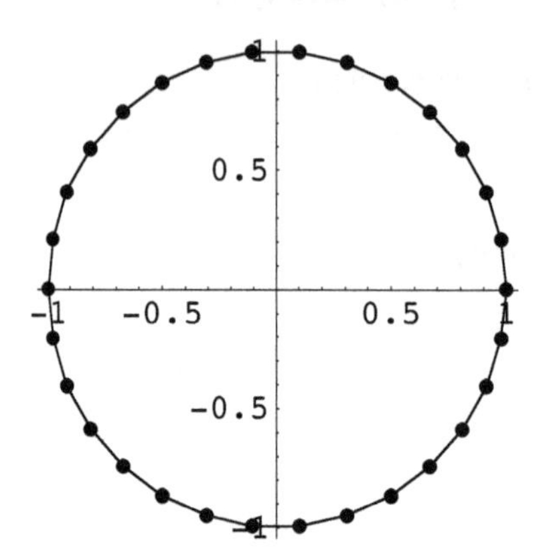

11.2 Contour integration

Let U be an open subset of $\mathbb{C}$ and let $\phi : [a,\ b] \to U$ be a smooth path. Let $f : U \to \mathbb{C}$ be continuous. Then the integral of f over the path ϕ is defined by the following equation. Note the notation – the function defining the path appears where the lower limit might usually be placed in the real case:

$$\int_\phi f(z)\, dz = \int_a^b f[\phi(t)]\, \phi'(t)\, dt \tag{11.12}$$

The real and imaginary parts of the integrand are now manifestly continuous functions of t, and so are integrable. Note that this looks a great deal like the ordinary substitution rule for changing variables in real integration. Indeed, this is a good way of remembering the formula! But here it is a *definition* of the integral over the path. Also, in complex analysis, we often tend to refer to integration over 'contours'. Remember - you should think of the word contour as synonymous with 'piecewise smooth path'. For a piecewise smooth path you just add up the integrals associated with each smooth piece.

■ Powers of z integrated around a circle centred on the origin

The best way of getting a grip on the definition of integration is to do a simple example. The example that will follow is at the same time one of the simplest and one of the most important that we can give. Let $f(z) = z^n$ and let $\phi(t) = r e^{it}$, for $t \in [0,\ 2\pi]$. The integer n can be positive, negative or zero. Then

$$f(\phi(t)) = r^n\, e^{int} \tag{11.13}$$

$$\phi'(t) = i r e^{it} \tag{11.14}$$

$$\int_\phi f(z)\, dz = i \int_0^{2\pi} r^{n+1}\, e^{i(n+1)t}\, dt \tag{11.15}$$

But the integral on the right side is zero unless $n = -1$, because the exponential function is periodic, with period 2π, and the values of the indefinite integral at the limits of integration cancel. When $n = -1$, the integral is $i \times 2\pi$. So around any circle centred on the origin, parametrized by the function ϕ,

$$\int_\phi \frac{1}{z}\, dz = 2\pi i \tag{11.16}$$

The fact that this is non-zero, while all other powers give zero, is surprisingly important, as you will see later.

11.3 Theorem 11.2: The fundamental theorem of calculus

This is stated as follows. *Let U be an open subset of* $\mathbb{C}$ *and let* $\phi : [a, b] \to U$ *be a contour (piecewise smooth path). Let* $f : U \to \mathbb{C}$ *be continuous. Suppose that there is a function* $\;: U \to \mathbb{C}$ *that is holomorphic on U* and $F'(z) = f(z)$. *Then*

$$\int_\phi f(z)\, dz = F(\phi(b)) - F(\phi(a)) \tag{11.17}$$

Proof: First note that we just need to consider smooth paths, as the result for piecewise smooth paths follows by adding up the pieces, and noting cancellations of intermediate terms (you might like to check this!). The rest follows by reducing the problem to the corresponding real fundamental theorem of calculus, applied to the real and imaginary parts of the second line of the following:

$$\begin{aligned} \int_\phi f(z)\, dz &\equiv \int_a^b f[\phi(t)]\, \phi'(t)\, dt = \int_a^b F'[\phi(t)]\, \phi'(t)\, dt \\ &= \int_a^b \frac{d}{dt} F[\phi(t)]\, dt = F(\phi(b)) - F(\phi(a)) \end{aligned} \tag{11.18}$$

There is one very important consequence: if there is such a function and ϕ is closed, so that $\phi(b) = \phi(a)$, then

$$\int_\phi f(z)\, dz = 0 \tag{11.19}$$

■ The trouble with $1/z$

We have already noted that

$$\int \frac{1}{z}\, dz = 2\pi i \neq 0 \tag{11.20}$$

when the path is any circle centred on the origin. It follows that there is no holomorphic function F defined on $\mathbb{C} - \{0\}$ with the property that $F'(z) = 1/z$. This is related to the many-valued nature of the logarithm function, and you will see this discussed more fully in the next chapter.

11.4 The value and length inequalities

There are two inequalities that are fundamental to further analysis of integration. The *value inequality* states that *if $g(t)$ is a continuous complex-valued function of the real variable t, then*

$$\left|\int_a^b g(t)\,dt\right| \leq \int_a^b |g(t)|\,dt \tag{11.21}$$

The proof is easy, for by setting

$$\int_a^b g(t)\,dt = R\,e^{i\theta} \tag{11.22}$$

where R is real and positive, we have, using the fact that $\text{Re}[z] \leqslant |z|$,

$$\begin{aligned}\left|\int_a^b g(t)\,dt\right| &= R = \int_a^b e^{-i\theta}\,g(t)\,dt = \int_a^b \text{Re}[e^{-i\theta}\,g(t)]\,dt \\ &\leqslant \int_a^b |e^{-i\theta}\,g(t)|\,dt = \int_a^b |g(t)|\,dt\end{aligned} \tag{11.23}$$

Next, the *length* of a smooth path ϕ is defined to be

$$L(\phi) = \int_a^b |\phi'(t)|\,dt \tag{11.24}$$

The length is the sum of several such terms if the path is piecewise smooth. There is an 'almost obvious' inequality that relates the size of an integral to the length of the path and a bound on the function being integrated. The length inequality is the following theorem: *Let*

$$M_f = \sup\{|f(z)| : z \in \phi^*\} \tag{11.25}$$

That is, M_f is the smallest real number such that

$$|f(z)| \leq M_f \tag{11.26}$$

for all $z \in \phi^*$. Then

$$\left|\int_\phi f(z)\,dz\right| \leq M_f\,L(\phi) \tag{11.27}$$

Proof: By the value inequality, and the definition of the path length:

$$\left|\int_{\phi} f(z)\,dz\right| = \left|\int_a^b f(\phi(t))\,\phi'(t)\,dt\right| \\ \leqslant \int_a^b |f(\phi(t))|\,|\phi'(t)|\,dt \leqslant \int_a^b M_f\,|\phi'(t)|\,dt = M_f\,L(\phi) \tag{11.28}$$

11.5 Theorem 11.3: Uniform convergence and integration

Now that we have a definition of integration, and are armed with the length inequality, a proper statement and proof about the integral of a uniformly convergent series of functions can be given. We have the following theorem. *Suppose that the sequence of functions $f_n(z)$ is continuous on a set U and that the series*

$$S(z) = \sum_{n=1}^{\infty} f_n(z) \tag{11.29}$$

is uniformly convergent. Then the series can be integrated term by term, over a piece-wise smooth path $\phi \in U$, i.e.

$$\int_{\phi} S(z)\,dz = \sum_{n=1}^{\infty} \int_{\phi} f_n(z)\,dz \tag{11.30}$$

Proof: Write the sum $S(z)$ as a partial sum and a remainder:

$$S(z) = \sum_{n=1}^{N} f_n(z) + \sum_{n=N+1}^{\infty} f_n(z) = S_n(z) + R_n(z) \tag{11.31}$$

Since $S(z)$ is continuous, so are $S_n(z)$ (it is a finite sum of continuous functions) and $R_n(z)$ (it is then the difference of a pair of continuous functions), so they can be integrated. Furthermore, we can expand out the integral in S_n, since it consists of finitely many terms:

$$\int_{\phi} S(z)\,dz = \int_{\phi} S_n(z)\,dz + \int_{\phi} R_n(z)\,dz = \sum_{n=1}^{N} \int_{\phi} f_n(z)\,dz + \int_{\phi} R_n(z)\,dz \tag{11.32}$$

Since the series is uniformly convergent, given any $\epsilon > 0$, we can find an integer N, independent of z, such that $n > N \Rightarrow |R_n(z)| < \epsilon$. Now let L be the length of ϕ, then, by the length inequality, we have

$$\left|\int_{\phi} R_n(z)\,dz\right| < \epsilon L \tag{11.33}$$

So

$$\left|\int_{\phi} S(z)\,dz - \sum_{n=1}^{N} \int_{\phi} f_n(z)\,dz\right| \tag{11.34}$$

can be made as small as we like, by taking N large enough. Hence the result. Theorem 11.2 will be used elsewhere. Note that uniform convergence is necessary, as the examples given in Chapter 9 made clear. A weaker statement of the theorem can also be given by mimicking the proof of the Weierstrass M-test and assuming some bounds on f_n, without mentioning the words 'uniform convergence' at all, but the statement and proof within the framework of uniform convergence are by far the cleanest.

11.6 ⚙ Contour integration and its perils in *Mathematica!*

Recall our definition of two piecewise linear paths in *Mathematica,* given by **psp** and **cpsp**:

```
{psp, cpsp}
```

$\{\{1, i, -1, -i\}, \{1, i, -1, -i, 1\}\}$

These lists can also be used to supply information to *Mathematica*'s contour integration routine. In the complex plane, the *Mathematica* functions **Integrate** and **NIntegrate** expect to receive a list consisting of the variable of integration and points on the contour. This makes use of the **Prepend** function

```
? Prepend
```

```
Prepend[expr, elem] gives expr with elem prepended.
```

```
ContourIntegral[expr_, vbl_, contour_] :=
    Integrate[expr, Prepend[contour, vbl]]
```

```
NContourIntegral[expr_, vbl_, contour_] :=
    NIntegrate[expr, Evaluate[ Prepend[contour, vbl]]]
```

```
NContourIntegral[1 / z, z, cpsp]
```

$1.11022\times 10^{-16} + 6.28319\,i$

```
Chop[%]
```

$6.28319\,i$

```
% / (2 Pi I)
```

1.

Here's another integral with the same result:

```
Chop[NContourIntegral[Sin[z] / z^2, z, cpsp]]
```

$6.28319\,i$

Imortant Warning: this type of integration is most reliably done by doing the calculation both symbolically *and numerically* in *Mathematica*, so that the path is explicitly fed in to the numerical integrator. When using symbolic integration, *Mathematica* will attempt to find a definite integral and apply the fundamental theorem of calculus to the end points of the contour. For closed contours where the integrand has a singularity within the contour, this can lead to problems, especially in older versions of *Mathematica*, such as the following pair, neither of which are correct:

```
N[ContourIntegral[Sin[z] / z^2, z, cpsp]]
```

$0. + 0\,i$

```
N[ContourIntegral[1 / z, z, cpsp]]
```

0.

In version 5, these do return the correct results, both numerically and symbolically.

```
N[ContourIntegral[Sin[z] / z^2, z, cpsp]]
```

$0. + 6.28319\,i$

```
N[ContourIntegral[1 / z, z, cpsp]]
```

$0. + 6.28319\,i$

```
ContourIntegral[Sin[z] / z^2, z, cpsp]
```

$-\text{Ci}(-1)^* + \text{Ci}(-1)$

```
ContourIntegral[1 / z, z, cpsp]
```

$2\,i\,\pi$

The lesson of this is to keep up with version upgrades, but always check the numerics against the analytics!

Exercises

11.1 Evaluate the integral

$$\int \bar{z}\, dz$$

over a circle of radius r centred on the origin. Is the integrand a holomorphic function of z?

11.2 Construct a parametrization of the perimeter of the square with vertices $1 + i,\ -1 + i,\ -1 - i,\ 1 - i$. Calculate the length of this path using the definition given in Section 11.4. Does this agree with your intuition? Evaluate the integral of z and of $\bar{z}$ around this path.

11.3 What is the geometrical interpretation of the closed path given by

$$z = \phi(\theta) = a\cos\theta + b\,i\sin\theta$$

for $0 \le \theta \le 2\pi$? Evaluate the integrals

$$\int \bar{z}\,dz \text{ and } \int z\,dz$$

over this path.

11.4 Evaluate the integral of $3z^2 + 2$ over each of the following paths:
(i) The semi-circular arc of radius 1 from $+1$ to -1 in the upper half-plane;
(ii) The straight line from $+1$ to -1.
(iii) The union of the three line segments from (a) $+1$ to $+1+i$, (b) from $+1+i$ to $-1+i$, (c) from $-1+i$ to -1.
What do you notice about these three answers? By considering the function $z^3 + 2z$ relate your answers to the Fundamental Theorem of Calculus.

11.5 Let $P(z)$ be a polynomial of degree $n \ge 2$ with leading term $a_0 z^n$. By writing $P(z)$ in the form

$$P(z) = a_0 z^n Q(z)$$

show that for $|z|$ large enough we can ensure that

$$|P(z)| \ge \frac{1}{2} |a_0 z^n|$$

Now let Γ_R be the semicircle in the upper half-plane extending from $z = R$ to $z = -R$. Use the length inequality to show that as $R \to \infty$,

$$\int_{\Gamma_R} \frac{1}{P(z)}\,dz \to 0$$

11.6 Let Γ be a path from the point $z_1 = 2$ to the point $z_2 = i$. Using the Fundamental Theorem of Calculus, calculate each of the following integrals:

$$\int_\Gamma \sin z\,dz$$

$$\int_\Gamma e^z \sin z\,dz$$

$$\int_\Gamma \cos^2(z)\sin z\,dz$$

11.7 Let $P(z)$ be any polynomial of degree n. Using the Fundamental Theroem of Calculus, show that

$$\int_{\Gamma} P(z)\, dz = 0$$

for any piecewise smooth *closed* path Γ.

11.8 ✵ Use *Mathematica* to evaluate the integral of z over the square with vertices $1+i$, $-1+i$, $-1-i$, $1-i$. How does the result compare with your answer from exercise 11.2?

11.9 ✵ Use *Mathematica* to evaluate the integral of $3z^2+2$ over each of the following polygonal paths:

(i) The straight line from $+1$ to -1.

(ii) The union of the three line segments from (a) $+1$ to $+1+i$, (b) from $+1+i$ to $-1+i$, (c) from $-1+i$ to -1.

How do your answers compare with those you calculated for Exercise 11.4?

11.10 ✵ Use *Mathematica* to calculate each of the integrals given in Exercise 11.6.

11.11 ✵ Recall the following definitions:

```
path[n_] := Table[Exp[2 * I * Pi * k / n], {k, 0, n}]

NContourIntegral[expr_, vbl_, contour_] :=
    NIntegrate[expr, Evaluate[ Prepend[contour, vbl]]]

ContourIntegral[expr_, vbl_, contour_] :=
    Integrate[expr, Prepend[contour, vbl]]
```

Calculate the values of

```
ContourIntegral[z, z, path[n]]
```

and

```
NContourIntegral[Conjugate[z], z, path[n]]
```

for a variety of increasing values of n. Useful starting choices might be $n = 5,\ 20,\ 100$. To what value does the latter appear to converge as n becomes very large?

12 Cauchy's theorem

Introduction

In this chapter we discuss the integration of holomorphic functions around closed paths. The main result, Cauchy's theorem, is that such integrals *vanish* when the function is holomorphic on and inside the path. This superficially null result has astonishing consequences when fully developed – in particular it will ultimately focus our attention on functions with singularities (i.e. functions that fail to be holomorphic at certain points), in order to get interesting results from integration. First, however, we need to explain Cauchy's theorem. There are several levels of explanation of this result, depending on how general one wishes to make the conditions of the theorem. At one level, the theorem can be explained by reference to Green's theorem in the plane (the planar form of Stokes' theorem), from elementary vector calculus, and this will be considered first. But the theorem can be stated and proved using fewer assumptions than are involved in the Stokes'/Green's theorems, and we shall consider such a case also.

The first part of this chapter assumes a basic familiarity with real line integrals of the form

$$\int A\,dx + B\,dy \tag{12.1}$$

defined on suitable curves in two real dimensions.

12.1 Green's theorem and the weak Cauchy theorem

To state these theorems properly we need a few topological definitions. Recall first that a 'contour' is shorthand for a piecewise-smooth path.

■ Topological preliminaries

An open set U is said to be *connected* if every pair of points in U can be joined by a polygonal path that lies entirely in U. An open connected set is called a *domain*. A *simply connected domain* D is a domain with the property that if C is any simple closed contour lying in D, then the domain interior to C is entirely within D. We may visualize simply connected domains as being open sets of the complex plane consisting of a single piece with no holes. A contour C with domain interior to C, called D', is understood to be 'traversed in the positive sense' if a hypothetical observer travelling along C always has D' to her left.

■ Theorem 12.1: Green's theorem in the plane

Let $A(x, y)$ and $B(x, y)$ be continuous and have continuous partial derivatives in a simply connected domain R of $\mathbb{C}$. Let C be a simple closed contour within R that is traversed in the positive sense. Furthermore, let D' be the domain interior to C. Then the following real line integral

$$\int_C A\,dx + B\,dy \tag{12.2}$$

obtained by traversing C in a positive sense, is equal to the following area integral over D':

$$\int_{D'} \left(\frac{\partial B}{\partial x} - \frac{\partial A}{\partial y}\right) dx\,dy \tag{12.3}$$

We take this result from elementary vector calculus, texts on which should be consulted for a proof.

■ Theorem 12.2: Green's theorem for a complex function

Suppose that a complex function $f(z) = u(x, y) + iv(x, y)$ is integrated over such a curve C, traversed in the positive sense. Then we can rewrite the left side of Green's theorem as follows.

$$\int_C f(z)\,dz = \int_C (u + iv)\,(dx + i\,dy) = \int_C (u + iv)\,dx + (-v + iu)\,dy \tag{12.4}$$

We set $A = u + iv$, $B = -v + iu$, and so, by Green's theorem applied to the real and imaginary parts, this is equal to

$$\begin{aligned} &\int_{D'} \left(\frac{\partial(-v + iu)}{\partial x} - \frac{\partial\,(u + iv)}{\partial y}\right) dx\,dy \\ &= i \int_{D'} \left(\frac{\partial(iv + u)}{\partial x} + i\frac{\partial\,(u + iv)}{\partial y}\right) dx\,dy \\ &= i \int_{D'} \left(\frac{\partial}{\partial x} + i\frac{\partial}{\partial y}\right)(u + iv)\,dx\,dy = 2i \int_{D'} \frac{\partial f}{\partial \bar{z}}\,dx\,dy \end{aligned} \tag{12.5}$$

where we have used the notation introduced in Chapter 10. So, to summarize, if $f(z)$ is continuous with continuous partial derivatives, then

$$\int_C f(z)\,dz = 2i \int_{D'} \frac{\partial f}{\partial \bar{z}}\,dx\,dy \tag{12.6}$$

■ Theorem 12.3: The weak form of Cauchy's theorem

Suppose that C, D′ and f(z) satisfy the conditions of Green's theorem and, furthermore, f(z) is holomophic. Since

$$\frac{\partial f}{\partial \bar{z}} = 0 \tag{12.7}$$

it follows that

$$\int_C f(z)\,dz = 0 \tag{12.8}$$

12.2 The Cauchy–Goursat theorem for a triangle

The form of Cauchy's theorem that we have stated, based on Green's theorem, requires that $f'(z)$ exists and is continuous. However, the requirement of continuity on the derivative is superfluous, and one can give a proof without this assumption. This stronger form is often called the Cauchy–Goursat theorem.

■ Relation to Green's theorem

The relationship of the Cauchy–Goursat theorem to the version based on Green's theorem is rather subtle. It will turn out that one can show that if a function is holomorphic, then it is in fact infinitely differentiable! This means that if a function is holomorphic, its derivatives are necessarily continuous. So why can't one use the version based on Green's theorem? The answer is that the result, that a function is infinitely differentiable provided it is holomorphic, and in particular, that the first derivatives are continuous, is a *consequence* of the full Cauchy–Goursat result. So the Cauchy–Goursat theorem is a stronger result than Green's theorem, which requires continuity of the first patial derivatives.

We shall now consider the proof of the stronger form of Cauchy's theorem. Note that we have not actually given a proof of Green's theorem – most elementary texts on vector calculus give one, and it is superseded by the following argument. A common proof of Green's theorem really just involves integrating the continuous functions $\partial B/\partial x$ and $\partial A/\partial y$, given in the statement of the theorem, along horizontal and vertical strips within the area of interest, and applying the real fundamental theorem of calculus. The proof below is a much more interesting exercise. We do it first for triangular contours, and then we elevate this result to one for more general 'star-shaped' regions.

■ Theorem 12.4: Cauchy–Goursat theorem for triangles

First we need some notation. Suppose that $a,\ b,\ c \in \mathbb{C}$. The triangle $T_{a,b,c}$ is defined by

$$T_{a,b,c} = \{z : z = a\lambda_1 + b\lambda_2 + c\lambda_3,\ \lambda_i \geq 0,\ \lambda_1 + \lambda_2 + \lambda_3 = 1\} \tag{12.9}$$

Suppose that $f: U \to \mathbb{C}$ is holomorphic on the open set U, and that $T_{a,b,c} \subset U$. This triangle has a boundary given by the piecewise smooth path

$$\phi_{a,b,c}(t) = \begin{cases} a + t(b-a) & 0 \leqslant t \leqslant 1 \\ b + (t-1)(c-b) & 1 \leqslant t \leqslant 2 \\ c + (t-2)(a-c) & 2 \leqslant t \leqslant 3 \end{cases} \tag{12.10}$$

Then the Cauchy–Goursat Theorem for a triangle states that, *given that* $f: U \to \mathbb{C}$ *is holomorphic on the open set* U, *with* $T_{a,b,c} \subset U$,

$$\int_{\phi_{a,b,c}} f(z)\, dz = 0 \tag{12.11}$$

■ Proof by contradiction

To prove this result, we assume that the result is not true, so we set

$$J = \int_{\phi_{a,b,c}} f(z)\, dz \neq 0 \tag{12.12}$$

and we aim to show that this leads to a contradiction. To prove this first requires some basic geometry. The length of the path $\phi_{a,b,c}$ is

$$L = |b-a| + |c-b| + |a-c| \tag{12.13}$$

We define the mid-points of the three sides by

$$a_1 = \frac{b+c}{2}; \; b_1 = \frac{a+c}{2}; \; c_1 = \frac{a+b}{2} \tag{12.14}$$

In this way the original triangle is split into four similar triangles, as shown below (the *Mathematica* routines for generating **plota**, etc., are given at the end of this chapter):

```
Show[plota]
```

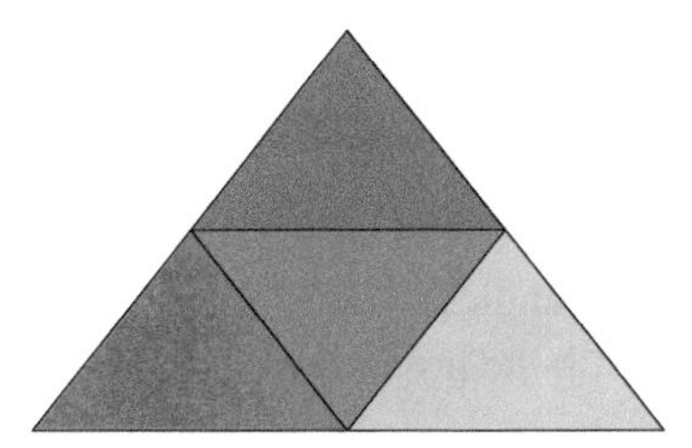

Now consider adding up the four contour integrals obtained by proceeding around each sub-triangle in a positive sense. Since there are two integrals around each interior line, one in each direction, these integrals over interior lines cancel, leaving just the integrals around the exterior lines. Hence we can write

$$
\begin{aligned}
J &= \int_{\phi_{a,b,c}} f(z)\,dz \\
&= \int_{\phi_{a,c_1,b_1}} f(z)\,dz + \int_{\phi_{c_1,b,a_1}} f(z)\,dz + \int_{\phi_{a_1,c,b_1}} f(z)\,dz + \int_{\phi_{a_1,b_1,c_1}} f(z)\,dz
\end{aligned}
\tag{12.15}
$$

By appplying the triangle inequality to this expression, and picking the sub-integral that is largest in magnitude – we call its boundary points $a^{(1)}$, $b^{(1)}$, $c^{(1)}$ – we have constructed a sub-triangle $T_{a^{(1)},b^{(1)},c^{(1)}}$ satisfying the following three conditions:

$$T_{a^{(1)},b^{(1)},c^{(1)}} \subset T_{a,b,c} \tag{12.16}$$

$$\left| \int_{\phi_{a^{(1)},b^{(1)},c^{(1)}}} f(z)\,dz \right| \geqslant \frac{|J|}{4} \tag{12.17}$$

$$L\,\phi_{a^{(1)},b^{(1)},c^{(1)}} = \frac{L}{2} \tag{12.18}$$

We can repeat this process, subdividing more finely:

```
Show[GraphicsArray[{{plotb, plotc}, {plotd, plote}}]]
```

In each case we pick the triangle where the integral is that of the four obtained with the largest magnitude. This defines a sequence of triangles $T_n = T_{a^{(n)},b^{(n)},c^{(n)}}$ with boundaries given by the images ϕ_n^* of the mappings $\phi_n(t)$, with the properties that

$$T_{a^{(n)},b^{(n)},c^{(n)}} \subset T_{a^{(n-1)},b^{(n-1)},c^{(n-1)}} \tag{12.19}$$

$$\left| \int_{\phi_{a^{(n)},b^{(n)},c^{(n)}}} f(z)\,dz \right| \geqslant \frac{|J|}{4^n} \tag{12.20}$$

$$L\,\phi_{a^{(n)},b^{(n)},c^{(n)}} = \frac{L}{2^n} \tag{12.21}$$

This sequence of triangles has a limiting point contained in all the triangles of the sequence. This can most easily be seen by exploiting the Cauchy convergence criteria discussed in Chapter 9, Section 9.2. Note first that if $n < p < q$, and $\alpha_q \in T_q$, then $\alpha_q \in T_p \subset T_n$, so that if $\alpha_q \in T_q$ also, the distance between α_q and α_p satisfies

$$|\alpha_p - \alpha_q| \leqslant L\,\phi_n = \frac{L}{2^n} \tag{12.22}$$

The same holds if $n < q < p$. So α_p is a Cauchy sequence, and there must be a limit point α such that $\alpha_p \to \alpha$. Furthermore, note that

$$|\alpha_p - \alpha| \leqslant L\,\phi_p = \frac{L}{2^p} \tag{12.23}$$

Finally the fact that the function $f(z)$ is differentiable at α is used to obtain a contradiction. Note that the differentiability condition can be written as

$$\begin{aligned} f(z) &= f(\alpha) + (z-\alpha)\,f'(\alpha) + \sigma(z-\alpha) \\ &= \frac{\partial}{\partial z}\left(zf(\alpha) + \frac{1}{2}f'(\alpha)(z-\alpha)^2\right) + \sigma(z-\alpha) \end{aligned} \tag{12.24}$$

so we deduce, from the Fundamental Theorem of Calculus, that when we integrate f around the closed triangle, the first two terms on the right side of this last equation integrate to zero, leaving us with

$$\int_{\phi_n} f(z)\,dz = \int_{\phi_n} \sigma(z-\alpha)\,dz \tag{12.25}$$

Now we apply the length inequality

$$\left|\int_{\phi_n} f(z)\,dz\right| = \left|\int_{\phi_n} \sigma(z-\alpha)\,dz\right| \leqslant L\,\phi_n \sup_{z\in\phi_n^*}|\sigma(z-\alpha)| \tag{12.26}$$

Now, given $\epsilon > 0$, there is a $\delta > 0$ such that if $|z-\alpha| < \delta$ then $|\sigma(z-\alpha)| < \epsilon\,|z-\alpha|$. Next pick n such that $L/2^n < \delta$:

$$\left|\int_{\phi_n} f(z)\,dz\right| \leqslant L\phi_n\,\epsilon\,L/2^n = \epsilon\,L^2/4^n \tag{12.27}$$

If we make the convenient choice $\epsilon = |J|/(2L^2) > 0$, we obtain

$$\left|\int_{\phi_n} f(z)\,dz\right| \leqslant \frac{1}{2}\frac{|J|}{4^n} \tag{12.28}$$

which contradicts the inequality that the integral has been constructed so that it is larger than $|J|/4^n$. So we cannot have that $|J| > 0$. Hence $J = 0$ and the result is proved.

12.3 The Cauchy–Goursat theorem for star-shaped sets

To elevate the result for a triangle to more general sets we need to make some suitable definitions. We have already seen that if there is a function $F(z)$ whose derivative is f, then the fundamental theorem of calculus tells us that the integral around any closed contour is zero. The converse is also true, as can be shown by actually constructing F as an integral of f. The details depend on what one assumes about the shape of the region of the complex plane under consideration. It is convenient to define a 'star-shaped' region as follows. Let U be an open subset of $\mathbb{C}$. We say that U is *star-shaped* about z_0 if, for all $z \in U$, the straight line

$$[z_0, z] \equiv \{z_0 + t(z - z_0) \mid 0 \leqslant t \leqslant 1\} \subset U \tag{12.29}$$

For example, a *convex* open set is, by definition, star-shaped about any of its points. So suppose that U is star-shaped about a, and that $z \in U$. We define (bear in mind w is just a dummy integration variables)

$$F(z) = \int_{[a,z]} f(w)\, dw \tag{12.30}$$

We can now state the theorem – note that we do not need all possible contours – all triangles will do.

■ Theorem 12.5: Existence of indefinite integral for star-shaped sets

Let f be a continuous complex valued function on U, open and star-shaped about a, and suppose that the integral of f around any triangle vanishes. Then $F(z)$ is holomorphic on U, with $F'(z) = f(z)$.

Proof: Fix $z \in U$, and let r be such that $|h| < r \Rightarrow z + h \in U$. So the line segments $[a, z]$, $[a, z + h]$, $[z, z + h]$ are all within U. So consider the closed triangular contour T from a to z to $z + h$ to a. The Cauchy–Goursat Theorem for a triangle establishes that:

$$\begin{aligned} 0 = \int_T f(w)\, dw &= \int_{[a,z]} f(w)\, dw + \int_{[z,z+h]} f(w)\, dw + \int_{[z+h,a]} f(w)\, dw \\ &= F(z) - F(z+h) + \int_{[z,z+h]} f(w)\, dw \end{aligned} \tag{12.31}$$

That is,

$$F(z+h) - F(z) = \int_{[z,z+h]} f(w)\, dw \tag{12.32}$$

Also, we can write

$$f(z) = (1/h) \int_{[z,z+h]} f(z)\, dw \tag{12.33}$$

$$\frac{F(z+h) - F(z)}{h} - f(z) = \frac{1}{h} \int_{[z,z+h]} [f(w) - f(z)]\, dw \tag{12.34}$$

and then

$$\begin{aligned} \left| \frac{F(h+z) - F(z)}{h} - f(z) \right| &= \frac{1}{|h|} \left| \int_{[z,z+h]} [f(w) - f(z)]\, dw \right| \\ &\leqslant \frac{1}{|h|} |h| \sup_{w \in [z,z+h]} |f(w) - f(z)| = \sup_{w \in [z,z+h]} |f(w) - f(z)| \end{aligned} \tag{12.35}$$

which can be made as small as we wish by continuity of f. Hence the result.

■ Cauchy–Goursat theorem for star-shaped and convex domains

We have actually done all the work necessary. The result is proved for triangles. Triangles were in fact all we needed to show the existence of a holomorphic function $F(z)$ with the property that $F'(z) = f(z)$. By the Fundamental Fheorem of Falculus applied to a simple closed path C within a star-shaped domain, the integral around C of $f(z)$ is zero. End of proof! Note that since any convex set is star-shaped about any of its points, the result has been proved for convex domains also.

One can consider making other proofs based on different assumptions. However, we have proved the theorem without assuming continuity of the derivatives, and the form of the theorem involving triangles is very direct and straightforward – differentiability leads directly to a contradiction unless the integral around a triangle is zero. Then the FTC elevates this to domains of a very general character. This therefore is as far as we shall go with relaxation of the conditions for Cauchy's theorem. Now it is time to look at some of the consequences of the result.

12.4 Consequences of Cauchy's theorem

It is useful to look at simple consequences of the result, and equivalent statements. The most important basic points are:

(1) Whether we can deform the contour defining an integral between two points A and B without changing the result for the integral.

(2) Whether we can shrink simple closed contours to integrals around a circle. (This is useful to do some explicit calculations.)

Consider the first point. Suppose we have two piecewise smooth paths C and C' from A to B, as shown below:

```
$DefaultFont = {"Helvetica", 12}; a = {1, 1}; b = {5, 5}; c = {3, 3};
l = Line[{a, b}]; c = Circle[c, 2 Sqrt[2], {Pi / 4, 5 * Pi / 4}];
pts = {PointSize[0.04], Map[Point, {a, b}]};
labels = {Text["A", {1.4, 1}],
   Text["B", {5.4, 5}], Text["C", {3.4, 3}], Text["C′", {1.4, 5}]} ;
Show[Graphics[{pts, l, c, labels}], AspectRatio -> 1]
```

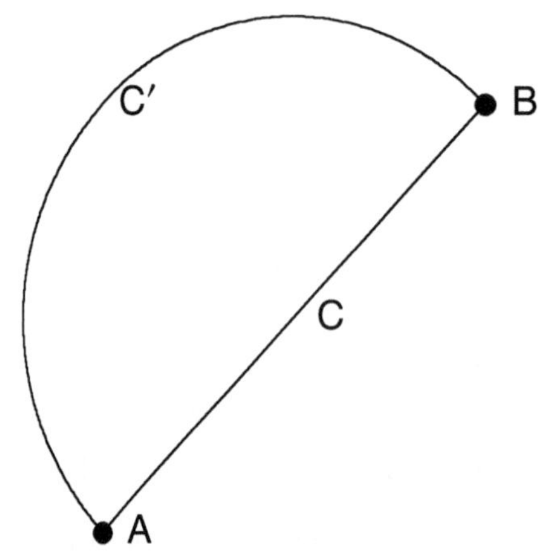

Let D denote the contour obtained by traversing C from A to B and then traversing backwards along C' from B to A. Clearly D is a closed contour, so that

$$\int_D f(z)\,dz = 0 \tag{12.36}$$

But

$$0 = \int_D f(z)\,dz = \int_C f(z)\,dz - \int_{C'} f(z)\,dz \tag{12.37}$$

Hence the integral from A to B is independent of the route taken. This of course links in to the notion that f possesses an 'anti-derivative' F.

■ Deformation to a circle

The last basic consequence of Cauchy's theorem involves seeing that we may replace quite complicated contours by circles around points of interest. Such points may arise quite naturally as points where the function being integrated may actually fail to be holomorphic.

Suppose that we have a piecewise smooth, positively traversed, simple, path $\phi(t)$ within an open set $U \subset \mathbb{C}$, and that f is holomorphic on $U - \{a\}$, where a is inside ϕ^*. To keep the proof straightforward we shall assume further that U is actually star-shaped about a. Note that we do not exclude the possibility that f is holomorphic on all of U – we just want to allow for the possibility of non-differentiability at a. Since U is open there is a number r such that the open disk

$$D(a, r) = \{z \in \mathbb{C} : |z - a| < r\} \tag{12.38}$$

is contained entirely within U. Suppose that $R < r$ and let γ be the positively traversed smooth path given by the circle radius R centred on a. What we want to establish is that

$$\int_\phi f(z)\, dz = \int_\gamma f(z)\, dz \tag{12.39}$$

Note first that the fact that the path ϕ is simple means it cannot intersect itself, so it loops around a just once. Now take a line L through a that intersects ϕ^* at c and d, and intersects γ at p and q. This line lies entirely within U because U is star-shaped about a. The geometry is as shown below:

```
$DefaultFont = {"Times", 14};
ϕ[t_] := {2 Cos[t] + 0.6 Sin[4 t], 2 Sin[t] + 0.6 Sin[2 t]};
γ[t_] := {Cos[t], Sin[t]};
data = {{AbsoluteThickness[0.1], Line[{{-2, -2}, {2, 2}}]}, {PointSize[0.03],
        Point[{-1 / Sqrt[2], -1 / Sqrt[2]}],
    Point[{ 1 / Sqrt[2],  1 / Sqrt[2]}],
        Point[{-1.4, -1.4}], Point[{ 1.83,  1.83}]},
        Text["γ", {1.2, 0}], Text["ϕ", {2.2, 0}],
        Text["p", {-0.7, -1.0}], Text["q", {0.7, 1.2}],
        Text["c", {-1.7, -1.4}], Text["d", {2.2, 1.9}]};
ParametricPlot[Evaluate[{ϕ[t], γ[t]}], {t, 0, 2 Pi}, AspectRatio -> 1,
    PlotRange -> {{-3, 3}, {-3, 3}},
 PlotStyle → AbsoluteThickness[0.1], Axes -> False, Epilog -> data]
```

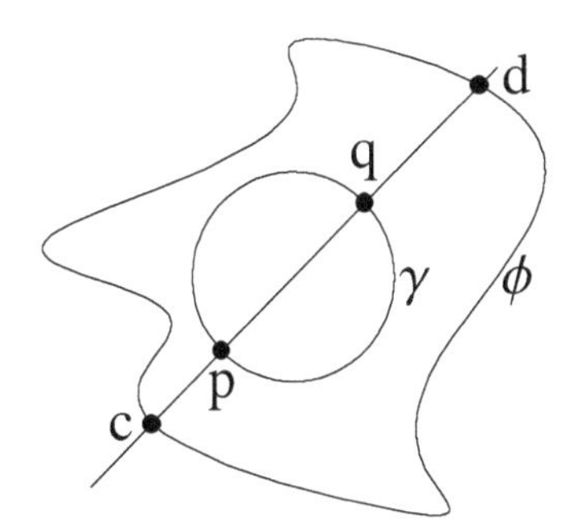

Now focus attention on one part of this geometry as shown below, where the line segment from q to d is split into two neighbouring lines L and L' as shown:

```
$DefaultFont = {"Times", 14};
ϕ[t_] := {2 Cos[t] + 0.6 Sin[4 t], 2 Sin[t] + 0.6 Sin[2 t]};
γ[t_] := {Cos[t], Sin[t]};
shift = 0.03;
data = {{AbsoluteThickness[0.1], Line[
     {{ 1/Sqrt[2] + shift,  1/Sqrt[2] - shift}, {1.83 + shift, 1.83 - shift}}],
        Line[{{ 1/Sqrt[2],  1/Sqrt[2]}, {1.83, 1.83}}]},
        {PointSize[0.01],
    Point[{ 1/Sqrt[2], 1/Sqrt[2]}], Point[{ 1.83,  1.83}]},
        Text["γ", {1.2, 0}], Text["ϕ", {2.2, 0}],
        Text["L", {1.6, 1.0}], Text["L′", {1.3, 1.6}],
        Text["q", {0.7, 1.2}], Text["d", {2.1, 2.0}]};
ParametricPlot[Evaluate[{ϕ[t], γ[t]}], {t, 0, 2 Pi}, AspectRatio -> 1,
    PlotRange -> {{-3, 3}, {-3, 3}},
 PlotStyle → AbsoluteThickness[0.1], Axes -> False, Epilog -> data]
```

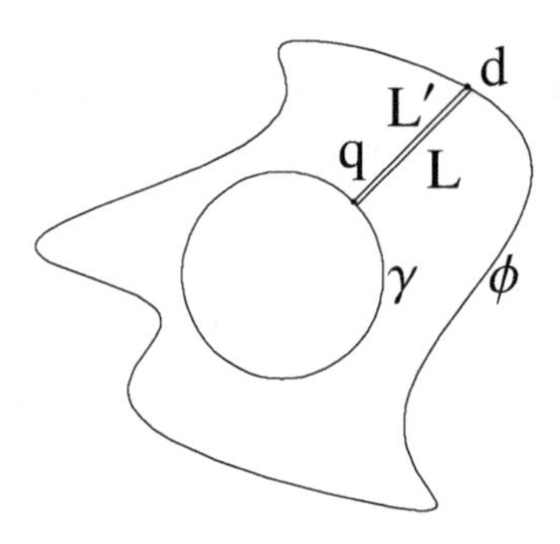

Consider the contour C defined by starting at $L' \cap \phi$, and traversing ϕ counter-clockwise until $L \cap \phi$ is reached. Then move along the line segment L from $L \cap \phi$ to $L \cap \gamma$, then around γ in a negative sense to $L' \cap \gamma$, then along L' from $L' \cap \gamma$ to $L' \cap \phi$. This is a closed contour and so

$$\int_C f(z)\,dz = 0 \tag{12.40}$$

but

$$\int_C f(z)\,dz = \int_\phi f(z)\,dz + \int_{L:d\to q} f(z)\,dz - \int_\gamma f(z)\,dz + \int_{L':q\to d} f(z)\,dz \tag{12.41}$$

Since the integrals along L and L' cancel, we deduce that

$$\int_\phi f(z)\,dz = \int_\gamma f(z)\,dz \tag{12.42}$$

So we can replace quite complicated contours by circles as long as they are simple and closed.

12.5 *Mathematica* pictures of the triangle subdivision

The production of iterated graphics in a straightforward manner can be done particularly elegantly in *Mathematica*. In particular, we can give a particularly elegant picture of the repeated subdivision of a triangle used in the proof of Cauchy's theorem:

```
Triangle[{a_, b_, c_}] :=
 Module[{edge = {{Re[a], Im[a]}, {Re[b], Im[b]},
           {Re[c], Im[c]}, {Re[a], Im[a]}}},
               {{Hue[Random[]], Polygon[edge]},
   {AbsoluteThickness[0.0025], Line[edge]}}]

Subdivide[{a_, b_, c_}] :=
    {{a, (a + b) / 2, (a + c) / 2},
  {(a + b) / 2, (b + c) / 2, (a + c) / 2},
  {(a + b) / 2, b, (b + c) / 2}, {(b + c) / 2, c, (a + c) / 2}}

Subdivide[ h_?MatrixQ] := Flatten[Map[Subdivide, h], 1]

RepeatedSubdivide[{a_, b_, c_}, n_] :=
 Nest[Subdivide, {a, b, c}, n]

checka = RepeatedSubdivide[{0, 2, 1 + I}, 1]
```

$$\begin{pmatrix} 0 & 1 & \frac{1}{2}+\frac{i}{2} \\ 1 & \frac{3}{2}+\frac{i}{2} & \frac{1}{2}+\frac{i}{2} \\ 1 & 2 & \frac{3}{2}+\frac{i}{2} \\ \frac{3}{2}+\frac{i}{2} & 1+i & \frac{1}{2}+\frac{i}{2} \end{pmatrix}$$

```
checkb = RepeatedSubdivide[{0, 2, 1 + I}, 2];
checkc = RepeatedSubdivide[{0, 2, 1 + I}, 3];
checkd = RepeatedSubdivide[{0, 2, 1 + I}, 4];
checke = RepeatedSubdivide[{0, 2, 1 + I}, 5];

ViewDivision[data_] :=
  Show[Graphics[Map[Triangle, data]]];
```

```
plota = ViewDivision[checka]
```

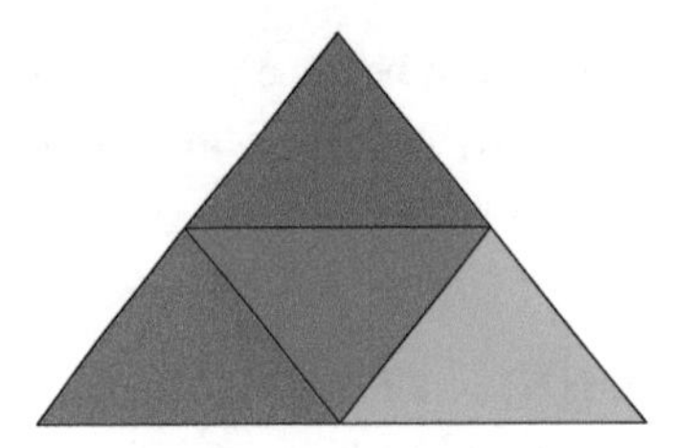

```
plotb = ViewDivision[checkb]
```

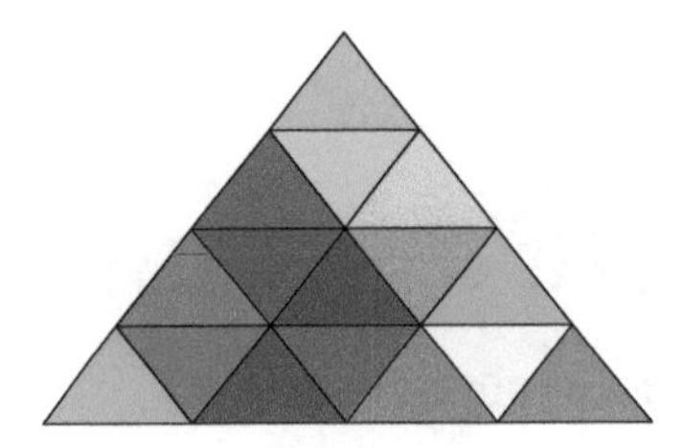

```
plotc = ViewDivision[checkc]
```

```
plotd = ViewDivision[checkd]
```

```
plote = ViewDivision[checke]
```

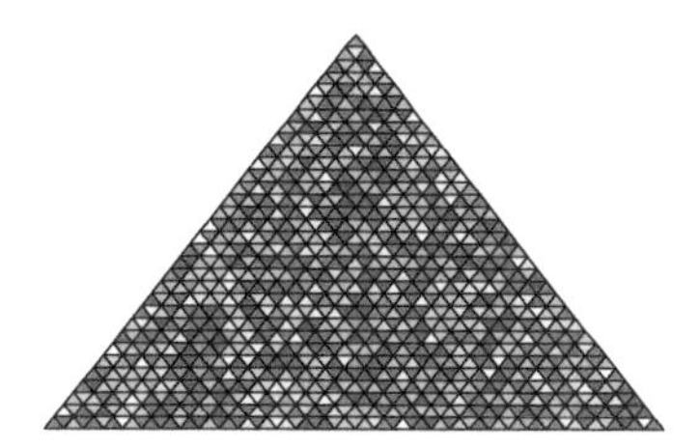

Exercises

12.1 Use Green's theorem in the plane for the case $f(z) = \bar{z}$ to show that if C has interior D', then

$$\int_C f(z)\,dz = 2iA(D')$$

where $A(D')$ is the area of the interior.

12.2 (The last part needs an understanding of moments of inertia in mechanics.) Use the complex form of Green's theorem in the plane for the case

$$f(z) = \frac{-i}{4}\bar{z}^2 z$$

to show that if C has interior D', then

$$\int_C f(z)\,dz = \int_{D'} |z|^2\,dx\,dy$$

Hence show, by evaluating a suitable integral around the unit circle, that the moment of inertia of a circular disk of radius r, and mass per unit area ρ, with respect to an axis normal to the disk through its centre (the polar moment), is

$$\frac{\pi\rho}{2}r^4$$

If the total mass is M, deduce that the polar moment of inertia is $Mr^2/2$.

12.3 A two-dimensional vector field $\underline{v} = (v_1(x, y), v_2(x, y), 0)$ is said to be irrotational if

$$\underline{\nabla} \times \underline{v} = \underline{0}$$

Show that this reduces to the constraint that

$$\frac{\partial v_2}{\partial x} = \frac{\partial v_1}{\partial y}$$

12.4 A two-dimensional vector field $\underline{v} = (v_1(x, y), v_2(x, y), 0)$ is said to be divergence-free, or solenoidal, if

$$\underline{\nabla}.\ \underline{v} = 0$$

Show that this reduces to the constraint that

$$\frac{\partial v_1}{\partial x} = -\frac{\partial v_2}{\partial y}$$

12.5 Show that if $f = v_1 - i v_2$ is holomorphic, then the two-dimensional vector field corresponding to f is both irrotational and divergence-free.

12.6 If a two-dimensional vector field has the property that its line integrals are independent of the path taken, then vector calculus theory tells us that there is a potential Φ with the property that

$$\underline{v} = \underline{\nabla}\ \Phi$$

How is this related to the existence of an anti-derivative for the corresponding function f? How do we interpret Cauchy's theorem in this context?

12.7 Calculate the integral

$$\int \frac{1}{z^2 + 2z + 2}\, dz$$

over the square with corners at $(0, 0)$, $(-2, 0)$, $(-2, -2)$, $(0, -2)$, to be traversed counter-clockwise. (Hint: factorize and do a partial fraction expansion on the denominator, and deform the square to a small circle around $-1 - i$.)

12.8 ❀ Explore the integrals of the following functions around a unit square centred on the origin, using *Mathematica*'s **`NIntegrate`** function. What is the pattern in the answers? What special cases are there?

$$z,\ |z|,\ \text{Conjugate}(z),\ J_0(z),\ J_0(|z|),\ \frac{1}{z},\ \frac{1}{z^2}$$

13 Cauchy's integral formula and its remarkable consequences

Introduction

In this chapter we introduce and prove the Cauchy integral formula and look at one part of the remarkable chain of consequences that results from it. The Cauchy integral formula is itself almost obvious from the deformation of an integral to a small circle, but its consequences are truly astonishing. First one establishes Taylor's theorem – in this case expressed as the fact that if a function is holomorphic, it is in fact infinitely differentiable, and can be represented as a power series involving all its derivatives. Formulae are also available for the derivatives. This is in manifest distinction to the real case, where being differentiable once is no guarantee of further differentiability, and being infinitely differentiable is no guarantee of having a power series! The formulae for the derivatives can be converted into useful inequalities (the Cauchy inequalities) for the scale of the derivatives, which allows a quick proof of the remarkable theorem of Liouville – that any bounded function that is holomorphic everywhere must be constant. Again, we explore the manifest distinctions with the real case of this odd result, and use it to establish a neat proof of the fundamental theorem of algebra – that any non-constant polynomial with complex coefficients has a root (and hence all its roots) in $\mathbb{C}$. This finally justifies the remark made in Chapter 1, Section 1.1 – that 'complex numbers are enough' – so we do not need to extend the number system any further to solve polynomial equations. Then we establish Morera's theorem, which is the converse to Cauchy's theorem.

13.1 The Cauchy integral formula

There are many levels of the statement of this result, depending on the type of contour one wishes to admit into the formula that results. We begin by establishing the simplest result, which is in fact sufficient for subsequent developments.

▪ Theorem 13.1: The Cauchy integral formula

Let U be an open subset of $\mathbb{C}$ and let f be analytic on U. Let $a \in U$ and suppose that $N_r(a) \subset U$. Let $R < r$ and let

$$\phi_R[t] = a + Re^{it} \tag{13.1}$$

for $0 \leqslant t \leqslant 2\pi$ be a circular path of radius R centred on a. Then

$$f(a) = \frac{1}{2\pi i} \int_{\phi_R} \frac{f(z)}{z-a}\, dz \tag{13.2}$$

Henceforth we shall abbreviate the name of this result to the 'CIF'.

■ Proof

We apply our deformation results from Chapter 12 to argue that if $0 < s < R$, we may replace our original circle by a smaller one of radius s:

$$\int_{\phi_R} \frac{f(z)}{z-a}\, dz = \int_{\phi_s} \frac{f(z)}{z-a}\, dz \tag{13.3}$$

Now $f(z)$ is differentiable at a, so given any $\epsilon > 0$, there is a $\delta : 0 < \delta < R$ such that, if $|z - a| < \delta$, then

$$\left| \frac{f(z) - f(a)}{z-a} - f'(a) \right| < \frac{\epsilon}{R} \tag{13.4}$$

(This is just the statement that the remainder term in the definition of differentiability goes to zero). So choose s such that $0 < s < \delta$. Then using the length inequality

$$\left| \int_{\phi_s} \left(\frac{f(z) - f(a)}{z-a} - f'(a) \right) dz \right| \leqslant \frac{\epsilon\, 2\pi\delta}{R} \leqslant 2\pi\epsilon \tag{13.5}$$

But also, we have that

$$\int_{\phi_s} f'(a)\, dz = 0 \tag{13.6}$$

$$\int_{\phi_s} \frac{f(a)}{z-a}\, dz = 2\pi i f(a) \tag{13.7}$$

$$\left| \frac{1}{2\pi i} \int_{\phi_s} \frac{f(z)}{z-a}\, dz - f(a) \right| \leqslant \epsilon \tag{13.8}$$

Since we can make ϵ as small as we wish, the result is proved.

■ Generalizations and winding numbers

First of all note that we can replace the circular path by a simple closed path, using the deformation theorem given in Chapter 12. Of particular importance is the case of a circle not necessarily centred on the point of interest.

We can also ask what happens for a path that is not necessarily simple. Suppose first that our original path is replaced by

$$\phi_{R,n}[t] = a + R\, e^{i t} \tag{13.9}$$

for $0 \leqslant t \leqslant 2n\pi$. Then a simple extension of the argument given above shows that

$$2\pi i n f(a) = \int_{\phi_{R,n}} \frac{f(z)}{z-a}\, dz \tag{13.10}$$

More generally still, given a path ϕ, we can define its *winding number* about a, $n(\phi, a)$, by the integral

$$2\pi i n(\phi, a) = \int_{\phi} \frac{1}{z-a}\, dz \tag{13.11}$$

It can be shown that this is well-defined, and obtain a result generalizing the CIF in the form

$$2\pi i n(\phi, a)\, f(a) = \int_{\phi} \frac{f(z)}{z-a}\, dz \tag{13.12}$$

We shall not go into details about such winding numbers in our main discussion – they are beyond the scope of this book.

13.2 Taylor's theorem

The following result gives **Theorem 13.2**: Taylor's theorem. *Let f be analytic on an open set $U \subset \mathbb{C}$. Then the following are true*:

(i) *f has derivatives of all orders at all points of U.*
(ii) *Suppose that $a \in U$, $r > 0$ and $\overline{N}_r(a) \subset U$, then*

$$f^{(n)}(a) = \frac{n!}{2\pi i} \int_{\phi_{r,a}} \frac{f(z)}{(z-a)^{n+1}}\, dz \tag{13.13}$$

where $\phi_{r,a}[t] = a + re^{it}, 0 \leqslant t \leqslant 2\pi$.
(iii) *If $|h| < r$,*

$$f(a+h) = \sum_{n=0}^{\infty} \frac{h^n}{n!} f^{(n)}(a) = f(a) + hf'(a) + \frac{1}{2}h^2 f''(a) + \ldots \tag{13.14}$$

and the series converges absolutely and uniformly for $h \in N_r(0)$. Before looking at the proof some remarks are in order.

This result goes some way to explaining our early interest in power series. It has turned out that any holomorphic function can be locally represented as a power series. We also know, from Chapter 9, that power series converge and are differentiable within their radius of convergence. Once one knows this result it is tempting to regard ‘holomorphic’ and ‘power series’ as synonymous. The term ‘analytic’ is used by some to mean ‘having a power series’, irrespective of whether the context is real or complex.

■ Differences from the real case

The theorem as stated is remarkable in the differences it implies between the consequences of real differentiability compared to complex differentiability. Consider first a function $f : \mathbb{R} \to \mathbb{R}$ defined by

$$f(x) = \begin{cases} 0 & x < 0 \\ x^2 & x \geqslant 0 \end{cases} \tag{13.15}$$

This function can be differentiated everywhere, including the origin (a first-principles application of the limit formula works there) to yield the function

$$f'(x) = \begin{cases} 0 & x < 0 \\ 2x & x \geqslant 0 \end{cases} \tag{13.16}$$

This new function is not differentiable at the origin. The following plots of the function (on the left) and its derivative (on the right) show the reason why one cannot differentiate twice:

```
plotf = Plot[If[x < 0, 0, x^2], {x, -1, 1},
DisplayFunction -> Identity, PlotRange -> {0, 1},
        PlotStyle -> Thickness[0.01]];
derplot = Plot[If[x < 0, 0, 2 x], {x, -1, 1},
DisplayFunction -> Identity, PlotRange -> {0, 1},
        PlotStyle -> Thickness[0.01]];
Show[GraphicsArray[{plotf, derplot}],
    DisplayFunction -> $DisplayFunction]
```

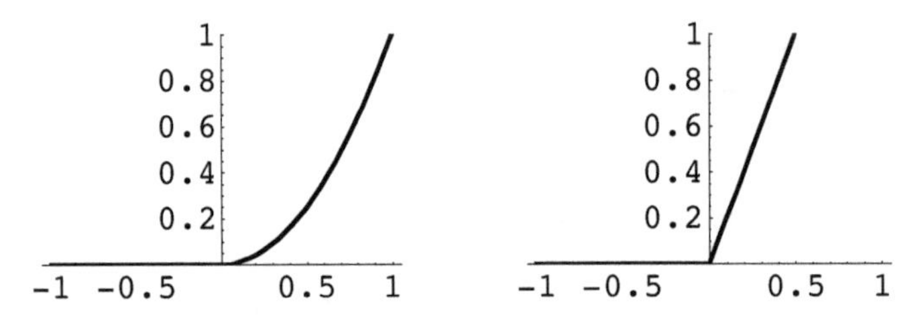

Another function illustrates the failure of the power series representation even when the function is *infinitely* differentiable. Let $g(x)$ be defined by

$$g(x) = \begin{cases} 0 & x = 0 \\ e^{-1/x^2} & x \neq 0 \end{cases} \tag{13.17}$$

It requires some work to see what is happening here, and the proofs of the following are discussed in the exercises. It turns out that $g(x)$ is infinitely differentiable everywhere, including the origin, but

$$g^{(n)}(0) = 0 \tag{13.18}$$

for all values of n. The function is incredibly flat in a neighbourhood of the origin:

```
Plot[If[x == 0, 0, Exp[-1 / x^2]], {x, -5, 5},
    PlotRange -> {0, 1}, PlotPoints -> 100,
        PlotStyle -> Thickness[0.005]]
```

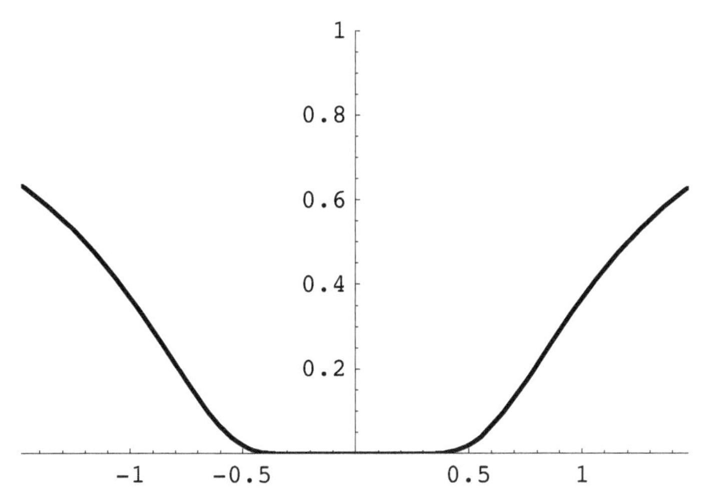

It is hopeless trying to make analogues of these functions that are differentiable at the origin. The function

$$g(z) = e^{-1/z^2} \tag{13.19}$$

actually possess one of the nastiest types of singularity, known as an *essential singularity*, at the origin. Not only does it blow up badly along the imaginary axis, but it also attains any non-zero value infinitely often in a neighbourhood of the origin! A demonstration of this is in one of the exercises. We will define and discuss this and other types of singularity in Chapter 14.

■ Proof of Taylor's theorem parts (i) and (ii)

How do we prove this result? First of all note that if f is once differentiable in U, we have the Cauchy integral formula. So parts (i) and (ii) are certainly true for $n = 0$. So we shall adopt an inductive approach, and assume that the result is true for a value of n, for all $a \in U$, and for each a we can find a range of values of r for which the result is true. Let h be such that $h \leqslant r/2$. So by our inductive hypothesis, we can write

$$f^{(n)}(a+h) = \frac{n!}{2\pi i} \int_{\phi_{r/2,a+h}} \frac{f(z)}{(z-(a+h))^{n+1}}\, dz \tag{13.20}$$

where $\phi_{r/2,a+h}[t] = a + h + (r/2)e^{it}$, $0 \leqslant t \leqslant 2\pi$. Why do we do this? Well, the point $a + h$ is at most $r/2$ from a, and a circle of radius $r/2$ centred on $a + h$ is therefore within

$N_r(a)$. So this expression is over a circular contour lying entirely within U. We now deform this contour to the original circular contour of radius r centred on a, to write

$$f^{(n)}(a+h) = \frac{n!}{2\pi i} \int_{\phi_{r,a}} \frac{f(z)}{(z-(a+h))^{n+1}} dz \tag{13.21}$$

It does not hurt to remind the reader why such a deformation is allowed – we apply Cauchy's theorem to the contour γ shown here. We have drawn it using *Mathematica* with illustrative values of a and h.

```
a = {0, 0}; h = {1 / 4, 0};
phione[t_] := {Cos[t], Sin[t]}; phitwo[t_] := h + 0.5 * phione[t];
gamma[t_] := 1.05 * phione[Pi / 160 + 0.9875 * t];
delta[t_] := h + 0.55 * phione[Pi / 160 + 0.9875 * t];
ParametricPlot[Evaluate[{phione[t], phitwo[t], gamma[t], delta[t]}],
    {t, 0, 2 Pi + 0.01}, AspectRatio -> 1, Axes -> False, Compiled -> False,
    PlotStyle ->
  {Thickness[0.005], Thickness[0.005], Thickness[0.01], Thickness[0.01]},
 Epilog -> {PointSize[0.03], Point[a], Point[a + h],
 Text["a", {0, -0.1}], Text["a+h", h + {0, -0.1}],
Thickness[0.015],     Line[{gamma[2 Pi], delta[2 Pi]}],
    Line[{gamma[0], delta[0]}]}]
```

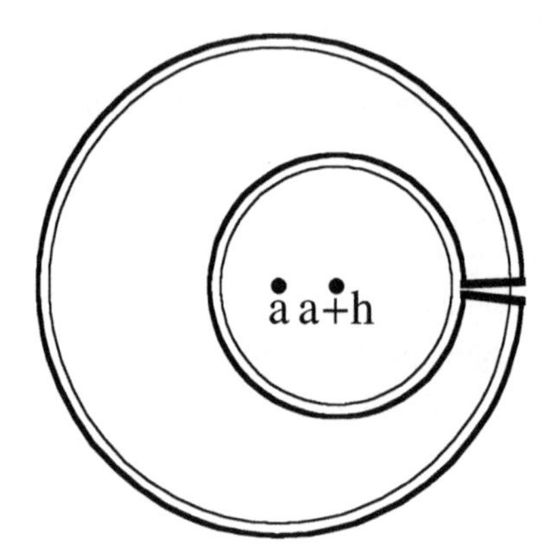

The outer circle is $\phi_{r,a}$ and the inner is $\phi_{a+h,r/2}$. The contour γ first traverses almost all of the outer circle anticlockwise, jumps to the inner circle, traverses it clockwise, then jumps back out to the outer circle. By Cauchy's theorem, we have

$$\int_{\gamma} \frac{f(z)}{(z-(a+h))^{n+1}} dz = 0 \tag{13.22}$$

Given that we can deform the contour to the same one used for $f^{(n)}$(a), we consider the definition of differentiability of $f^{(n)}$, with the proposed integral expression for $f^{(n+1)}$, by examining

$$\sigma = f^{(n)}(a+h) - f^{(n)}(a) - h\,\frac{(n+1)!}{2\pi i} \int_{\phi_{r,a}} \frac{f(z)}{(z-a)^{n+2}} dz \tag{13.23}$$

To show that the nth derivative is differentiable, with $f^{(n+1)}$ given by the same integral formula, it is sufficient to show that σ is $O(h^2)$, so that $\sigma/h \to 0$, as $h \to 0$. But we can write

$$\sigma = \frac{n!}{2\pi i}\int_{\phi_{r,a}} f(z)\left[\frac{1}{(z-(a+h))^{n+1}} - \frac{1}{(z-a)^{n+1}} - \frac{h(n+1)}{(z-a)^{n+2}}\right]dz \tag{13.24}$$

So we need to estimate the bracketed term in the integrand. The astute reader may well have noticed by now that the fact that the bracketed term is of the right order is nothing more than the observation that $1/(z-a)^{n+1}$ is differentiable, with the obvious derivative, but we can check it explicitly. First, putting it all over a common denominator, we have

$$\begin{aligned}&\frac{1}{(z-(a+h))^{n+1}} - \frac{1}{(z-a)^{n+1}} - \frac{(n+1)h}{(z-a)^{n+2}} = \\ &\frac{(z-a)^{n+2} - (z-(a+h))^{n+1}\,(z-a+(n+1)h)}{(z-a)^{n+2}(z-(a+h))^{n+1}}\end{aligned} \tag{13.25}$$

Next, by the binomial theorem, we can write

$$(z-(a+h))^{n+1} = (z-a)^{n+1} - (n+1)\,h\,(z-a)^n + R \tag{13.26}$$

where R is $O(h^2)$. On insertion of this the term in the numerator in $(z-a)^{n+2}$ is cancelled, leaving us with

$$\frac{(z-a)^n h^2 (n+1)^2 - (z-a+h(n+1))R}{(z-a)^{n+2}\,(z-(a+h))^{n+1}} \tag{13.27}$$

So everything in the numerator is $O(h^2)$. Hence we have shown that $f^{(n)}$ is differentiable, with $f^{(n+1)}$ also given by the integral formula. This proves parts (i) and (ii) of the theorem, by induction.

■ Proof of Taylor's theorem part (iii)

Now we know that $\overline{N}_r(a) \subset U$. In fact, we can stretch this closed disk a little, and state that there is an s with $s > r$ and $N_s(a) \subset U$. This follows from the fact that there is a certain minimum non-zero distance from a point on the circle $\phi_r(a)$ to points not in U. This is proved properly in the appendix (to this proof) on the disk-stretching lemma and requires some results from formal analysis. This lemma can be used to extend the scope of the Cauchy integral formula somewhat, but for no real gain. Here, however, the lemma is needed, so we first assume it and then prove it.

Proceeding on this assumption, let $r < q < s$, and suppose that $|h| \leqslant r$. By the CIF, bearing in mind that, having stretched the disk, $a + h$ is inside ϕ_q, and applying an expansion to the denominator,

$$\begin{aligned} f(a+h) &= \frac{1}{2\pi i}\int_{\phi_\sigma} \frac{f(z)}{(z-(a+h))}\,dz \\ &= \frac{1}{2\pi i}\int_{\phi_q} f(z)\left[\frac{1}{(z-a)}+\ldots+\frac{h^n}{(z-a)^{n+1}}+\frac{h^{n+1}}{(z-a)^{n+1}(z-a-h)}\right]dz \\ &= f(a)+h\,f'(a)+\ldots+\frac{h^n}{n!}f^{(n)}(a) \\ &\quad+\frac{h^{n+1}}{2\pi i}\int_{\phi_q} f(z)\left[\frac{1}{(z-a)^{n+1}(z-a-h)}\right]dz \end{aligned} \tag{13.28}$$

So let's look at the size of the remainder term. Suppose that $|f(z)| \leqslant M$ on the path over the image of ϕ_q. Then

$$\begin{aligned} &\left|\frac{1}{2\pi i}h^{n+1}\int f(z)\left[\frac{1}{(z-a)^{n+1}(-a-h+z)}\right]dz\right| \leqslant \\ &\frac{r^{n+1}}{2\pi}\left[\frac{M}{(q-r)\,q^{n+1}}\right]2\pi q = \frac{M\,r\left(\frac{r}{q}\right)^n}{q-r} \end{aligned} \tag{13.29}$$

and this tends to zero as $n \to \infty$.

■ Absolute and uniform convergence

The establishment of this result is based on an analysis similar to that established for the remainder. The $(n+1)$th term in the series is

$$\frac{h^n}{n!}f^{(n)}(a) = \frac{h^n}{2\pi i}\int_{\phi_q}\frac{f(z)}{(z-a)^{n+1}}\,dz \tag{13.30}$$

so taking absolute values,

$$\left|\frac{h^n}{n!}f^{(n)}(a)\right| \leqslant \frac{r^n}{2\pi}\left[\frac{M}{q^{n+1}}\right]2\pi\sigma = M\left(\frac{r}{q}\right)^n \tag{13.31}$$

and by comparison with a geometric series, this is absolutely convergent, and. by the Weierstrass M-test, is uniformly convergent.

■ Appendix to Taylor's theorem: the disk-stretching lemma

Let $U \subset \mathbb{C}$, U open, but $U \neq \mathbb{C}$, and let $d(z)$ denote the shortest distance from z to a point not in U. Formally, we let

$$d(z) = \inf\{|z-w|;\, w \notin U\} \tag{13.32}$$

This is a continuous function on $\mathbb{C}$, and is only zero if $z \notin U$, because if $z \in U$, then, by the definition of an open set, $N_\epsilon(z) \subset U$, for some ϵ, hence $d(z) \geqslant \epsilon$. Now suppose the closed disk $\overline{N}_r(a)$ lies within U. This closed disk is a closed and bounded subset of $\mathbb{C}$, and we assume a result from basic analysis that a continuous function defined on such a subset is bounded on such a subset and attains its bounds. In particular, $d(z)$ attains its lower bound on $\overline{N}_r(a)$. This value is necessarily greater than zero, we call it $d_{\min}$. It then it follows that $N_{r+d_{\min}}(a) \subset U$.

All we have done is to prove carefully that, along the circle, the minimum distance to a point not in U is greater than zero, so we can stretch the circle out a little.

13.3 The Cauchy inequalities

If we take the estimate used to establish the uniform convergence of the Taylor series:

$$\left| \frac{h^n}{n!} f^{(n)}(a) \right| \leqslant M\left(\frac{r}{q}\right)^n \tag{13.33}$$

and set $h = r$, we obtain the Cauchy inequalities:

$$\left| f^{(n)}(a) \right| \leqslant \frac{M\, n!}{q^n} \tag{13.34}$$

Recall that M is the maximum value of $|\, f(z)\,|$ on the image of ϕ_σ.

13.4 Liouville's theorem

Theorem 13.3: *Suppose that f is holomorphic on all of* $\mathbb{C}$. Liouville's theorem states that *if f is bounded on $\mathbb{C}$ then f is constant.*

The proof of this involves a simple application of the $n = 1$ Cauchy inequality. Suppose that $|\, f(z)\,| < m$ for all $z \in \mathbb{C}$. Let $a \in \mathbb{C}$, and suppose that $s > 0$. Let M be the maximum value of $|\, f(z)\,|$ on a circle of radius s centred on a. Then, choosing the $n = 1$ case from the Cauchy inequalities gives:

$$|f'(a)| \leqslant \frac{M}{s} \leqslant \frac{m}{s} \tag{13.35}$$

Now since s is arbitrary, and so can be as large as we choose, it follows that

$$f'(a) = 0 \tag{13.36}$$

This is true for all a, so $f'^{(z)} = 0$. So $f(z)$ is constant.

■ Differences from the real case

A reminder about the real case is in order here, to make the point about just how surprising Liouville's theorem is! There are plenty of differentiable, indeed infinitely differentiable functions, that are bounded on all of □. For example, if we consider one of the simple trigonometric functions, we have, for all $x \in$ □,

$$|\sin(x)| \leqslant 1 \tag{13.37}$$

But if we extend this function into the complex plane, the function sin(z) grows in the imaginary direction, and is unbounded as one approaches $\pm i\infty$. It is nicely frustrating to play this game with any bounded real function. As a second example, consider

$$f(x) = \frac{1}{a^2 + x^2} \tag{13.38}$$

This again is bounded for x real, but if we extend this function (holomorphically) into the complex, by considering

$$f(z) = \frac{1}{a^2 + z^2} \tag{13.39}$$

it inevitably blows up at $z = \pm ia$. This notion can be neatly extended to prove the next important result, about roots of polynomials.

13.5 The fundamental theorem of algebra

Theorem 13.4: *Suppose that $P(z)$ is a non-constant polynomial. Then there is a complex number w such that $P(w) = 0$.*

The idea of the proof is very simple, now that we are armed with Liouville's theorem, for we just consider

$$f(z) = \frac{1}{P(z)} \tag{13.40}$$

and suppose that there are no points w where $P(w) = 0$. Under this assumption f is continuous. We have to show carefully that we get a contradiction unless P is constant. We do this by demonstrating that, under our assumptions, f is bounded, and this begins by writing out P in the form

$$P(z) = a_0 + a_1 z + \ldots + a_n z^n = z^n \left(a_n + \frac{a_{n-1}}{z} + \ldots + \frac{a_0}{z^n}\right) \tag{13.41}$$

Let's consider the absolute value of that part of the bracketed term not involving a_n. This is just

$$\left|\frac{a_{n-1}}{z} + \ldots + \frac{a_0}{z^n}\right| \tag{13.42}$$

By choosing a sufficiently large value for R, we can arrange that if $|z| > R$, then

$$\left|\frac{a_{n-1}}{z} + \ldots + \frac{a_0}{z^n}\right| \leqslant \left|\frac{a_n}{2}\right| \tag{13.43}$$

It follows by the triangle inequality that if $|z| > R$, then

$$|P(z)| \geqslant \frac{1}{2} R^n \, |a_n| \tag{13.44}$$

and hence that

$$f(z) \leqslant \frac{2}{R^n \, |a_n|} \tag{13.45}$$

So we have shown that f is bounded on $|z| > R$. Since it is continuous on the closed and bounded set $|z| \leqslant R$, it is bounded there also. Hence it is bounded for all z. Hence, by Liouville, it must be a constant. This contradicts the assumption that P is non-constant. This finally justifies the remark made in Chapter 1, Section 1.1 – that 'complex numbers are enough' – so we do not need to extend the number system any further to solve polynomial equations.

■ Remarks and visualization

Once one has found one root the fact that there are n roots follows by factorization. A careful analysis is given in Exercise 13.3. There may be multiple roots at a given point. At this stage you may wish to revisit the discussion in Chapter 1. Remember that a polynomial, indeed linear, equation with integer coefficients requires a number system larger than the integers. A general linear equation will have a solution that is rational. A quadratic equation with rational coefficients may have solutions that are rational or real, but in general has a complex solution. A general polynomial equation with complex coefficients always has all its solutions within the set of complex numbers – we do not need to extend our concept of numbers any further. Of course, people have invented higher-order number systems – the Quaternions of Hamilton being a classic example. But the requirement of polynomial equation solving does not necessitate their introduction.

There is a particularly neat way of visualizing the roots in the complex plane with *Mathematica*, which provides an informal view of the fundamental theorem of algebra. The idea is to plot minus the log of the absolute value of the polynomial. The following plot is a variant of an approach that was originally given in the 'Solving the Quintic' poster (Adamchik and Trott, 1994). For example, consider the polynomial $q(z) = z^5 - 1$. We can plot the logarithm of its absolute value:

```
q[z_] := z^5 - 1; Plot3D[-Log[Abs[q[x + I y]]],
 {x, -1.5, 1.5}, {y, -1.5, 1.5}, PlotPoints → 50]
```

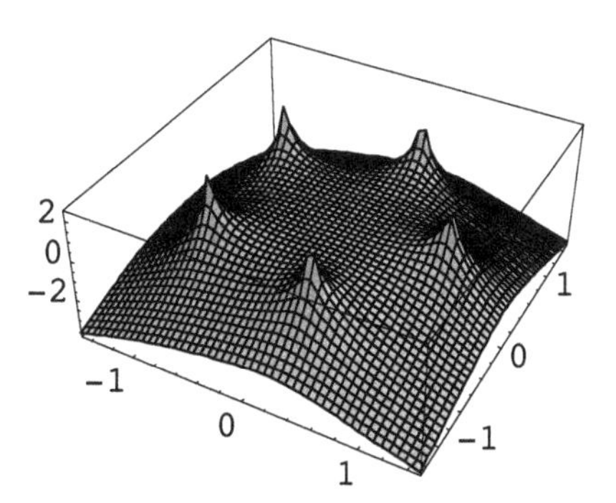

The five zeroes stand out as peaks in this plot.

13.6 Morera's theorem

This is the converse to Cauchy's theorem, and states that if, given a function f, its integral around all suitable closed paths is zero, then the function is holomorphic. It hinges critically on the first part of Taylor's theorem. What we do is to build a function F that is the integral of the given function f, and show that it is differentiable on a suitable region, provided that certain integrals vanish, with a derivative that is our original function f. It follows, from the fact that we can differentiate as often as we wish, that f is itself differentiable.

■ Theorem 13.5: Morera's theorem

Let f be a continuous function defined on an open set $U \subset \mathbb{C}$. Suppose that

$$\int_{\phi} f(z)\, dz = 0 \tag{13.46}$$

for any piecewise-smooth closed curve ϕ. Then f is holomorphic in U.

Proof: It is sufficient (as in our proof of the Cauchy–Goursat Theorem), to consider triangular paths $\phi_{a,b,c}$ for triangles $T_{a,b,c} \subset U$. With this in mind, let $z \in U$, and suppose, given that U is open, that $N_r(z) \subset U$, and $w \in N_r(z)$. Let χ_w be the straight-line path from z to w, and define F by the integral

$$F(w) = \int_{\chi_w} f(\zeta)\, d\zeta \tag{13.47}$$

Now let ψ be the straight line joining w and $w + h$, for $w + h \in N_r(z)$. Clearly,

$$F(w+h) - F(w) = \int_{\psi} f(\zeta)\, d\zeta \tag{13.48}$$

and, making a subtraction that will lead to the right expression for the derivative

$$F(w+h) - F(w) - h f(w) = \int_{\psi} [f(\zeta) - f(w)]\, d\zeta \tag{13.49}$$

The right side of this is an integral over a path of a length that goes to zero, of a function that goes to zero, and hence is $o(|h|)$ – to prove this formally, given $\epsilon > 0$, we can find a $\delta > 0$ such that $|f(\zeta) - f(w)| < \epsilon$ for $|\zeta - w| < \delta$. So if $|h| < \delta$,

$$\left| \int_{\psi} [f(\zeta) - f(w)]\, d\zeta \right| < \epsilon |h| \tag{13.50}$$

The remainder is $o(|h|)$, so F is differentiable with derivative f. Since we can differentiate again, f must be analytic.

13.7 The mean-value and maximum modulus theorems

Let's now go back to the CIF in the form

$$f(z_0) = \frac{1}{2\pi i}\int_{\phi_R} \frac{f(z)}{z - z_0}\,dz \tag{13.51}$$

We can deduce some other important results directly from the CIF. These other results do not rely on Taylor's Theorem. To establish the main result we first rewrite the CIF using an explicit parametrization on the circle:

$$z = z_0 + Re^{it}$$

Then Eq. (13.51) becomes

$$f(z_0) = \frac{1}{2\pi}\int_0^{2\pi} f(z_0 + Re^{it})\,dt \tag{13.52}$$

and we have established **Theorem 13.6** (the mean-value theorem): *the value of a holomorphic function f at a point is the average of its values on a circle centred at that point.* We can deduce from this that

$$|f(z_0)| \le \frac{1}{2\pi}\int_0^{2\pi} |f(Re^{it} + z_0)|\,dt \tag{13.53}$$

It follows that we can deduce **Theorem 13.7** (maximum modulus theorem): *for a non-constant function $f, |f(z_0)|$ cannot be a maximum of the modulus of f.* The proof relies on the fact that if the result is not true then Eq. (13.53) leads to a contradiction (see Exercise 13.4 for details). This observation is the starting point for a chain of theorems involving the maximum of the modulus.

Exercises

13.1 (Harder) Consider the function defined by:

$$g(x) = \begin{cases} 0 & x = 0 \\ e^{-1/x^2} & x \ne 0 \end{cases}$$

Show, using the first-principles definition of differentiability, that g has derivatives of all orders at $x = 0$, and that $g^{(n)}(0) = 0$ for all positive integers n.

13.2 Consider the complex function defined for $z \ne 0$ by

$$g(z) = e^{-\frac{1}{z^2}}$$

Find all solutions, with w □ 0 specified, of the equation

$$w = g(z)$$

in the set $|z| < \varepsilon$, where ε is given. (Hint: use the multi-valued nature of the logarithm function to analyse this problem.) Does this function tend to a limit as $z \to 0$? Does it tend to infinity?

13.3 Suppose that $P(z)$ is a polynomial of degree n and that $P(w) = 0$. By considering the identity

$$z^k - w^k = (z - w)(z^{k-1} + z^{k-2} w + \ldots + z w^{k-2} + w^{k-1})$$

for $1 \leqslant k \leqslant n$ show that $P(z)$ can be written as $(z - w) Q(z)$ for some polynomial Q of degree $n - 1$. Deduce that P has exactly n zeroes, counting repeated roots as making multiple contributions.

13.4 Suppose that f is holomorphic and that $|f(z)|$ has a maximum at $z = z_0$. Show that Eq. (13.53) implies that $|f|$ must be constant. Deduce from the Cauchy–Riemann equations, and the constancy of $|f|^2$, that if this holds on any open disk or connected set, f must be constant on such a set.

13.5 Use the formula for the derivatives supplied by Taylor's theorem (Eq. 13.13) to deduce the value of

$$\int \frac{e^{6z}}{z^4} \, dz$$

where the integral is taken over a unit circle centred on the origin.

13.6 Use Taylor's theorem to give a *rigorous* proof of the fact that if $f(z)$ is holomorphic, then, if we write $f(z) = u(x, u) + i\, v(x, y)$

$$\frac{\partial^2 u}{\partial x^2} + \frac{\partial^2 u}{\partial y^2} = 0$$

and similarly for v.

13.7 Let $P(z)$ be a polynomial of degree n and suppose that its zeroes are at $z = z_k$ for $k = 1, \ldots, n$. Show that

$$\frac{P'(z)}{P(z)} = \sum_{k=1}^{n} \frac{1}{z - z_k}$$

13.8 Let C be a simple closed positively traversed contour, and suppose that $P(z)$ is a polynomial of degree n with no zeroes on C. Use the result of Exercise 13.7 to show that the number of zeroes of P inside C, counting multiplicities, is

$$\frac{1}{2\pi i}\int \frac{P'(z)}{P(z)}\,dz$$

13.9 For each of the following real differentiable and bounded functions explain how the complex generalization shown fails to satisfy the conditions for Liouville's theorem:

$$\cos(x) \longrightarrow \cos(z)$$

$$\frac{1}{(a^2+x^2)^2} \longrightarrow \frac{1}{(a^2+z^2)^2}$$

$$\frac{1}{(a^2+x^2)^2} \longrightarrow \frac{1}{(a^2+|z|^2)^2}$$

13.10 ✵ By replacing the function **Plot3D** in Section 13.5 with **ContourPlot**, establish another method by which the locations of roots may be identified graphically and hence give approximate numerical values for all the roots of

```
q[z_] := z^5 - 3 z - 1
```

13.11 ✵ Using the values you found in the previous exercise as starting values for **FindRoot**, find accurate values for all roots of $q(z) = 0$. Check your results using **NSolve**.

13.12 ✵ Use Exercise 13.8 and a polygonal approximation to a circle to calculate the number of zeroes of the polynomial

$$8z^3 + 46z - 36z^2 - 15$$

inside each of the three circles $|z| = 1, 2, 3$.

14 Laurent series, zeroes, singularities and residues

Introduction

In this chapter we introduce the generalization of a Taylor series to the Laurent series, and discuss the consequences of the existence of a Laurent series. We also makes some definitions that allow us to get a grip on the concept of a singularity of a complex function, and introduce a definition and classification of isolated singularities. The notion of a reside at an isolated singularity is introduced, and various devices for calculating residues are introduced for simple functions of interest. This chapter is fundamental to the development of the theory of integration, and the evaluation of Fourier and Laplace transforms and their inverses. It also highlights the importance of seeking out and investigating singularities.

14.1 The Laurent series

Taylor's theorem relates to functions holomorphic on a disk within some open set. Now we make a hole in the disk and consider functions holomorphic on an annulus. So let U be the annulus

$$U = \{z : \delta_1 < |z - z_0| < \delta_2\} \tag{14.1}$$

where

$$0 \leqslant \delta_1 < \delta_2 \leqslant \infty \tag{14.2}$$

Note that the annulus can have a hole that is just the point z_0, and its outer limit can be infinity. Now we can state the theorem:

■ Theorem 14.1: The Laurent series

Suppose that f is holomorphic on the annulus U. Then

$$f(z) = \sum_{n=-\infty}^{\infty} a_n (z - z_0)^n \tag{14.3}$$

$$= \ldots + \frac{a_{-2}}{(z - z_0)^2} + \frac{a_{-1}}{z - z_0} + a_0 + a_1 (z - z_0) + a_2 (z - z_0)^2 + \ldots \tag{14.4}$$

where the second line is just there to emphasize that the series contains *negative* powers of $(z - z_0)$. *Furthermore, if $\delta_1 < r_1 < r_2 < \delta_2$, the series converges absolutely and* *uniformly in the set* $\{z : r_1 \leqslant |z - z_0| \leqslant r_2\}$, *and the coefficients are all given by the formula*

$$2\pi i\, a_n = \int_{\phi_r} \frac{f(z)}{(z - z_0)^{n+1}}\, dz \tag{14.5}$$

where $\delta_1 < r < \delta_2$, $\phi_r = z_0 + r\, e^{i\,t}$, for $0 \leqslant t \leqslant 2\pi$. In particular, note that the constant term is given by what would be the Cauchy Integral Formula (CIF) for $f(z_0)$, if the function were holomorphic 'all the way in':

$$2\pi i a_0 = \int_{\phi_r} \frac{f(z)}{(z - z_0)}\, dz \tag{14.6}$$

and also, for the negative terms:

$$2\pi i a_{-1} = \int_{\phi_r} f(z)\, dz \tag{14.7}$$

$$2\pi i a_{-n} = \int_{\phi_r} f(z)\,(z - z_0)^{n-1}\, dz \tag{14.8}$$

Note that these all vanish if the function is holomorphic in a disk, rather than just an annulus – then the result reduces to Taylor's theorem, with the corresponding formula for the coefficients.

■ Proof of the theorem

Suppose that $z \in U$ can be written as $z = z_0 + h$, and pick $r_1,\ r_2$ such that

$$0 \leqslant \delta_1 < r_1 < |h| < r_2 < \delta_2 \tag{14.9}$$

How we proceed depends a little on which version of Cauchy's theorem we use. Given that we have the Cauchy–Goursat result for star-shaped regions, it is logical to divide up the annular region into n blocks, each of which is star-shaped, so that we can apply the result.

```
a = {0, 0};
phione[t_] := {Cos[t], Sin[t]}; phitwo[t_] := 2 * phione[t];
inner[t_] := 0.5 * phione[t]; outer[t_] := 2.5 * phione[t];
lines = Table[Line[{{Cos[θ], Sin[θ]}, {2 Cos[θ], 2 Sin[θ]}}],
   {θ, Pi / 16, 31 Pi / 16, Pi / 8}];
ParametricPlot[Evaluate[{phione[t], phitwo[t], inner[t], outer[t]}],
    {t, 0, 2 Pi}, AspectRatio -> 1, Axes -> False, Compiled -> False,
 PlotRange → {{-3, 3}, {-3, 3}}, PlotRegion -> {{0., 1}, {0., 1}},
    PlotStyle -> {Thickness[0.01], Thickness[0.01],
   Thickness[0.005], Thickness[0.005]},
 Epilog -> {PointSize[0.03], Point[a], Point[{1.5, 0}],
            Text["z₀", {0, -0.2}], Text["z", {1.7, -0.1}],
            Text["δ₁", {0, -0.6}], Text["r₁", {0, -1.1}],
            Text["r₂", {0, -2.1}], Text["δ₂", {0, -2.6}],
            Thickness[0.01], lines}]
```

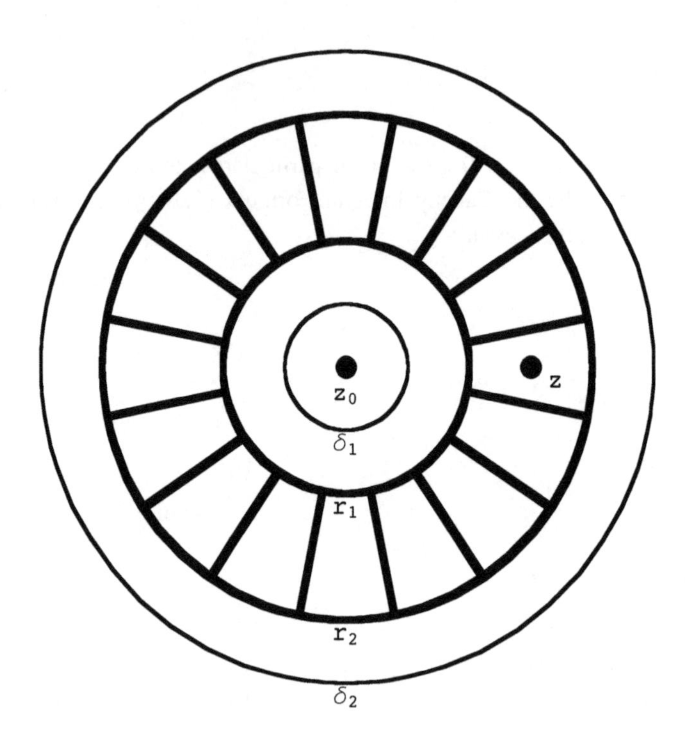

Let the anticlockwise contour bounding each block be denoted γ_i, for $i = 1, \ldots, n$, with $i = 1$ giving the block containing z. By the CIF, using w as the integration variable, we have

$$f(z) = f(z_0 + h) = \frac{1}{2\pi i} \int_{\gamma_1} \frac{f(w)}{(w - (z_0 + h))} dw \tag{14.10}$$

and by applying Cauchy's theorem to each of the other blocks, if $i > 1$,

$$0 = \frac{1}{2\pi i} \int_{\gamma_i} \frac{f(w)}{(w - (z_0 + h))} dw \tag{14.11}$$

Now we add together all these results. The integrals over radial lines cancel in pairs, leaving two integrals:

$$f(z_0 + h) = \frac{1}{2\pi i} \int_{\phi_{r_2}} \frac{f(w)}{(w - (z_0 + h))} dw - \frac{1}{2\pi i} \int_{\phi_{r_1}} \frac{f(w)}{(w - (z_0 + h))} dw \tag{14.12}$$

We have dealt with one of these already, in Taylor's theorem. The integral over ϕ_{r_2} can be expanded precisely as in part (ii) of Taylor's theorem in Section 13.2, with coefficients given by the same formula. This leaves us to deal with

$$\mathcal{J} = \frac{-1}{2\pi i} \int_{\phi_{r_1}} \frac{f(w)}{(w - (z_0 + h))} \, dw = \frac{1}{2\pi i} \int_{\phi_{r_1}} \frac{f(w)}{(h - (w - z_0))} \, dw \tag{14.13}$$

We treat the denominator in the integrand by writing it, with $v = (w - z_0)/h$, as

$$\frac{1}{h(1-v)} = \frac{1 - v^n}{h(1-v)} + \frac{v^n}{h(1-v)} = \frac{1}{h}(1 + v + v^2 + v^{n-1}) + \frac{v^n}{h(1-v)} \tag{14.14}$$

Putting this back in terms of $(w - z_0)/h$ gives

$$\frac{1}{h} + \frac{w - z_0}{h^2} + \frac{(w - z_0)^2}{h^3} + \ldots + \frac{(w - z_0)^{n-1}}{h^n} + \frac{(w - z_0)^n}{h^n (h - (w - z_0))} \tag{14.15}$$

Insertion of this expansion gives

$$\mathcal{J} = \sum_{j=1}^{n} \frac{1}{2\pi i} \int_{\phi_{r_1}} \frac{f(w)(w - z_0)^{j-1}}{h^j} \, dw + \frac{1}{2\pi i} \int_{\phi_{r_1}} \frac{f(w)(w - z_0)^n}{h^n (h - (w - z_0))} \, dw \tag{14.16}$$

With a_{-j} as defined in the statement of the theorem, we can write this as

$$\mathcal{J} = \sum_{j=1}^{n} \frac{a_{-j}}{(z - z_0)^j} + R_n(h) \tag{14.17}$$

For the remainder, we have

$$| R_n(h) | = \left| \frac{1}{2\pi i} \int_{\phi_{r_1}} \frac{f(w)\,(w - z_0)^n}{h^n (h - (w - z_0))} \, dw \right| \tag{14.18}$$

Suppose that $\delta_1 < \sigma < r_1$ and that $| f(z) | < M$ on the path ϕ_{r_1}. Then

$$|R_n(h)| \leq \frac{1}{2\pi} \frac{M\sigma^n}{r_1^n (r_1 - \sigma)} 2\pi\sigma = \frac{M\sigma}{r_1 - \sigma} \left(\frac{\sigma}{r_1} \right)^n \tag{14.19}$$

which tends to zero uniformly as $n \to \infty$, for $|z| \geqslant r_1$. By a similar analysis, looking at a general terms in the series, it follows that

$$\left| \frac{a_{-n}}{(z - z_0)^n} \right| \leqslant M \left(\frac{\sigma}{r_1} \right)^n \tag{14.20}$$

and so the convergence is absolute.

▪ Uniqueness of the coefficients

The fact that the convergence is uniform leads to a simple proof that the coefficients in the Laurent expansion are unique. Suppose that we have some other expansion

$$f(z) = \sum_{n=-\infty}^{\infty} b_n (z - z_0)^n \tag{14.21}$$

that applies within the annulus. Inspection of the positive and negative n terms in this expansion reveals that each is a power series, in $(z - z_0)$ and $1/(z - z_0)$ respectively, and hence is uniformly convergent in the interior and exterior of a pair of disks that overlap in the annulus. Now take the formula

$$a_n = \frac{1}{2\pi i} \int_{\phi_r} \frac{f(z)}{(z - z_0)^{n+1}} \, dz \tag{14.22}$$

and insert the expansion in terms of b_m. Uniformity of the convergence implies that we can swap the order of integration and summation. Evaluation of the integral of each term then gives

$$a_n = b_n \tag{14.23}$$

14.2 Definition of the residue

The residue is just the term a_{-1}, with $n = -1$ in the Laurent series. Note that

$$2\pi i a_{-1} = \int f(z) \, dz \tag{14.24}$$

so that the residue measures the extent to which Cauchy's theorem fails. In fact, there is rather more to it than that – by getting a value for a_{-1}, using independent methods, we can actually evaluate contour integrals with very little work – more on this later!

14.3 Calculation of the Laurent series

The uniqueness result for the coefficients in the Laurent series may seem to be something of a formal dead end, but it is in fact very useful. It means that if we can find *some* method of extracting the coefficients, we know we have found the right ones. There is no one right way of getting a Laurent series. Usually we just need to know a few terms (for residue applications – see below), and it may often be a matter of identifying a handful of the negative terms, or perhaps bolting together known Taylor or binomial expansions with other functions or changes of variable. The following examples are illustrative of the kind of approaches generally taken. **`TraditionalForm`** is used for *Mathematica* output.

■ Laurent example 1

In this case we take a simple function and multiply it by a negative power of z. We consider

$$f(z) = \frac{\sin(z)}{z^3} \tag{14.25}$$

In this case the route is plain – we take the known Taylor series for sin(z) and divide it by z^3. The series for sin(z) is well known, and can be written down without or with any help from *Mathematica*:

```
Series[Sin[z],{z,0,9}]
```

$$z - \frac{z^3}{6} + \frac{z^5}{120} - \frac{z^7}{5040} + \frac{z^9}{362880} + O(z^{10})$$

We just divide this by z^3, leading to the following, which can also be obtained directly with *Mathematica*:

```
Series[Sin[z] / z^3, {z, 0, 4}]
```

$$\frac{1}{z^2} - \frac{1}{6} + \frac{z^2}{120} - \frac{z^4}{5040} + O(z^5)$$

■ Laurent example 2

Frequently, in dealing with rational functions, the binomial theorem can be applied to get expansions about various different points. Depending on the point chosen for the centre of the annulus, we may get a Laurent expansion or just a Taylor expansion. Consider

$$f(z) = \frac{1}{z(1-z)} \tag{14.26}$$

A good start here is to do a partial fraction expansion, either through pen and paper algebra or with the *Mathematica* function **Apart**, with some tidying up of signs by use of **Cancel**:

```
?Apart
```

```
Apart[expr] rewrites a rational expression as
  a sum of terms with minimal denominators. Apart[expr,
  var] treats all variables other than var as constants.
```

```
Cancel[Apart[1/ (z (1 - z))]]
```

$$\frac{1}{z} + \frac{1}{1-z}$$

The binomial theorem can be applied to the second term, to get the expansion

$$\frac{1}{z} + 1 + z + z^2 + z^3 + \ldots \tag{14.27}$$

We can proceed directly with *Mathematica* to obtain:

```
Series[1 / (z (1 - z)), {z, 0, 3}]
```

$$\frac{1}{z} + 1 + z + z^2 + z^3 + O(z^4)$$

This series has a radius of convergence of one. We can also apply the binomial theorem, or *Mathematica*, at other points. For example, expanding this expression about $z = 2$ leads to:

```
Series[1 / (z (1 - z)), {z, 2, 6}]
```

$$-\frac{1}{2} + \frac{3(z-2)}{4} - \frac{7}{8}(z-2)^2 + \frac{15}{16}(z-2)^3 - \frac{31}{32}(z-2)^4 + \frac{63}{64}(z-2)^5 - \frac{127}{128}(z-2)^6 + O((z-2)^7)$$

This also has a radius of convergence of one. We can also expand about infinity. If we put $z = 1/w$, we obtain

$$\frac{w^2}{w-1} \tag{14.28}$$

which we can expand in powers of w using the binomial theorem, for $|w| < 1$, to obtain:

$$-w^2 - w^3 - w^4 - \ldots = -\frac{1}{z^2} - \frac{1}{z^3} - \frac{1}{z^4} - \ldots \tag{14.29}$$

which is valid for $|z| > 1$. This latter series can also be obtained directly with *Mathematica*:

```
Series[1 / (z (1 - z)), {z, Infinity, 6}]
```

$$-\left(\frac{1}{z}\right)^2 - \left(\frac{1}{z}\right)^3 - \left(\frac{1}{z}\right)^4 - \left(\frac{1}{z}\right)^5 - \left(\frac{1}{z}\right)^6 + O\left(\left(\frac{1}{z}\right)^7\right)$$

■ Laurent example 3

Sometimes it may be convenient to combine a known Taylor series with the binomial expansion, especially if just a few terms are required. For example, consider

$$f(z) = \frac{1}{\sin z} = \operatorname{cosec}(z) \tag{14.30}$$

The series for $\sin(z)$ begins

$$z - \frac{z^3}{6} + \frac{z^5}{120} + \ldots \tag{14.31}$$

We can write its reciprocal in the form

$$\frac{1}{z\left(1-\frac{z^2}{6}+\frac{z^4}{120}+\ldots\right)} \tag{14.32}$$

Now we make a binomial expansion of the denominator, obtaining

$$\frac{1}{z}\left(1+\frac{z^2}{6}+\frac{7\,z^4}{360}+\ldots\right)=\frac{1}{z}+\frac{z}{6}+\frac{7\,z^3}{360}+\ldots \tag{14.33}$$

A direct approach with *Mathematica* leads to the same result:

```
Series[Csc[z], {z, 0, 3}]
```

$$\frac{1}{z}+\frac{z}{6}+\frac{7\,z^3}{360}+O(z^4)$$

■ Laurent example 4

Here is a more interesting example:

$$f(z)=e^{\frac{1}{z}} \tag{14.34}$$

We know the series for e^w in the form

$$1+w+\frac{w^2}{2}+\frac{w^3}{6}+\frac{w^4}{24}+O(w^5)$$

We just substitute $w = 1/z$:

$$1+\frac{1}{z}+\frac{1}{2}\left(\frac{1}{z}\right)^2+\frac{1}{6}\left(\frac{1}{z}\right)^3+\frac{1}{24}\left(\frac{1}{z}\right)^4+O\left(\left(\frac{1}{z}\right)^5\right)$$

A similar approach is also required in managing *Mathematica*. The obvious attempt fails, with an interesting message:

```
Series[Exp[1 / z], {z, 0, 3}]
```

— *Series::esss : "Essential singularity encountered in* $e^{\frac{1}{z}+O(z^4)}$.

$$e^{\frac{1}{z}}$$

So we proceed as before:

```
Series[Exp[w], {w, 0, 4}]
```

$$1+w+\frac{w^2}{2}+\frac{w^3}{6}+\frac{w^4}{24}+O(w^5)$$

Now we make a substitution:

```
% /. w -> 1 / z
```

$$1 + \frac{1}{z} + \frac{1}{2}\left(\frac{1}{z}\right)^2 + \frac{1}{6}\left(\frac{1}{z}\right)^3 + \frac{1}{24}\left(\frac{1}{z}\right)^4 + O\left(\left(\frac{1}{z}\right)^5\right)$$

and recover the same result. Note that this last series has infinitely many negative terms. This is related to *Mathematica*'s complaint about essential singularities and will be discussed in Section 14.5.

14.4 Definitions and properties of zeroes

The Laurent series allows us to get a formal grip on various types of singularity in a complex function. However, before exploiting this fully we need to get a grip on various global properties of complex functions, and we need to develop some definitions and theorems to do so. It is illuminating to explore not just singularities, but zeroes also, in order to see the similarities and differences between the two types of object. Naively, one might imagine that a function f becomes singular whenever $1/f$ has a zero, but matters are rather more complicated than this. We shall see, eventually, that certain types of singularity are indeed so associated with zeroes, but not all are.

■ Connected sets and domains revisited

We have already met one definition of a connected (open) set, in Section 12.1. A result we shall take from topology gives us an equivalent characterization: An open subset $U \subset \mathbb{C}$ is *connected* if it *cannot* be written as

$$U = U_1 \bigcup U_2 \tag{14.35}$$

where $U_1 \bigcap U_2 = \phi$ (the sets are disjoint), with U_1 and U_2 non-empty and open. A *domain* is a connected open set in $\mathbb{C}$. Note that this is distinct from the elementary notion of the *domain of definition* of a function. On domains we can make some useful statements about the behaviour of functions. So let U be a domain, and let f be holomorphic on U and *not identically zero*. Then, for each $z \in U$, there is a value of n such that $f^{(n)} \neq 0$. (Note that this is not true for the infinitely differentiable real case, as the example $f(x) = e^{-1/x^2}$ makes clear.)

To prove this result, let

$$A = \{z \in U : f^{(n)} = 0 \text{ for all } n = 0, 1, 2, 3, \ldots\} \tag{14.36}$$

Now if $z \in A$, we can find a neighbourhood $N_r(z) \subset U$. If $w \in N_r(z)$, then $f(w) = 0$ on using the Taylor series expansion about z. So $N_r(z) \subset A$, so A is open. Now let

$$B = \{z \in U : \text{there is an } n \text{ such that } f^{(n)}(z) \neq 0\} \tag{14.37}$$

Now B is open by the continuity of $f^{(n)}$. So we have written

$$U = A \bigcup B, \quad A \bigcap B = \phi \tag{14.38}$$

Since f is not identically zero, B is non-empty, so by connectedness A must be empty.

■ Theorem 14.2: The zeroes of a holomorphic function are isolated

The previous result allows us to prove an important theorem, that the zeroes of a function holomorphic on a domain, and not identically zero, are isolated. We state this carefully thus: *Let* $N = \{z : f(z) = 0\}$. *Then if* $z \in N$, *there is a* $\delta > 0$ *such that*

$$N_\delta(z) \bigcap N = \{z\} \tag{14.39}$$

That is, we can find a neighbourhood around z *on which the function is non-zero except at* z *itself.*

Proof: Suppose that $f(z) = 0$, and pick the smallest n such that $f^{(n)} \neq 0$. There is a neighbourhood of z on which there is a Taylor series

$$\begin{aligned} f(w) &= \sum_{j=n}^{\infty} \frac{f^{(j)}(z)(w-z)^j}{j!} \\ &= (w-z)^n \sum_{j=0}^{\infty} \frac{f^{(n+j)}(z)(w-z)^j}{(n+j)!} = (w-z)^n\, g(w) \end{aligned} \tag{14.40}$$

where $g(w)$ is holomorphic (it is defined by a power series), and $g(z) = f^{(n)}(z)/n! \neq 0$. By continuity of g, there is a neighbourhood of z on which $g(w) \neq 0$, and hence on which $f(w) \neq 0$.

14.5 Singularities

A function can fail to be holomorphic in a variety of ways. A large class of interesting cases can be managed by exploring those singularities that are isolated in the same sense that zeroes are necessarily isolated. This excludes cases such as $\text{Log}(z)$ or $\sqrt{z}$ near the origin, where we have to introduce branch cuts just to have a well-defined function, but does include a large and very important set of possibilities.

■ Definition of an isolated singularity

Suppose that we have an open $U \subset \mathbb{C}$ and that $z \in U$. We define $U^* = U - \{z\}$. We say that a function f that is holomorphic on U^* has an *isolated singularity* at z. Since U is open, $N_r(z) \subset U$ for some r, and we define $N_r{}^*(z) = N_r(z) \bigcap U^*$. This is just a punctured disk, which is a special case of an annulus, so we can write down a Laurent expansion:

$$f(z) = \sum_{n=-\infty}^{-1} a_n(z - z_0)^n + \sum_{n=0}^{\infty} a_n(z - z_0)^n \tag{14.41}$$

■ Classification of isolated singularities

The Laurent series allows us to classify isolated singularities into three types:

(1) The point z is said to be a *removable singularity* if $a_{-n} = 0$ for all $n > 0$.
(2) If $a_{-N} \neq 0$ but $a_{-n} = 0$ for all $n > N$ then z is a *pole of order* N.
(3) If infinitely many negative terms are present, then z is an *isolated essential singularity*.

The first situation may seem puzzling – students often argue that it is not really a singularity in this case! The reason we introduce this case is that there may be very good reasons for excluding the centre of the annulus from the domain of definition of the function, in terms of the way we write down the form of the function. It may only turn out on close inspection that the excluded point is actually OK, and that we may then extend the domain of definition to include the centre. An excellent example of this is the function

$$f(z) = \frac{1}{(z+\pi)\sin(z)} - \frac{1}{\pi z} \tag{14.42}$$

Clearly the origin $z = 0$ is a bad point from the point of view of working out this definition. But this point is in fact a removable singularity. Work it out yourself by computing the Laurent series for this function in an annulus centred on the point $z = 0$.

■ Characterizations of isolated singularity types

The three types of isolated singularity may be characterized in other ways besides the behaviour of the terms in the Laurent expansion. One of these has a special name:

▪ Theorem 14.3: Riemann removable singularities theorem

This may be stated as follows. *If z is an isolated singularity of $f(z)$ then it is a removable singularity if and only if there is an $r > 0$ such that $f(z)$ is bounded on $N_r{}^*(z)$.* Note that one way is obvious - if the function indeed has a Taylor series it is bounded on a neighbourhood of z. For the other way, consider the negative terms in the Laurent series:

$$a_{-n} = \frac{1}{2\pi i} \int_{\phi_r} f(z)(z - z_0)^{n-1}\, dz \tag{14.43}$$

Now f is bounded on a circle centred on z_0 of radius s, for any s with $0 < s < r$, so suppose that $|f| < M$ for $|z| < r$.

$$|a_{-n}| \leqslant \frac{1}{2\pi} M s^{n-1} 2\pi s = M s^n \tag{14.44}$$

Since s can be as small as we wish, $a_{-n} = 0$.

▪ Characterization of poles and essential singularities

We can also assert that:

(1) $f(z)$ has a pole at z_0 if and only if $|f(w)| \to \infty$ as $w \to z_0$;
(2) z_0 is an isolated essential singularity if and only if for each W and each $\epsilon > 0$, there is a value of z such that $|z - z_0|$ and $|f(z) - W| < \epsilon$.

It is easiest to discuss these together. Suppose first that f has a pole of order k. Then the Laurent series is

$$f(w) = \frac{a_{-k}}{(w-z)^k} + \frac{a_{-k+1}}{(w-z)^{k-1}} + \ldots = \frac{1}{(w-z)^k}(a_{-k} + a_{-k+1}(w-z) + \ldots) \tag{14.45}$$

where $a_{-k} \neq 0$. This can be written as

$$f(w) = \frac{h(w)}{(w-z)^k} \tag{14.46}$$

where h has a removable singularity. We remove it, defining $h(z) = a_{-k} \neq 0$. Then as $w \to z$, $h(w) \to a_{-k}$ and $|f| \to \infty$. Now suppose that the conditions of (2) apply. Then f is not bounded, nor does $|f| \to \infty$, so f does not have a removable singularity nor does it have a pole. It must therefore have an isolated essential singularity.

Now suppose that f has an isolated essential singularity and that the result of (2) fails. Therefore there is a W for which the result fails, and an $\epsilon > 0$ such that if $0 < |z - z_0| < \epsilon$, then $|f(z) - W| \geqslant \epsilon$. We can make up a new function

$$g(z) = \frac{1}{f(z) - W} \tag{14.47}$$

that is holomorphic on the punctured disk, and we know that $|g(z)| \leqslant 1/\epsilon$. So by the Riemann removable singularity theorem, g can be extended holomorphically to the whole disk. It has a Taylor series, and can therefore be written as

$$g(z) = (z - z_0)^k h(z) \tag{14.48}$$

where $k \geqslant 0$ and $h(z_0) \neq 0$. Reorganizing, we get f as

$$f(z) = W + \frac{1}{g(z)} = W + \frac{1}{h(z)(z - z_0)^k} \tag{14.49}$$

If $k = 0$, f has a manifest removable singularity, whereas if $k > 0$ it has an explicit pole, which is a contradiction with f having an isolated essential singularity. So there is no W

for which the result fails.

What this result says is that f gets close to any given point. Sometimes you can be more explicit, as in the example of $f(z) = e^{1/z}$. Suppose that

$$W = e^{1/z} \tag{14.50}$$

then

$$z = \frac{1}{i\left(2\pi k + \text{Arg}_p[w]\right) + \log(|W|)} \tag{14.51}$$

and, except for $W = 0$, we see that there are infinitely many solutions of Eq. (14.50) in an arbitrarily small neighbourhood of the origin.

■ Pictures of singularities

We can use *Mathematica*'s three-dimensional plot routines to get a better grip on the behaviour of the singular functions and the way they are characterized. For a removable singularity there is nothing to see, as we merely fill in the hole where the function is defined. For poles and isolated essential singularities there is more to it. We begin by recalling the definition of **ComplexPlot3D**:

```
ComplexPlot3D[func_, xrange_, yrange_, options___] :=
Plot3D[
  {Abs[func[x + I y]], Hue[N[(Pi + Arg[func[x + I y]]) / (2 Pi)]]},
  xrange, yrange, options]
```

You will find the following more interesting if you view the material in colour, and you might like to turn off the mesh by setting it to **False**! In the following examples we shall use *Mathematica*'s pure function notation, where **#** stands for the argument of the function, and the function is terminated by **&**. If you are using *Mathematica* technologies beyond version 5.2, see the enclosed CD and on-line supplement.

▪ Simple pole

In this case the function $f(z) = 1/z$ blows up at the singularity at the origin, and there is a variation of phase (colour) in the function of 2π as one loops around the singularity:

```
ComplexPlot3D[1 / # &, {x, -1, 1}, {y, -1, 1},
 PlotRange -> {0, 40}, PlotPoints -> 40, Mesh -> True]
```

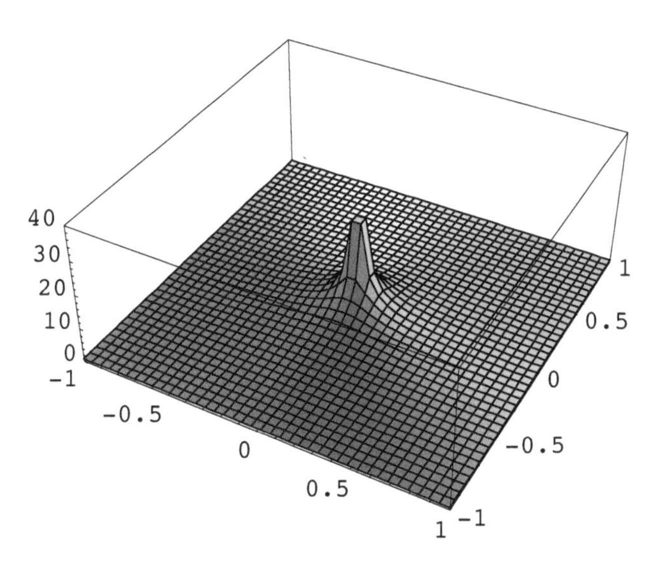

■ Double pole

In this case the function $f(z) = 1/z^2$ blows up more strongly at the singularity, and there is a variation of phase (colour) in the function of 4π as one loops the singularity:

```
ComplexPlot3D[1 / #^2 &, {x, -1, 1}, {y, -1, 1},
 PlotRange -> {0, 80}, PlotPoints -> 40, Mesh -> True]
```

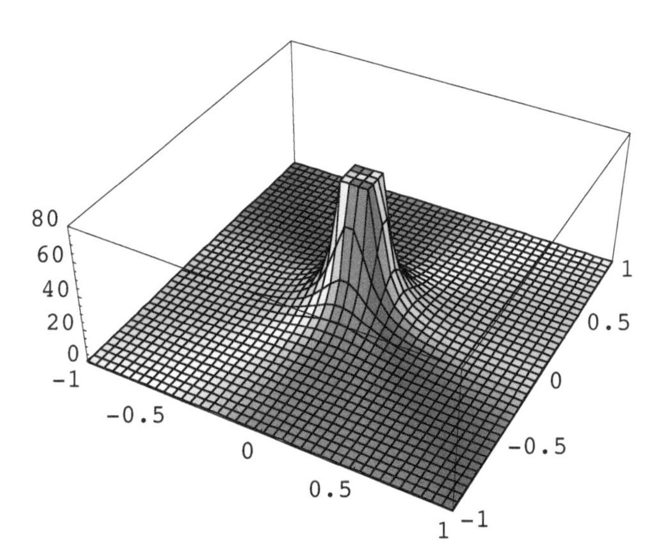

■ Isolated essential singularity

Now the function e^{-1/z^2} jumps all over, with rapid variations in phase around the singular point:

```
ComplexPlot3D[(Exp[-1 / #^2] &), {x, -1, 1},
 {y, -1, 1}, PlotRange -> {0, 80}, PlotPoints -> 60,
   ViewPoint -> {2, 0, 1}, Mesh -> True]
```

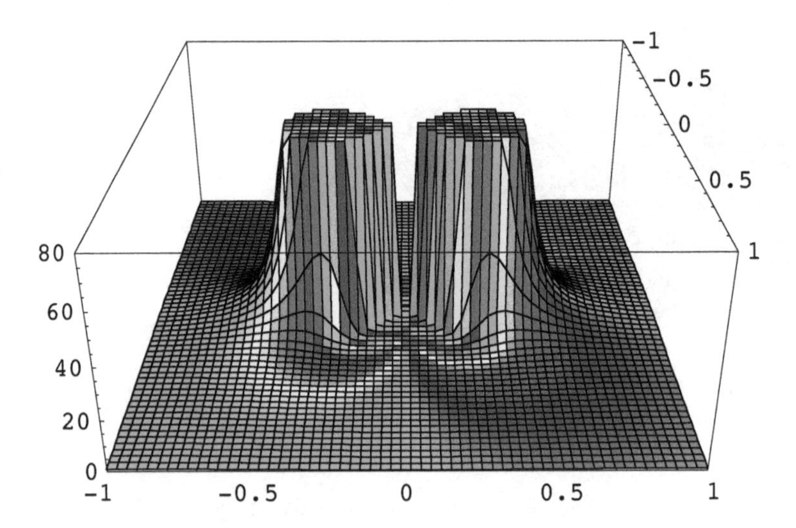

14.6 Computing residues

To get a value for a contour integral we shall need an independent way of calculating a_{-1}. The following results are immensely useful. There are standard results for poles of a given order, but for isolated essential singularities we have to try a variety of approaches.

■ Residues for poles

Let $f(z)$ have a pole of order m at $z = z_0$. Then the residue of $f(z)$ at z_0 is given by

$$(m-1)!\, a_{-1} = \lim_{z \to z_0} \frac{\mathrm{d}^{m-1}((z-z_0)^m f(z))}{\mathrm{d}z^{m-1}} \tag{14.52}$$

In particular, for a simple pole (a pole of order one)

$$a_{-1} = \lim_{z \to z_0} ((z-z_0) f(z)) \tag{14.53}$$

And, for a pole of order two

$$a_{-1} = \lim_{z \to z_0} \frac{\mathrm{d}}{\mathrm{d}z} ((z-z_0)^2 f(z)) \tag{14.54}$$

▪ Proof

If $f(z)$ has a pole of order m then

$$f(z) = \sum_{n=-m}^{\infty} a_n (z - z_0)^n \tag{14.55}$$

so that

$$(z - z_0)^m f(z) = \sum_{n=-m}^{\infty} a_n (z - z_0)^{n+m} \tag{14.56}$$

By relabelling the summation we have

$$(z - z_0)^m f(z) = \sum_{k=0}^{\infty} a_{k-m} (z - z_0)^k \tag{14.57}$$

Now we just differentiate $m - 1$ times, to obtain

$$(m-1)!\, a_{-1} + m!\, a_0 (z - z_0) + O(z - z_0)^2 \tag{14.58}$$

Taking the limit gives the desired result.

■ Rational and related functions

To identify the order of the pole of a function $f(z) = P(z)/Q(z)$, we first make sure that P and Q have had any common factors cancelled, and then we locate the zeroes of the denominator Q. If Q is a polynomial, the order of each pole is then the multiplicity of the zeroes of Q. For more general denominators, we do a Taylor series expansion. If $Q(z_0) = 0$ but $Q'(z_0) \neq 0$, then z_0 is a simple pole. If $Q(z_0) = Q'(z_0) = 0$ but $Q''(z_0) \neq 0$, it is a double pole, etc. Note also that for a simple pole, if we can pull out a linear factor in the denominator, writing

$$f(z) = \frac{P(z)}{(z - z_0)\, q(z)} \tag{14.59}$$

with $P(z_0) \neq 0 \neq q(z_0)$, then the residue is just

$$\frac{P(z_0)}{q(z_0)} = \frac{P(z_0)}{Q'(z_0)} \tag{14.60}$$

which amounts to a useful application of L'Hospital's rule.

14.7 Examples of residue computations

■ Residue example 1

$$f(z) = \frac{1}{a^2 + z^2} \tag{14.61}$$

In this case the denominator factorizes explicitly as

$$(z - ia)(z + ia) \tag{14.62}$$

Now the simple pole formula can be applied directly. For example, working at $z = ia$, multiplying by the factor $(z - ia)$ just amounts to covering up the term $(z - ia)$ in the denominator generating the singularity, leaving

$$\frac{1}{z + ia} \tag{14.63}$$

to be evaluated in the limit as $z \to ia$, which now amounts to just substituting $z = ia$ in what is left, yielding $1/(2ia) = -i/(2a)$. Similarly, the other singularity at $z = -ia$ gives a residue of $1/(-2ia) = i/(2a)$.

■ *Mathematica*-assisted calculation

There are two ways of using *Mathematica* to extract values of residues. The first is to apply the **Residue** function, and the other is to use **Limit**. We shall explore both of these in subsequent examples, beginning here with a direct application of **Residue**:

```
? Residue
```

```
Residue[expr, {x, x0}] finds the residue of expr at the point x = x0.
```

```
Clear[a, f];
f[z_] := 1/(a^2 + z^2)
```

```
Residue[f[z], {z, I a}]
```

$$-\frac{i}{2a}$$

```
Residue[f[z], {z, -I a}]
```

$$\frac{i}{2a}$$

■ Residue example 2

$$f(z) = \frac{1}{(a^2 + z^2)^2} \tag{14.64}$$

In this case the denominator factorizes explicitly as

$$(z - ia)^2 (z + ia)^2 \tag{14.65}$$

So we see a pair of double poles. Again, working first at $z = ia$, we multiply by the square of the factor $z - ia$, and we must calculate

$$\frac{d}{dz}\frac{1}{(z+ia)^2} = -\frac{2}{(ia+z)^3} \tag{14.66}$$

at the point $z = ia$. This evaluates to $-i/(4a^3)$. The other pole's residue is $i/(4a^3)$.

■ *Mathematica*-assisted calculation

```
f[z_] := 1/(a^2 + z^2)^2
```

```
Residue[f[z], {z, I a}]
```

$$-\frac{i}{4a^3}$$

```
Residue[f[z], {z, -I a}]
```

$$\frac{i}{4a^3}$$

■ Residue example 3

Next we consider a more complicated rational function:

$$f(z) = \frac{z^2 - 2z}{(z+1)^2(z^2+4)} \tag{14.67}$$

In this case the denominator factorizes explicitly as

$$(z+1)^2(z+2i)(z-2i) \tag{14.68}$$

and we have a double pole at -1, and a pair of simple poles at $\pm 2i$. To deal with the double pole, first cover up the singular factor, leaving

$$\frac{(z^2-2z)}{(z^2+4)} \tag{14.69}$$

The derivative of this is

$$\frac{2z-2}{z^2+4} - \frac{2z(z^2-2z)}{(z^2+4)^2} \tag{14.70}$$

and evaluation of this at $z = -1$ gives us $-14/25$. For the simple pole at $z = 2i$, we cover up the term $(z - 2i)$ in the denominator, leaving

$$\frac{z^2-2z}{(z+1)^2(z+2i)} \tag{14.71}$$

and evaluation of this at $z = 2i$ gives $(7+i)/25$. The other value is $(7-i)/25$.

■ *Mathematica*-assisted calculation

```
Clear[f];
f[z_] := (z^2 - 2 z) / ((z + 1)^2 (z^2 + 4))
```

We can use the **Residue** function, or take limits using the formulae for single and double poles:

```
Residue[f[z], {z, -1}]
```

$$-\frac{14}{25}$$

```
Residue[f[z], {z, 2 I}]
```

$$\frac{7}{25} + \frac{i}{25}$$

```
Residue[f[z], {z, -2 I}]
```

$$\frac{7}{25} - \frac{i}{25}$$

As an alternative, we can use the built-in **Limit** function (you might like to check the other two limits):

```
Limit[D[(z + 1)^2 f[z], z], z -> -1]
```

$$-\frac{14}{25}$$

■ Residue example 4

$$f(z) = e^{\frac{1}{z}} \tag{14.72}$$

This time there is no alternative but to use the known series for the exponential function:

$$e^{1/z} = 1 + \frac{1}{z} + \frac{1}{2 z^2} + \frac{1}{6 z^3} + \ldots \tag{14.73}$$

We just extract the coefficient of $1/z$ as the residue, which is therefore 1.

■ *Mathematica*-assisted calculation

Residue is not happy at essential singularities:

```
Residue[Exp[1 / x], {x, 0}]
```

$$\text{res}\left(e^{\frac{1}{x}}, \{x, 0\}\right)$$

But you can use **Series** to extract the term in $1/z$ directly:

```
Series[Exp[w], {w, 0, 3}]
```

$$1 + w + \frac{w^2}{2} + \frac{w^3}{6} + O(w^4)$$

```
% /. w -> 1/z
```

$$1 + \frac{1}{z} + \frac{1}{2}\left(\frac{1}{z}\right)^2 + \frac{1}{6}\left(\frac{1}{z}\right)^3 + O\left(\left(\frac{1}{z}\right)^4\right)$$

Another option is to compute the residue by numerical integration using the **NResidue** function that is in the **NumericalMath** package:

```
Needs["NumericalMath`NResidue`"]
```

This is best used by controlling the radius of the circle used to do the integration:

```
NResidue[Exp[1/z], {z, 0}, Radius → 1]
```

$$1. + 4.16334 \times 10^{-17}\, i$$

At an essential singularity the small radius used by default, 1/100, can produce numerical errors:

```
NResidue[Exp[1/z], {z, 0}]
```

— *NIntegrate::ploss :*
"Numerical integration stopping due to loss of precision. Achieved neither the requested PrecisionGoal nor AccuracyGoal; suspect one of the following: highly oscillatory integrand or the true value of the integral is 0. If your integrand is oscillatory on a (semi−)infinite interval try using the option Method−>Oscillatory in NIntegrate.

$$-4.37399 \times 10^{25} - 1.60089 \times 10^{24}\, i$$

Less extreme singularities are happy with the default radius employed by **NResidue**:

```
NResidue[1/z, {z, 0}]
```

1.

This is all fine provided there is no other singularity very close to the one under investigation. Careful analysis is then needed. You might like to consider the example of

```
Exp[1/z] + 1/(z - 1/1000)
```

to see the difficulties that can arise.

■ Residue example 5

The following example raises a number of issues. Consider the function

$$f(z) = \frac{e^z}{\sin^2(z)} \tag{14.74}$$

The denominator is periodic, and vanishes at $z = n\pi$. Its periodicity, given that it is the square of $\sin(z)$, is also π. We let $z = n\pi + w$, and note that

$$f(z) = \frac{e^{\pi n + w}}{\sin^2(\pi n + w)} = e^{n\pi} f(w) \tag{14.75}$$

It is sufficient to work out the residue at $z = 0$, since the residues elsewhere are given by the residue at 0 multiplied by $e^{n\pi}$. At $z = 0$, we have a double pole, as $\sin^2(z) \sim z^2$ near the origin. In contrast to our previous examples, we do not have an explicit factor we can pull out before the differentiation. We can now proceed in one of two ways — apply the formula anyway, or do a manual series approach. In this case the latter is easier, for we can write

$$\begin{aligned}\frac{e^z}{\sin^2(z)} &= \frac{e^z}{\left(z - \frac{z^3}{6} + ..\right)^2} = \frac{e^z}{z^2\left(1 - \frac{z^2}{6} + ..\right)^2} \\ &= \frac{\left(1 + z + \frac{z^2}{2} + ..\right)\left(1 + \frac{z^2}{3} + ...\right)}{z^2} = \frac{1}{z^2} + \frac{1}{z} + ...\end{aligned} \tag{14.76}$$

and identify the residue at $z = 0$ as $+1$. So at $z = n\pi$ the residue is $e^{n\pi}$. Note the manner in which we have treated the function $\sin(z)$. We look at a special point that is easy to manage ($z = 0$) and about which we have a known series. Then the behaviour elsewhere is deduced from the periodic character of the function. This type of analysis extends to functions such as $\cos(z)$, $\sinh(z)$ and $\cosh(z)$.

■ *Mathematica*-assisted calculation

Care must be taken here. There is no problem in identifying what happens at the origin:

```
Clear[f];
f[z_] := Exp[z]/Sin[z]^2

Residue[f[z], {z, 0}]
```

1

Other *specific* values can be checked, by either method:

```
Residue[f[z], {z, Pi}]
```

e^{π}

```
Limit[D[(z - 2 Pi)^2 Exp[z] / Sin[z]^2, z], z -> 2 Pi]
```

$e^{2\pi}$

But you must *not* substitute general values of n, as *Mathematica* does not know that n is necessarily an integer, and fails to spot the singularity in $\sin(z)$:

```
Limit[D[(z - 3 Pi)^2 Exp[z] / Sin[z]^2, z], z -> 3 Pi]
```

$e^{3\pi}$

```
Residue[f[z], {z, n Pi}] /. IntegerQ[n] → True
```

0

What you can do, and what may be very helpful, is to get *Mathematica* to help you spot a pattern:

```
Table[Residue[f[z], {z, n*Pi}], {n, -3, 3}]
```

$\{e^{-3\pi}, e^{-2\pi}, e^{-\pi}, 1, e^{\pi}, e^{2\pi}, e^{3\pi}\}$

Exercises

14.1 Suppose that f and g are functions that are holomorphic in a punctured disc centred on $z = a$, and that they have poles of order m and n respectively at $z = a$. State the nature of the singularity of the following functions:

$$f(z) + g(z); \qquad f(z) \times g(z); \qquad f(z)/g(z)$$

Take care to include *all* possibilities. What, if anything, can be said about the case when both have essential singularities at $z = a$? (For this last part you may find it helpful to consider the function $\exp(1/z)$ and variations of it.)

14.2 Locate and classify all the singularities of the following functions:

$$f(z) = \frac{1}{(z+2)(z^2+1)^2(z-1)^3}$$

$$g(z) = \frac{1}{(z+\pi)\sin(z)} - \frac{1}{\pi z}$$

$$h(z) = e^{\frac{1}{z(z-1)}}$$

$$k(z) = \frac{1}{z\left(e^{\frac{1}{z}} + 1\right)}$$

What is special about the origin $z = 0$ in the last of these four examples?

14.3 For each of the following functions:

(i) locate all the singularities;
(ii) classify each of the singularities;
(iii) calculate the residues at each of the singularities.

$$f(z) = \frac{1}{(a^2 + z^2)(b^2 + z^2)}; \quad g(z) = \frac{1}{(a^2 + z^2)^2 (b^2 + z^2)^2}$$

$$h(z) = \frac{1}{z^6 + 1}; \quad k(z) = \frac{e^z}{\cos(z)}; \quad m(z) = \frac{1}{\cosh(z)}$$

$$p(z) = \frac{e^z}{\sinh^2(z)}; \quad q(z) = \frac{\cot(z)\coth(z)}{z^3}; \quad r(z) = z^3 e^{\frac{1}{z^2}}$$

14.4 Prove, using the Laurent series and the formula for the coefficients, that

$$e^{\frac{1}{2} a(z-1/z)} = \sum_{n=-\infty}^{\infty} J_n(a)\, z^n$$

where

$$J_n(a) = \frac{1}{2\pi} \int_0^{2\pi} \cos(n\theta - a\sin(\theta))\, d\theta$$

(Hint: use a unit circle in the formula for the coefficients.)

14.5 Find all of the zeroes of the polynomial

$$z^4 + 3z^3 - 15z - 7z^2 + 18$$

enlisting *Mathematica*'s help if you get stuck. Hence calculate the residues of the function

$$\frac{z^2 + 1}{z^4 + 3z^3 - 15z - 7z^2 + 18}$$

at each of its singularities.

14.6 ❄ Using the functions **`Residue`**, **`Series`**, **`Limit`** and **`NResidue`**, find as many ways as you can of using *Mathematica* to check your answers to Exercise 14.3. In the case of the functions containing a periodic component, such as $\cos(z)$, you should restrict attention to considering particular points, and try to infer the general pattern from those, in accordance with the comments made at the end of Example 5 in Section 14.7.

14.7 ☸ Use the function **ComplexPlot3D** to visualize the functions of Exercise 14.3.

14.8 ☸ Although it does not distinguish between singularities and zeroes so effectively, the function **ContourPlot** used in conjunction with **Abs** gives a very revealing picture of singularities. For example, try out:

```
ContourPlot[Abs[1/Cosh[(x + I y)]],
  {x, -2, 2}, {y, -20, 20}, PlotPoints -> 120];
```

and explore the other functions in Exercise 14.3 in a similar way.

14.9 ☸ Use *Mathematica* to find out what the function J_n is in Exercise 14.4 (Hint: consider the cases $n = 0, 1$).

15 Residue calculus: integration, summation and the argument principle

Introduction

In this chapter we introduce the methods by which certain types of definite integral may be evaluated. Similar methods may be used to sum certain types of infinite series. The approach has many applications, and will be considered again in Chapter 16, in applications to Fourier transforms, and in Chapter 17, on Laplace transforms. We begin by establishing the Residue theorem, which relates a contour integral to the residues of the integrand at its various singularities. Then we explore how various types of real integral can be transformed into contour integrals, and then evaluated by an analysis of their singularities. Finally we take a brief look at the summation of series by residue methods.

Mathematica can play various roles in this part of the theory related to the evaluation of integrals by the calculus of residues. It can just be there to help with the algebra in calculating residues. You can use the functions **Residue** and **NResidue** to work out the residues directly. Finally you can use **Integrate** and **NIntegrate** to do a direct calculation of the answer. In this last case considerable care is required. The symbolic treatment of general integrals is an evolving (black) art and the results, mostly in the way they are displayed and the full details of conditions for the results to hold, will vary from version to version of the software. This matters particularly when the integrand contains parameters. In general, the most recent version is the most reliable, but not necessarily the quickest. In all cases it is safest to use *Mathematica* to evaluate the integrals in various different ways and check for consistency. This author's advice is to compare the results from:

(1) **Integrate** with **GenerateConditions -> True** (the default setting);
(2) **Integrate** with specific **Assumptions** about parameters;
(3) **Integrate** with parameters set to be specific exact values;
(4) **NIntegrate** with parameters set to be specific exact values.

You must not make the mistake of trying to cut corners by setting **GenerateConditions -> False**, as this may give misleading answers.

15.1 The residue theorem

Theorem 15.1. *Let* $S = \{z_1, z_2, \ldots, z_n\}$ *be a set of points in an open convex subset* U *of* $\mathbb{C}$. *Suppose that* $f(z)$ *is holomorphic on* $U - S$. *Let* ϕ *be a piecewise smooth path whose image is contained in* $U - S$, *i.e. it avoids the singularities, and let* $n(\phi, z_i)$ *be the winding number of the path about* z_i. *Then*

$$\int_{\phi} f(z)\,dz = 2\pi i \sum_{i=1}^{n} n(\phi, z_i)\,\mathrm{Res}(f, z_i) \tag{15.1}$$

■ Proof for simple case

Suppose first that there is just one singularity at $z = a$, and that our contour winds around a just once. We consider a contour as shown inside which the function is holomorphic:

```
h = {1 / 2, 0}; phione[t_] := {Cos[t], Sin[t]};
gamma[t_] :=  phione[Pi / 160 + 0.9875 * t];
delta[t_] :=  h + 0.15 * phione[Pi / 40 + 0.95 * t];
ParametricPlot[Evaluate[{gamma[t], delta[t]}], {t, 0, 2 Pi + 0.01},
 AspectRatio -> 1, Axes -> False, Compiled -> False,
    PlotStyle -> {Thickness[0.01], Thickness[0.01]},
 Epilog -> {PointSize[0.03], Point[h], Text["a", h + {0, -0.07}],
Thickness[0.01],
   Line[{gamma[2 Pi], delta[2 Pi]}], Line[{gamma[0], delta[0]}]}]
```

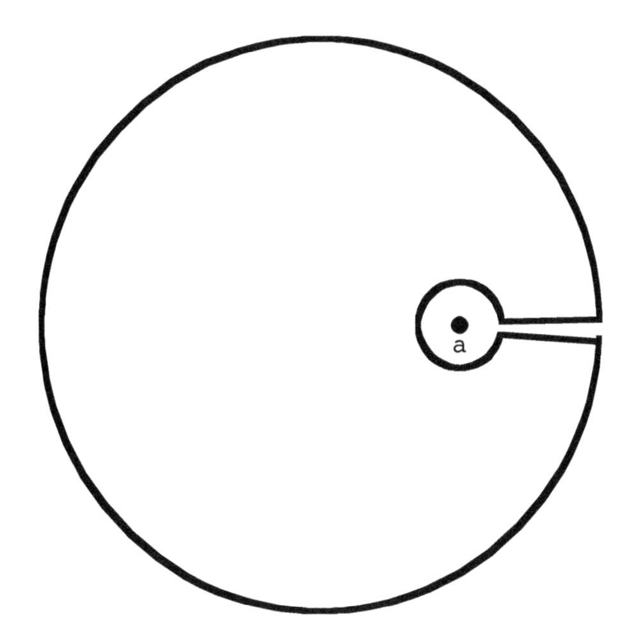

Cauchy's theorem tells us that the integral around the entire contour is zero. Hence we can relate the integral around the outer boundary to that around a small circle centred on the singular point. Now we use the Laurent expansion about a, and the formula

$$2\pi i a_{-1} = \int f(z)\,dz \tag{15.2}$$

applied on the small circle. When there are several singularities we make as many indentations in our original contour as there are singularities, and we add up the results. When the contour winds around any singularity more than once, the contribution is multiplied by the winding number for that singularity.

15.2 Applying the residue theorem

We proceed directly to exploring the means by which this result can be used to evaluate integrals. We shall explore 3 routes to the evaluation:

(1) Traditional mathematical approach;
(2) *Mathematica*-assisted approach;
(3) Direct evaluation within *Mathematica*.

Several examples will be considered that illustrate many different choices of contour. The examples considered are chosen to cover most of the basic cases of interest.

■ Aside: a graphical definition

In order to draw our contour integrals in *Mathematica*, we shall need some definitions of graphical objects frequently encountered. We introduce symbols for poles of various orders.

```
SimplePole[{x_, y_}, r_] :=
  {Disk[{x, y}, 0.6 * r], Circle[{x, y}, r]};
DoublePole[{x_, y_}, r_] := {Disk[{x, y}, 0.6 * r],
   Circle[{x, y}, r], Circle[{x, y}, 0.8 * r]};
TriplePole[{x_, y_}, r_] :=
  {Disk[{x, y}, 0.6 * r], Circle[{x, y}, r],
   Circle[{x, y}, 0.8 * r], Circle[{x, y}, 0.9 * r]};
HighPole[{x_, y_}, r_] := {Disk[{x, y}, 0.6 * r],
  Table[Circle[{x, y}, (0.6 + k / 20) * r], {k, 1, 8}]}
EssSing[{x_, y_}, r_] := Disk[{x, y}, r]

Show[Graphics[
    {SimplePole[{1, 1}, 1], DoublePole[{5, 1}, 1],
     TriplePole[{1, 5}, 1], HighPole[{5, 5}, 1],
     EssSing[{3, 3}, 1]}, AspectRatio -> 1]]
```

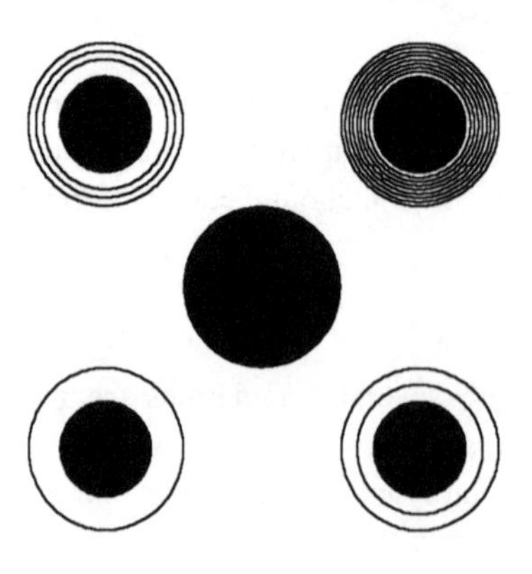

15.3 Trigonometric integrals

We are interested here in integrals of the form

$$\int_0^{2\pi} G(\cos(t), \sin(t))\, dt \tag{15.3}$$

The integration from 0 to 2π suggests a circular contour that may be parametrized by t. What we wish to do is to reverse the process by which a complex contour integral is reduced to a standard real integral by the introduction of a parametrization. We wish to take the given real integral and represent it by a contour integral.

This particular example illustrates the need to take care that the mapping into the complex plane is holomorphic – a common mistake is to represent the trigonometric functions as obvious non-holomorphic functions.

■ Example 1

$$\int_0^{2\pi} \frac{1}{8\cos^2(t) + 1}\, dt \tag{15.4}$$

In this case, we introduce a circular contour parametrized by t, obtained by setting

$$z = e^{it}, \ \cos(t) = \frac{1}{2}(e^{it} + e^{-it}) \tag{15.5}$$

The cosine function can be represented in two ways, when z is on the unit circle:

$$\cos(t) = \frac{1}{2}(z + \bar{z}) = \frac{1}{2}\left(z + \frac{1}{z}\right) \tag{15.6}$$

It is vital that the second and *holomorphic* form is chosen for the mechanics of the residue theorem to operate properly. Also, do not worry that the singular function $1/z$ has been introduced – in this context 'singularities are good' – they are what help us evaluate the integral! For subsequent manipulations with *Mathematica* we introduce the following function:

```
cosof[z] := (z + 1 / z) / 2
```

Next, we need to make the change of variable $t \to z$ under the integral, obtained by mapping the differential according to the rule:

$$dt \to -\frac{i\,dz}{z} \tag{15.7}$$

■ Traditional mathematical approach

By multiplying the integrand top and bottom by z, and then expanding the denominator, the integrand reduces to:

$$-\frac{iz}{2z^4 + 5z^2 + 2} \tag{15.8}$$

The denominator is a quadratic in z^2, and by application of the formula for a quadratic, or by factorization by inspection, the roots are easily found to be $-1/2$ and -2. We do not care about the square roots of -2, since they are outside the unit circle, but we do want the other two square roots of $-1/2$. These are

$$\{-\frac{i}{\sqrt{2}}, \frac{i}{\sqrt{2}}\} \tag{15.9}$$

There are four roots on the imaginary axis, as shown here, together with the contour that is the unit circle:

```
ci =
    {SimplePole[{0, Sqrt[2]}, 0.1],
        SimplePole[{0, -Sqrt[2]}, 0.1],
        SimplePole[{0, 1/Sqrt[2]}, 0.1],
        SimplePole[{0, -1/Sqrt[2]}, 0.1],
        Thickness[0.01], Circle[{0, 0}, 1]};
cplot = Show[Graphics[ci, AspectRatio -> Automatic,
            Axes -> True, Ticks -> None, PlotRange -> {{-2, 2}, {-2, 2}}]]
```

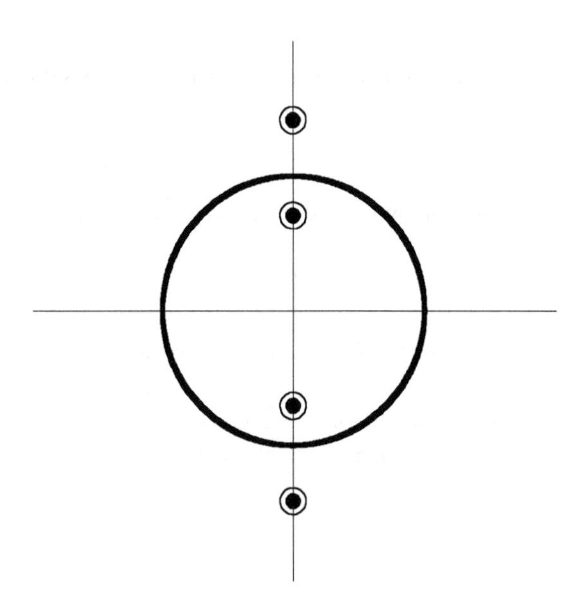

Since we know the roots of the denominator, the integrand can be factorized about the two roots of interest and the residues computed by the method of covering up the singular factor. Here it is easy to use the differentiation formula for a simple pole. So set

$$P(z) = -iz\ ; \quad Q(z) = 2z^4 + 5z^2 + 2 \tag{15.10}$$

The function giving the residues of the function $P(z)/Q(z)$ is then just:

$$\text{Resfunc}(z) = \frac{P(z)}{Q'(z)} = -\frac{iz}{8z^3 + 10z} \tag{15.11}$$

Evaluating this at the two poles gives

$$\text{Resfunc}\left(-\frac{i}{\sqrt{2}}\right) = -\frac{i}{6} \tag{15.12}$$

$$\text{Resfunc}\left(\frac{i}{\sqrt{2}}\right) = -\frac{i}{6} \tag{15.13}$$

so that the value of the integral is just

$$2\pi i\left(-\frac{i}{6} - \frac{i}{6}\right) = \frac{2\pi}{3} \tag{15.14}$$

■ *Mathematica*-assisted approach

First we define the integrand for *Mathematica* and ask for its simplification:

```
integrand = -(I/(z*(8*cosof[z]^2 + 1)))
```

$$-\frac{i}{z\left(2\left(z+\frac{1}{z}\right)^2+1\right)}$$

```
simp = Simplify[integrand]
```

$$-\frac{i z}{2 z^4 + 5 z^2 + 2}$$

The singularities of the integrand are given by the zeroes of the polynomial denominator, so the denominator is defined as a polynomial expression, and its roots found:

```
poly = Denominator[simp]
```

$$2 z^4 + 5 z^2 + 2$$

```
z /. Solve[poly == 0, z]
```

$$\left\{-\frac{i}{\sqrt{2}}, \frac{i}{\sqrt{2}}, -i\sqrt{2}, i\sqrt{2}\right\}$$

Note that the denominator is a fourth-order polynomial and we have found four distinct roots – there are no multiple roots. So the four roots give four simple poles. Two are obviously inside the unit circle. We can use the differentiation method used above, or take limits directly. We shall ask *Mathematica* to do the latter. For each of these, we multiply by the singular factor and take the limit using the `Limit` function:

```
resone = Simplify[Limit[(z + I/√2) integrand, z → -I/√2]]
```

$$-\frac{i}{6}$$

```
restwo = Simplify[
  Limit[(z - I / Sqrt[2]) integrand, z -> I / Sqrt[2]]]
```

$$-\frac{i}{6}$$

The value of the integral is given by $2\pi i$ times the sum of the residues:

```
2 π I (resone + restwo)
```

$$\frac{2\pi}{3}$$

You might like to look at using **Residue** as another method.

■ Direct *Mathematica* evaluation check

This can be done directly using *Mathematica*'s definite integration routine. Here the *Mathematica* input is shown in **InputForm**:

```
Integrate[1/(8*Cos[t]^2 + 1), {t, 0, 2*Pi}]
```

$$\frac{2\pi}{3}$$

In some older versions of *Mathematica* you might have got a more complicateed answer. In such circumstances you should use **Simplify** or **FullSimplify** to boil the result down to simplest form.

■ Use of *Mathematica*'s numerical residue function

When residues are tricky, perhaps because of a very high order pole or essential singularity, there is a numerical function available as a package, which we now load:

```
Needs["NumericalMath`NResidue`"]
```

Here is a reminder of how to find what functions have been added:

```
?NumericalMath`NResidue`*
```

NumericalMath`NResidue`

```
NResidue Radius
```

Here is how to find out what each one does:

```
?NResidue
```

```
NResidue[expr, {x, x0}] uses NIntegrate to
  numerically find the residue of expr near the point x = x0.
```

```
?Radius
```

```
Radius is an option to NResidue that specifies the
  radius of the circle on which the integral is evaluated.
```

Let's apply this to the problem at hand:

```
nresone = NResidue[integrand, {z, I/Sqrt[2]}]
```

$-1.43928\times 10^{-17} - 0.166667\, i$

```
nrestwo = NResidue[integrand, {z, -I/Sqrt[2]}]
```

$-9.45966\times 10^{-18} - 0.166667\, i$

```
2 Pi I * (nresone + nrestwo)
```

$2.0944 - 1.49869\times 10^{-16}\, i$

```
Chop[%]
```

2.0944

```
2 Pi/3 // N
```

2.0944

Sometimes, because of the numerical method used, very small numbers may appear in the final or, as here, intermediate results. *Mathematica* does a numerical integration around a small circle to work out the residue! It is often useful to use the **Chop** function to remove machine-precision glitches:

```
Chop[nresone]
```

$-0.166667\, i$

This example illustrates that you have many options involving varying degrees of intervention by *Mathematica.*

■ Another trigonometric example with double pole

It is worth doing another example of this type, to see explicitly how to cope with an appearance of the sine function and how to manage double poles. In this case we shall do a standard mathematical evaluation and just check the result against that arising from the *Mathematica* integrator. The integral of interest is, for $a > b > 0$,

$$\int_0^{2\pi} \frac{\sin^2(t)}{a + b\cos(t)}\, dt \tag{15.15}$$

The ingredients we need are:

$$z = e^{it} \tag{15.16}$$

$$\cos(t) = \frac{1}{2}\left(z + \frac{1}{z}\right) \tag{15.17}$$

$$\sin(t) = \frac{1}{2i}\left(z - \frac{1}{z}\right) \tag{15.18}$$

$$dt = -\frac{i\,dz}{z} \tag{15.19}$$

The integrand is therefore

$$\frac{i(z - \frac{1}{z})^2}{4z(a + \frac{1}{2}b(z + \frac{1}{z}))} \tag{15.20}$$

and, after a little simplification, this becomes

$$\text{integrand} = \frac{i\,(z^2 - 1)^2}{2\,z^2\,(b\,z^2 + 2\,a\,z + b)} \tag{15.21}$$

There is manifestly a double pole at the origin, and a pair of poles at the roots of the polynomial

$$\text{poly} = bz^2 + 2az + b \tag{15.22}$$

These roots are the list:

$$\left\{\frac{-a - \sqrt{a^2 - b^2}}{b}, \frac{\sqrt{a^2 - b^2} - a}{b}\right\} \tag{15.23}$$

Clearly the first of these lies outside the unit circle, since $a > b$. Inspection of the quadratic reveals that the product of the roots is one, so the second must be inside. So let

$$z_1 = \frac{-a - \sqrt{a^2 - b^2}}{b} \tag{15.24}$$

$$z_2 = \frac{\sqrt{a^2 - b^2} - a}{b} \tag{15.25}$$

As an example, with $a = 2$, $b = 1$, the roots are as shown:

```
ci = {DoublePole[{0, 0}, 0.1],
SimplePole[{-2 - Sqrt[3], 0}, 0.1],
SimplePole[{-2 + Sqrt[3], 0}, 0.1],
Thickness[0.01], Circle[{0, 0}, 1]};
cplot = Show[Graphics[ci, AspectRatio -> Automatic,
   Axes -> True, Ticks -> None, PlotRange -> {{-4, 4}, {-2, 2}}]]
```

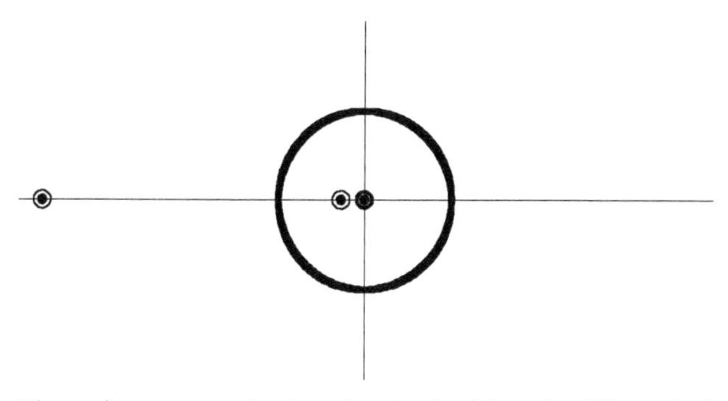

The residue at zero is given by the double-pole differentiation formula. First we differentiate the integrand times the singular factor:

$$\text{deriv} = \frac{\text{d}(z^2 \text{ integrand})}{\text{d}z} = \frac{2iz(z^2 - 1)}{bz^2 + 2az + b} - \frac{i(2a + 2bz)(z^2 - 1)^2}{2(bz^2 + 2az + b)^2} \tag{15.26}$$

Now the limit at the double pole at the origin is taken:

$$\lim_{z \to 0} \text{deriv} = -\frac{ia}{b^2} \tag{15.27}$$

For the other simple pole at z_2 we can use the polynomial-ratio formula, or cover up the factor and take limits:

$$\lim_{z \to z_2} [(z - z_2) \text{ integrand}] = \frac{ib^2\left(\frac{\left(\sqrt{a^2-b^2} - a\right)^2}{b^2} - 1\right)^2}{2\left(\sqrt{a^2 - b^2} - a\right)^2 \left(2\,a + 2\left(\sqrt{a^2 - b^2} - a\right)\right)} \tag{15.28}$$

This simplifies to

$$\frac{i\sqrt{a^2 - b^2}}{b^2} \tag{15.29}$$

Adding this to the residue at the origin and multiplying by $2\pi i$ gives the value of the integral as:

$$2i\left(-\frac{ia}{b^2}+\frac{i\sqrt{a^2-b^2}}{b^2}\right)\pi \tag{15.30}$$

and this simplifies to

$$\frac{2\pi\left(a-\sqrt{a^2-b^2}\right)}{b^2} \tag{15.31}$$

■ Checks against *Mathematica*

A *Mathematica*-assisted calculation can be done along the same lines as considered previously. In this case we look at just checking the results against *Mathematica*'s symbolic integrator. Now in our example we were assuming that $a > b > 0$. Just using **`Integrate`** without feeding in these assumptions leads to the following, where we have added a timing calculation and are using a fresh kernel session. This sort of result can drive *Mathematica* beginners crazy!

```
Timing[Integrate[Sin[t]^2/(a + b*Cos[t]), {t, 0,
2*Pi}]]
```

$$\Big\{44.1\text{ Second, If}\Big[b \neq a \bigwedge a+b \neq 0 \bigwedge \left(\text{Re}\left(\frac{a}{b}\right) \geq 1 \bigvee \text{Re}\left(\frac{a}{b}\right)+1 \leq 0 \bigvee \left(\text{Im}\left(\frac{a}{b}\right) \neq 0 \bigwedge \text{Im}\left(\frac{a+b}{b}\right) \neq 0\right)\right),$$

$$\frac{2\left(\pi a-\sqrt{b^2-a^2}\,\log\left(\frac{a-b}{\sqrt{b^2-a^2}}\right)+\sqrt{b^2-a^2}\,\log\left(\frac{b-a}{\sqrt{b^2-a^2}}\right)\right)}{b^2},$$

$$\text{Integrate}\left[\frac{\sin^2(t)}{a+b\cos(t)}, \{t, 0, 2\pi\}, \text{Assumptions} \to \left(\text{Re}\left(\frac{a}{b}\right)+1>0 \bigwedge \text{Re}\left(\frac{a}{b}\right)<1 \bigwedge \left(\text{Im}\left(\frac{a}{b}\right)=0 \bigvee \text{Im}\left(\frac{a+b}{b}\right)=0\right)\right) \bigvee b=a \bigvee a+b=0\right]\Big]\Big\}$$

Clearly the integrator has taken some time to work out some very complicated conditions for this integral to make sense for all possible complex a and b. Let us now feed our assumptions about a and b to **`Integrate`**:

```
Timing[Integrate[Sin[t]^2/(a + b*Cos[t]), {t, 0,
2*Pi}, Assumptions -> a>b>0]]
```

$$\left\{16.8\text{ Second}, \frac{2\left(a-\sqrt{a^2-b^2}\right)\pi}{b^2}\right\}$$

Note that supplying an inequality for a or b implicitly says that they must be real, as general complex numbers are not ordered! Finally, a warning: in some versions of

Mathematica it is not safe to just use **GenerateConditions -> False**. You might like to see what happens in this example if you try this. Again, if your version of *Mathematica* supplies a more complicated answer, use **Simplify** and **FullSimplify** on it.

15.4 Semicircular contours

A large class of real integrals involve integration from negative infinity to positive infinity. Many of these may be treated by adding a large semicircular completion contour in either the upper or lower half-plane. One then makes an argument that the contribution of the semicircle can be ignored, and proceeds to evaluate the residues in one of the half-planes. There are two levels of the theorem that allow us to discard the semicircular contribution. The first version makes strong assumptions about the decay of the integrand at infinity, and is easy to prove using simple inequalities. The second version, usually known as Jordan's lemma, makes weaker assumptions and is harder to prove. We shall defer a treatment of this until we consider Fourier and Laplace transforms, which is the main area of application of Jordan's lemma. Note that an integral from zero to infinity can be written as one from minus infinity to infinity if the integrand is even – otherwise special methods are required. Also, the nature of the process requires that integrals from minus infinity to infinity be interpreted in a special way – this leads us to introduce the notion of the 'Cauchy Principal Value'.

We motivate the discussion by working with a particular example, which is the integral, for $a \neq b$, and a and b real and positive:

$$\int_0^{\infty} \frac{1}{(a^2 + x^2)(b^2 + x^2)}\, dx \tag{15.32}$$

The integrand is an even function of x, so that the integral under consideration may be written as

$$\frac{1}{2}\int_{-\infty}^{\infty} \frac{1}{(a^2 + x^2)(b^2 + x^2)}\, dx \tag{15.33}$$

This is not an integral over a closed contour. We need to add a piece, in the form of a large semicircle in either the upper or lower half plane. To check that this gives zero we need the following definitions and theorem.

■ Cauchy principal values and semicircle theorem I

To understand the evaluation of integrals along the entire real axis by the use of contour methods, we need to appreciate their definition more carefully. Normally, a doubly infinite integral would be defined by requiring the existence of the double limit (i.e. R and S tend to infinity independently):

$$\int_{-\infty}^{\infty} f(x)\, dx = \lim_{R\to\infty} \lim_{S\to\infty} \int_{-R}^{S} f(x)\, dx \tag{15.34}$$

In our discussion of contour integrals, it is convenient to relax this definition to the Cauchy Principal Value, usually denoted by putting a 'P' before the integral:

$$P\int_{-\infty}^{\infty} f(x)\, dx = \lim_{R\to\infty} \int_{-R}^{R} f(x)\, dx \tag{15.35}$$

This can make a real difference to whether an integral exists! For example, the function

$$f(x) = \frac{a}{\pi(a^2 + x^2)} \tag{15.36}$$

occurs in elementary probability, for real $a > 0$. It is called the 'Cauchy Distribution', and is a probability density function (p.d.f.) by virtue of the fact that it is non-negative and

$$\int_{-\infty}^{\infty} \frac{a}{\pi\,(a^2 + x^2)}\, dx = 1 \tag{15.37}$$

The mean value of the associated random variable is given by the double limit of

$$\int_{-R}^{S} \frac{x\,a}{\pi(a^2 + x^2)}\, dx \tag{15.38}$$

which evaluates to

$$\frac{a \log(a^2 + S^2)}{2\pi} - \frac{a \log(a^2 + R^2)}{2\pi} \tag{15.39}$$

Clearly, if we let R and S approach infinity independently the answer can be any value we like! However, if we set $R = S$ the answer is zero for all finite R, and hence the principal value is zero. This makes sense – the p.d.f. is an even function so the expected value of the random variable should be zero, since positive and negative values of the same magnitude carry the same weight in the p.d.f. Or, to put it another way, the integrand for the expected value is an odd function, so we should expect the total area under the curve to be zero.

When we work with principal value integrals, the contour along the real axis from $-R$ to R is closed by attaching a large semicircle. The question arises as to whether we should use the upper or lower half-plane. For ratios of polynomials, it does not matter, provided certain conditions are satisfied. For integrands containing trigonometric or exponential functions, it is critical that the right choice is made, and this is also very important for the development of the theory of Fourier and Laplace transforms.

Our first semicircle theorem is the following. The proof comes with the statement! **Theorem 15.1**: *suppose $f(z)$ satisfies the condition*

$$|f(z)| \leq \frac{M}{R^k} \tag{15.40}$$

on the contour Γ, parametrized by

$$z = Re^{i\theta} \tag{15.41}$$

for either $0 \leqslant \theta \leqslant \pi$ *(upper half-plane) or* $\pi \leqslant \theta \leqslant 2\pi$ *(lower half-plane), where* $k > 1$ *and* M *are constants. Then, by the length inequality,*

$$\left| \int_\Gamma f(z)\,dz \right| \leq \frac{M\pi R}{R^k} = \frac{\pi M}{R^{k-1}} \tag{15.42}$$

because the arc-length of Γ *is* πR*. This vanishes as* $R \to \infty$*, provided* $k > 1$. For ratios of polynomials, it is clearly more than sufficient that the degree of the denominator is 2 or more greater than that of the numerator.

▪ Back to our integral

$$f(z) = \frac{P(z)}{Q(z)}; \quad P(z) = 1; \quad Q(z) = 2(a^2 + z^2)(b^2 + z^2) \tag{15.43}$$

The denominator clearly has zeroes of multiplicity one at the points

$$\text{poles} = \{ia, -ia, ib, -ib\} \tag{15.44}$$

and these are simple poles. There are two in the upper half plane, at $\{ia, ib\}$. In this case we can appeal to the rule for just differentiating the denominator, which uses the function

$$q(z) = \frac{\mathrm{dQ}(z)}{\mathrm{dz}} = 4z(a^2 + z^2) + 4z(b^2 + z^2) \tag{15.45}$$

The residues are, first at ia,

$$\frac{1}{q(ia)} = -\frac{i}{4a(b^2 - a^2)} \tag{15.46}$$

and then at ib,

$$\frac{1}{q(ib)} = -\frac{i}{4b(a^2 - b^2)} \tag{15.47}$$

$$\tag{15.48}$$

So the total value of the integral is

$$\begin{aligned} 2\pi i\left(\frac{1}{q(ib)} + \frac{1}{q(ia)}\right) &= 2i\left(-\frac{i}{4b(a^2 - b^2)} - \frac{i}{4a(b^2 - a^2)}\right)\pi \\ &= \frac{\pi}{2ab(a+b)} \end{aligned} \tag{15.49}$$

Clearly the intermediate steps break down if $a = b$. In this case there is a double pole, and the result is obtained by setting

$$g(z\_) := \frac{1}{2(a^2 + z^2)^2} \tag{15.50}$$

and evaluating the integral as

$$2\pi i \left(\lim_{z \to i a} \frac{d((z - i a)^2 g(z))}{d z} \right) = \frac{\pi}{4a^3} \tag{15.51}$$

■ **Final check against *Mathematica***

We can of course check the result using *Mathematica* directly. You can get a very complicated answer if you do not feed in assumptions about a and b, but let's do it right first time now:

```
Integrate[1/((x^2 + a^2)*(x^2 + b^2)),{x, 0,
Infinity}, Assumptions -> {a>0, b>0}]
```

$$\frac{\pi}{2 b a^2 + 2 b^2 a}$$

This, of course, confirms our residue calculation. Note that *Mathematica* is telling us that this answer is valid even if $a = b$. The condition that a and b be distinct was introduced to force the poles to be simple in our residue calculation. The result holds when $a = b$, but the detailed residue calculation must be done using the results for double poles. You should do this calculation.

15.5 Semicircular contour: easy combinations of trigonometric functions and polynomials

Our first semicircle theorem is sufficient to cope with products of trigonometric functions and rational functions, provided that we pick a semicircle in the correct half-plane, and the polynomial component in the denominator grows sufficiently rapidly at infinity. As an example, consider:

$$\int_{-\infty}^{\infty} \frac{\sin(x)}{x^2 + x + 1}\, dx \tag{15.52}$$

We cannot complete the contour in either the lower or the upper half-plane as things stand, because the sine function blows up in each case. The trick is to consider the integral as the imaginary part of

$$\int_{-\infty}^{\infty} \frac{e^{ix}}{x^2 + x + 1}\, dx \tag{15.53}$$

Now we can complete in the upper half-plane, since the numerator is less than unity in absolute value there, and the denominator sends the integrand to zero sufficiently fast for our semicircle theorem to apply. Now, where are the poles? These are at the zeroes of the denominator. This can be found by use of the formula for a quadratic, leading to the factorization. Replacing x by the complex variable z, we have

$$z^2 + z + 1 = \left(z + \frac{1}{2} + \frac{i\sqrt{3}}{2}\right)\left(z + \frac{1}{2} - \frac{i\sqrt{3}}{2}\right) \tag{15.54}$$

The pole in the upper half plane is at

$$z = -\frac{1}{2} + \frac{i\sqrt{3}}{2} \tag{15.55}$$

So the residue can be evaluated as

$$\lim_{z \to -\frac{1}{2} + \frac{i\sqrt{3}}{2}} \frac{e^{iz}}{z + \frac{1}{2} + \frac{i\sqrt{3}}{2}} = -\frac{i e^{\frac{1}{2}(-i-\sqrt{3})}}{\sqrt{3}} \tag{15.56}$$

Therefore, taking the real and imaginary parts as

$$\frac{2e^{-\sqrt{3}/2}\pi\cos(\frac{1}{2})}{\sqrt{3}} - \frac{2ie^{-\sqrt{3}/2}\pi\sin(\frac{1}{2})}{\sqrt{3}} \tag{15.57}$$

we obtain the desired result as:

$$-\frac{2e^{-\sqrt{3}/3}\pi\sin(\frac{1}{2})}{\sqrt{3}} \tag{15.58}$$

Note that at the same time we have also derived, by taking real parts, the result that

$$\int_{-\infty}^{\infty} \frac{\cos(x)}{x^2 + x + 1}\, dx = \frac{2e^{-\sqrt{3}/2}\pi\cos(\frac{1}{2})}{\sqrt{3}} \tag{15.59}$$

It is quite commonplace to obtain the value of a further integral in addition to the value of the integral originally sought.

▪ Direct *Mathematica* evaluation

This time there are no assumptions to be set. The answer for the sine integral comes back in the form we want right away, but the cosine integral needs a little simplification.

```
Integrate[{Sin[x], Cos[x]}/(x^2 + x + 1),
{x, -Infinity, Infinity}]
```

$$\left\{-\frac{2\,e^{-\frac{\sqrt{3}}{2}}\,\pi\sin(\frac{1}{2})}{\sqrt{3}}, \frac{(1 + e^{i})\,e^{-\sqrt[6]{-1}}\,\pi}{\sqrt{3}}\right\}$$

```
Simplify[ComplexExpand[%[[2]]]]
```

$$\frac{2\,e^{-\frac{\sqrt{3}}{2}}\,\pi\cos(\frac{1}{2})}{\sqrt{3}}$$

The details of the output may depend on the version you are using, but the answers can be reconciled with such simplification.

15.6 Mousehole contours

▪ Theorem 15:2

*Suppose that $f(z)$ has a **simple** pole at a, with residue σ. Let $\phi_\epsilon(t)$ be the path*

$$\phi_\epsilon(t) = a + \epsilon\, e^{it}, \quad \alpha \leqslant t \leqslant \beta \tag{15.60}$$

Then

$$\lim_{\epsilon \to 0} \int_{\phi_\epsilon} f(z)\, dz = (\beta - \alpha) i \sigma \tag{15.61}$$

This is typically used to obtain the contributions of small semicircular indentations in linear contours. Note that it only makes sense for simple poles. The proof is easy – just write out the Laurent series

$$f(z) = g(z) + \frac{\sigma}{z - a} \tag{15.62}$$

where $g(z)$ is holomorphic, and evaluate the integral

$$\begin{aligned} \int_{\phi_\epsilon} f(z)\, dz &= \int_{\phi_\epsilon} \frac{\sigma}{z - a}\, dz + \int_{\phi_\epsilon} g(z)\, dz \\ &= \int_\alpha^\beta \frac{\sigma i \epsilon e^{it}}{\epsilon e^{it}}\, dt + \int_{\phi_\epsilon} g(z)\, dz = (\beta - \alpha) i \sigma + \int_{\phi_\epsilon} g(z)\, dz \end{aligned} \tag{15.63}$$

Taking limits sends the latter integral to zero, leaving the desired result. The classic type of application of the Mousehole result is exemplified by the evaluation of

$$\int_0^\infty \frac{x - \sin(x)}{x^3}\, dx \tag{15.64}$$

We first note that the integrand is even, and so consider half the value of

$$\int_{-\infty}^\infty \frac{x - \sin(x)}{x^3}\, dx \tag{15.65}$$

First we need to consider how to put the integrand in complex form. Our first guess might be to regard the real expression as the imaginary part of

$$f(z) = \frac{iz - e^{iz}}{z^3} \tag{15.66}$$

but at the origin this now has a triple pole. One further correction does the trick:

$$f(z) = \frac{iz + 1 - e^{iz}}{z^3} \tag{15.67}$$

Expansion of the exponential in the numerator shows that

$$f(z) = \frac{1}{2z} + \frac{i}{6} - \frac{z}{24} + \ldots \tag{15.68}$$

There is now only one singularity in f, at the origin, where there is a simple pole with residue 1/2. Now we can apply our semicircle theorem in the upper half-plane, as the denominator is degree two greater than the polynomial part of the numerator, and the exponential part of the numerator is well behaved. We therefore use the contour shown for evaluation:

```
ci = {SimplePole[{0, 0}, 0.25], Thickness[0.01], Circle[{0, 0}, 0.5, {0, Pi}],
Circle[{0, 0}, 5, {0, Pi}],
   Line[{{0.5, 0}, {5, 0}}], Line[{{-0.5, 0}, {-5, 0}}],
        Text["C₁", {-2.5, -0.3}], Text["C₂", {2.5, -0.3}],
        Text["C₃", {0.4, 0.7}], Text["C₄", {2.5, 4}]};
cplot =
 Show[Graphics[ci, AspectRatio -> Automatic, Axes -> True, Ticks -> None,
           PlotRange -> {{-6, 6}, {-1, 6}}]]
```

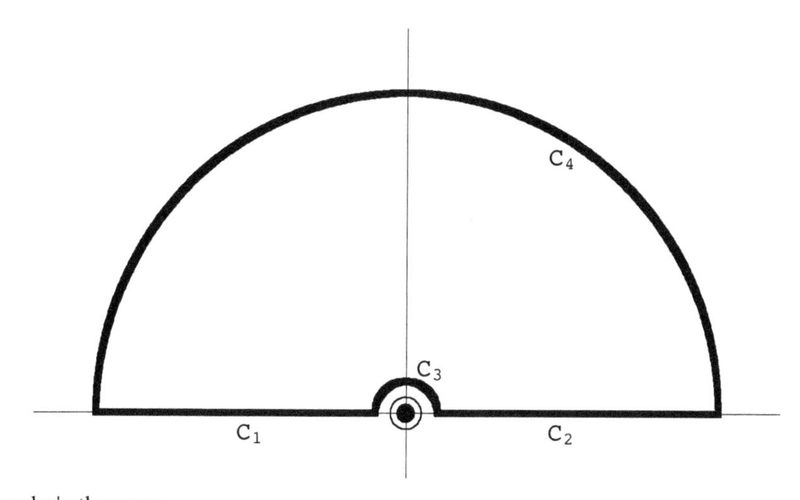

By Cauchy's theorem

$$\int_{C_1} f(z)\,dz + \int_{C_2} f(z)\,dz + \int_{C_3} f(z)\,dz + \int_{C_4} f(z)\,dz = 0 \tag{15.69}$$

But in the limit, we can ignore the contribution of C_4. The integral we want is the combination of C_1 and C_2, and this is

$$J = \int_{C_1} f(z)\,dz + \int_{C_2} f(z)\,dz = -\int_{C_3} f(z)\,dz \tag{15.70}$$

But this is now the integral over a small semicircle, traversed clockwise, and so is

$$-(-i\pi)\,\mathrm{res}(f, 0) = \frac{i\pi}{2} \tag{15.71}$$

The imaginary part of this is just $\pi/2$, and so our original integral is just $\pi/4$.

■ Final check against *Mathematica*

```
Integrate[(x - Sin[x]) / x^3, {x, 0, Infinity}]
```

$$\frac{\pi}{4}$$

15.7 Dealing with functions with branch points

There are numerous integrands involving functions such as z^α and Log(z) for which special methods are required. All of these involve keeping careful track of the arguments (in the sense of the phases or 'args' of the variables) of the functions involved. There are several ways of managing integrals involving these type of functions and sometimes one actually deliberately introduces the log function in order to handle a certain type of problem. Sometimes a contour such as the following does quite well, where are branch cut is made down the negative imaginary axis:

```
ci = {Thickness[0.01], Circle[{0, 0}, 0.5, {0, Pi}],
Circle[{0, 0}, 5, {0, Pi}],
   Line[{{0.5, 0}, {5, 0}}], Line[{{-0.5, 0}, {-5, 0}}],
         Text["C1", {-2.5, -0.3}], Text["C2", {2.5, -0.3}],
               Text["C3", {0.4, 0.7}], Text["C4", {2.5, 4}],
         {Thickness[0.007],
     Dashing[{0.01, 0.01}], Line[{{0, 0}, {0, -3}}]}};
cplot = Show[Graphics[ci, AspectRatio -> Automatic,
               Axes -> False, Ticks -> None, PlotRange -> {{-6, 6}, {-3, 6}}]]
```

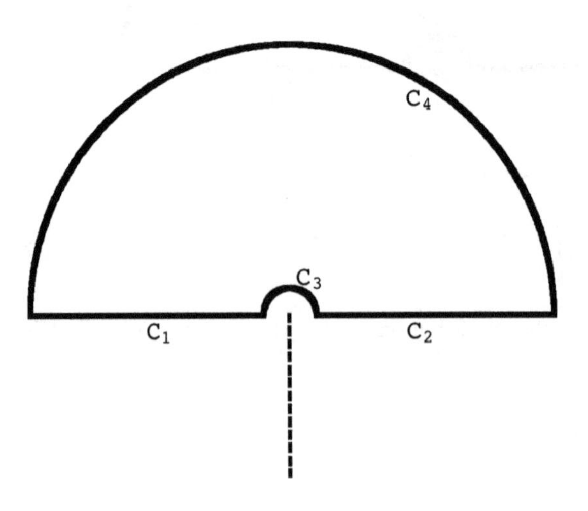

These contours are good when one wishes to avoid unnecessary evaluation of residues in the lower half-plane. Sometimes however, there may be poles on the negative real axis, and unless they are simple, this method will fail. A good general purpose contour for coping with a variety of functions with branch points is the following contour:

```
ci = {Thickness[0.01], Circle[{0, 0}, 0.5, {Pi / 5, 2 Pi - Pi / 5}],
Circle[{0, 0}, 5, {Pi / 55, 2 Pi - Pi / 55}],
Text["C1", {2.5, 0.7}], Text["C2", {2.5, -0.7}],
   Text["C3", {0.4, 0.7}], Text["C4", {2.5, 4}],
Line[{{0.5 Cos[Pi / 5], 0.5 Sin[Pi / 5]}, {5 Cos[Pi / 55], 5 Sin[Pi / 55]}}],
Line[{{0.5 Cos[Pi / 5], -0.5 Sin[Pi / 5]}, {5 Cos[Pi / 55], -5 Sin[Pi / 55]}}],
{Thickness[0.005], Dashing[{0.01, 0.01}],    Line[{{0, 0}, {8, 0}}]}};
cplot = Show[Graphics[ci, AspectRatio -> Automatic,
   Axes -> False, Ticks -> None, PlotRange -> {{-6, 6}, {-6, 6}}]]
```

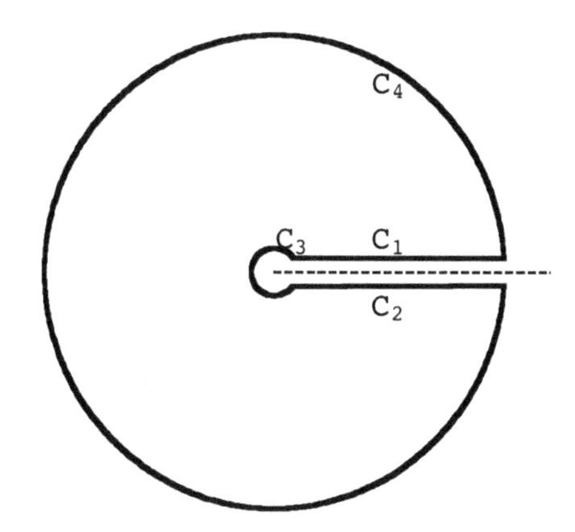

In this case the branch cut is taken along the positive real axis. We consider some applications to illustrate the power of this method. The first gives us a powerful general theorem. We consider an integral of the form

$$\int_0^{\infty} f(x)\,dx \tag{15.72}$$

where f has no branch points, and no poles on the positive real axis or at zero, and goes to zero sufficiently fast at infinity that we neglect integrals over arcs of large circles. We do not require that f has any even symmetry. The trick is to consider

$$\int f(z)\log(z)\,dz \tag{15.73}$$

over the contour consisting of C_1, C_2, C_3, C_4. On the contour just above the real axis, the contribution from C_1 is just

$$\int_0^{\infty} f(x)\log(x)\,dx \tag{15.74}$$

Just below the real axis, we get a contribution

$$\int_{\infty}^{0} f(x)\,(\log(x) + 2\pi i)\,dx \tag{15.75}$$

Now the combination of these two is just

$$-2\pi i \int_{0}^{\infty} f(x)\,dx \tag{15.76}$$

Finally, we can check that the integral over the small semicircle goes to zero as its radius goes to zero. Application of the residue theorem then shows that

$$\int_{0}^{\infty} f(x)\,dx = -\sum \mathrm{Res}[\log(z)\,f(z)] \tag{15.77}$$

where the sum is over the singularities within the contour. This is a strange and beautiful result!

■ Example using branch method

We will consider the following integral using the method just outlined. Can you think of another (simpler) approach using a wedge that will work for this particular type of function? (See also Exercise 15.7.)

$$\int_{0}^{\infty} \frac{1}{x^3 + 1}\,dx \tag{15.78}$$

The denominator is zero at the points

$$z = \left\{e^{\frac{i\pi}{3}},\, e^{i\pi},\, e^{\frac{5i\pi}{3}}\right\} \tag{15.79}$$

and so can be factorized over these roots, but we do not actually do this, as it will be easier to use the differentiation theorem to extract the residue. We need to consider $\mathrm{Log}(z)$ times this function. The singularities are on the unit circle, where $\mathrm{Log}(z) = i\,\mathrm{Arg}(z)$, so if the function is

$$g(z) = \frac{\log(z)}{(z^3 + 1)} \tag{15.80}$$

the residue at any pole on the unit circle is the numerator divided by the derivative of the denominator, i.e. on the unit circle

$$\mathrm{Res}(z) = \frac{i\,\mathrm{Arg}(z)}{3z^2} \tag{15.81}$$

The three pieces, in order of that given in the list in Eq. (15.79), are

$$\frac{\pi i}{3\left(3e^{\frac{2i\pi}{3}}\right)} = -\frac{i\pi}{18} + \frac{\pi}{6\sqrt{3}} \tag{15.82}$$

$$\frac{\pi i}{3e^{2 i \pi}} = \frac{i\pi}{3} \tag{15.83}$$

$$\frac{5\pi i}{3\left(3e^{\frac{10 i \pi}{3}}\right)} = -\frac{5i\pi}{18} - \frac{5\pi}{6\sqrt{3}} \tag{15.84}$$

The negative of the sum of these is just

$$\frac{2\pi}{3\sqrt{3}} \tag{15.85}$$

This is the desired result, which we can check directly with *Mathematica*:

```
Integrate[1/(x^3 + 1), {x, 0, Infinity}]
```

$$\frac{2\pi}{3\sqrt{3}}$$

If you obtain something different in your current version of *Mathematica*, use the simplification tools.

■ More than one branch structure

When the function being integrated has a branch point, and one then adds a logarithm, matters are more complicated. Let's look at the following integral:

$$\int_0^\infty \frac{\sqrt{x}\,\log(x)}{(x+1)^2}\,dx \tag{15.86}$$

Now this can be treated in a variety of ways – another option is given in the exercises. Here we proceed directly, so we consider the function

$$f(z) = \frac{\sqrt{z}\,\log(z)}{(z+1)^2} \tag{15.87}$$

integrated over the same contour. This time the contribution from C_1 is the integral we want, and that from C_2 is

$$-\int_0^\infty \frac{e^{i\pi}\sqrt{x}\,(\log(x) + 2\pi i)}{(x+1)^2}\,dx \tag{15.88}$$

These two add up to

$$\int_0^\infty \frac{\sqrt{x}\,(2\log(x) + 2\pi i)}{(x+1)^2}\,dx \tag{15.89}$$

Now there is one double pole at $z = -1$, so to get the residue we need to evaluate the derivative at $z = -1$, of

$$\sqrt{z}\,\log(z) \tag{15.90}$$

and the derivative is

$$\frac{\log(z)}{2\sqrt{z}} + \frac{1}{\sqrt{z}} \tag{15.91}$$

Putting $z = e^{i\pi}$ we obtain

$$\frac{i\pi}{2\,e^{\frac{i\pi}{2}}} + \frac{1}{e^{\frac{i\pi}{2}}} = -i + \frac{\pi}{2} \tag{15.92}$$

Multiplication by $2\pi i$ and application of the residue theorem gives

$$\int_0^\infty \frac{\sqrt{x}\,(2\log(x) + 2\pi i)}{(x+1)^2}\,dx = 2\pi i\left(\frac{\pi}{2} - i\right) = 2\pi + i\pi^2 \tag{15.93}$$

Now we just take real and imaginary parts, to obtain the following pair of results:

$$\int_0^\infty \frac{\sqrt{x}\,\log(x)}{(x+1)^2}\,dx = \pi \tag{15.94}$$

$$\int_0^\infty \frac{\sqrt{x}}{(x+1)^2}\,dx = \frac{\pi}{2} \tag{15.95}$$

■ *Mathematica* **checks**

We can check these directly with *Mathematica*:

```
Integrate[(Sqrt[x]*Log[x])/(1 + x)^2, {x, 0, Infinity}]
```

π

```
Integrate[Sqrt[x]/(1 + x)^2, {x, 0, Infinity}]
```

$\frac{\pi}{2}$

15.8 Infinitely many poles and series summation

There are two further classes of integral that are important, both involving integrands with infinitely many poles. In the first case we wish to pick the contour to just contain a small number poles, in order to evaluate an integral. In the second case we wish to take a sequence of contours neatly avoiding a sequence of poles, and carefully add up an infinite series. We consider three representative problems.

■ An integral involving a periodic function

$$J = \int_{-\infty}^{\infty} \frac{x}{\sinh(x)}\, dx \tag{15.96}$$

In this case we consider the function

$$f(z) = \frac{z}{\sinh(z)} \tag{15.97}$$

The denominator vanishes when $z = i n\pi$, for n an integer, and these will generate simple poles in the absence of any cancellations. The potential pole at the origin is nullified by the presence of z in the numerator, so we consider non-zero positive and negative integer values for n. If we tried some kind of semicircular contour we would have all kinds of problems since there are infinitely many poles in either half-plane. The point to note here is that the denominator is periodic in the imaginary direction:

$$\sinh(z + i\pi) = -\sinh(z) \tag{15.98}$$

So if we use a rectangular contour going along the real axis and back along the real axis displaced by a multiple of $i n \pi$, we can arrange that the two integrals are closely related. We use the following contour:

```
ci = {SimplePole[{0, Pi}, 0.25],
   SimplePole[{0, 2 Pi}, 0.25], SimplePole[{0, -Pi}, 0.25],
Thickness[0.01], Circle[{0, Pi}, 0.5, {Pi, 2 Pi}],
Line[{{-9, 0}, {9, 0}}], Line[{{-9, Pi}, {-0.5, Pi}}],
Line[{{0.5, Pi}, {9, Pi}}],
   Line[{{9, 0}, {9, Pi}}], Line[{{-9, 0}, {-9, Pi}}],
Text["C₁", {2.5, -0.3}], Text["C₂", {9.7, Pi / 2}], Text["C₄", {1, 2.8}],
Text["C₃", {4.5, 3.8}], Text["C₅", {-4.5, 3.8}], Text["C₆", {-9.6, Pi / 2}]};
cplot = Show[Graphics[ci, AspectRatio -> Automatic,
Axes -> True, Ticks -> None, PlotRange -> {{-10, 10}, {-4, 7}}]]
```

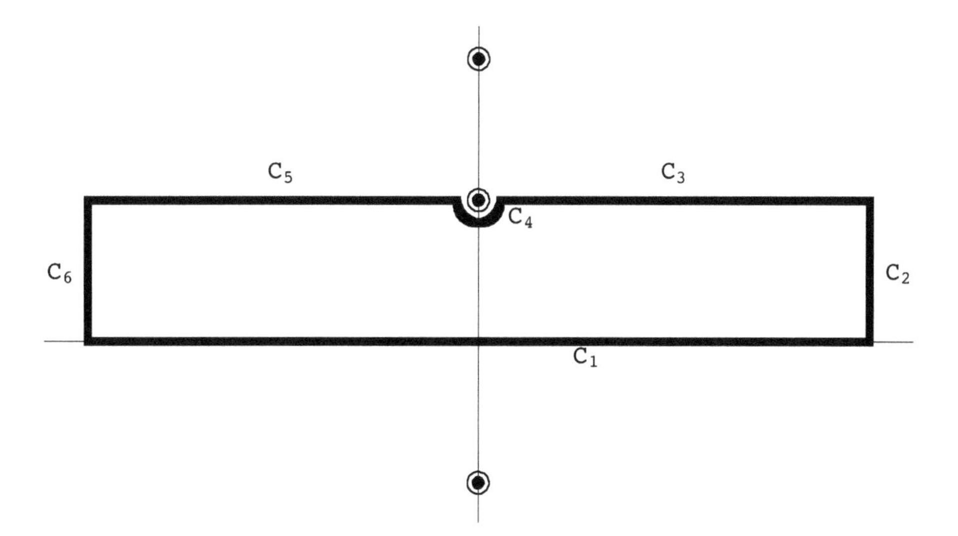

By Cauchy's theorem:

$$\begin{aligned}&\int_{C_1} f(z)\,dz + \int_{C_2} f(z)\,dz + \int_{C_3} f(z)\,dz \\ &+ \int_{C_4} f(z)\,dz + \int_{C_5} f(z)\,dz + \int_{C_6} f(z)\,dz = 0\end{aligned} \tag{15.99}$$

Now we can ignore the contributions from C_2 and C_6 in the limit that the ends of the rectangle go to infinity, as the denominator blows up exponentially. For the three horizontal sections, we can write

$$\begin{aligned}&\int_{C_1} f(z)\,dz + \int_{C_3} f(z)\,dz + \int_{C_5} f(z)\,dz \\ &= \int_{-\infty}^{\infty} \frac{x}{\sinh(x)}\,dx + P\int_{-\infty}^{\infty} \frac{x+i\pi}{\sinh(x)}\,dx = \int_{-\infty}^{\infty} \frac{2x}{\sinh(x)}\,dx = 2J\end{aligned} \tag{15.100}$$

where the imaginary principal value integral is zero because the integrand is odd. Finally we need to deal with the upturned mousehole contour. The residue at $i\pi$ is the limit

$$\lim_{z\to i\pi} \frac{z(z-i\pi)}{\sinh(z)} = \lim_{z\to i\pi} \frac{2z-i\pi}{\cosh(z)} = -i\pi \tag{15.101}$$

where the central step is L'Hospital's rule, and we note that

$$\cosh(i\pi) = \cos(\pi) = -1 \tag{15.102}$$

So Cauchy's theorem and the mousehole theorem give us

$$2J + (-i\pi)(-i\pi) = 0 \tag{15.103}$$

so $J = \pi^2/2$.

■ *Mathematica* check

```
Integrate[x / Sinh[x], {x, -Infinity, Infinity}]
```

$$\frac{\pi^2}{2}$$

■ Summation type 1

Let $\gamma(n)$ be an even rational function of n, holomorphic except at the origin, satisfying, for large $|z|$, $|\gamma(z)| = O(|z|^{-k})$ for $k > 1$, and we wish to evaluate

$$\sum_{n=1}^{\infty} \gamma(n) \tag{15.104}$$

We consider

$$f(z) = \pi\gamma(z)\cot(\pi z) \tag{15.105}$$

First of all note that the poles of $f(z)$ are the poles of $\gamma(z)$ and also at the poles of $\cot(\pi z)$. The residues of $f(z)$ at the poles of $\cot(\pi z)$ are just $\gamma(n)$. Now we pick a square contour with vertices at

$$\left(N+\frac{1}{2}\right)(1+i);\ \left(N+\frac{1}{2}\right)(-1+i);\ \left(N+\frac{1}{2}\right)(-1-i);\ \left(N+\frac{1}{2}\right)(1-i); \tag{15.106}$$

It is a simple task (that you should do) to show that $\cot(\pi z)$ is bounded on such a square (note that this square goes between the singularities), so by the behaviour of f for large z, by letting $N \to \infty$ we deduce that the integral over this square tends to zero, and hence that

$$2\sum_{n=1}^{\infty}\gamma(n) = -\sum_{\text{poles of }\gamma}\text{res}[\pi\gamma(z)\cot(\pi z)] \tag{15.107}$$

▪ **Example**

$$\sum_{n=1}^{\infty}\frac{1}{n^2} = \frac{\pi^2}{6} \tag{15.108}$$

This is proved by taking $\gamma(n) = 1/n^2$ and computing the residue of $\pi\cot(\pi z)/z^2$ at the origin as $-\pi^2/3$.

Mathematica can do many sums of this type

```
Sum[1 / n^2, {n, 1, Infinity}]
```

$$\frac{\pi^2}{6}$$

■ Summation type 2

Let $\gamma(n)$ be an even rational function of n, holomorphic except at the origin, satisfying, for large $|z|$, $|\gamma(z)| = O(|z|^{-k})$ for $k > 1$, and we wish to evaluate:

$$\sum_{n=1}^{\infty}(-1)^n\gamma(n) \tag{15.109}$$

We consider

$$f(z) = \pi\gamma(z)\text{cosec}(\pi z) \tag{15.110}$$

and proceed exactly as before, noting that the residue of $f(z)$ at the poles of $\text{cosec}(\pi z)$ is now $(-1)^n\gamma(n)$.

■ Example

By a similar argument as before, it follows that

$$\sum_{n=1}^{\infty} \frac{(-1)^n}{n^2} = -\frac{\pi^2}{12} \tag{15.111}$$

This is proved by taking $\gamma(n) = 1/n^2$ and computing the residue of $\pi \text{cosec}(\pi z)/z^2$ at the origin as $\pi^2/6$.

```
Sum[(-1)^n/n^2, {n, 1, Infinity}]
```

$$-\frac{\pi^2}{12}$$

By considering sums from negative infinity to infinity we can obtain other types of results – see the last part of Exercise 15.10 for an example. But some series cannot be summed by these methods – try Exercise 15.19 if you want to get an idea of the depth of the water you can rapidly find yourself in! We also have the famous Mittag-Leffler expansion theorem, which can be thought of as a kind of fancy partial fraction expansion based on the poles of functions, rather than on a simple factorization of a denominator. The form of this theorem is described below.

Theorem 15.3: Mittag-Leffler *Suppose that $f(z)$ only has singularities at simple poles $a_1, a_2, \ldots$ arranged in increasing absolute value, with residues $b_1, b_2, \ldots$ at each of these poles. Suppose further that $f(z)$ is less than M in magnitude on a sequence of circles of radius R_m,* with $R_m \to \infty$ *as $m \to \infty$, with these circles not passing through any of the poles, and M is independent of m.* Then

$$f(z) = f(0) + \sum_{n=1}^{\infty} b_n \left(\frac{1}{a_n} + \frac{1}{z - a_n} \right) \tag{15.112}$$

The proof is left as Exercise 15.11 for more enterprising students. Some simple applications are given in Exercise 15.12. It may surprise you to learn that *Mathematica* is perfectly capable of adding up series of this type, to recover the original functions. This is very helpful, particularly for checking answers, and some examples are given in Exercise 18.

15.9 The argument principle and Rouché's theorem

■ The argument principle

Suppose that $f(z)$ is a meromorphic function defined on and inside a simple closed contour C, with no zeroes or poles on C itself. (You are reminded that *meromorphic* means holomorphic in the finite plane apart from poles.) Consider the integral

$$J = \frac{1}{2\pi i} \int_C \frac{f'(z)}{f(z)}\, dz \tag{15.113}$$

We can consider this integral in two ways.

▪ *J* as a change in argument

If we parametrize C by t, with $a \leqslant t \leqslant b$ we can write

$$J = \frac{1}{2\pi i} \int_C \frac{f'(z(t))}{f(z(t))} z'(t)\, dt = \frac{1}{2\pi i} \int_C \frac{\partial \log(f(z(t)))}{\partial t}\, dt \tag{15.114}$$

and so

$$J = \frac{1}{2\pi i} \Big[\log(f(z(t)))\Big]_{t=a}^{t=b} \tag{15.115}$$

which is the change in the value of the logarithm of f as we move round the path. Decomposing the complex logarithm we obtain

$$J = \frac{1}{2\pi i} \Big[\log(|f(z(t))|) + i\, \mathrm{Arg}(f(z(t)))\Big]_{t=a}^{t=b} = \frac{1}{2\pi} \Big[\mathrm{Arg}(f(z(t)))\Big]_{t=a}^{t=b} \tag{15.116}$$

so that J is just the change in the argument of f around the path, divided by 2π. This is normally written without reference to the parameter as

$$J = \frac{1}{2\pi} \Big[\mathrm{Arg}(f(z))\Big]_C \tag{15.117}$$

Another interpretation of this view of J is in terms of winding numbers. If we let the path D denote the image of C under the mapping $z \to w = f(z)$, then

$$J = \frac{1}{2\pi i} \int_D \frac{1}{w}\, dw \tag{15.118}$$

is the number of times w winds around the origin, since it is $1/2\pi$ times the change in the argument of $w = f(z)$ as z progresses around C and w progresses around D.

▪ *J* as number of zeroes minus number of poles

Suppose that in a neighbourhood of a point z_i within C we can write

$$f(z) = (z - z_i)^{n_i} g(z) \tag{15.119}$$

where g is both holomorphic and non-zero in a neighbourhood of z_i, and n_i is a non-zero integer. This is a zero of of order n_i if n_i is positive, and a pole of order $-n_i$ if n_i is negative. Clearly

$$f'^{(z)} = g(z)\, n_i\, (z - z_i)^{n_i - 1} + g'^{(z)}\, (z - z_i)^{n_i} \tag{15.120}$$

and so, in a neighbourhood of z_i, we can see that

$$\frac{f'(z)}{f(z)} = \frac{n_i}{z - z_i} + h(z) \tag{15.121}$$

where $h(z) = g'^{(z)} / g(z)$ is holomorphic in this same neighbourhood. Applying the calculus of residues we see immediately that

$$J = \sum_{i=1}^{k} n_i \tag{15.122}$$

where k is the total number of locations of the zeroes or poles. This result is usually written as

$$J = N - P \tag{15.123}$$

where N is the total number of zeroes, including multiplicies, and P is the total number poles, each pole weighted by its orders. The argument principle is then summarized by an equation combining these observations into the form:

$$J = \frac{1}{2\pi i} \int_C \frac{f'(z)}{f(z)} dz = \frac{1}{2\pi} [\mathrm{Arg}(f(z))]_C = N - P \tag{15.124}$$

■ Theorem 15.4: Rouché's theorem

Suppose that $f(z)$ and $g(z)$ are holomorphic inside and on a simple closed contour C. Then Rouché's theorem states that if $|g(z)| < |f(z)|$ on C then $f(z)$ and $f(z) + g(z)$ have the same number of zeroes inside C, (where multiplicities are included).

To prove this result we set $h(z) = g(z)/f(z)$, and observe that the assumption of the theorem is that $h(z) < 1$ on C. We let N denote the number of zeros of $g(z)/f(z)$, and M the number of zeroes of $f(z)$. By the argument principle and the fact that f and g are holomorphic, and hence have no poles inside C, we have

$$N = \frac{1}{2\pi i} \int_C \frac{f'(z) + g'^{(z)}}{f(z) + g(z)} dz \; ; \quad M = \frac{1}{2\pi i} \int_C \frac{f'(z)}{f(z)} dz \tag{15.125}$$

We now write down the difference, $N - M$, and show that this is zero under the assumptions given:

$$N - M = \frac{1}{2\pi i} \int_C \frac{f'(z) + g'^{(z)}}{f(z) + g(z)} dz - \frac{1}{2\pi i} \int_C \frac{f'(z)}{f(z)} dz \tag{15.126}$$

Because $g(z) = f(z)\, h(z)$ we can reorganize this as

$$N - M$$
$$= \frac{1}{2\pi i} \int_C \frac{f'(z) + f'(z)\, h(z) + f(z)\, h'(z)}{f(z)\,(1 + h(z))} dz - \frac{1}{2\pi i} \int_C \frac{f'(z)}{f(z)} dz$$

$$= \frac{1}{2\pi i}\int_C \frac{f'(z)(1+h(z)) + f(z)\,h'(z)}{f(z)(1+h(z))}\,dz - \frac{1}{2\pi i}\int_C \frac{f'(z)}{f(z)}\,dz$$

$$= \frac{1}{2\pi i}\int_C \frac{f'(z)(1+h(z))}{f(z)(1+h(z))} + \frac{f(z)\,h'(z)}{f(z)(1+h(z))}\,dz - \frac{1}{2\pi i}\int_C \frac{f'(z)}{f(z)}\,dz$$

$$= \frac{1}{2\pi i}\int_C \frac{h'(z)}{(1+h(z))}\,dz$$

Now we note that since $h(z) < 1$ on C the integrand of this last expression can be expanded term by term:

$$N - M = \frac{1}{2\pi i}\int_C h'(z)(1 - h(z) + h^2(z) - h^3(z) + \ldots)\,dz \qquad (15.128)$$

The series is uniformly convergent and can be integrated term by term, each term giving zero. Hence $N = M$.

An example using Rouché's theorem

Consider the polynomial

$$P(z) = z^8 - 4z^3 + 24 \qquad (15.129)$$

We can find out about the zeroes of this function in various ways, with or without *Mathematica*. We can use *Mathematica* to work out directly the integral involved in the argument principle. We can try to apply Rouché's theorem using 'pen and paper'. Finally, we can just ask *Mathematica* to tell us where the roots are, possibly symbolically, and if necessary, numerically.

Approach 1: application of the argument principle with *Mathematica*

$$P(z) = z^8 - 4z^3 + 24 \qquad (15.130)$$

```
P[z_] := z^8 - 4 z^3 + 24;
q[z_] = D[P[z], z];
Q[z_] := q[z]
```

Let's define a function to investigate the argument principle applied over a circle. Note that we will use numerical integration and then round the result. This is important because if we used **`Integrate`**, this would try to find an analytical expression (involving the logarithm in many examples) and this would not reliably pick up the correct branch of the argument.

```
RootCount[r_] :=
 Round[1 / (2 Pi I) NIntegrate[Q[r Exp[I t]] / P[r Exp[I t]]
    I r Exp[I t], {t, 0, 2 Pi}]]
```

The following produces some warnings, which arise quite rightly because on the first (inner) integral the correct result is zero:

```
Map[RootCount[#] &, {1, 2}]
```

– *NIntegrate::ncvb :*
NIntegrate failed to converge to prescribed accuracy after 7 recursive bisections in t near t = 1.98804.

{0, 8}

So we know that all eight roots lie between the circles of radius one and two.

■ Approach 2: applying Rouché

Let $f(z) = 24$ and $g(z) = z^8 - 4z^3$. On the inner circle where $|z| = 1$ we note $|f(z)| = 24$. Furthermore

$$|g(z)| = |z^8 - 4z^3| \leq |4z^3| + |z^8| \leqslant 5 < 24 = |f(z)| \tag{15.131}$$

So by Rouché's theorem the sum $f + g = P$ has the same number of roots as does f inside the circle $|z| = 1$, i.e., none. Now consider the outer circle $|z| = 2$. This time we choose $f(z) = z^8$ and $g(z) = 24 - 4z^3$. On the outer circle we have

$$|g(z)| = |24 - 4z^3| \leq 24 + |4z^3| \leqslant 24 + 32 = 56 < 128 = 2^8 = |f(z)| \tag{15.132}$$

so that $P = f + g$ has the same number of roots inside as does f, i.e. 8.

■ ✵ Approach 3: brute-force root-finding

The **Solve** function does not yield analytical results:

```
z /. Solve[z^8 - 4 z^3 + 24 == 0, z]
```

{Root[#1^8 – 4 #1^3 + 24 &, 1], Root[#1^8 – 4 #1^3 + 24 &, 2],
Root[#1^8 – 4 #1^3 + 24 &, 3], Root[#1^8 – 4 #1^3 + 24 &, 4],
Root[#1^8 – 4 #1^3 + 24 &, 5], Root[#1^8 – 4 #1^3 + 24 &, 6],
Root[#1^8 – 4 #1^3 + 24 &, 7], Root[#1^8 – 4 #1^3 + 24 &, 8]}

But the numerical results are easily found:

```
rootlist = z /. NSolve[z^8 - 4 z^3 + 24 == 0, z]
```

{–1.37849 – 0.670517 i, –1.37849 + 0.670517 i,
–0.569794 – 1.26996 i, –0.569794 + 1.26996 i, 0.57115 – 1.47273 i,
0.57115 + 1.47273 i, 1.37713 – 0.465032 i, 1.37713 + 0.465032 i}

and we can check their absolute values:

```
Map[Abs, rootlist]
```

{1.53291, 1.53291, 1.39193, 1.39193, 1.5796, 1.5796, 1.45353, 1.45353}

■ A graphical representation

The results above can be summarized by the following *Mathematica* plot showing the two circles and the roots:

```
circlesnroots = ParametricPlot[
  {{2 Cos[t], 2 Sin[t]}, {Cos[t], Sin[t]}},
  {t, 0, 2 Pi}, AspectRatio → 1,
  Epilog → {PointSize[0.03],
    Map[Point[{Re[#], Im[#]}] &, rootlist]} ]
```

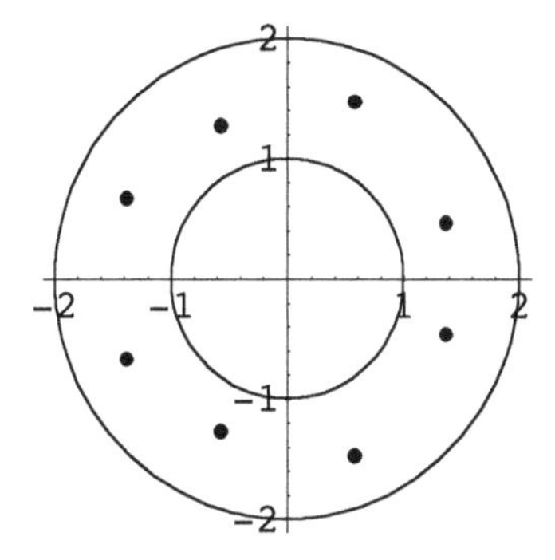

■ Other methods for root isolation in *Mathematica*

Some further help for counting and locating roots of polynomials in *Mathematica*, beyond merely using **Solve** to locate them in the low degree case, is provided by the **RootIsolation** package in the algebra area. This can be loaded in the usual way, and the relevant function is **CountRoots**:

```
Needs["Algebra`RootIsolation`"];
?CountRoots
```

```
CountRoots[f, {x,a,b}] computes the number of roots (
  multiplicities counted) of a univariate polynomial f[x] in
  the interval (a,b) (for complex numbers a,b interval (a,b) is
  the open rectangle (or open line segment or point) of which
  a is the lower-left vertex and b is the upper-right vertex).
```

In other words, this function is already set up to do the work for us using rectangular intervals!

```
CountRoots[z^8 - 4 z^3 + 24, {z, 0, 2}]
```

0

```
CountRoots[z^8 - 4 z^3 + 24, {z, -2, 2}]
```

0

```
CountRoots[z^8 - 4 z^3 + 24, {z, -1.5 - I, 1.5 + I}]
```

4

```
CountRoots[z^8 - 4 z^3 + 24, {z, -2 - 2 I, 2 + 2 I}]
```

8

```
Show[Graphics[
  {{RGBColor[0, 1, 0], Rectangle[{-2, -2}, {2, 2}]},
   {RGBColor[0, 0, 1], Rectangle[{-1.5, -1},
     {1.5, 1}]}, {PointSize[0.03], RGBColor[0, 0, 0],
    Map[Point[{Re[#], Im[#]}] &, rootlist]}},
  Frame → True, AspectRatio → 1]]
```

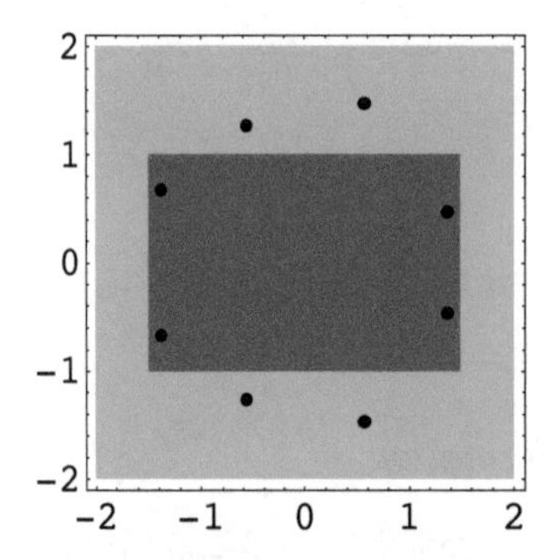

We can overlay this with the circles plot as well:

```
Show[%, circlesnroots]
```

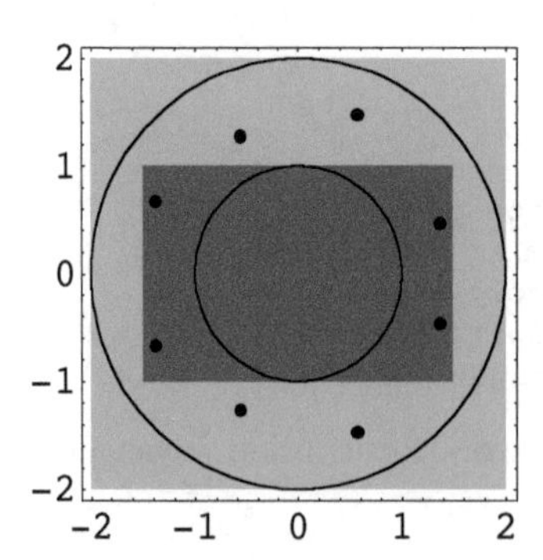

▪ Projects

Write down some polynomials and investigate them using the techniques described above. Two examples to start you off are given in Exercise 15.20.

Exercises

15.1 Evaluate, for a and b real, with $a > b > 0$, the integral

$$\int_0^{2\pi} \frac{1}{a + b\cos(\theta)}\, d\theta$$

Also evaluate using residue methods the integral:

$$\int_0^{2\pi} \frac{1}{(5 - 2\cos(\theta))^2}\, d\theta$$

How might the second result be deduced from the first?

15.2 Evaluate using residue methods the integral

$$\int_0^{2\pi} \cos^6(\theta)\, d\theta$$

and explain how the result can also be obtained using DeMoivre's Theorem.

15.3 Evaluate using residue methods each of the three integrals

$$\int_{-\infty}^{\infty} \frac{1}{(x^2+3)^2}\, dx, \quad \int_{-\infty}^{\infty} \frac{1}{(x^2+3)^3}\, dx, \quad \int_0^{\infty} \frac{1}{(x^2+3)^3}\, dx$$

15.4 Evaluate using residue methods each of the three integrals

$$\int_{-\infty}^{\infty} \frac{\cos(x)}{(x^2+3)^2}\, dx, \quad \int_{-\infty}^{\infty} \frac{\cos(x)}{(x^2+3)^3}\, dx, \quad \int_0^{\infty} \frac{\cos(x)}{(x^2+3)^3}\, dx$$

15.5 Using the Mousehole result, evaluate the integral

$$\int_0^{\infty} \frac{\sin(x)}{x}\, dx$$

15.6 Consider again the integral

$$\int_0^{\infty} \frac{\sqrt{x}\,\log[x]}{(x+1)^2}\, dx$$

Use the substitution $x = y^2$ to show that this is equivalent to

$$4\int_0^{\infty} \frac{y^2 \log[y]}{(y^2+1)^2}\, dy$$

and hence evaluate the integral.

15.7 Using methods similar to that employed in Section 15.7, evaluate

$$\int_0^\infty \frac{1}{x^5 + 1}\, dx$$

Using a similar approach but without logs, also evaluate:

$$\int_0^\infty \frac{1}{x^6 + 1}\, dx$$

Also show how these can be done using a wedge-shaped contour. To figure out where the wedge should be, consider the angle about the origin over which the integrand is periodic.

15.8 For $0 < \alpha < 1$, evaluate the integral

$$\int_0^\infty \frac{x^{\alpha-1}}{x+1}\, dx$$

15.9 Evaluate the integral

$$\int_0^\infty \frac{\cosh(\alpha\, x)}{\cosh(x)}\, dx$$

using a suitable rectangular contour.

15.10 Evaluate the following four infinite sums:

$$\sum_{n=1}^{\infty} \frac{1}{n^4}, \quad \sum_{n=1}^{\infty} \frac{1}{n^6}$$

$$\sum_{n=1}^{\infty} \frac{1}{a^2 + n^2}, \quad \sum_{n=-\infty}^{\infty} \frac{(-1)^n}{a^2 + n^2}$$

15.11 (Harder) By considering the function

$$g(z) = \frac{f(z)}{z - \zeta}$$

and integrating over the sequence of circles given in the statement of the theorem, prove the Mittag-Leffler expansion theorem.

15.12 Assuming the Mittag-Leffler theorem, derive generalized partial fraction expansions for cosec(z), sec(z), tan(z), cot(z). How may these be used directly to infer their equivalents for hyperbolic functions?

❄ In the following exercises with *Mathematica* please note the suggestions in the introduction for cross-checking the output of **`Integrate`**. In all cases where there are parameters with assumptions about them, try both the deafult setting of **`GenerateConditions -> True`**, and making use of **`Assumptions`** settings. Avoid the use of **`GenerateConditions -> False`**.

15.13 ✵ Check the results of Exercises 15.1 through 15.5 using *Mathematica*. Note that you may need to use algebra functions such as **PowerExpand** and **Simplify** to get obvious agreement.

15.14 ✵ Use *Mathematica* to directly evaluate the following integral without making any assumptions about a, and then repeat the process with some suitable assumptions.

$$\int_0^\infty \frac{1}{(a^2 + x^2)^2}\, dx$$

Simplify the results using **Simplify**, **PowerExpand** etc. as necessary, and compare your answer with that obtained by residue methods.

15.15 ✵ Use *Mathematica* to check the results obtained in Exercise 15.7.

15.16 ✵ Use *Mathematica* to check the results obtained in exercise 9. You might find the use of a combination of **TrigExpand** and **Simplify** to be helpful.

15.17 ✵ *Mathematica*'s built-in function **Sum** can treat a large set of infinite sums symbolically. Use this function to confirm your answers to Exercise 15.10.

15.18 ✵ *Mathematica* can also check results of Mittag-Leffler type. Use the **Sum** function to evaluate, for example

$$8\, z \sum_{n=1}^{\infty} \frac{1}{(2\, n - 1)^2\, \pi^2 - 4\, z^2} \quad \text{and} \quad 8\, z \sum_{n=1}^{\infty} \frac{1}{(2\, n - 1)^2\, \pi^2 + 4\, z^2}$$

and confirm as many of your answers to Exercise 15.12 as you can.

15.19 ✵ (More of an investigative project) Find out what happens when you try to sum

$$\sum_{n=1}^{\infty} \frac{1}{n^3}$$

by the methods introduced in this chapter. Try looking up ‘Apery's constant’, and ‘Zeta(3)’ on the internet and in the literature to see what this question leads to. *Mathematica* is aware of these issues, and gives symbolic results, for example:

```
Sum[1 / n^3, {n, 1, Infinity}]
```

$\zeta(3)$

15.20 ✵ Using the principle of the argument, Rouché's theorem and *Mathematica*'s root isolation and find techniques, investigate the location of the zeroes of

$$p(z) = z^4 - z^3 + 4$$
$$q(z) = z^8 - z^3 + 2$$

16 Conformal mapping I: simple mappings and Möbius transforms

Introduction

Complex functions have an elegant interpretation in terms of mappings of the complex plane into itself. We explored this briefly in Chapter 8. Now we wish to study the geometrical aspects in rather more detail. Our plan is as follows. First, we shall literally play with *Mathematica* to get a feel for what some simple mappings do to simple regions. Next we shall look at the property of 'conformality' – that holormorphic functions, when interpreted as mappings, preserve angles between curves at most points. Then we shall explore the relationship between the geometry of circles and lines and a special class of mappings called Möbius transforms.

This chapter is the foundation for several that follow. In particular, in Chapter 19 we shall explore the application of conformal mapping to problems in physics in 2-dimensional regions. Chapter 23 will explore how some of this material may be generalized to higher dimensions. Chapter 21 will look at how conformal maps, and the Schwarz–Christoffel transformation in particular, can be managed numerically. Chapter 23 will also reveal the real physics underlying the Möbius transform when it is seen in terms of Einstein's theory of special relativity.

16.1 ✵ Recall of visualization tools

Our first goal is to use *Mathematica* to explore some simple mappings. We shall do so by loading the **ComplexMap** Package and making a pair of additional functions, **CartesianMap** and **PolarMap**. Let's load the package and ask about the first of these functions:

```
Needs["Graphics`ComplexMap`"];
?Graphics`ComplexMap`CartesianMap
```

```
"CartesianMap[f, {x0, x1, (dx)}, {y0, y1, (dy)}] plots the
  image of the cartesian coordinate lines under the function
  f. The default values of dx and dy are chosen so that the
  number of lines is equal to the value of the option Lines.
```

We extend the functionality of the **ComplexMap** package by defining a pair of functions that show us the effect of the conformal map both 'before' and 'after'. Clearly we can define both Cartesian and polar versions. In each case two plots are created. The first has the identity mapping, while the second has the function incorporated. This makes it very straightforward to see the effect of a mapping on a region bounded by Cartesian or polar coordinate lines:

```
PolarConformal[func_, radial_, polar_, options___] :=
Show[GraphicsArray[{PolarMap[# &, radial, polar,
      options, DisplayFunction -> Identity],
    PolarMap[func, radial, polar,
      options, DisplayFunction -> Identity]
       }], DisplayFunction -> $DisplayFunction];
```

```
CartesianConformal[func_,
   xrange_, yrange_, options___] :=
Show[GraphicsArray[{
    CartesianMap[# &, xrange, yrange,
      options, DisplayFunction -> Identity],
    CartesianMap[func, xrange, yrange,
      options, DisplayFunction -> Identity]
       }], DisplayFunction -> $DisplayFunction];
```

Functions to plot complex regions defined by inequalities

In recent versions of *Mathematica*, there is another wonderful package, **Inequality-Graphics**, that helps with the visualization of mappings:

```
Needs["Graphics`InequalityGraphics`"];
```

```
? ComplexInequalityPlot
```

```
ComplexInequalityPlot[ineqs, {z, zmin, zmax}] plots the the region defined
  by ineqs within the box bounded by {Re[zmin], Im[zmin]} and {Re[zmax],
  Im[zmax]}. The functions that occur within the inequality need to
  be real valued functions of a complex argument, e.g. Abs, Re and Im.
```

```
ComplexInequalityPlot[ 1 / 2 ≤ Abs[z] ≤ 1, {z} ]
```

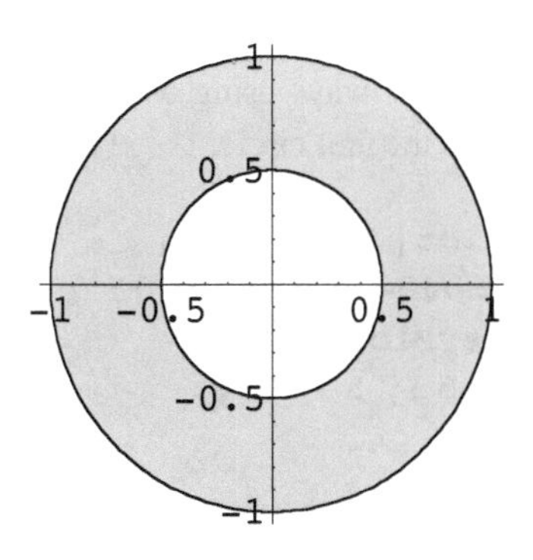

16.2 A quick tour of mappings in *Mathematica*

The purpose of this section is to give you a feel for what happens to regions in the complex plane when we apply a mapping associated with an elementary function.

■ A simple example of a Möbius transform

It is easiest to begin with a simple example of a Möbius transform. We shall look at these in greater generality shortly. Consider the mapping $W(z)$ given by the formula

```
W[z_] = (z - I)/(z + I);
```

The corresponding inverse mapping is readily found by solving for z in terms of $w = W[z]$:

$$-\frac{i(w+1)}{w-1} \tag{16.1}$$

You can of course get *Mathematica* to do it:

```
z /. Solve[W[z] == w, z]
```

$$\{-\frac{i(w+1)}{w-1}\}$$

Let's look first at the images of four points under the mapping:

```
{W[0], W[I], Limit[W[z], z -> Infinity], W[-I]}
```

– *Power::infy : "Infinite expression* $\frac{1}{0}$ *encountered.*

{−1, 0, 1, ComplexInfinity}

Consider also the real axis in z-space. Any real point is equidistant from $+i$ and $-i$ so $|w| = 1$ if z is real. So the real axis gets mapped to the unit circle. If z is in the upper half-plane the modulus of $W[z]$ will be less than one, so the upper half-plane is mapped into the interior of the unit circle. We can look at this in various ways using our functions. First, let's check that the upper half-plane gets mapped to the unit circle:

```
zUpperHalfPlane = ComplexInequalityPlot[ Im[z] ≥ 0,
   {z, -5 - 5 I, 5 + 5 I }, DisplayFunction → Identity ];
wOfzUpperHalfPlane = ComplexInequalityPlot[
   Im[-I (w + 1) / (w - 1)] ≥ 0, {w, -5 - 5 I, 5 + 5 I },
   DisplayFunction → Identity ];
Show[
 GraphicsArray[{zUpperHalfPlane, wOfzUpperHalfPlane}],
 DisplayFunction → $DisplayFunction ]
```

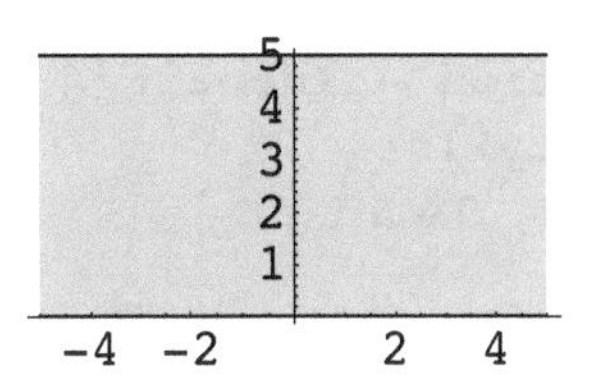

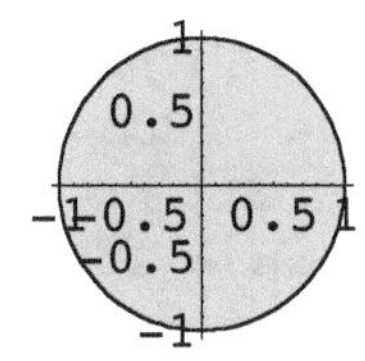

If we want to look at the Cartesian grid image, we can do that with **CartesianConformal**. Note that this time we get, in the right-hand image, just that part of the unit circle mapped to by the plotted rectangle:

```
CartesianConformal[((# - I) / (# + I)) &, {-5, 5}, {0, 5},
 Lines -> 30, PlotPoints -> 40, PlotRange -> All]
```

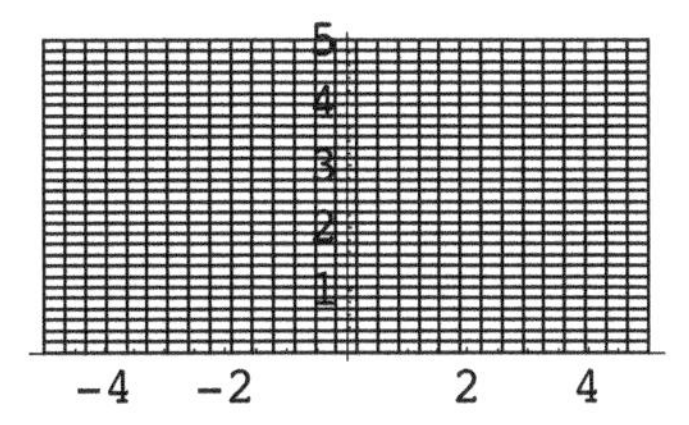

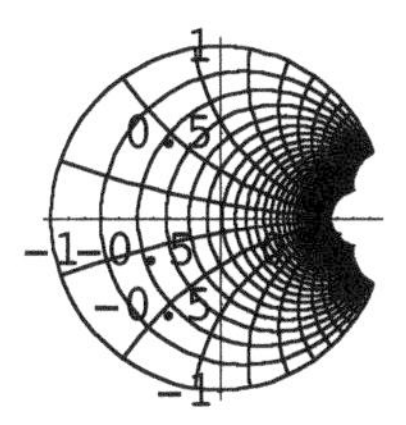

We can also view this using **PolarConformal** applied to the unit circle applied to the inverse mapping:

```
PolarConformal[(I * (# + 1) / (1 - #)) &, {0, 1},
 {0, 2 Pi}, Lines -> 20, PlotPoints -> 40,
 PlotRange → {{-2, 2}, {-2, 2}}]
```

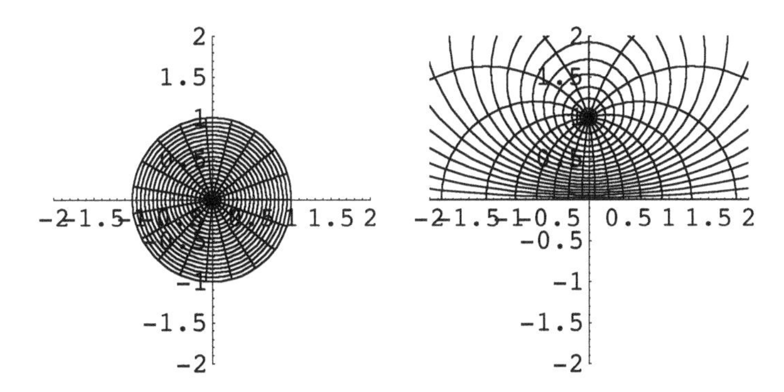

We can look at the inverse mapping in other ways. This is given by Eq. (16.1)

$$z = -\frac{i(w+1)}{w-1}$$

In this case we see that a complex point equidistant from +1 and -1 will get mapped to the unit circle. That is, the imaginary axis in w-coordinates gets mapped to the unit circle in z-coordinates. Let's visualize this as before:

```
LeftHalfwPlane = ComplexInequalityPlot[ Re[w] ≤ 0,
   {w, -5 - 5 I, 5 + 5 I }, DisplayFunction → Identity ];
zOfLeftHalfwPlane = ComplexInequalityPlot[
   Re[(z - I) / (z + I)] ≤ 0, {z, -5 - 5 I, 5 + 5 I },
   DisplayFunction → Identity ];
Show[
 GraphicsArray[{LeftHalfwPlane, zOfLeftHalfwPlane}],
 DisplayFunction → $DisplayFunction ]
```

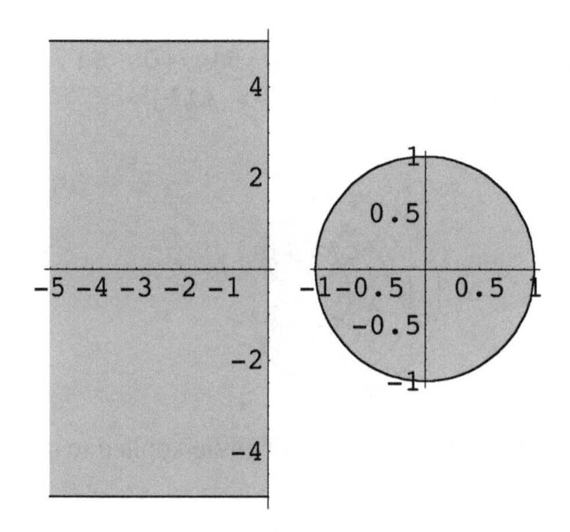

■ Power mappings

Let's take a look at the square and the cube mapping

```
PolarConformal[#^2 &, {0, 1}, {0, Pi / 2},
 Lines -> 20, PlotPoints -> 40, PlotRange -> All]
```

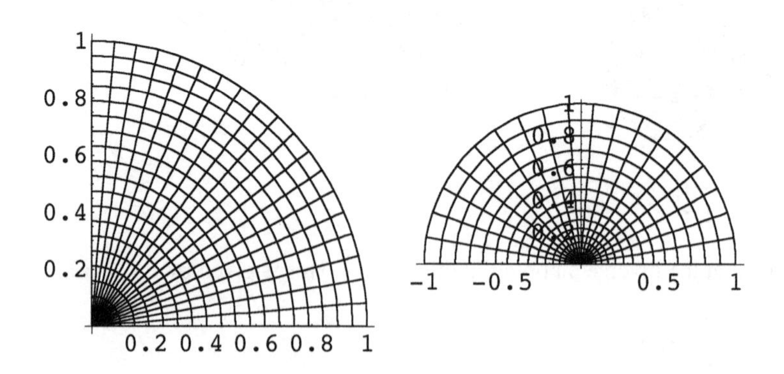


```
PolarConformal[#^3 &, {0, 1}, {0, Pi / 3},
 Lines -> 20, PlotPoints -> 40, PlotRange -> All]
```

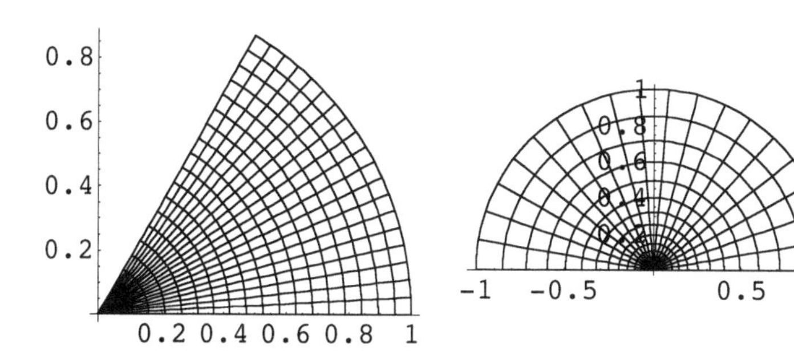

■ Fractional powers

By taking suitable fractional powers the effect of a power mapping can be undone:

```
PolarConformal[#^(1/2) &, {0, 1}, {0, Pi},
 Lines -> 20, PlotPoints -> 40, PlotRange -> All]
```

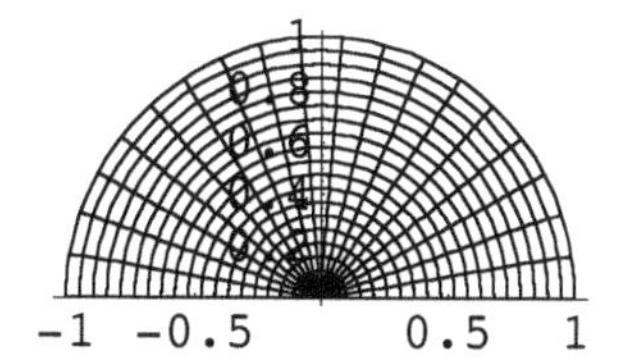

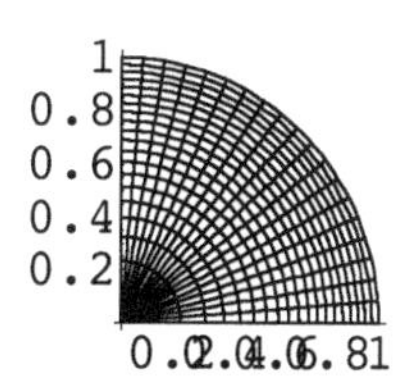

This gives us the idea of opening up a wedge to a half-plane, or vice-versa:

```
zFirstQuadrant =
  ComplexInequalityPlot[Re[z] ≥ 0 && Im[z] ≥ 0,
   {z, -5 - 5 I, 5 + 5 I}, DisplayFunction → Identity];
wOfzFirstQuadrant = ComplexInequalityPlot[
   Re[w^2] ≥ 0 && Im[w^2] ≥ 0, {w, -5 - 5 I, 5 + 5 I},
   DisplayFunction → Identity];
Show[
 GraphicsArray[{zFirstQuadrant, wOfzFirstQuadrant}],
 DisplayFunction → $DisplayFunction]
```

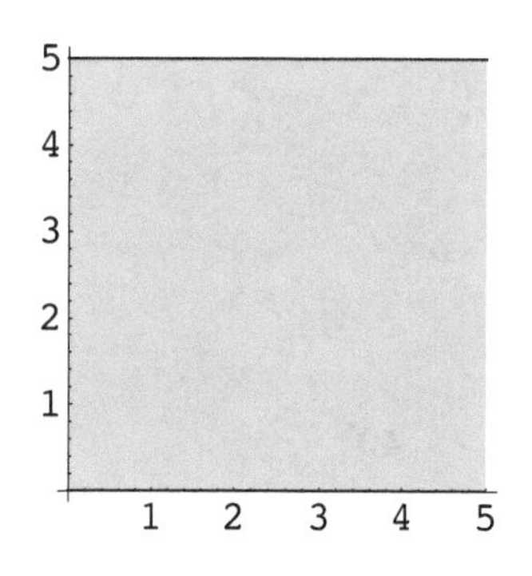

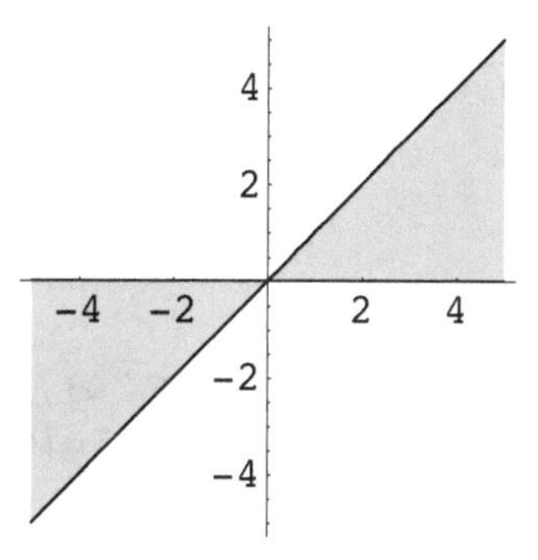

Next we look at some other mappings.

■ Rotations

```
PolarConformal[Exp[I Pi / 6] * # &,
 {0, 1}, {-Pi / 6, Pi / 6}, Lines -> 20,
 PlotPoints -> 40, PlotRange -> All]
```

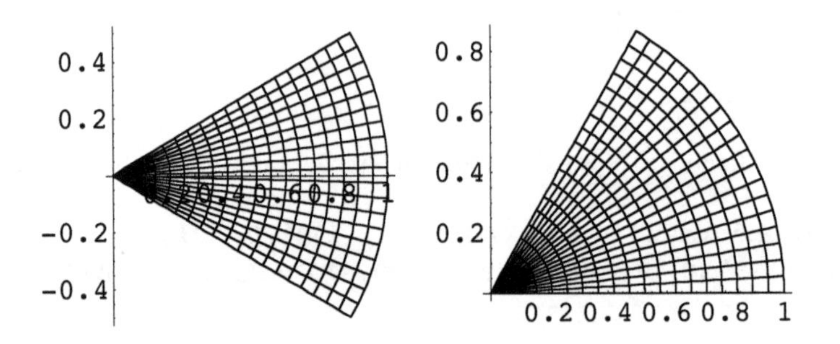

■ Exponential mappings

```
CartesianConformal[Exp, {0, 1}, {0, Pi / 2},
 Lines -> 20, PlotPoints -> 40, PlotRange -> All]
```

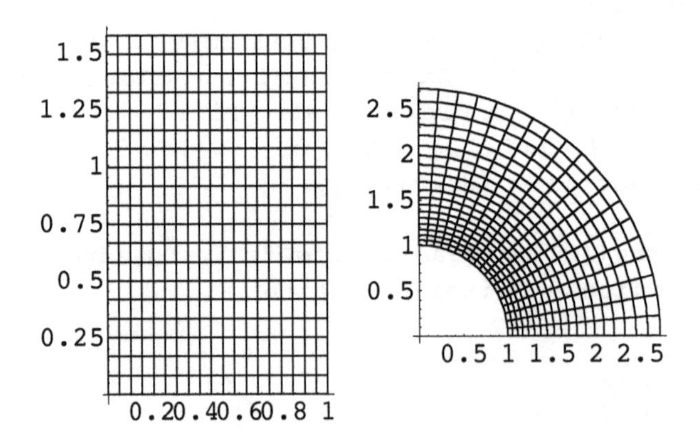

■ Log mappings

```
PolarConformal[Log, {E^2, E^4},
 {0, Pi / 2}, Lines -> 20, PlotPoints -> 40]
```

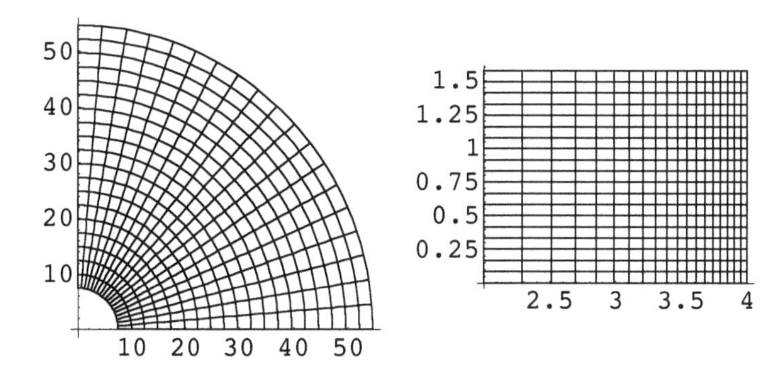

■ Joukowski aerofoil mapping

There is a very famous conformal mapping that captures several interesting features of aerofoils. The Joukowski mapping can be used to investigate various problems related to fluid flow, by transforming the region exterior to a circle to that exterior to an aerofoil shape:

```
Joukowski[x_, y_, R_, vbl_, θ_] :=
Exp[I θ] * ((vbl + x + I y) +
    (-x + Sqrt[R^2 - y^2])^2 / (vbl + x + I y))
```

■ Varying the thickness

When both x and y are zero the circle is mapped into a real line segment. As x changes the mapped region transforms from an aerofoil-shaped region into a plane. The first frame of an animation demonstrating this is shown in the text in the output below – the electronic version contains the full movie. If you are using *Mathematica* technologies beyond version 5.2 see the enclosed CD and on-line supplement as well.

```
Do[PolarConformal[Joukowski[x, 0, 2, #, 0] &,
  {2, 5}, {0, 2 Pi}, Lines -> 10, PlotPoints -> 40,
PlotRange -> {{-6, 6}, {-6, 6}}], {x, 0.5, 0.0, -0.05}]
```

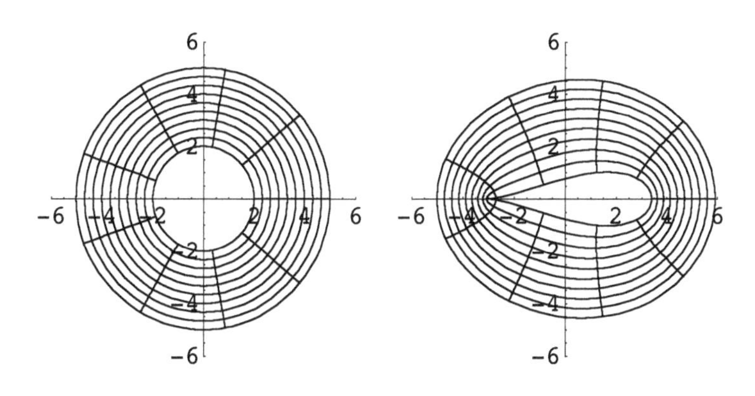

▪ Varying the camber

Now, by varying y, the camber of the aerofoil can be adjusted:

```
Do[PolarConformal[Joukowski[0.5, y, 2, #, 0] &,
  {2, 5}, {0, 2 Pi}, Lines -> 10, PlotPoints -> 40,
PlotRange -> {{-6, 6}, {-6, 6}}], {y, 1, -1, -0.1}]
```

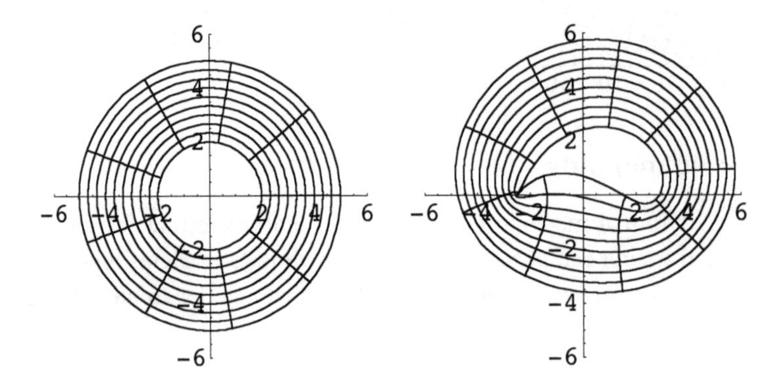

▪ Varying the angle of attack

A simple rotation can also be applied:

```
Do[PolarConformal[Joukowski[0.5, 0.5, 2, #, θ] &,
  {2, 5}, {0, 2 Pi}, Lines -> 10, PlotPoints -> 40,
PlotRange -> {{-6, 6}, {-6, 6}}], {θ, 0, Pi / 4, Pi / 16}]
```

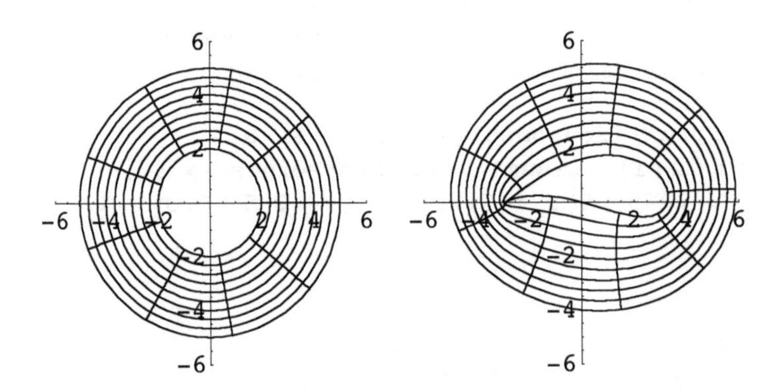

16.3 The conformality property

You should now take a careful look at the pictures in the previous section and ask yourself the question: At what angles do the curves (in the right-hand plots) intersect? It should not take long to convince yourself that the answer is, usually, a right angle. Apart from special points, such as the trailing edge of the aerofoil, the mappings have preserved the property that the original (left-hand plot) Cartesian or polar coordinate curves intersect at a right angle. This occurs in spite of the fact that globally the shapes of regions are transformed dramatically – locally, two curves that intersected at a right angle usually intersect at a right-angle after the mapping has been applied. What is happening here – and why do things occasionally go wrong? The answer lies in the conformality property of holomorphic mappings. Consider two curves $\phi_1(t)$ and $\phi_2(t)$ with the property that

$$\phi_1[0] = \phi_2[0] = z_0 \tag{16.2}$$

That is, they intersect at z_0, and we make the convenient choice that this corresponds to $t = 0$ for both curves. Suppose that we apply a holomorphic mapping

$$w = f(z) \tag{16.3}$$

to a region of the complex plane containing a neighbourhood of z_0. We get a pair of curves

$$\begin{aligned} \psi_1(t) &= f(\phi_1(t)) \\ \psi_2(t) &= f(\phi_2(t)) \end{aligned} \tag{16.4}$$

such that $\psi_1(0) = \psi_2(0) = w_0 = f(z_0)$. The question that we wish to pose and answer is : What is the angle between the curves ϕ_1 and ϕ_2 at $t = 0$, and how is it related to the angle between the curves ψ_1 and ψ_2 at $t = 0$? To define the angle between the curves ϕ_1 and ϕ_2, we must assume that the tangents to the curves exist at $t = 0$. A point, z_1, on ϕ_1 in a neighbourhood of z_0 is given by

$$z_1 = \phi_1(\Delta t) = \phi_1(0) + \Delta t\,\dot{\phi}_1(0) + O(\Delta t^2) \simeq z_0 + +\Delta t\,\dot{\phi}_1(0) \tag{16.5}$$

Similarly, a point z_2 on ϕ_2 in a similar neighbourhood is given by

$$z_2 = \phi_2(\Delta t) = \phi_2(0) + \Delta t\,\dot{\phi}_2(0) + O(\Delta t^2) \simeq z_0 + +\Delta t\,\dot{\phi}_2(0) \tag{16.6}$$

The angle between the ϕ curves at $t = 0$ is given by the limit as $\Delta t \to 0$ of

$$\mathrm{Arg}\left(\frac{z_1 - z_0}{z_2 - z_0}\right) \tag{16.7}$$

and this is then given by

$$\mathrm{Arg}\left(\frac{\dot{\phi}_1(0)}{\dot{\phi}_2(0)}\right) \tag{16.8}$$

Similarly, the angle between the ψ curves is

$$\text{Arg}\left(\frac{\dot{\psi}_1(0)}{\dot{\psi}_2(0)}\right) \tag{16.9}$$

But by the chain rule, for $i = 1, 2$,

$$\dot{\psi}_i(0) = f'(z_0)\,\dot{\phi}_i(0) \tag{16.10}$$

so we can assert that provided $f'(z_0) \neq 0$, the angles between the two set of curves are identical. Matters go wrong when the derivative vanishes.

16.4 The area-scaling property

The effect of the derivative being non-zero or zero may also be interpreted in terms of areas of infinitesimal regions. The following argument assumes that you have taken a basic multi-variable calculus course. If we consider our mappings as real mappings

$$\begin{aligned} u &= u(x, y) \\ v &= v(x, y) \end{aligned} \tag{16.11}$$

then the area magnification factor is given by the Jacobian

$$J = \left|\frac{\partial(u, v)}{\partial(x, y)}\right| \tag{16.12}$$

Exercise 16.3 encourages you to use the Cauchy–Riemann equations to show that this is equal to

$$|f'(z)|^2 \tag{16.13}$$

So areas are locally rescaled by a conformal mapping, and are squashed to zero where the derivative vanishes, which is also where the angle-preserving property fails.

16.5 The fundamental family of transformations

There is a simple family of transforms that lead up to the Möbius transform. The members of this simple family are:

(1) translation by a: $w = z + a$;
(2) rotation by θ: $w = z e^{i\theta}$;
(3) stretching by a (real): $w = az$;
(4) inversion: $w = 1/z$;
(5) the general linear transformation, for a, b complex: $w = az + b$.

By composing all of these maps, we get the Möbius transform, sometimes also called the bilinear or fractional transform:

$$w = \frac{az+b}{cz+d} \tag{16.14}$$

16.6 Group properties of the Möbius transform

We can get *Mathematica* do to quite a bit of the work involved here. First, we wish to show that a Möbius transform is invertible. So define the mapping in *Mathematica* thus:

```
Clear[z, w, W];
W[z] = (a z + b) / (c z + d);
```

We require that $ad \neq bc$, because of the derivative being:

```
Simplify[D[W[z], z]]
```

$$\frac{a\,d - b\,c}{(d + c\,z)^2}$$

The inversion is trivial:

```
z /. Solve[W[z] == w, z][[1]]
```

$$\frac{d\,w - b}{a - c\,w}$$

Let's define another Möbius transform as follows:

```
Q[z_] = (α z + β) / (γ z + δ);
```

Now we compose the two transforms:

```
Simplify[Q[W[z]]]
```

$$\frac{b\,\alpha + a\,z\,\alpha + d\,\beta + c\,z\,\beta}{b\,\gamma + a\,z\,\gamma + d\,\delta + c\,z\,\delta}$$

To make it completely obvious what has happened, we group the terms with and without a z in both numerator and denominator:

```
Collect[Numerator[%], z] / Collect[Denominator[%], z]
```

$$\frac{b\,\alpha + d\,\beta + z\,(a\,\alpha + c\,\beta)}{b\,\gamma + d\,\delta + z\,(a\,\gamma + c\,\delta)}$$

So the composition of two Möbius transforms is also a Möbius transform, and the inverse of a Möbius transform is also a Möbius transform. Given that there is an obvious identity mapping that is also a Möbius transform, it is now clear that Möbius transforms form a *group* under composition. (If you know precisely what a group is you should check the remaining group requirement of associativity yourself.). This is a very interesting group for all sorts of reasons, of both a mathematical and physical character. For now we content ourselves with a few mathematical observations, but in Chapter 23 the true significance is revealed as a complex representation of the Lorentz group of special relativity.

16.7 Other properties of the Möbius transform

If you have done any algebra the fact that Möbius transforms form a group may have caused you to wonder whether this group is in any sense a *symmetry group* – is there some object or set of objects that are left invariant by Möbius transforms? There are in fact at least two 'yes' answers to this question. We shall give one answer here and another in Chapter 23. Even if you do not know what a group is, the observations made here a rather ctitical in understanding the effect of Möbius transforms, and in understanding how to build one with specific properties.

■ A simple characterization of circles and lines

What we wish to establish is that Möbius transformations map the set of all circles and lines into itself. To establish this it is necessary to have a characterization of circles and lines that facilitates this observation. There is more than one way of treating circles and lines to make this straightforward, and we will take an algbraic route. Consider first a circle, centre α and radius r. The equation of this circle is

$$|z-a|^2 = r^2 > 0 \tag{16.15}$$

which we may expand as

$$z\bar{z} - a\bar{z} - \bar{a}z + (|a|^2 - r^2) = 0 \tag{16.16}$$

More flexibly, with A and C real, we consider an equation of the form

$$Az\bar{z} + B\bar{z} + \bar{B}z + C = 0 \tag{16.17}$$

This is equivalent to:

$$z\bar{z} + (B/A)\,\bar{z} + (\bar{B}/A)z + C/A = 0 \tag{16.18}$$

which is a circle with centre $-B/A$ and radius-squared given by

$$\frac{B^2 - AC}{A^2} \tag{16.19}$$

So the quadratic equation Eq. (16.17) with A and C real and $B^2 > AC$ represents a circle. For a line, whose real description is

$$ax + by = c \tag{16.20}$$

we just have to note that this may be rewritten in complex form as the condition

$$B\bar{z} + \bar{B}z + C = 0 \tag{16.21}$$

with C real. So combining the cases of A zero and non-zero we see that the set of circles and lines is characterized by the single condition

$$Az\bar{z} + B\bar{z} + \bar{B}z + C = 0 \tag{16.22}$$

where A and C are real and $B^2 > AC$. It is a line if $A = 0$ and is a circle otherwise. This is sometimes referred to as the equation of a *circline*.

■ *Mathematica* implementation

Let's define the circline condition symbolically. For what follows we remind the reader that complex conjugation is denoted in *Mathematica* **InputForm** by **Conjugate** and in output by a star in recent versions. (We have used a bar in the text).

```
Circline[z_, A_, B_, C_] :=
  A z Conjugate[z] + B Conjugate[z] + Conjugate[B] z + C;
```

Let's also introduce a general symbolic characterization of a Möbius transform:

```
Möbius[z_, a_, b_, c_, d_] := (a z + b) / (c z + d)
```

Let's take the equation of a circline and apply it to the Möbius transform of z:

```
Circline[Möbius[z, a, b, c, d], A, B, C]
```

$$C + \frac{(b + a\,z)\,B^*}{d + c\,z} + \frac{B\,(b + a\,z)^*}{(d + c\,z)^*} + \frac{A\,(b + a\,z)\,(b + a\,z)^*}{(d + c\,z)\,(d + c\,z)^*}$$

First, we simplify it:

```
Simplify[%]
```

$$\frac{(A\,(b + a\,z) + B\,(d + c\,z))\,(b + a\,z)^* + (C\,(d + c\,z) + (b + a\,z)\,B^*)\,(d + c\,z)^*}{(d + c\,z)\,(d + c\,z)^*}$$

Let's inspect the denominator and numerator:

```
Denominator[%]
```

$$(d + c\,z)\,(d + c\,z)^*$$

The denominator is just

$$|d + c\,z|^2 \tag{16.23}$$

and the numerator is:

```
Numerator[%%]
```

$$(A\,(b+a\,z)+B\,(d+c\,z))\,(b+a\,z)^{*}+(C\,(d+c\,z)+(b+a\,z)\,B^{*})\,(d+c\,z)^{*}$$

Now, if **Möbius[z,a,b,c,d]** satisfies the equation of a circline, this numerator must vanish. What condition does this give on z? This is just another quadratic expression involving z and its complex conjugate. Let's shorten and simplify it a little:

```
% /. {Conjugate[b + a z] -> b̄ + ā z̄,
  Conjugate[B] -> B̄, Conjugate[d + c z] -> d̄ + c̄ z̄}
```

$$(A\,(b+a\,z)+B\,(d+c\,z))\,(\overline{b}+\overline{a}\,\overline{z})+(C\,(d+c\,z)+(b+a\,z)\,\overline{B})\,(\overline{d}+\overline{c}\,\overline{z})$$

```
Collect[%, {z, z̄}]
```

$$A\,b\,\overline{b}+B\,d\,\overline{b}+C\,d\,\overline{d}+b\,\overline{B}\,\overline{d}+(A\,b\,\overline{a}+B\,d\,\overline{a}+C\,d\,\overline{c}+b\,\overline{B}\,\overline{c})\,\overline{z}+ z\,(a\,A\,\overline{b}+B\,c\,\overline{b}+c\,C\,\overline{d}+a\,\overline{B}\,\overline{d}+(a\,A\,\overline{a}+B\,c\,\overline{a}+c\,C\,\overline{c}+a\,\overline{B}\,\overline{c})\,\overline{z})$$

Let's make some checks. The coefficients not involving z, and multiplying $z\overline{z}$ are real, and the coefficent of z is the complex conjugate of that of $\overline{z}$. So we set:

```
A' = a A ā + B c ā + c C c̄ + a B̄ c̄;
B' = A b ā + B d ā + C d c̄ + b B̄ c̄;
B̄' = a A b̄ + B c b̄ + c C d̄ + a B̄ d̄;
C' = A b b̄ + B d b̄ + C d d̄ + b B̄ d̄;
```

We need to check the sign of the primed version of $B^2 - AC$. This is given by:

```
Simplify[B' B̄' - A' C']
```

$$-(b\,c-a\,d)\,(A\,C-B\,\overline{B})\,(\overline{b}\,\overline{c}-\overline{a}\,\overline{d})$$

This is $B^2 - AC$ times the square of the absolute value of $bc - ad$, so is positive if the original $B^2 - AC$ was! So the circline equation is preserved in that z also satisfies a circline equation. As an alternative, note that since a Möbius transformation is the composition of various translations, rotations, rescalings and inversions, it is sufficient to check that the circline equation is preserved by each of these separately.

■ Working with three given points

Sometimes one wishes to find a Möbius transform mapping three given points into three known image points. This is easily accomplished with *Mathematica:*

```
Clear[MöbiusCreator, za, zb, zc, wa, wb, wc]
```

```
MöbiusCreator[ z_,
  { za_, zb_, zc_}, {wa_, wb_, wc_}] :=
Module[{soln},
    soln = Solve[{wa == Möbius[za, a, b, c, 1] ,
wb == Möbius[zb, a, b, c, 1] ,
wc == Möbius[zc, a, b, c, 1] }, {a, b, c}];
        Möbius[z, a, b, c, 1] /. soln[[1]]]

MöbiusCreator[z, {0, -I, -1}, {I, 1, 0}]
```

$$\frac{i z + i}{1 - z}$$

This works fine in *Mathematica* for finite points. If you know that particular values of z must end up at zero or infinity, this makes the construction of the numerator and denominator straightforward, up to a scaling in each, and then a third point can be used to fix the overall scale. Let's look at our initial example in this light, but now backwards. Suppose we want the points

$$\{-1, 0, 1\} \tag{16.24}$$

to map to

$$\{0, i, \infty\} \tag{16.25}$$

The numerator must be proportional to $(z + 1)$, in order for -1 to map to zero. Similarly, the denominator must be proportional to $(z - 1)$. So the map must be of the form

$$w = \frac{\alpha(z + 1)}{z - 1} \tag{16.26}$$

and we deduce that

$$\alpha = -i \tag{16.27}$$

in order for the the image of zero to be correct.

■ Comments

A large number of transformations may be effected by combining Möbius transforms with simple power functions and other elementary functions. The action of Möbius transformations is best handled by understanding the circline property and exploring the behaviour of particular points. When trying to map one region into one seemingly unrelated, a good policy is to try to break down the mapping into a sequence of simple mappings that get there step by step. *Above all, practice is required to obtain a feeling for what works*! In the exercises below, this idea of composing several simpler mappings is exploited several times – these questions should be attempted in order as more complex questions may use the result of the one previous.

16.8 More about ComplexInequalityPlot

It will help with the exercises if you understand how to represent combinations of inequalities. For example, if you wish to bound one function by two values of its absolute value, or real or imaginary parts, you can just write down a double inequality:

```
ComplexInequalityPlot[ 1 ≤ Abs[z - I] ≤ Sqrt[2] ,
 {z, -5 - 5 I, 5 + 5 I }]
```

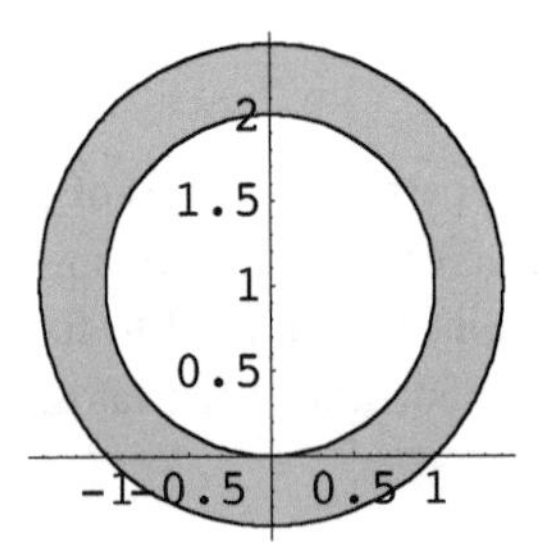

If you need to combine bounds on different functions, you can use the logical AND denoted && in *Mathematica:*

```
ComplexInequalityPlot[ 1 ≤ Abs[z - I] ≤ Sqrt[2]
  && Abs[z + I] ≤ Sqrt[2], {z, -5 - 5 I, 5 + 5 I }]
```

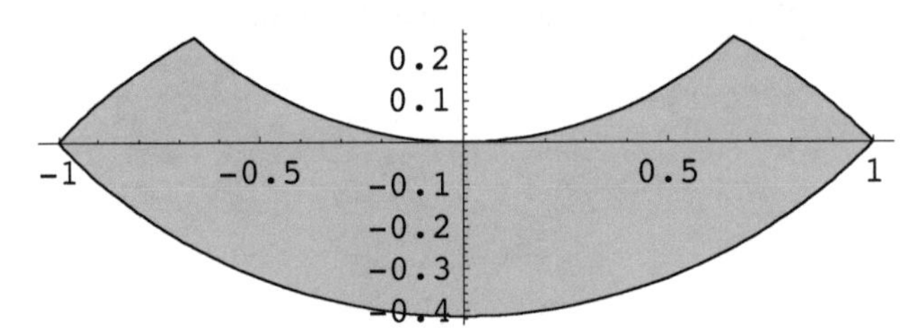

The logical OR, given in *Mathematica* by ||, can also be used:

```
ComplexInequalityPlot[
 Abs[z] ≤ 1/2 || Abs[z - 1] ≥ 2, {z, -5 - 5 I, 5 + 5 I }]
```

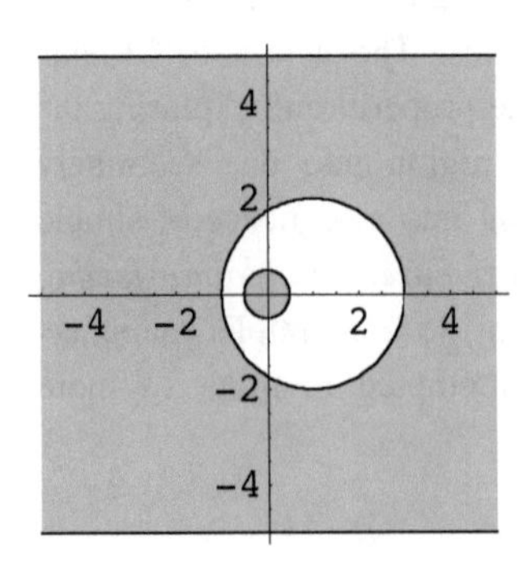

Exercises

In Exercises 16.3-10, you are encouraged to use the **CartesianConformal** and **ComplexInequalityPlot** functions to verify any drawings you have made using *Mathematica*, preferably after you have thought about them with pen and paper.

16.1 What happens to the angles between curves through the origin under the mapping $w = z^2$?

16.2 What happens to the angles between curves through the origin under the mapping $w = z^n$, for $n = 3, 4, 5, \ldots$ etc.?

16.3 Use the Cauchy–Riemann equations to show that the Jacobian is given by

$$J = |f'(z)|^2$$

16.4 What is the image of the quadrant

$$0 \leqslant \text{Arg}(z) \leqslant \frac{\pi}{2}$$

under the mapping $w = z^2$?

16.5 What is the image of the region

$$0 \leq \text{Arg}(z) \leq \frac{\pi}{n}$$

under the mapping $w = z^n$, for n a positive integer?

16.6 What is the image of the real axis under the mapping

$$w = \frac{z - i}{z + i}$$

and what is the image of the upper half-plane under the same mapping?

16.7 Find a mapping that takes the wedge

$$-\pi/6 \leq \text{Arg}(z) \leq \pi/6$$

into the upper half-plane with the real axis included. [Uses Exercise 16.2]

16.8 Find a mapping that takes the wedge

$$-\pi/6 \leq \text{Arg}(z) \leq \pi/6$$

into the closed disk bounded by the unit circle. [Uses Erercises 16.6, 16.7]

16.9 Find a mapping that takes the wedge

$$-\pi/4 \leq \text{Arg}(z) \leq \pi/4$$

into the closed disk bounded by the unit circle. [Similar to Exercise 16.8]

16.10 Consider the region defined by the inequalities

$$|z - i| < \sqrt{2} \qquad |z + i| < \sqrt{2}$$

Draw the region, and verify that the points $+1$ and -1 lie on its boundary, and that zero is in its interior. Find a Möbius transform that takes the points $\{1, -1, 0\}$ to $\{0, \infty, 1\}$. Hence find a mapping that takes the given region into the interior of the unit circle. [Uses Exercise 16.9]

16.11 ✵ Using the **CartesianConformal** and **ComplexInequalityPlot** functions, and trying particular values of a, explore the action of the mapping

$$w = \sin\left(\frac{\pi z}{a}\right)$$

on the region

$$\text{Im}(z) \geq 0; -\frac{a}{2} < \text{Re}(z) < \frac{a}{2}$$

16.12 ✵ Using the **CartesianConformal** function and **ComplexInequality-Plot** functions, and trying particular values of a, explore the action of the mapping

$$w = \cos\left(\frac{\pi z}{a}\right)$$

on the region

$$\text{Im}(z) \geq 0; 0 < \text{Re}(z) < a$$

16.13 ✵ Find a Möbius transform that maps the cube roots of unity, $\{\omega, \omega^2, 1\}$, into $\{1, 0, -1\}$, where

$$\omega = e^{2\pi i/3}$$

17 Fourier transforms

Introduction

In this chapter we shall explore the notion of a 'transform' of a function, where an integral mapping is used to construct a 'transformed' function out of an original function. The continuous Fourier transform is one of a family of such mappings, which also includes the Laplace transform and the discrete Fourier transform. The Laplace transform will be discussed in Chapter 18. Numerical methods for the discrete Fourier transform and for the inversion of Laplace transforms will be given in Chapter 21.

What is the point of such transforms? Perhaps the most important lies in the solution of linear differential equations. Here the operation of a transform can convert differential equations into algebraic equations. In the case of an ordinary differential equation (ODE), one such transform can produce a single algebraic condition that can be solved for the transform by elementary means, leaving one with the problem of inversion – the means by which the transformed solution is turned into the function that is desired. In the case of a partial differential equation (PDE), for example in two variables, one transform can be used to reduce the PDE into an ODE, which may be solved by standard methods, or, perhaps, by the application of a further transform to an algebraic condition. Again one proceeds to a solution of the transformed problem. One or more inversions is required to obtain the solution.

In probability theory the transform of a distribution is called the characteristic function of the distribution. All the moments of the distribution are easily obtained from this characteristic function (which is just a complex form of the moment-generating function), and the Fourier inversion theorem guarantees that the density function can be recovered from the characteristic function.

This chapter will not constitute a comprehensive investigation of Fourier transform calculus. What we shall do is focus on the relevance of contour integration to calculate transforms and their inverses – the material developed in Chapter 15 is critical and should be reviewed now if you are not already familiar with it. Although this will be our focus, along the way we shall give a brief tour of distributions and the delta-function – enough to get a proper feel for the inversion theorem in its proper setting and explore some of the applications to differential equations.

Another approach to managing Fourier transforms is simply to compile a list of functions, transforms and their inverses, and some rules. The method of contour integration is then the method of last resort. Of course, contour integration has to be used to compile some of these tables!

Finally, note that the focus here will be on the full complex form of the transform, using the complex exponential function. Some elementary texts use sine and cosine trasforms – the view here is that the use of the complex form is much cleaner, and can always be re-expressed in sine/cosine terms when absolutely necessary, for those functions posessing the necessary symmetry.

17.1 Definition of the Fourier transform

Their are many different definitions of the Fourier transform in the literature, all of which are related by numerical normalization factors involving 2π and -1. Try not to let this worry you – you just need to be careful about compiling information from different sources. You should also be aware, if you are using a very old version of *Mathematica*, (versions 3.x or earlier) that the default conventions in *Mathematica* are different in version 4.x or later from those in earlier versions. Another difference is that from version 4.x onwards the transforms are in the kernel rather than in an add-on package, as used to be the case. The definition used here as the default is the following. Given a function $f(x)$ defined for $-\infty < x < \infty$, we set

$$\hat{f}(\omega) = \frac{1}{\sqrt{2\pi}} \int_{-\infty}^{\infty} f(x)\, e^{i\omega x}\, dx \tag{17.1}$$

■ *Mathematica* implementation

The conventions used here are consistent with a large body of common use, particularly in modern physics and also with *Mathematica*'s defaults. *Mathematica* has a number of Fourier-related functions. You can find out what is present by the following query:

```
? *Fourier*
```

If you execute that command you will see several functions to try out – what we want here is to get at the details of the definition of immediate interest:

```
? FourierTransform
```

```
FourierTransform[expr, t, ω] gives the symbolic Fourier
  transform of expr. FourierTransform[expr, {t1, t2, ... }, {ω1,
  ω2, ... }] gives the multidimensional Fourier transform of expr.
```

So let's ask about the options, thereby getting the default values:

```
Options[FourierTransform]
```

```
{Assumptions :→ $Assumptions,
  GenerateConditions → False, FourierParameters → {0, 1}}
```

Note the appearance of **Assumptions** and **GenerateConditions**. Given that **Fourier** is essentially an integration, whenever you supply symbolic parameters, you should supply explicit assumptions about the nature of the parameters. This is for the same reasons as discussed in Chapter 15. Here, becasue the default is **GenerateConditions -> False** , it is especially important. The list **FourierParameters** is given in the form $\{a, b\}$, and supplies the general result:

$$\hat{f}(\omega) = \sqrt{\frac{|b|}{(2\pi)^{1-a}}} \int_{-\infty}^{\infty} f(t) e^{ib\omega t} \, dt \tag{17.2}$$

Some common choices for $\{a, b\}$ are `{0, 1}` (default; modern physics), `{1, -1}` (pure mathematics; systems engineering), `{-1, 1}` (classical physics), `{0, -2 Pi}` (signal processing).

17.2 An informal look at the delta-function

This section is not meant to be rigorous. It contains a sketch of harder material that you may wish to skip for now. Our goal is to give enough information to see how the inversion and convolution theorems may be proved when put in the most appropriate setting. The point is that a full description of transform calculus makes use of a set of objects that is rather bigger than the set of all functions – this is the set of all *distributions*. One can think about distributions in various ways. Here first is an algebraic view. Let's forget about functions for a moment and think about ordinary real vectors in n real dimensions – this constitutes a real vector space V. You can think about ordinary real space where $n = 3$ if it helps. Such a vector space has, naturally associated with it, a dual space, V' consisting of all linear mappings from V to $\mathbb{R}$. The dual space V' *has the same dimension*, n, as V. A very explicit representation can be given. If $v \in V$, and has components with respect to a basis e_i that are v^i, there is a dual basis f^i of V' with respect to which a vector $w \in V'$ has components w_i. The linear mapping to $\mathbb{R}$ is given by

$$\sum_{i=1}^{n} w_i v^i \tag{17.3}$$

Given a notion of *distance* in V, an explicit isomorphim between V and its dual can be set up. The distance is usually expressed in terms of a metric function g_{ij}–in Euclidean space with a standard Cartesian basis it would be 1 if $i = j$ and zero otherwise. The mapping to $\mathbb{R}$ obtained from a vector v and its metric-induced dual is

$$\sum_{i=1}^{n} \sum_{j=1}^{n} v^i v^j g_{ij} \tag{17.4}$$

and in the Euclidean case it is just the standard inner product

$$\sum_{i=1}^{n} v^i v^i \tag{17.5}$$

What does all this have to do with functions? Now consider the set of all continous real functions $f(x)$ defined on some interval $\alpha \leqslant x \leqslant \beta$, which is possibly infinite in extent. Given a particular function f and another function $g(x)$, we can define an element of the dual space associated with g by the mapping L_g, by the formula

$$L_g[f] = \int_\alpha^\beta g(x) f(x)\, dx \tag{17.6}$$

(It is not important here to consider the convergence issues when one or both of α, β are infinite.) So any function can be associated with an element of the dual space by this device – a function can be converted into a mapping from functions to the reals by the use of integration. Does this mean that the dual space is equivalent to the set of functions?

The answer is a resounding NO! The dual space is much bigger than the original function space (which is already infinite in dimension), and it is called the set of 'distributions'. How can we see that it is bigger? The answer takes us straight to the notion of a 'delta-function', which is not really a function at all! Pick a point a in the interval on which the functions are defined, and set

$$\Delta_a[f] = f(a) \tag{17.7}$$

This is a perfectly valid linear mapping that does not arise through the integration of f against any other continuous function. What we do now is to invent a representation of such distributions that *looks like* an ordinary function, and call it

$$\delta(x - a) \tag{17.8}$$

the *delta-function, or δ-function, with support at a*. It has the following properties:

$$\delta(x - a) = 0 \text{ if } x \neq a \tag{17.9}$$

$$\int_\alpha^\beta f(x)\delta(x - a)\, dx = f(a) \tag{17.10}$$

provided $\alpha < a < \beta$. You can think of $\delta(x - a)$ as being an 'infinite spike' located at a that integrates to unity.

■ The δ-function as the limit of a sequence of functions

Although this object is not really a function at all, it is very convenient to think of it in terms of the limit of a sequence of functions. This 'limit' is not the limit in any of the senses of classical analysis – it is only to be applied under integration. It does not actually matter what particular sequence of functions is employed, provided that they are continuous and integrate to unity over the entire real line. Let's introduce two sequences that do the job. First, let's define Δ_1 as follows.

```
Δ1[ϵ_, x_, a_] :=
  1 / (ϵ Sqrt[2 Pi]) Exp[- (x - a) ^2 / (2 ϵ^2)]
```

In standard mathematical notation (we have also converted the output) it is

```
TraditionalForm[Δ1[ϵ, x, a]]
```

$$\frac{e^{-\frac{(x-a)^2}{2\epsilon^2}}}{\sqrt{2\pi}\,\epsilon} \tag{17.11}$$

So this is just a Gaussian function. It integrates to unity provided ϵ is real and positive:

```
Integrate[Δ1[ϵ, x, a],
 {x, -Infinity, Infinity}, Assumptions → ϵ > 0]
```

1

Students of probability or statistics will recognize this as the density function associated with a normal distribution with mean a and standard deviation ϵ. The second function we shall introduce is

```
Δ2[ϵ_, x_, a_] := ϵ/Pi/((x - a)^2 + ϵ^2)
```

In mathematical notation (the output is converted here also) this is

$$\frac{\epsilon}{\pi\,((x-a)^2+\epsilon^2)} \tag{17.12}$$

```
Integrate[Δ2[ϵ, x, a], {x, -Infinity, Infinity},
 Assumptions → {Im[a] == 0, ϵ > 0}]
```

1

Students of probability or statistics will recognize this as the density function associated with a Cauchy distribution centred on a, parametrized by ϵ. Both of these functions have the property that they integrate to unity, and are peaked at a. (You might like to plot these functions using *Mathematica*.) As the parameter ϵ tends to zero, these functions become more strongly peaked at a. The idea is that their limiting form is precisely that of a δ-function. You can get a better grip on this by considering integrating either function against a 'test-function' f. We can write (from now on the integration range is fixed as the entire real line):

$$\begin{aligned}
&\int_{-\infty}^{\infty} f(x)\Delta_i(\epsilon, x, a)\,dx \\
&= \int_{-\infty}^{\infty} (f(x)-f(a))\Delta_i(\epsilon, x, a)\,dx + \int_{-\infty}^{\infty} f(a)\,\Delta_i(\epsilon, x, a)\,dx \\
&= f(a)\int_{-\infty}^{\infty} \Delta_i(\epsilon, x, a)\,dx + \int_{-\infty}^{\infty} (f(x)-f(a))\Delta_i(\epsilon, x, a)\,dx \\
&= f(a) + \int_{-\infty}^{\infty} (f(x)-f(a))\Delta_i[\epsilon, x, a]\,dx
\end{aligned} \tag{17.13}$$

Now consider what happens as $\epsilon \to 0$. The last integral has an integrand that is zero at $x = a$, because of the factor $f(x) - f(a)$, but the Δ-function concentrates itself at this point, becoming zero elsewhere. Some careful analysis shows that this latter term tends to zero, leaving us with just $f(a)$ when $\epsilon = 0$. So these Δ-functions do have a limit that is a δ-function, when all 'limits' are taken assuming an integration is being carried out.

Having got a grip on the δ-function, we need to say what it has to do with Fourier transforms. The concept we are after is to define the Fourier transform of unity, i.e,

$$\frac{1}{\sqrt{2\pi}} \int_{-\infty}^{\infty} 1\, e^{i\omega x}\, dx \tag{17.14}$$

This does not exist at all in the usual sense, but it has a very simple interpretation once distributions are introduced. The way to get at this is very simple. We replace '1' in the integral by a function whose limit is unity. There are several choices, but the one that is most convenient is to consider, for $\epsilon > 0$,

$$\frac{1}{\sqrt{2\pi}} \int_{-\infty}^{\infty} e^{-\epsilon|x|}\, e^{i\omega x}\, dx \tag{17.15}$$

This can be done by pen-and-paper in two pieces, or we can get *Mathematica* to sort it out. Note that we have said that the imaginary part of ω mustg be less than ϵ in magnitude. Think about why this must be true for the integral in Eq. (17.15) to converge.

```
1/Sqrt[2 Pi]Integrate[Exp[-є*Abs[x]]*Exp[I*ω*x],
  {x, -Infinity, Infinity}, Assumptions → {є>0, -є <
Im[ω] < є}]
```

$$\frac{\sqrt{\frac{2}{\pi}}\,\epsilon}{\epsilon^2 + \omega^2}$$

which is precisely

$$\sqrt{2\pi}\, \Delta_2(\epsilon, \omega, 0) \tag{17.16}$$

That is, we have established that

$$\frac{1}{\sqrt{2\pi}} \int_{-\infty}^{\infty} e^{-\epsilon|x|} e^{i\omega x}\, dx = \sqrt{2\pi}\, \Delta_2(\epsilon, \omega, 0) \tag{17.17}$$

Under any subsquent integration over ω, we can let $\epsilon \to 0$, and hence assert that, *as a distribution*,

$$\int_{-\infty}^{\infty} e^{i\omega x} dx = 2\pi\delta(\omega) \tag{17.18}$$

Eq. (17.18) gives the fundamental link between the Fourier transform and the δ-function.

■ *Mathematica*'s δ-function

In *Mathematica* the δ-function is associated with the physicist P. Dirac, its key inventor. Its definition is available within the kernel in version 4 or later. See the Calculus packages if you are using an older version of *Mathematica*.

```
? DiracDelta
```

```
DiracDelta[x] represents the Dirac delta
  function δ(x). DiracDelta[x1, x2, ... ] represents
  the multidimensional Dirac delta function δ(x1, x2, …).
```

```
DiracDelta[x]
```

$\delta(x)$

Mathematica understands the key link between Fourier transforms and the δ- function:

```
FourierTransform[1, t, ω]
```

$\sqrt{2\pi}\,\delta(\omega)$

17.3 Inversion, convolution, shifting and differentiation

We are now in a position to give informal distributional proofs of the key results – the inversion theorem and the convolution theorem. These are stated with our definitional convention given in Eq. (17.1).

■ Theorem 17.1: the inversion theorem

Suppose that Eq. (17.1) *holds, i.e.*

$$\hat{f}(\omega) = \frac{1}{\sqrt{2\pi}} \int_{-\infty}^{\infty} f(x) e^{i\omega x}\, dx$$

Then the inversion theorem states that

$$f(x) = \frac{1}{\sqrt{2\pi}} \int_{-\infty}^{\infty} \hat{f}(\omega) e^{-i\omega x}\, d\omega \tag{17.19}$$

Assuming the distributional result from Eq. (17.18) and that some reordering of integrals is possible, we can give a simple proof. We have:

$$\begin{aligned} \frac{1}{\sqrt{2\pi}} \int_{-\infty}^{\infty} \hat{f}(\omega) e^{-i\omega x}\, d\omega &= \frac{1}{2\pi} \int_{-\infty}^{\infty} \left(\int_{-\infty}^{\infty} f(y) e^{i\omega y}\, dy \right) e^{-i\omega x}\, d\omega \\ = \int_{-\infty}^{\infty} \left(\frac{1}{2\pi} \int_{-\infty}^{\infty} e^{i\omega(y-x)}\, d\omega \right) f(y)\, dy &= \int_{-\infty}^{\infty} \delta(y-x)\, \phi(y)\, dy = f(x) \end{aligned} \tag{17.20}$$

■ Mathematica inversion

```
? InverseFourierTransform
```

```
InverseFourierTransform[expr, ω, t] gives the symbolic inverse Fourier
  transform of expr. InverseFourierTransform[expr, {ω1, ω2, ... }, {t1,
  t2, ... }] gives the multidimensional inverse Fourier transform of expr.
```

```
InverseFourierTransform[Sqrt[2 Pi] DiracDelta[ω], ω, t]
```

1

■ Theorem 17.2: the convolution theorem

Suppose we have two transforms:

$$\hat{f}(\omega) = \frac{1}{\sqrt{2\pi}} \int_{-\infty}^{\infty} f(x) e^{i\omega x}\, dx;\ \hat{g}(\omega) = \frac{1}{\sqrt{2\pi}} \int_{-\infty}^{\infty} g(x) e^{i\omega x}\, dx \tag{17.21}$$

The convolution of f with g, $[f * g](x)$, is defined by,

$$[f * g](x) = \int_{-\infty}^{\infty} f(y)\, g(x-y)\, dy \tag{17.22}$$

If $h(x) = [f * g](x)$, then the Fourier transform of the convolution is

$$\hat{h}(\omega) = \sqrt{2\pi}\ \hat{f}(\omega)\, \hat{g}(\omega) \tag{17.23}$$

Here is the proof.

$$\begin{aligned} \sqrt{2\pi}\ \hat{h}(\omega) &= \int_{-\infty}^{\infty} e^{i\omega x} h(x)\, dx \\ &= \int_{-\infty}^{\infty} e^{i\omega x} \int_{-\infty}^{\infty} \left(\frac{1}{\sqrt{2\pi}} \int_{-\infty}^{\infty} \hat{f}(p) e^{-ipy} dp \right) \left(\frac{1}{\sqrt{2\pi}} \int_{-\infty}^{\infty} \hat{g}(q) e^{-iq(x-y)} dq \right) dy\, dx \\ &= \frac{1}{(2\pi)} \int_{-\infty}^{\infty} \int_{-\infty}^{\infty} \int_{-\infty}^{\infty} \int_{-\infty}^{\infty} dp\, dq\, dy\, dx\, e^{i\omega x} e^{-ipy}\, e^{-iq(x-y)} \hat{f}(p) \hat{g}(q) \end{aligned} \tag{17.24}$$

Doing the y-integration reduces this to

$$\int_{-\infty}^{\infty} \int_{-\infty}^{\infty} \int_{-\infty}^{\infty} dp\, dq\, dx\, e^{i\omega x}\, e^{-iqx}\, \hat{f}(p) \hat{g}(q) \delta(q-p) \tag{17.25}$$

We now use the δ-function to do the q integration, leaving us with

$$\int_{-\infty}^{\infty} \int_{-\infty}^{\infty} dp\, dx\, e^{i(\omega-p)x}\, \hat{f}(p) \hat{g}(p) \tag{17.26}$$

Now we integrate by x, to obtain

$$2\pi \int_{-\infty}^{\infty} dp \delta(\omega - p) \hat{f}(p)\hat{g}(p) = 2\pi \hat{f}(\omega)\hat{g}(\omega) \quad (17.27)$$

So we have established that

$$\hat{h}(\omega) = \sqrt{2\pi} \hat{f}(\omega) \hat{g}(\omega) \quad (17.28)$$

That is, the Fourier transform of the convolution is essentially (up to a normalization) the product of the transforms. Similarly, if we have the product of two functions in x terms, the Fourier transform of such a product can be written as the convolution of the transforms.

■ The shift and scaling theorems

Theorem 17.3: the shift theorem. This is simply the observation that *if*

$$h(x) = e^{ixa} f(x) \quad (17.29)$$

then

$$\hat{h}(\omega) = \hat{f}(a + \omega) \quad (17.30)$$

The proof is left to you as Exercise 17.1. Exercise 17.2 asks you to prove:
Theorem 17.4: **the scaling theorem**. This is the result that *if*

$$h(x) = f(x/a) \quad (17.31)$$

then

$$\hat{h}(\omega) = a\hat{f}(\omega a) \quad (17.32)$$

Note also that the Fourier transform is linear. If

$$h(x) = \alpha f(x) + \beta g(x) \quad (17.33)$$

then the transform of h satisfies

$$\hat{h}(\omega) = \alpha \hat{f}(\omega) + \beta \hat{g}(\omega) \quad (17.34)$$

■ Theorem 17.5: the differentiation theorem

For applications in mathematical physics the critial observation is the manner in which differentiation with respect to x becomes simple multiplication by $(-i\omega)$ on the transform. *Let*

$$h(x) = f'(x) \quad (17.35)$$

then

$$\hat{h}(\omega) = -i\omega\hat{f}(\omega) \quad (17.36)$$

To prove this, we note that this is a point at which we must be a little less cavalier about the class of functions we are dealing with. From the definition

$$\hat{h}(\omega) = \frac{1}{\sqrt{2\pi}} \int_{-\infty}^{\infty} f'(x)\, e^{i\omega x}\, dx \tag{17.37}$$

and we need to be able to write:

$$\hat{h}(\omega) = \frac{1}{\sqrt{2\pi}} \int_{-\infty}^{\infty} \left(\frac{\mp(f(x)\, e^{i\omega x})}{\mp x} - i\omega f(x)\, e^{i\omega x} \right) dx \tag{17.38}$$

and use integration to kill the first term – this requires of course that $f \to 0$ as $x \to \pm\infty$. Then we obtain

$$\hat{h}(\omega) = -i\,\omega \frac{1}{\sqrt{2\pi}} \int_{-\infty}^{\infty} f(x) e^{i\omega x}\, dx = -i\omega \hat{f}(\omega) \tag{17.39}$$

Provided higher derivatives tend to zero at $\pm\infty$ and the derivatives of the function remain integrable (continuous will do), repeated application of this result can be used to show that the Fourier transform of the nth derivative is given by

$$(\widehat{f^{(n)}})[\omega] = (-i\omega)^n\, \hat{f}(\omega) \tag{17.40}$$

Note that if the opposite sign convention is employed for the exponent, the right side of this becomes $(i\omega)^n\, \hat{f}(\omega)$. There is a corresponding inverse result that multiplication by x corresponds to differentiation with respect to ω – see Exercise 17.3 for the details.

17.4 Jordan's lemma: semicircle theorem II

It is evident that we need general methods for computing integrals of the form

$$\int_{-\infty}^{\infty} f(x) e^{i\omega x}\, dx$$

A key result for the evaluation of such integrals is Jordan's lemma, which gives very useful conditions under which the integration region may be completed by a large semicircle in the upper (or lower) half-plane (UHP or LHP), and hence evaluated by the calculus of residues. Let's work with the upper half-plane version, where we assume that $\omega > 0$.

Theorem 17.6: Jordan's lemma. *Consider the semicircular path*

$$\begin{gathered} \Phi_R[t] = R e^{it} \\ 0 \le t \le \pi \end{gathered} \tag{17.41}$$

and let M_R be the maximum (formally, the supremum) of $|f(z)|$ on the image of Φ_R in $\mathbb{C}$. Suppose that $\omega > 0$ and that as $R \to \infty$,

$$M_R \to 0 \tag{17.42}$$

Then Jordan's lemma states that

$$\int_{\Phi_R} f(z)\, e^{i\omega z}\, dz \to 0 \tag{17.43}$$

as $R \to \infty$.

▪ Comments

(1) The condition on f is very weak – we just need that the function tends to zero.
(2) If $\omega > 0$ then as the imaginary part of z becomes large the integrand is exponentially damped – this is why we need to associate positive ω with the *upper* half-plane. If $\omega < 0$ there is an obvious corresponding result for the *lower* half-plane.
(3) Once we have this result the answer for the integral for $\omega > 0$ is just

$$2\pi i \sum_{\text{UHP}} \text{Res}[f(z) e^{i\omega z}] \tag{17.44}$$

and for $\omega < 0$ it is

$$-2\pi i \sum_{\text{LHP}} \text{Res}[f(z) e^{i\omega z}] \tag{17.45}$$

Note the additional minus sign in Eq. (17.44) – we traverse the LHP contour clockwise. These can give quite different functional forms for the answer, and this is important.

▪ Proof of Jordan's lemma

Let's first write the integration in terms of an integral over the path parameter t. We apply the integration inequality from Section 11.4:

$$\left| \int_{\Phi_R} F(z) e^{i\omega z}\, dz \right| \leqslant M_R \int_0^{\pi} \left| e^{i\omega R e^{it}}\, i R e^{it} \right| dt \tag{17.46}$$

But

$$\left| e^{i\omega R e^{it}}\, i R e^{it} \right| = R |e^{i\omega R(\cos(t) + i\sin(t))}| = R e^{-\omega R \sin(t)} \tag{17.47}$$

so that

$$\left| \int_{\Phi_R} F(z) e^{i\omega z}\, dz \right| \leqslant R M_R \int_0^{\pi} e^{-\omega R \sin(t)}\, dt = 2 R M_R \int_0^{\pi/2} e^{-\omega R \sin(t)}\, dt \tag{17.48}$$

Now we have the inequality, valid for $0 \leqslant t \leqslant \pi/2$,

$$\sin(t) \geqslant \frac{2t}{\pi} \tag{17.49}$$

(some plots with *Mathematica*, and the concave nature of the sine function in this range of t, should quickly convince you of this) and so we can state that

$$\left| \int_{\Phi_R} F(z) e^{i\omega z}\, dz \right| \leqslant 2RM_R \int_0^{\pi/2} e^{-2\omega R t/\pi}\, dt = 2RM_R \left(\frac{\pi(1 - e^{-R\omega})}{2R\omega} \right) \tag{17.50}$$

$$\left| \int_{\Phi_R} F(z) e^{i\omega z}\, dz \right| \leqslant \frac{\pi M_R (1 - e^{-R\omega})}{\omega} \leqslant \frac{\pi M_R}{\omega} \tag{17.51}$$

which tends to zero as required.

17.5 Examples of transforms

A large number of Fourier transforms can now be generated by combining together:

(1) the basic distributional results;
(2) the shift, scaling and differentiation theorems;
(3) applications of Jordan's lemma and the calculus of residues;
(4) other contours invented for various special cases;
(5) *Mathematica* calculations.

It must be appreciated that one or a combination of these methods must be employed.

■ An example using the basic distributional properties

What is the Fourier transform of $f(x) = \sin(a\, x)$? This, by definition, is

$$\hat{f}(\omega) = \frac{1}{\sqrt{2\pi}} \int_{-\infty}^{\infty} \sin(ax) e^{i\omega x}\, dx \tag{17.52}$$

This is best approached by writing the integrand in terms of pure exponential functions:

$$\begin{aligned} \hat{f}(\omega) &= \frac{1}{\sqrt{2\pi}} \int_{-\infty}^{\infty} \left(\frac{e^{iax} - e^{-iax}}{2i} \right) e^{i\omega x}\, dx \\ &= \frac{1}{2i} \frac{1}{\sqrt{2\pi}} \int_{-\infty}^{\infty} (e^{i(\omega + a)x} - e^{i(\omega - a)x})\, dx \\ &= -i\pi \frac{1}{\sqrt{2\pi}} (\delta(a + \omega) - \delta(\omega - a)) = i\sqrt{\frac{\pi}{2}} (\delta(\omega - a) - \delta(a + \omega)) \end{aligned} \tag{17.53}$$

■ Checking with *Mathematica*

```
FourierTransform[Sin[a t], t, w]
```

$$i\sqrt{\frac{\pi}{2}}\, \delta(w - a) - i\sqrt{\frac{\pi}{2}}\, \delta(a + w)$$

See Exercise 17.4 for the cosine function analogue.

■ Example using Jordan's lemma

Let's consider the function we have used before in defining δ-functions:

$$\frac{\epsilon}{\pi(x^2+\epsilon^2)} \tag{17.54}$$

We can now evaluate its transform very quickly, for this function tends to zero for large z, and hence we can apply Jordan's lemma immediately. For $\omega > 0$ the transform is given by

$$\frac{1}{\sqrt{2\pi}}\, 2\pi i \sum_{\text{UHP}} \text{Res}\Big[\frac{\epsilon}{\pi(z^2+\epsilon^2)}\, e^{i\omega z}\Big] = \frac{1}{\sqrt{2\pi}} \sum_{\text{UHP}} \text{Res}\Big[\frac{2i\epsilon\, e^{i\omega z}}{(z-i\epsilon)(z+i\epsilon)}\Big] \tag{17.55}$$

In the UHP there is only one pole, at $z = i\epsilon$, and it is simple. So we cover up the one singular factor and evaluate the remaining expression to obtain

$$\frac{1}{\sqrt{2\pi}}\, e^{-\epsilon\omega} \tag{17.56}$$

For $\omega < 0$ a similar calculation can be done in the LHP (Exercise 17.5) to obtain $e^{\epsilon\omega}$. Thus the answer for all real ω (it integrates to $1/\sqrt{2\pi}$ if ω=0) is just

$$\frac{1}{\sqrt{2\pi}}\, e^{-\epsilon|\omega|} \tag{17.57}$$

■ An inverse transform using Jordan's lemma

Consider the transform function

$$\hat{f}(\omega) = \frac{i}{\omega + ia} \tag{17.58}$$

What is the associated $f(x)$? It is given by the inversion formula Eq. (17.19)

$$f(x) = \frac{i}{\sqrt{2\pi}} \int_{-\infty}^{\infty} \frac{1}{\omega + ia}\, e^{-i\omega x}\, d\omega \tag{17.59}$$

Note that now we are integrating over ω and the parameter in Jordan's Lemma is $-x$. The non-exponential part of the integral tends to zero at infinity so we can go ahead and apply the lemma. For $x > 0$ we must complete in the *lower* half-plane. There is one simple pole and we get the answer

$$f(x) = \frac{i}{\sqrt{2\pi}}\,(-2\pi i)\, e^{-i(-ia)x} = \sqrt{2\pi}\, e^{-ax} \tag{17.60}$$

But for $x < 0$ we must complete in the upper half-plane, where there are no poles at all! Hence the answer is then zero.

■ Computational check

Note that *Mathematica* gets all this straight, though it helps to be told the sign of a:

```
InverseFourierTransform[
 I / (ω + I a), ω, x, Assumptions → a > 0]
```

$$e^{-a x} \sqrt{2 \pi}\, \theta(x)$$

```
InverseFourierTransform[
 I / (ω + I a), ω, x, Assumptions → a < 0]
```

$$-e^{-a x} \sqrt{2 \pi}\, \theta(-x)$$

The function θ, denoted in *Mathematica* by **UnitStep**, takes care of the two cases – here is a plot of it:

```
Plot[UnitStep[x], {x, -3, 3},
 PlotStyle -> Thickness[0.01]]
```

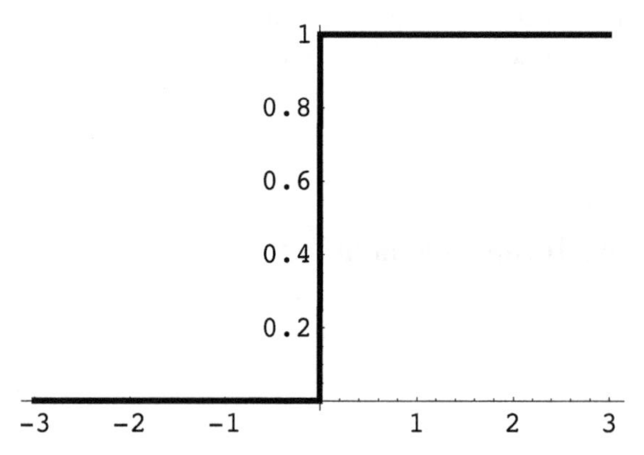

Note the following, which is often useful and follows from the definition of the δ-function:

```
D[UnitStep[x], x]
```

$$\delta(x)$$

■ A special contour for Gaussian functions

The transforms of some functions require special treatment. Consider the Gaussian function

$$\frac{e^{-(x-\mu)^2/(2\sigma^2)}}{\sigma\sqrt{2\pi}} \tag{17.61}$$

We have written it this way in order to make the link with the characteristic function for the normal distribution with mean μ and standard deviation σ. The Fourier transform of this is

$$\hat{f}(\omega) = \frac{1}{2\pi\sigma}\int_{-\infty}^{\infty} e^{-(x-\mu)^2/(2\sigma^2)}\, e^{i\omega x}\, dx \tag{17.62}$$

The term in the exponential is

$$\begin{aligned}
i\omega x - \frac{(x-\mu)^2}{2\sigma^2} &= -\frac{x^2}{2\sigma^2} + \left(\frac{\mu}{\sigma^2} + i\omega\right)x - \frac{\mu^2}{2\sigma^2}\\
&= \frac{-x^2 - \mu^2 + 2x(\mu + i\sigma^2\omega)}{2\sigma^2}\\
&= \frac{-(x-(\mu+i\sigma^2\omega))^2 - \mu^2 + (\mu + i\sigma^2\omega)^2}{2\sigma^2}\\
&= \frac{-(x-(\mu+i\sigma^2\omega))^2}{2\sigma^2} + i\mu\omega - \frac{\sigma^2\omega^2}{2}\\
&= \frac{-p^2}{2\sigma^2} + i\mu\omega - \frac{\sigma^2\omega^2}{2}
\end{aligned} \tag{17.63}$$

where p is the complex shifted variable

$$p = x - (i\omega\sigma^2 + \mu) \tag{17.64}$$

We can make a real change of variables to eliminate μ, but what do we do about the imaginary shift? We can write the result so far as

$$e^{i\omega\mu - \sigma^2\omega^2/2}\, J \tag{17.65}$$

where the quantity J is

$$J = \frac{1}{2\pi\sigma}\int_{-\infty - i\omega\sigma^2}^{\infty - i\omega\sigma^2} e^{-q^2/(2\sigma^2)}\, dx \tag{17.66}$$

Now consider a rectangular contour obtained by taking the contour in the definition of J, and adding a piece coming backwards along the real axis, joined at both ends, to form a rectangle. We observe:

(1) there is no contribution from the vertical contours, as the integrand tends to zero;
(2) there are no poles inside the rectangle.

By Cauchy's theorem, the total integral must be zero. Hence we note that

$$J = \frac{1}{2\pi\sigma}\int_{-\infty}^{\infty} e^{-q^2/(2\sigma^2)}\, dx = \frac{1}{\sqrt{2\pi}}\,\frac{1}{\sqrt{2\pi}\,\sigma}\int_{-\infty}^{\infty} e^{-q^2/(2\sigma^2)}\, dx = \frac{1}{\sqrt{2\pi}} \tag{17.67}$$

and hence that

$$\hat{f}(\omega) = \frac{1}{\sqrt{2\pi}} e^{i\omega\mu - \sigma^2 \omega^2/2} \tag{17.68}$$

This is almost the 'characteristic function' for the normal distribution, which omits the factor $1/\sqrt{2\pi}$.

■ Checking with *Mathematica*

```
FourierTransform[
 1 / (Sqrt[2 Pi] σ) Exp[ - (x - μ) ^2 / (2 σ^2)],
 x, ω, Assumptions → {σ > 0, Im[μ] == 0}]
```

$$\frac{e^{i\,\mu\,\omega - \frac{\sigma^2\,\omega^2}{2}}}{\sqrt{2\,\pi}}$$

17.6 Expanding the setting to a fully complex picture

Complex numbers actually play other roles in the management of Fourier transforms. It is not just a matter of finding the values of integrals using Jordan's lemma and applying the calculus of residues. In fact, we really need to see transforms as being defined for complex values of ω. Why should we bother with this? In fact it allows us to cope with transforms of a rather larger class of functions, and furthermore to regard Laplace transforms and Fourier transforms as being related in a rather trivial fashion, by a 90-degree rotation in the complex plane. Furthermore, the application of Jordan's lemma requires that the function being integrated is actually holomorphic – when is this so?

In this subsection the independent variable will be taken to be t rather than x. This is partly to make the link with Laplace transforms easier, as we shall be concerned here mainly with Fourier transforms of functions that are identically zero for $t < 0$.

As a motivating example, consider the function

$$f(t) = \begin{cases} 0; & t < 0 \\ e^{at}; & t > 0 \end{cases} \tag{17.69}$$

where we make no particular requirement on the sign of the real parameter a. The Fourier transform is just

$$\frac{1}{\sqrt{2\pi}} \int_0^\infty e^{(a+i\omega)t}\, dt \tag{17.70}$$

This integral is given by evaluating the difference in the values at the limits of the indefinite integral

$$\frac{1}{\sqrt{2\pi}} \frac{e^{t(a+i\omega)}}{a + i\omega} \tag{17.71}$$

Under what circumstances does the limit of this, for $t \to \infty$, exist and equal zero? We need the real part of

$$a + i\omega \tag{17.72}$$

to be less than zero, i.e.

$$a < \mathrm{Im}(\omega) \tag{17.73}$$

Under these circumstances the transform exists and equals

$$-\frac{1}{\sqrt{2\pi}} \frac{1}{(a + i\omega)} \tag{17.74}$$

Mathematica can also calculate this:

```
FourierTransform[UnitStep[t] Exp[a t], t, ω]
```

$$-\frac{1}{\sqrt{2\pi}\,(a + i\,\omega)}$$

So it is convenient to allow the transform variable ω to be complex. The transform exists provided the imaginary part is large enough to kill the exponential growth in the function being transformed.

The inversion theorem must be adjusted accordingly – we need to do the inversion integration by integating along any horizontal contour *above* $\mathrm{Im}[\omega] = a$. We shall not give a proof of the following remark, but it turns out that this type of behaviour is absolutely typical for the transforms of functions that are zero for $t < 0$. More specifically, let $f(t)$ be zero for $t < 0$ and satisfy a condition that

$$|f(t)| \leqslant K e^{at} \tag{17.75}$$

for $t > 0$ and some real $K > 0$ and real a. Then the Fourier transform exists and is a holomorphic function of ω for $\mathrm{Im}(\omega) > a$ – the upper half-plane above a – see Dettman (1984) for a discussion of this, and other related properties. The inversion takes place along a horizontal contour in the half-plane above a. More generally still, for functions that are not zero for $t < 0$, there will typically be a strip in which the transform is holomorphic. See Exercise 17.7 for an example.

17.7 Applications to differential equations

The most elegant applications of Fourier transforms are those to simple partial differential equations involving a space variable that extends over the entire real line.

■ The diffusion equation

Consider the equation, for $-\infty < x < \infty$ and $\tau > 0$,

$$\frac{\partial^2 u(x, \tau)}{\partial x^2} = \frac{\partial u(x, \tau)}{\partial \tau} \tag{17.76}$$

with the initial condition that $u(x, 0) = f(y)$. We write the spatial Fourier transform as

$$\hat{u}(\omega, \tau) = \frac{1}{\sqrt{2\pi}} \int_{-\infty}^{\infty} e^{i\omega x} u(x, \tau)\, dx \tag{17.77}$$

with the inversion relation

$$u(x, \tau) = \frac{1}{\sqrt{2\pi}} \int_{-\infty}^{\infty} e^{-i\omega x} \hat{u}(\omega, \tau)\, d\omega \tag{17.78}$$

Now, as u satisfies the diffusion equation, $\hat{u}$ satisfies the ordinary differential equation

$$\frac{\partial \hat{u}(\omega, \tau)}{\partial \tau} + \omega^2 \hat{u}(\omega, \tau) = 0 \tag{17.79}$$

This has the solution

$$\hat{u}(\omega, \tau) = \hat{u}(\omega, 0)\, e^{-\omega^2 \tau} \tag{17.80}$$

So, using our formula for u at general times,

$$u(x, \tau) = \frac{1}{\sqrt{2\pi}} \int_{-\infty}^{\infty} e^{-i\omega x} \hat{u}(\omega, 0)\, e^{-\omega^2 \tau}\, d\omega \tag{17.81}$$

Substituting for the value of $\hat{u}$ at $\tau = 0$

$$u(x, \tau) = \frac{1}{2\pi} \int_{-\infty}^{\infty} e^{-i\omega x} \int_{-\infty}^{\infty} e^{i\omega y} u(y, 0)\, dy\, e^{-\omega^2 \tau}\, d\omega \tag{17.82}$$

Reorganizing gives

$$u(x, \tau) = \int_{-\infty}^{\infty} f(y) H(x - y, \tau)\, dy \tag{17.83}$$

where

$$H(z, \tau) = \frac{1}{2\pi} \int_{-\infty}^{\infty} e^{-i\omega z - \omega^2 \tau}\, d\omega \tag{17.84}$$

We have already analysed such Gaussians and can identify

$$H(z, \tau) = \frac{e^{-z^2/(4\tau)}}{2\sqrt{\tau\pi}} \tag{17.85}$$

This can be checked in Exercise 17.8. So the answer is the convolution of the initial data with this Gaussian function – the *Green's function* for the heat equation. See Exercises 17.9 and 17.10 for some further worked applications.

■ Laplace's equation in a half-plane

This follows a very similar pattern to the heat equation. Consider the equation, for $-\infty < x < \infty$ and $y > 0$,

$$\frac{\partial^2 u(x, y)}{\partial x^2} + \frac{\partial^2 u(x, y)}{\partial y^2} = 0 \tag{17.86}$$

with the boundary condition that $u(x, 0) = f(y)$ and the requirement that $u \to 0$ for large $r = \sqrt{x^2 + y^2}$. We write the spatial Fourier transform as

$$\hat{u}(\omega, y) = \frac{1}{\sqrt{2\pi}} \int_{-\infty}^{\infty} e^{i\omega x}\, u(x, y)\, dx \tag{17.87}$$

with the inversion relation

$$u(x, y) = \frac{1}{\sqrt{2\pi}} \int_{-\infty}^{\infty} e^{-i\omega x}\, \hat{u}(\omega, y)\, d\omega \tag{17.88}$$

Now, as u satisfies the Laplace equation, $\hat{u}$ satisfies the ordinary differential equation

$$\frac{\partial^2 \hat{u}(\omega, y)}{\partial y^2} = \omega^2\, \hat{u}(\omega, \tau) \tag{17.89}$$

With the condition that the answer tends to zero for large y, this has the solution

$$\hat{u}(\omega, y) = \hat{u}(\omega, 0)\, e^{-|\omega| y} \tag{17.90}$$

So using our formula for u at general y:

$$u(x, y) = \frac{1}{\sqrt{2\pi}} \int_{-\infty}^{\infty} e^{-i\omega x}\, \hat{u}(\omega, 0)\, e^{-|\omega| y}\, d\omega \tag{17.91}$$

Substituting for the value of $\hat{u}$ at $y = 0$:

$$u(x, \tau) = \frac{1}{2\pi} \int_{-\infty}^{\infty} e^{-i\omega x} \int_{-\infty}^{\infty} e^{i\omega y}\, u(y, 0)\, dy\, e^{-|\omega| y}\, d\omega \tag{17.92}$$

Reorganizing gives

$$u(x, y) = \int_{-\infty}^{\infty} f(t) H(x - t, y)\, dt \tag{17.93}$$

where now

$$H(q, \tau) = \frac{1}{2\pi} \int_{-\infty}^{\infty} e^{-i\omega q - |\omega| y}\, d\omega = \frac{y}{\pi (q^2 + y^2)} \tag{17.94}$$

by our previous result for the transform of the Cauchy distribution. Hence

$$u(x, y) = \frac{y}{\pi} \int_{-\infty}^{\infty} \frac{f(t)}{((x-t)^2 + y^2)} \, dt \qquad (17.95)$$

This formula is known as the *Poisson integral formula for a half-plane*. We shall discuss it again in Chapter 19, where it arises from another approach. See Exercise 17.11 for a simple application. The answer to Exercise 17.11 should look like:

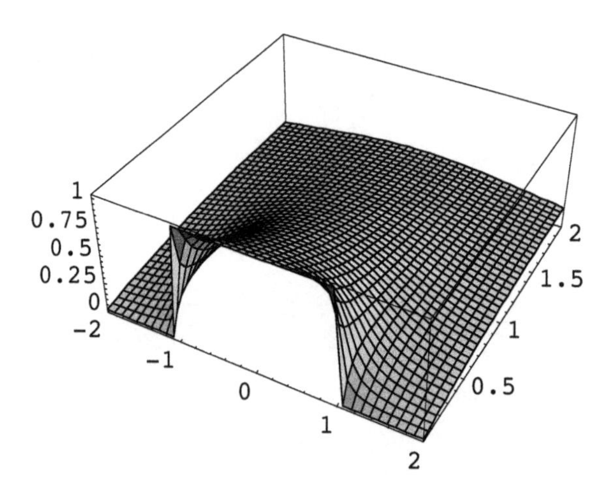

17.8 Specialist applications and other *Mathematica* functions and packages

In this book we have focused on those aspects of Fourier transforms that link most naturally either to complex analysis and the calculus of residues, or to the default functionality built in to *Mathematica*.

We will not pursue here the variant transforms that make use of the sine and cosine functions, though their applications are very important in particular fields. In situations where $f(x)$ has a particular symmetry, e.g. being odd or even in x, it makes sense to encode this symmetry within the transform, and use the sine or cosine versions. These transforms and their inverses are available within *Mathematica*.

```
? *Fourier*
```

System`

Fourier	FourierSinTransform	InverseFourierCosTransform
FourierCosTransform	FourierTransform	InverseFourierSinTransform
FourierParameters	InverseFourier	InverseFourierTransform

For example, in the case of the sine transform, we have:

```
?FourierSinTransform
```

```
FourierSinTransform[expr, t, ω] gives the symbolic Fourier sine
  transform of expr. FourierSinTransform[expr, {t1, t2, ... }, {ω1,
  ω2, ... }] gives the multidimensional Fourier sine transform of expr.
```

This gives the result of the integration

$$\sqrt{\frac{2}{\pi}} \int_0^\infty f(t)\sin(\omega t)\, dt \tag{17.96}$$

Such transforms are very useful in particular fields, but do not have the elegant simplicity, in complex analytical terms, of the basic complex exponential form.

When one wishes to work numerically there is a package for doing so:

```
Needs["Calculus`FourierTransform`"]
```

You can find out what is in it, in version 4 or later, by executing the following command:

```
?Calculus`FourierTransform`*
```

The long result is suppressed here, but the most important ones here are **NFourierTransform** and its inverse, which allow numerical estimation of the transform and its imverse. We will be more interested in the kernel function **Fourier**, which will be discussed in Chapter 20.

Appendix 17: older versions of *Mathematica*

Prior to version 4 the Fourier transform itself was located within an add-on package. This can be loaded as follows:

```
Needs["Calculus`FourierTransform`"]
```

Let's find out what we have loaded (note that the returned list of functions is specific to version 3):

```
?Calculus`FourierTransform`*
```

```
FourierCosSeriesCoefficient  InverseFourierSinTransform
FourierCosTransform          InverseFourierTransform
FourierExpSeries             NFourierCosSeriesCoefficient
FourierExpSeriesCoefficient  NFourierExpSeries
FourierFrequencyConstant     NFourierExpSeriesCoefficient
FourierOverallConstant       NFourierSinSeriesCoefficient
FourierSample                NFourierTransform
FourierSinSeriesCoefficient  NFourierTrigSeries
FourierSinTransform          NInverseFourierTransform
FourierTransform             $FourierFrequencyConstant
FourierTrigSeries            $FourierOverallConstant
InverseFourierCosTransform
```

So we have a great many functions to try out – what we want here is to get at the details of the definition:

```
? FourierTransform
```

```
FourierTransform[expr, t, w] gives a function of w, which is the
  Fourier transform of expr, a function of t.  It is defined by
  FourierTransform[expr, t, w] = FourierOverallConstant * Integrate[
  Exp[FourierFrequencyConstant I w t] expr, {t, -Infinity, Infinity}].
```

The default settings actually defined, in earlier versions, are the following:

$$\hat{f}(\omega) = \int_{-\infty}^{\infty} f(x)\, e^{i\omega x}\, dx \tag{17.97}$$

So let's ask about the options, thereby getting the default values:

```
Options[FourierTransform]
```

```
{FourierFrequencyConstant → 1,
 FourierOverallConstant → 1,
 DefiniteIntegral → False, Assumptions → {}}
```

The first two of these gives is the *old* form of the list now entitled **FourierParameters**. Agreement with the version 4 defaults can be obtained by putting a factor of $1/\sqrt{2\pi}$ in front of the integral. This can be set in *Mathematica* by setting an option with:

```
FourierOverallConstant -> 1 / Sqrt[2 Pi]
```

or, permanently within a kernel session, by setting the default:

```
$FourierOverallConstant = 1 / Sqrt[2 Pi]
```

Mathematically, we are then asserting that the transform is now

$$\hat{f}(\omega) = \frac{1}{\sqrt{2\pi}} \int_{-\infty}^{\infty} f(x)\, e^{i\omega x}\, dx \tag{17.98}$$

The other main variation is to insert a minus sign in the exponent, which can be achieved in *Mathematica* by setting an option

```
FourierFrequencyConstant -> -1
```

or, permanently within a session, by setting

```
$FourierFrequencyConstant = -1
```

If both these variations were to be used, we would have, for example, the new definition

$$\hat{f}(\omega) = \frac{1}{\sqrt{2\pi}} \int_{-\infty}^{\infty} f(x)\, e^{-i\omega x}\, dx \tag{17.99}$$

Exercises

In the following the use of *Mathematica* should be regarded as an optional extra useful for verifying answers found by 'pen and paper'.

17.1 Prove the shift theorem, that if

$$h(x) = e^{ixa} f(x)$$

then

$$\hat{h}(\omega) = \hat{f}(a + \omega)$$

17.2 Prove the scaling theorem, that if

$$h(x) = f(x/a)$$

then

$$\hat{h}(\omega) = a\hat{f}(\omega/a)$$

17.3 Show that the Fourier transform of

$$x^n f(x)$$

is given by

$$\left(-i \frac{\partial}{\partial \omega}\right)^n \hat{f}(\omega)$$

17.4 Show that the Fourier transform of

$$\cos(ax)$$

is given by

$$\sqrt{\frac{\pi}{2}}\, (\delta(\omega - a) + \delta(a + \omega))$$

and use *Mathematica* to confirm your answer.

17.5 Calculate the Fourier transform of

$$\frac{\epsilon}{\pi(x^2 + \epsilon^2)}$$

for $\omega < 0$. (Hint: Use Jordan's lemma in the LHP.)

17.6 Use Jordan's lemma to calculate the inverse transforms of the following:

$$\frac{1}{a^2+\omega^2}; \quad \frac{1}{(a^2+\omega^2)^2}; \quad \frac{1}{(a^2+\omega^2)(b^2+\omega^2)}$$

Calculate the answers for both $x > 0$ and $x < 0$. ❀ In each case verify your answers using *Mathematica*'s **InverseFourierTransform** function.

17.7 (Care required!) Consider the function $f(t)$ defined piecewise, for a, b real, but not necessarily positive, by

$$f(t) = \begin{cases} e^{-a t}; & t > 0 \\ e^{+b t}; & t < 0 \end{cases}$$

For what values of a, b and ω does the Fourier transform of this function exist? When it exists, how must the inversion be done? Show that the inversion recovers the given function when the transform exists.

17.8 ❀ Use *Mathematica* to calculate the inverse Fourier transform of

$$e^{-\omega^2 \tau}$$

17.9 At time zero ($\tau = 0$), an infinite rod is at temperate $T_0 \delta(x)$. The temperature at later times satisfies the heat equation

$$\kappa \frac{\partial^2 u}{\partial x^2} = \frac{\partial u}{\partial \tau}$$

Find the temperature everywhere in the rod for $\tau > 0$. For physicists: If the speed of light is c, evaluate the temperature at the point $2c\tau$. What does this imply about the speed of heat transfer compared to the speed of light? Does this make sense? What do you think might be wrong, if anything, with the heat equation?

17.10 At time zero ($\tau = 0$), an infinite rod is at temperate T_0 in the region $-a < x < a$, and is at zero temperature elsewhere. The temperature at later times satisfies the heat equation

$$\kappa \frac{\partial^2 u}{\partial x^2} = \frac{\partial u}{\partial \tau}$$

Find the temperature everywhere in the rod for $\tau > 0$.

17.11 ❀ A function u satisfies Laplace's equation in the region $-\infty < x < \infty$ and $y > 0$. On the line $y = 0$ it is given by $u = 1$ if $-a < x < a$, and is zero elsewhere. Find the solution for $y > 0$. Use *Mathematica* to plot the result for $a = 1$ and confirm that the function looks like that given in Section 17.8.

18 Laplace transforms

Introduction

There is a transform that is closely related to a special case of the Fourier transform, known as the Laplace transform. While the Laplace transform is very similar, historically it has come to have a separate identity, and one can often find separate tables of the two sets of transforms. Furthermore, it is very appropriate to make a separate assessment of both its inversion, and its applications to differential equations. In the latter context, Laplace transforms are particularly useful when dealing with ODEs and PDEs defined on a half-space – in this setting its differential properties are slightly different from the Fourier transform due to the influence of the boundary.

The goal of this chapter is to define the Laplace transform and explain the basic results and links to complex variable theory. It should be appreciated that there is an extensive knowledge base of known transforms and their inverses. Sadly, many of the excellent books of tables of transforms are old and hard to find if not actually out of print. You might like to check if your library has copies of the old works by Erdelyi. One notable exception is the extraordinarily comprehensive series of books by Prudnikov, Brychkov and Marichev, in which volumes 4 and 5 (Prudnikov *et al*, 1998, 2002) give tables of transforms and their inverses. Although these are expensive, they are in print, are a worthwhile addition to any library and a must for serious professional use. For a further supply of problems and another friendly introduction to the theory, the text by Spiegel (1965) is hard to beat and readily available. You should also note that you can make quite extensive tables of your own with *Mathematica*!

18.1 Definition of the Laplace transform

It is conventional to regard the independent variable as t, (which is often a time coordinate). For the transform variable we shall use the symbol s, though books and papers will often use p or q. We are only interested in functions $f(t)$ defined for

$$t \geqslant 0 \tag{18.1}$$

The Laplace transform of $f(t)$, denoted $\tilde{f}(s)$, or $[Lf](s)$ when we want to avoid any confusion, is defined by the integral

$$\tilde{f}(s) = \int_0^\infty f(t)\, e^{-ts}\, dt \tag{18.2}$$

In terms of the Fourier transform $\hat{f}(\omega)$, it is readily noted that on the class of functions that are zero for $t < 0$, the Laplace transform satisfies the relationship

$$\tilde{f}(s) = \int_0^\infty f(t)\, e^{it(is)}\, dt = \sqrt{2\pi}\, \hat{f}(is) \tag{18.3}$$

and is therefore easily seen as the Fourier transform in complex coordinates rotated by a right angle in the complex plane. Fortunately, everybody seems to agree about the conventions for Laplace transforms so there are no further parameters to worry about. The package definition of Laplace transforms that has to be used prior to version 4 is consistent with the definition within the kernel in versions 4.x or later.

The class of functions that can be Laplace-transformed, as in the case of the Fourier transform, depends on whether one regards s as a real variable (with complex numbers only being introduced to interpret the inversion integral as a contour integration problem), or as an essentially complex variable. The latter view is much more useful, so we require that f is integrable on any finite interval, and that f is bounded by a simple exponential function for t large; i.e. there are real constants M, s_0, T, such that for $t > T$,

$$|f(t)| \leqslant M e^{s_0 T} \tag{18.4}$$

Then the Laplace transform exists in at least the region

$$\text{Re}(s) > s_0 \tag{18.5}$$

■ Example

To make the domain of definition clear, let us consider the exponential function itself, as we did with the Fourier transform, where we found an *upper half-plane* (UHP) of convergence. Since $\omega = i s$, an upper-half-plane of convergence should translate to a *right half-plane* (RHP) of convergence in the Laplace coodinate s. We can work this out directly again for the Laplace transform. We set, for $t \geqslant 0$, $f(t) = e^{at}$, so that

$$\tilde{f}(s) = \int_0^\infty e^{t(a-s)}\, ds \tag{18.6}$$

This will converge (as a function) if and only if

$$\text{Re}(s) > \text{Re}(a) \tag{18.7}$$

and then the transform evaluates to

$$f(s) = \frac{1}{s-a} \tag{18.8}$$

Note that this has a simple pole at $s = a$. As a complex function, it is now defined as a holomorphic function *everywhere except* $s = a$. There is certainly a half-plane of convergence, but once evaluated the transform can be defined on a larger region – this type of behaviour facilitates the inversion.

18.2 Properties of the Laplace transform

The Laplace transform clearly inherits many properties from the Fourier transform, such as linearity. It is useful to have separate characterizations of such results as the shift theorem and differentiation results – the latter is slightly different due to boundary terms.

■ Theorem 18.1: the shift and scaling theorems

Let $g(t) = e^{-at} f(t)$, *then* (the proof is Exercise 18.1),

$$\tilde{g}(s) = \tilde{f}(a+s) \tag{18.9}$$

Also, if $g(t) = f(t/a)$, *then* (Exercise 18.2)

$$\tilde{g}(s) = a\tilde{f}(a\,s) \tag{18.10}$$

■ Theorem 18.2: the differentiation theorem

When we proved the differentiation identity for Fourier transforms, we freely assumed that integration by parts was possible and that there were no boundary terms. When the integral is cut off at zero, the integration by parts introduces boundary terms. In the Laplace case, where the boundary is always present, we need to make this explicit. We shall look at this explicitly for the first and second derivatives. Exercise 18.3 asks you to look at the general case by the method of induction. Consider the Laplace transform of $f'(t)$: This is (we use the longer notation to avoid confusing the tilde and $'$ notations)

$$\begin{aligned} [L f'](s) &= \int_0^\infty f'(t)\, e^{-st}\, dt \\ &= \int_0^\infty \left(\frac{\partial (f(t)\, e^{-st})}{\partial t} + s e^{-st}\, f(t) \right) dt \\ &= s \int_0^\infty f(t) e^{-st}\, dt - f(0) \\ &= s\tilde{f}(s) - f(0) \end{aligned} \tag{18.11}$$

If we apply this result again, with $f(s) = [L g''](s)$, we obtain

$$[L g''](s) = s^2\, \tilde{g}(s) - s g(0) - g'(0) \tag{18.12}$$

For higher derivatives, see Exercise 18.3. We shall find these results very useful in building in initial conditions to the solution of ordinary differential equations.

■ Other identities and the holomorphic property

What about the Laplace transform of $t\, f(t)$, or $t^n\, f(t)$? This is related to the derivative of the Laplace transform, and we can show quite easily that the Laplace transform is actually differentiable and hence holomorphic in a suitable region. Let's proceed naively at first. We have

$$\frac{\partial \tilde{f}(s)}{\partial s} = \frac{\partial\left(\int_0^\infty f(t)\, e^{-st}\, dt\right)}{\partial s} \tag{18.13}$$

Now *assuming* that we can differentiate under the integral, this is just

$$\frac{\partial \tilde{f}(s)}{\partial s} = -\int_0^\infty t f(t)\, e^{-st}\, dt \tag{18.14}$$

so that if this differentiation can be carried out, we see that the required Laplace transform of $t\, f(t)$ is

$$-\frac{\partial \tilde{f}(s)}{\partial s} \tag{18.15}$$

If the Laplace transform is holomorphic, this procedure may be repeated as many times as we like, so that the Laplace transform of $t^n f(t)$ is

$$(-1)^n \frac{\partial^n \tilde{f}(s)}{\partial s^n} \tag{18.16}$$

But is the Laplace transform actually differentiable as a complex function? The answer is yes, and now that we know what the derivative must look like, if it exists, we can prove it. So consider

$$\left| \frac{\tilde{f}(s+h) - \tilde{f}(s)}{h} + \int_0^\infty t f(t)\, dt \right| \tag{18.17}$$

If we can show that this tends to zero as $h \to 0$ we will have shown that the transform is differentiable. We can use the shift theorem to simplify this to

$$\left| \int_0^\infty \left(\frac{e^{-ht} - 1}{h} + t \right) e^{-st}\, f(t)\, dt \right| \tag{18.18}$$

But the exponential function has a series expansion, so we can write

$$\frac{e^{-ht} - 1}{h} + t = \sum_{n=2}^{\infty} \frac{(-ht)^n}{h n!} \tag{18.19}$$

This is of order h and so it is just a matter of tidying up on some analysis. Note that

$$\left| \int_0^\infty \left(\frac{e^{-ht} - 1}{h} + t \right) e^{-st} f(t)\, dt \right| \leq \left| \int_0^\infty \sum_{n=2}^{\infty} \frac{(-ht)^n}{hn!} e^{-st} f(t)\, dt \right| \\ \leq |h| \int_0^\infty t^2 e^{t|h|} |e^{-st} f(t)|\, dt \tag{18.20}$$

Now assuming that the transform exists for $\mathrm{Re}(s) > s_0$, let μ be the real positive constant defined by

$$2\mu = \mathrm{Re}(s) - s_0 \tag{18.21}$$

and constrain h so that

$$|h| < \mu \tag{18.22}$$

Then we have

$$\mathrm{Re}(h + s) > \mu + s_0 \tag{18.23}$$

and so

$$|h| \int_0^\infty t^2\, e^{t|h|}\, |e^{-st} f(t)|\, dt \leq |h| \int_0^\infty t^2\, e^{-(s_0 + \mu)t}\, |f(t)|\, dt \tag{18.24}$$

The integral converges by virtue of Eq. (18.4), and so this term tends to 0 as $h \to 0$. So the Laplace transform is differentiable, and hence holomorphic in the right half-plane of convergence.

■ *Mathematica* and the Laplace transform

Many such transforms can easily be evaluated by standard integration, using substitutions and tables of integrals. *Mathematica* has a pair of built in functions to carry out the transform and its inverse. Prior to version 4.x you had to load a package to carry out the computation. It was in the Calculus area and called **LaplaceTransform**. So with much older versions you need to type (most readers will *not* need to do this)

```
Needs["Calculus`LaplaceTransform`"]
```

before proceeding. Let's find out about the key function of interest:

```
?LaplaceTransform
```

```
LaplaceTransform[expr, t, s] gives the Laplace transform of
  expr. LaplaceTransform[expr, {t1, t2, ... }, {s1, s2, ... }]
  gives the multidimensional Laplace transform of expr.
```

Here are two examples:

```
LaplaceTransform[Exp[a t], t, s]
```

$$\frac{1}{s-a}$$

```
LaplaceTransform[Sin[a t],
 t, s, Assumptions → Im[a] == 0]
```

$$\frac{a}{a^2+s^2}$$

Several other standard calculations are given in Exercise 18.4. In each case you should check your answers against those of *Mathematica.*

■ Making tables of transforms with *Mathematica*

The results of Exercise 18.3 can be extended somewhat using the capabilities built in to *Mathematica.* In fact, you can generate entire tables of transforms in a few moments. Here is an extended set. The output style has been converted to that of a displayed equation, and an asumption that a and b are real has been applied. You might like to make your own tables from *Mathematica*, and print them out.

```
funclist = {1, t, t^2, t^α, t^α Exp[-a t],
   Sin[a t], Cos[a t], Exp[-a t] Cos[b t],
   Exp[-a t] Cos[b t], Sinh[a t], Cosh[a t],
        Erf[a Sqrt[t]]};
tranlist = Map[LaplaceTransform[#, t, s,
     Assumptions → {Im[a] == 0, Im[b] == 0}] &, funclist];
TraditionalForm[TableForm[Transpose[
   {funclist, tranlist}]]]
```

$$\begin{array}{ll} 1 & \frac{1}{s} \\ t & \frac{1}{s^2} \\ t^2 & \frac{2}{s^3} \\ t^\alpha & s^{-\alpha-1}\,\Gamma(\alpha+1) \\ e^{-a t}\, t^\alpha & (a+s)^{-\alpha-1}\,\Gamma(\alpha+1) \\ \sin(a t) & \frac{a}{a^2+s^2} \\ \cos(a t) & \frac{s}{a^2+s^2} \\ e^{-a t}\cos(b t) & \frac{a+s}{b^2+(a+s)^2} \\ e^{-a t}\cos(b t) & \frac{a+s}{b^2+(a+s)^2} \\ \sinh(a t) & \frac{a}{s^2-a^2} \\ \cosh(a t) & \frac{s}{s^2-a^2} \\ \mathrm{erf}(a\sqrt{t}) & \frac{|a|}{s\sqrt{a^2+s}} \end{array} \tag{18.25}$$

18.3 The Bromwich integral and inversion

The Fourier inversion result can be rotated by a right angle to get the Laplace transform inversion integral, sometimes called the *Bromwich integral*. For f continuous, whose transform exists, we have

$$f(t) = \frac{1}{2\pi i} \int_{c-i\infty}^{c+i\infty} \tilde{f}(s)\, e^{st}\, ds \tag{18.26}$$

where $c > s_0$. That is, we take a vertical infinite line within the right half-plane of convergence (to the right of any singularities), and integrate along this contour. The details of the derivation from the Fourier result are straightfoward and given in Exercise 18.5. You should try to understand now why Eq. (18.26) and the choice of c *guarantee* that $f(t) = 0$ for $t < 0$. The explanation is given in Section 18.4.

■ Use of a 'knowledge base' for inversion

One can avoid the direct use of this contour integration for a large number of cases. For example, suppose that we arrive at a transform given by

$$\tilde{f}(s) = \frac{a+s}{b^2 + (a+s)^2} \tag{18.27}$$

How is one going to determine $f(t)$? Quite clearly the quickest answer is to consult the table given by Eq. (18.25), from which the result may be deduced. A great many simple cases may be deduced by such methods, and you should not hesitate to use such methods. You can of course also ask *Mathematica*:

```
InverseLaplaceTransform[(a + s) / (b^2 + (a + s)^2), s, t]
```

$e^{-at} \cos(b\, t)$

Mathematica also knows how to make use of the rules:

```
InverseLaplaceTransform[19! / (s + I a)^20, s, t]
```

$t^{19} (\cos(a\, t) - i \sin(a\, t))$

18.4 Inversion by contour integration

The Bromwich integral can, when other methods fail, be computed by direct integration. We employ, as you might have guessed, a rotated version of Jordan's lemma to deal with the matter. As always, inspection of the exponentials in the integrand is a good guide to when and where the semicircles of integration can be added. So let's look at the inversion integral:

$$f(t) = \frac{1}{2\pi i} \int_{c-i\infty}^{c+i\infty} \tilde{f}(s)\, e^{st}\, ds \tag{18.28}$$

When $t < 0$ the integrand decays strongly as the real part of s becomes large and positive – this is a useful reminder that the rotated Jordan's lemma must be applied to a large semicircle in the right-half-plane. Since the transform is holomorphic in this region, the answer from integrating over the closed contour (obtained by joining this semicircle to the Bromwich contour) is identically zero, by Cauchy's theorem. Hence $f(t) \equiv 0$ for $t < 0$. When $t > 0$ matters are rather different. If $f(s)$ contains only a finite number of poles, we can add a large semicircle in the left half-plane and apply Jordan's lemma again and use the calculus of residues to deduce that:

$$f(t) = \sum_{\text{LHP}} \text{Res}[\tilde{f}(s)\, e^{st}] \tag{18.29}$$

where the LHP is the region $s < s_0 < c$. When $\tilde{f}(s)$ has a branch cut, one of two situations apply. If the branch cut is made between two finite points, and Jordan's lemma can be applied, we get a contribution from Eq. (18.29) plus a contribution from winding around the branch cut, which may or may not be expressible in terms of simple functions. If the branch cut goes off to infinity, the contour cannot be closed and one must use special methods to obtain the answer, and numerical methods may be needed.

▪ Inversion example 1

Consider again the function

$$\tilde{f}(s) = \frac{a+s}{b^2 + (a+s)^2} = \frac{a+s}{(s+a+ib)\,(s+a-ib)} \tag{18.30}$$

The denominator of this expression is zero at the two points

$$s = -a \pm ib \tag{18.31}$$

and the transform has a simple pole at each point. Eq. (18.29) gives us

$$f(t) = \frac{(a+(-a-ib))\, e^{(-a-ib)\,t}}{((-a-ib)+a-ib)} + \frac{(a+(-a+ib))\, e^{(-a+ib)\,t}}{((-a+ib)+a+ib)} \tag{18.32}$$

which simplifies to

$$\frac{1}{2}\, e^{-(a+bi)\,t}\,(1+e^{2ibt}) = e^{-at}\cos(bt) \tag{18.33}$$

as before. Some more examples of this type are given within Exercise 18.6. Although they can be done by other methods, those examples are also valuable practice for the application of the calculus of residues.

▪ Inversion example 2

This is a much more subtle example. Suppose the transform is

$$\frac{1}{\sqrt{a^2 + s^2}} \tag{18.34}$$

This has no poles, but has a branch cut extending between $\pm ia$. But Jordan's lemma applies, so we begin by adding a large semicircle in the left half-plane. Now we deform this until it wraps completely around the branch cut. We split the integral into two pieces:

(1) piece one takes us from $-ia$ to $+ia$ on the right side of the branch cut. We write $s = iu$ and write the contribution to the integral as

$$\frac{1}{2\pi} \int_{-a}^{a} \frac{e^{iut}}{\sqrt{a^2 - u^2}} \, du \tag{18.35}$$

(2) piece two takes us from $+ia$ to $-ia$ down the left side of the branch cut. The signs of the integrand need careful management. As we move around a small circle from the top right of the right side of the full contour to the top left, to begin this second peice, we pick up a minus sign from the square root. But we are also integrating *downwards*. So we get another copy of Eq. (18.35) with the same overall sign. So we deduce that

$$f(t) = \frac{1}{\pi} \int_{-a}^{a} \frac{e^{iut}}{\sqrt{a^2 - u^2}} \, du = \frac{1}{\pi} \int_{-a}^{a} \frac{\cos(ut)}{\sqrt{a^2 - u^2}} \, du \tag{18.36}$$

where the last step applies as the denominator of the integrand is even. What is this function? For this you may need to consult a table of integrals related to special functions. Here we can just ask *Mathematica* to tell us the answer in terms of Bessel functions:

```
Integrate[1 / Pi Cos[u t] / Sqrt[a^2 - u^2],
 {u, -a, a}, Assumptions → {a > 0, t > 0}]
```

$J_0(a\,t)$

You can run another check as well:

```
LaplaceTransform[BesselJ[0, a t], t, s]
```

$$\frac{1}{\sqrt{a^2 + s^2}}$$

▪ ✵ Inversion example 3

$$\tilde{f}(s) = e^{-x\sqrt{s}} \tag{18.37}$$

This is a very important case, which we shall ask *Mathematica* for:

```
InverseLaplaceTransform[Exp[-x Sqrt[s]], s, t]
```

$$\frac{e^{-\frac{x^2}{4t}}\, x}{2\sqrt{\pi}\, t^{3/2}}$$

A closely related case that is relatively straightforward to manage by contour integration is given in Exercise 18.7. A discussion of the case here is given in the next section and in Exercise 18.10.

18.5 Convolutions and applications to ODEs and PDEs

There are many applications of Laplace transforms to differential equations. Many of these involve the convolution theorem for Laplace transforms, which we now state.

■ Theorem 18.3: the convolution theorem for Laplace transforms

Let $f(t)$ and $g(t)$ be functions with Laplace transforms $f(s)$ and $g(s)$ respectively. We know that as Fourier transforms, with f and g *identically zero* for negative values of their arguments, the convolution theorem holds – we have

$$\tilde{f}(s) = \hat{f}(i s), \qquad \tilde{g}(s) = \hat{g}(i s) \tag{18.38}$$

$$h(t) = [f * g](t) = \int_{-\infty}^{\infty} f(\tau)\, g(t-\tau)\, d\tau \tag{18.39}$$

$$\hat{h}(\omega) = \sqrt{2\pi}\, \hat{f}(\omega)\, \hat{g}(\omega) \tag{18.40}$$

Hence, on sorting out the factors of $\sqrt{2\pi}$, we can assert immediately that the convolution theorem for Laplace transforms holds: *if* f, g, h *are as above, then*

$$\tilde{h}(s) = \tilde{f}(s)\, \tilde{g}(s) \tag{18.41}$$

The important point is to notice that the limits of the convolution in the definition of h are not in fact infinite. Since $f(\tau)$ vanishes for $\tau < 0$, and $g(t-\tau)$ vanishes for $\tau > t$, the integral for the convolution h is actually

$$h(t) = [f * g](t) = \int_0^t f(\tau)\, g(t-\tau)\, d\tau \tag{18.42}$$

The relevance of this to differential equations is that we can often think of the differential equation as having two components – a differential operator, and initial and/or boundary conditions and sources. The same operator may be associated with many different boundary conditions and sources, and the transform approach reveals that the solution to the problem is a convolution of a function associated with the operator, and functions associated with the particular sources and/or boundary conditions.

■ Second-order linear ODEs

Let's consider a family of linear systems of the form

$$a \frac{\partial^2 y(t)}{\partial t^2} + b \frac{\partial y(t)}{\partial t} + c y(t) = f(t) \tag{18.43}$$

We Laplace-transform the entire equation, to obtain

$$a(s^2 \tilde{y}(s) - s y(0) - y'(0)) + b(s \tilde{y}(s) - y(0)) + c \tilde{y}(s) = \tilde{f}(s) \tag{18.44}$$

We reorganize this to

$$(a s^2 + b s + c) \tilde{y}(s) = \tilde{f}(s) + a s y(0) + a y'(0) + b y(0) \tag{18.45}$$

So the transform is, immediately,

$$\tilde{y}(s) = \frac{\tilde{f}(s) + a s y(0) + a y'(0) + b y(0)}{a s^2 + b s + c} \tag{18.46}$$

We can see that the differential operator is characterized by the denominator, while the source f and the initial conditions are lumped together in the numerator. It is clear that if we can find the inverse Laplace transform of

$$\frac{1}{a s^2 + b s + c} \tag{18.47}$$

then the solution for any collection of source terms and initial conditions is a convolution of the inverse of this function and the inverse of the numerator. This picture is quite useful for developing general theory, but in practice we wish to consider practical examples. Here is one. Some more for you to try are given in Exercise 18.8.

$$\frac{\partial^2 y(t)}{\partial t^2} + 2 \frac{\partial y(t)}{\partial t} + y(t) = 0 \tag{18.48}$$

with $y(0) = 0$, $y'(0) = 1$. Clearly,

$$\tilde{y}(s) = \frac{1}{s^2 + 2 s + 1} = \frac{1}{(s+1)^2} \tag{18.49}$$

There is a double pole at $s = -1$, and we need to evaluate

$$y(t) = \sum_{s=-1} \text{Res}\left[\frac{1}{(s+1)^2} e^{s t}\right] = \frac{\partial e^{st}}{\partial s} \bigg|_{s=-1} = t e^{-t} \tag{18.50}$$

There are many ways of checking the answer. Here is one direct alternative that uses the *Mathematica* function **`DSolve`**:

```
y[t] /. DSolve[{y''[t] + 2 y'[t] + y[t] == 0,
    y[0] == 0, y'[0] == 1}, y[t], t][[1]]
```

$e^{-t}\, t$

■ Systems of first-order ODEs

Another common and important problem is that of an ordinary differential equation for a vector $y_i(t)$:

$$\frac{\partial y_i[t]}{\partial t} = \sum_{j=1}^{n} A_{ij}\, y_j[t] + f_i[t] \tag{18.51}$$

or, more compactly, as a matrix equation:

$$\dot{y} = A y + f \tag{18.52}$$

When the matrix A consists of constants, clearly the same approach may be applied:

$$s\,\tilde{y}(s) - \tilde{y}(0) = A\tilde{y}(s) + \tilde{f}(s) \tag{18.53}$$

This is rearranged to give

$$(sI - A)\tilde{y}(s) = \tilde{f}(s) + \tilde{y}(0) \tag{18.54}$$

and then inverted as a matrix equation to yield

$$\tilde{y}(s) = (sI - A)^{-1}(\tilde{f}(s) + \tilde{y}(0)) \tag{18.55}$$

This is a particularly elegant description as the poles in this function are given neatly as the eigenvalues of the matrix A. Again, this observation is useful for developing general theory, particularly in Control Theory applications. Sometimes, especially for small n, we can proceed more directly. As an example, consider the system with

$$\frac{\partial y_1(t)}{\partial t} = y_1(t) - y_2(t);\ \frac{\partial y_2(t)}{\partial t} = y_2(t) - y_1(t);\ y_1(0) = 1;\ y_2(0) = 0 \tag{18.56}$$

Here we have $f = 0$, the matrix A is

$$\begin{pmatrix} 1 & -1 \\ -1 & 1 \end{pmatrix} \tag{18.57}$$

and the vector $y(0)$ is

$$\begin{pmatrix} 1 \\ 0 \end{pmatrix} \tag{18.58}$$

The solution for the transformed vector is

$$\begin{pmatrix} s-1 & 1 \\ 1 & s-1 \end{pmatrix}^{-1} \begin{pmatrix} 1 \\ 0 \end{pmatrix} \tag{18.59}$$

which evaluates to

$$\begin{pmatrix} \frac{s-1}{s^2-2s} \\ -\frac{1}{s^2-2s} \end{pmatrix} \tag{18.60}$$

This may be inverted piece by piece, using any of the methods, to obtain

$$\begin{pmatrix} \frac{1}{2} + \frac{e^{2t}}{2} \\ \frac{1}{2} - \frac{e^{2t}}{2} \end{pmatrix} \tag{18.61}$$

As usual, this can be checked with *Mathematica*:

```
{x[t], y[t]} /. DSolve[{D[x[t], t] == x[t] - y[t],
D[y[t], t] == y[t] - x[t], x[0] == 1, y[0] == 0},
  {x[t], y[t]}, t]
```

$$\left(\frac{1}{2}(1+e^{2t}) \quad \frac{1}{2}(1-e^{2t}) \right)$$

You should not be put off learning the analytical methods just because *Mathematica* can do it directly more quickly – here we want to use *Mathematica* as a checking tool, so the example given was simple enough to be done many different ways. Some examples of the transform method applied to systems of ODEs are given in Exercise 18.9.

■ The heat equation with zero initial data and a boundary condition

Let's consider the heat equation

$$\frac{\partial^2 u(x, \tau)}{\partial x^2} = \frac{\partial u(x, \tau)}{\partial \tau} \tag{18.62}$$

in the region

$$x > a, \ \tau > 0 \tag{18.63}$$

with the following boundary and initial conditions:

$$u(a, \tau) = g(\tau); \quad u(x, 0) = 0 \text{ for } x \geq a \tag{18.64}$$

We also demand that at infinity

$$\lim_{x\to\infty} u(x, \tau) = 0 \tag{18.65}$$

This type of problem has many interpretations. The simplest is that of a semi-infinite rod whose temperature is initially zero, but where one end is subsequently as a temperature $g(\tau)$. This is a good place to use a time-based Laplace transform, as the initial data are zero, making the Laplace transform of the time-derivative simple. We set

$$\tilde{u}(x, s) = \int_0^\infty u(x, \tau)\, e^{-s\tau}\, d\tau \tag{18.66}$$

Then the Laplace transform of the diffusion equation gives us

$$\frac{\partial^2 \tilde{u}(x, s)}{\partial x^2} = s\, \tilde{u}(x, s) \tag{18.67}$$

The solution of this vanishing at infinity is

$$\tilde{u}(x, s) = \tilde{u}(a, s)\, e^{-\sqrt{s}\,(x-a)} = \tilde{g}(s)\, e^{-\sqrt{s}\,(x-a)} \tag{18.68}$$

where the Laplace transform of the boundary condition is

$$\tilde{g}(p) = \int_0^\infty g(\tau)\, e^{-p\tau}\, d\tau \tag{18.69}$$

It follows that the solution is the convolution of $g(\tau)$ with the inverse Laplace transform of $e^{-\sqrt{p}\,(x-a)}$, which we have already found. The solution is therefore just

$$u(x, \tau) = \frac{1}{2\sqrt{\pi}} \int_0^\tau \frac{e^{-\frac{(x-a)^2}{4(\tau-t)}}\,(x-a)\, g(t)}{(\tau-t)^{3/2}}\, dt \tag{18.70}$$

By a change of variables, this may be reorganized to

$$u(x, \tau) = \frac{1}{2\sqrt{\pi}} \int_{1/\tau}^\infty \frac{e^{-\frac{s(x-a)^2}{4}}\,(x-a)\, g(\tau - 1/s)}{s^{1/2}}\, ds \tag{18.71}$$

Let's work out a particular example, with $g = 1$. We let $q = (x-a) > 0$.

```
Integrate[(Exp[-(1/4)*s q^2] * q) / (2 Sqrt[Pi s]),
  {s, 1/τ, Infinity}, Assumptions → {q > 0, τ > 0}]
```

$$\operatorname{erfc}\left(\frac{q}{2\sqrt{\tau}}\right)$$

In our problem we have $x > a$, so that we can now write down the result as:

$$\operatorname{erfc}\left(\frac{x-a}{2\sqrt{\tau}}\right) = 1 - \operatorname{erf}\left(\frac{x-a}{2\sqrt{\tau}}\right) \tag{18.72}$$

Note that $\operatorname{erf}(0) = 0$ and $\operatorname{erf}(\infty) = 1$, so this solution is unity when $x = a$ and vanishes when $\tau = 0$, or $x \to \infty$. The integral for a general g, in the form we have given, can be extremely awkward to manage (see Exercise 18.10). Here is a plot of the result for $a = 0$.

```
Plot3D[Erfc[x / (2 Sqrt[t])], {x, 0, 2}, {t, 0.0001, 1}]
```

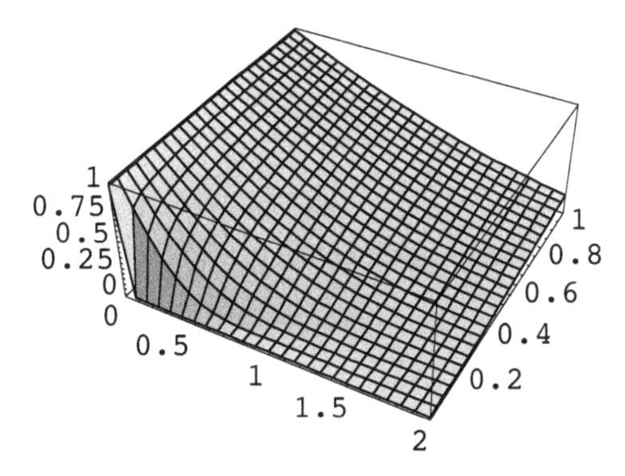

18.6 Conformal maps and Efros's theorem

In Chapter 16 we discussed conformal maps. The pedagogical applications of conformal mapping to potential theory exploit the chain rule for holomporhic functions and the preservation of certain types of boundary condition under conformal maps (e.g wedges, Joukowski aerofoilds). This will be discussed in Chapter 19. The question we will discuss here is what happens when we try to marry conformal mapping to Laplace transform theory? We can consider this in two different ways. First, if we have a Laplace transform pair $f(t)$ and $\tilde{f}(s)$, what function of time is associated with $\tilde{f}(\mu(s))$, where $\mu(s)$ is a holomorphic function associated with a conformal mapping? Second, if we have a (presumably real) mapping $t \to \tau(t)$, what is the Laplace transform of $f(\tau(t))$? Considering the first case, transforms with a structural form of the type

$$\tilde{f}(\mu(s)) \quad \text{or} \quad G(s)\,\tilde{f}(\mu(s))$$

where $\mu(s)$ and $G(s)$ are simple functions are commonplace. For example, in diffusion theory and related financial applications it is very common to generate transforms where μ is a simple square root function.

■ Theorem 18.4: Efros's theorem

According to the series of volumes by Prudnikov *et al* (1998, 2002), the relevant generalized multiplication theorem was written down by A. M. Efros in 1935. *Suppose that* $f(t)$ *has Laplace transform* $\tilde{f}(s)$ *and that we can find a function* $g(t,\ \zeta)$ *such that*

$$\int_0^\infty e^{-st}\, g(t,\,\zeta)\, dx = G(s)\, e^{-\zeta\mu(s)} \tag{18.73}$$

then let

$$H(t) = \int_0^\infty f(\zeta)\, g(t, \zeta)\, d\zeta \tag{18.74}$$

Then the Laplace transform of $H(t)$ is given by

$$\begin{aligned}\tilde{H}(t) &= \int_0^\infty \left(\int_0^\infty f(\zeta)\, g(t, \zeta)\, d\zeta\right) e^{-st}\, dx = G(s) \int_0^\infty f(\zeta)\, e^{-\zeta\mu(s)}\, d\zeta \\ &= G(s)\tilde{f}(\mu(s))\end{aligned} \tag{18.75}$$

Note that Eq. (18.75) contains the proof as well as the statement of the result.

▪ Corollary: the ordinary convolution theorem

If we set $\mu(s) = s$ in the above, then

$$\int_0^\infty e^{-st}\, g(t, \zeta)\, dx = G(s)\, e^{-\zeta s} \tag{18.76}$$

So $g(t, \zeta) = g(t - \zeta)$ and is zero for $t < \zeta$. Then

$$H(t) = \int_0^t f(\zeta)\, g(t - \zeta)\, d\zeta \tag{18.77}$$

with Laplace transform

$$G(s)\, \tilde{f}(s) \tag{18.78}$$

so that Efros's theorem is indeed a generalization of the usual convolution theorem we stated as Theorem 18.3.

▪ An example

Chapter 1 of both volumes 4 and 5 of Prudnikov *et al* (1998, 2002) contain several examples. Here is another example relevant to diffusion and finance theory.

$$\begin{aligned}\mu(s) &= \sqrt{\nu^2 + 2s} \\ G(s) &= \frac{1}{s + \gamma}\end{aligned} \tag{18.79}$$

$$\begin{aligned}g(t, \zeta) = e^{-\gamma t} \Bigg(& e^{-\sqrt{\nu^2 - 2\gamma}\, \zeta}\, \Phi\!\left(\frac{-\zeta + \sqrt{2}\, t \sqrt{\frac{\nu^2}{2} - \gamma}}{\sqrt{t}}\right) + \\ & e^{+\sqrt{\nu^2 - 2\gamma}\, \zeta}\, \Phi\!\left(\frac{-\zeta - \sqrt{2}\, t \sqrt{\frac{\nu^2}{2} - \gamma}}{\sqrt{t}}\right)\Bigg)\end{aligned} \tag{18.80}$$

here Φ is the cumulative normal distribution function.

■ An application of our example

Let's consider a simple case where $\gamma = \nu = 0$ and $\tilde{f}(s) = e^{-(x-a)s/\sqrt{2}}$. The transform is then

$$\frac{e^{-(x-a)\sqrt{s}}}{s} \tag{18.81}$$

which is appropriate to the PDE for the heat equation considered in Section 18.5. For this case

$$g(t, \zeta) = 2\,\Phi\left(\frac{-\zeta}{\sqrt{t}}\right) = \mathrm{erf}\left(-\frac{\zeta}{\sqrt{2t}}\right) + 1 = 1 - \mathrm{erf}\left(\frac{\zeta}{\sqrt{2t}}\right) \tag{18.82}$$

and this times the inverse transform of $\tilde{f}(s)$, with the inverse considered as a function of ζ, must be integrated with respect to ζ. But the inverse transform of

$$\tilde{f}(s) = e^{-\frac{(x-a)s}{\sqrt{2}}} \tag{18.83}$$

is just

$$\delta\left(\zeta - \frac{x-a}{\sqrt{2}}\right) \tag{18.84}$$

so the integration reduces to setting $\zeta = (x-a)/\sqrt{2}$ and we are led once again to the solution of the PDE in the form

$$1 - \mathrm{erf}\left(\frac{x-a}{2\sqrt{t}}\right) \tag{18.85}$$

■ Changes of time coordinate

These are not usually regarded as complex conformal maps (but in a fundamental symmetric Fourer transform picture they could be). What do we do about changes of time variables? What does this induce on the transform? Symmetry suggests a reverse form of Efros's result, with transforms that are generalized convolutions. An interesting example is a result written down at the very end of Chapter 1 of Prudnikov *et al* (2002). Let $f(t)$ have Laplace transform $\tilde{f}(s)$, then

$$\int_0^\infty \tilde{f}(s)\, J_p(a\, s)\, ds \tag{18.86}$$

is the Laplace transform, with coordinate p, of

$$f(a \sinh(t)) \tag{18.87}$$

How does this come about? Let's see how by establishing a simple and closely related result. Let $f(t)$ be given with transform

$$\tilde{f}(s) = \int_0^\infty e^{-st} f(t)\, dt \tag{18.88}$$

Consider now the generalized convolution, with Laplace argument p, and where we replace t by v:

$$\int_0^\infty \tilde{f}(s)\, e^{-s} I_p(s)\, ds = \int_0^\infty f(v)\, dv \int_0^\infty e^{-sv}\, e^{-s} I_p(s)\, ds \tag{18.89}$$

If we let $u = v + 1$, this integral is equal to (using a standard result for the transform of a modified Bessel function)

$$\begin{aligned} &\int_0^\infty f(v)\, dv \int_0^\infty e^{-su}\, I_p(s)\, ds \\ &= \int_0^\infty \frac{f(v)}{\sqrt{u^2-1}\,\left(u+\sqrt{u^2-1}\right)^s}\, dv = \int_0^\infty \frac{f(v)\, e^{-s\cosh^{-1}(u)}}{\sqrt{u^2-1}}\, dv \end{aligned} \tag{18.90}$$

and we can now carry out the inversion $s \to t$, and obtain, writing $u = v + 1$ explicitly now:

$$\begin{aligned} &\int_0^\infty \frac{f(v)\, \delta[t - \cosh^{-1}(v+1)]}{\sqrt{u^2-1}}\, dv \\ &= \int_0^\infty f(v)\, \delta[\cosh(t) - (v+1)]\, dv = f(\cosh[t] - 1) \end{aligned} \tag{18.91}$$

Which means that

$$\int_0^\infty F(q) e^{-q} I_p(q)\, dq \tag{18.92}$$

is the Laplace transform of

$$f(\cosh(t) - 1) \tag{18.93}$$

and we have figured out how to invert a Laplace transform whose variable is given by the index of a modified Bessel function, provided it appears under an integral with a function that is the Laplace transform of another function. The exponential is in there for technical reasons, as $\cosh(0) = 1$. The example with the Bessel function and the change of variables involving $\sinh(t)$ can be proved in a similar way. This is left to you as Exercise 18.11. Other examples can be worked out on a case by case basis.

Exercises

Those exercises containing a supplementary *Mathematica* component in addition to a pen and paper exercise are denoted [❋].

18.1 Prove the shift theorem, that if $g(t) = e^{-at} f(t)$, then $\tilde{g}(s) = \tilde{f}(a+s)$.

18.2 Prove the scaling theorem, that if $g(t) = f(t/a)$, then $\tilde{g}(s) = a\,\tilde{f}(a\,s)$.

18.3 Show that the Laplace transform of the nth derivative of f satisfies:

$$[L\,f^{(n)}]\,(s) = s^n\,\tilde{f}(s) - s^{n-1}\,f(0) - \ldots - f^{(n-1)}(0)$$

(Hint: Use one integration by parts in the definition of $[L\,f^{(n)}]\,(s)$, and use induction.)

18.4 [❀] By direct integration, show that the Laplace transforms of the functions in the left column are those in the right. You may assume that n is an integer for those cases that involve it. In each case, don't forget to use the shift theorem whenever possible, and check your answers with *Mathematica*.

1	$\frac{1}{s}$
t	$\frac{1}{s^2}$
t^n	$s^{-n-1}\,(\Gamma\,(n+1))$
$e^{-a\,t}\,t^n$	$(a+s)^{-n-1}\,(\Gamma\,(n+1))$
$\sin\,(a\,t)$	$\frac{a}{a^2+s^2}$
$\cos\,(a\,t)$	$\frac{s}{a^2+s^2}$
$e^{-a\,t}\,(\cos\,(b\,t))$	$\frac{a+s}{b^2+(a+s)^2}$

18.5 Given the Fourier inversion theorem, and the relationship

$$\tilde{f}(s) = \sqrt{2\pi}\,\hat{f}(i\,s)$$

between the Laplace and Fourier transforms, explain the Laplace inversion formula given by Equation (18.26).

18.6 Using the calculus of residues and Jordan's lemma, evaluate the inverse Laplace transforms of

$$\frac{1}{(s^2+1)}\,;\quad \frac{1}{(s^4+1)}\,;\quad \frac{1}{s\,(s+2)\,(s+4)}\,;\quad \frac{1}{(s^2+4)\,(s^2+9)}$$

18.7 [❀] (Harder) Evaluate, by contour integration, the inverse Laplace transform of

$$\frac{1}{\sqrt{s+1}}$$

and check your answer with *Mathematica*. You need to consider carefully how to modify the semicircular contour to treat the branch point at $s = -1$.

18.8 Using the Laplace transform method, find the solutions of

$$a\frac{\partial^2 y(t)}{\partial t^2} + b\frac{\partial y(t)}{\partial t} + c\,y(t) = f(t)$$

with the following specializations:

(i) $a = 1, b = 0, c = 1, f(t) = 0, y(0) = 1, y'(0) = 0;$
(ii) $a = 1, b = 0, c = 1, f(t) = \cos(t), y(0) = 1, y'(0) = 0;$
(iii) $a = 1, b = 6, c = 13, f(t) = 0, y(0) = 0, y'(0) = 1;$
(iv) $a = 1, b = 6, c = 13, f(t) = \sin(2t), y(0) = 0, y'(0) = 1.$

18.9 Using the Laplace transform method as applied to matrix systems, find the solutions of the following systems of ODEs:

(i) $\dot{x} = y, \dot{y} = -\omega^2 x, y(0) = 0, x(0) = 1.$
(ii) $\dot{x} = -2x, \dot{y} = 2x - 3y, \dot{z} = 3y, x(0) = 1, y(0) = 0, z(0) = 0.$

18.10 [❀] (Harder) Consider the transform inversion integral for the heat equation in the form

$$q(x, t) = \frac{1}{2\pi i} \int_{c-i\infty}^{c+i\infty} e^{-\sqrt{p}\,(x-a)}\, e^{p\,t}\, dp$$

Deform the contour to wrap around the negative real axis and taking care of the branch points, show that

$$q(x, t) = \frac{1}{\pi} \int_0^{\infty} e^{-pt} \sin\left(\sqrt{p}\,(x-a)\right) dp$$

Use *Mathematica* to evaluate this integral and check that it agrees with inversion example 3 of Section 18.4.

18.11 (Harder) Let $f(t)$ have Laplace transform $\tilde{f}(s)$. Prove that

$$\int_0^{\infty} \tilde{f}(s) J_p(as)\, ds$$

is the Laplace transform, with coordinate p, of

$$f(a \sinh(t))$$

18.12 ❀ Check your solutions to Exercise 18.8 using *Mathematica*'s **DSolve** function.

18.13 ❀ Check your solutions to Exercise 18.9 using *Mathematica*'s **DSolve** function.

19 Elementary applications to two-dimensional physics

Introduction

Many problems in applied mathematics reduce to a problem in potential theory, requiring the solution, ϕ, known as the *potential*, of Laplace's equation:

$$\nabla^2 \phi = 0 \tag{19.1}$$

Holomorphic functions of a complex variable play a critical role in the solution of this and related equations. In this chapter we shall explore the solution of Laplace's equation in two dimensions, when the solution is particularly straighthforward, and solutions may also be turned into other solutions via the use of conformal mapping. The matter will be revisited in Chapters 23 and 24, whence we shall see how to manage Laplace's equation and the Helmholtz equation in three dimensions, and the wave equation in four dimensions. Solutions of Eq. (19.1) have applications to several areas, including:

(1) steady-state diffusion;
(2) gravitation;
(3) electrostatics;
(4) magnetostatics;
(5) fluid dynamics.

Our first task will be to briefly review each of these topics, to see how the *potential* arises. Students interested in one particular physical application may wish to focus on the topic of most interest in Section 19.1. Then we shall look at the solution of Eq. (19.1) by holomorphic methods and illustrate many of them with *Mathematica*. The last section of this chapter looks briefly at generalizations of potential theory – in particular, the biharmonic equation and its application to viscous flow.

19.1 The universality of Laplace's equation

Consider first the time-dependent diffusion equation with constant diffusivity κ. This is, for a dependent variable T, which we may conceptualize as temperature:

$$\frac{\partial T}{\partial t} = \kappa \nabla^2 T \tag{19.2}$$

It is immediately evident that, when the temperature is time-independent, T satisfies Laplace's equation. The situation with gravity is even more straightforward. In the absence of matter, the gravitational field strength is the gradient of a field Φ satisfying

Laplace's equation. The story with electromagnetism is a little more subtle. In the time-independent case, Maxwell's equations reduce to (we are using simplified units):

$$\underline{\nabla}.\underline{E} = \rho \tag{19.3}$$

$$\underline{\nabla}.\underline{B} = 0 \tag{19.4}$$

$$\underline{\nabla} \times \underline{E} = \underline{0} \tag{19.5}$$

$$\underline{\nabla} \times \underline{B} = \underline{J} \tag{19.6}$$

where ρ is the charge density and $\underline{J}$ is the current density (usually these source terms are multiplied by appropriate electromagnetic constants, but these constants vary with convention and are of no consequence here). For the electric field $\underline{E}$, Eq. (19.5) holds everywhere, so that we may always write

$$\underline{E} = \underline{\nabla}\phi \tag{19.7}$$

for an electrostatic potential ϕ. Then Eq. (19.3) implies that ϕ satisfies Poisson's equation

$$\nabla^2\phi = \rho \tag{19.8}$$

and, in the absence of charges, Laplace's equation is obtained once more. The magnetic case is a little more subtle. Eq. (19.6) says that, away from any current sources, the magnetic field is curl-free and hence again the gradient, at least locally, of a potential, and Eq. (19.4) then guarantees that this magnetostatic potential satisfies Laplace's equation. However, the presence of currents at various points in space causes global problems. Suppose for example we have a current-carrying wire located at the origin, and consider the integral of $\underline{B}$ around a closed circle C centred on the origin, with interior A. By Stoke's theorem, we have, from Eq. (19.6),

$$\int_C \underline{B}.d\underline{r} = \int_A \underline{\nabla} \times \underline{B}.d\underline{S} = \int_A \underline{J}.d\underline{S} = I \tag{19.9}$$

where I is the current in the wire. If the magnetic field was strictly the gradient of a potential, this integral would have to be zero. What happens is that globally the potential is multi-valued. We shall see how this is elegantly managed by the multi-valued nature of certain complex functions!

The case of fluid dynamics is very rich in structure, and we shall explore this topic in detail in a section of its own. To introduce the matter, we note that the conservation of fluid mass leads to the equation

$$\frac{\partial \rho}{\partial t} + \underline{\nabla}.(\rho\,\underline{v}) = 0 \tag{19.10}$$

where $\underline{v}$ is the velocity field and ρ the fluid density. If the fluid is incompressible in the sense that ρ must be independent of both position and time, we deduce that

$$\underline{\nabla}.\underline{v} = 0 \tag{19.11}$$

In general, fluid mechanics involves the solution of the Navier-Stokes equations, as discussed in Section 19.7. The simplification we need here is the case of zero vorticity, i.e.

$$\underline{\nabla} \times \underline{v} = \underline{0} \tag{19.12}$$

in which case we can again write the velocity field as the gradient of a potential, which by the incompressibility condition must then satisfy Laplace's equation. We shall look at the special properties of fluids with regard to stream functions in Sections 19.3 and 19.7. A fluid flow satisfying Eq. (19.12) is said to be *irrotational.*

Our goal in this Chapter is to explore various approaches to solving Laplace's equation in *two* dimensions. This is not as silly as it sounds – although the physical world is three-dimensional there may be cases of exact or approximate translational (or other) symmetry that reduces the problem to one that is essentially two-dimensional.

19.2 The role of holomorphic functions

Let's remind ourselves of some facts about holomorphic functions. First of all, we know that a holomorphic function is infinitely differentiable, so that if we write

$$f(z) = f(x + i y) = \phi(x, y) + i \psi(x, y) \tag{19.13}$$

then ϕ and ψ may be differentiated twice. We also know (from real analysis) that the mixed second derivatives are symmetric – for example,

$$\frac{\partial^2 \phi}{\partial x \partial y} = \frac{\partial^2 \phi}{\partial y \partial x} \tag{19.14}$$

and similarly for ψ. The Cauchy–Riemann equations then guarantee (Exercise 19.1 reminds you of this!) that

$$\frac{\partial^2 \phi}{\partial x^2} + \frac{\partial^2 \phi}{\partial y^2} = 0 \tag{19.15}$$

and similarly for ψ. So the real and imaginary parts of a holomorphic function satisfy Laplace's equation.

■ Rewriting Laplace's equation

The fact the holomorphic functions satisfy the Laplace equation can be made rather more transparent by re-introducing the operators

$$\frac{\partial f}{\partial z} = \frac{1}{2}\left(\frac{\partial f}{\partial x} - i \frac{\partial f}{\partial y}\right) \tag{19.16}$$

$$\frac{\partial f}{\partial \bar{z}} = \frac{1}{2}\left(\frac{\partial f}{\partial x} + i \frac{\partial f}{\partial y}\right) \tag{19.17}$$

and noting that

$$\nabla^2 f = 4 \frac{\partial^2 f}{\partial z \partial \bar{z}} \tag{19.18}$$

This makes it clear that if f is a function of a complex variable, then it satisfies Laplace's equation if:

(1) it is holomorphic, i.e. a function only of z and not $\bar{z}$;
(2) it is anti-holomorphic, i.e. a function only of $\bar{z}$ and not z.

If you feel that this notational slickness might be hiding something, Exercise 19.2 encourages you to check in detail that an anti-holomorphic function satisfies the Laplace equation.

■ The significance of the chain rule and boundary conditions

The chain rule has a very special signifiance here, because it allows us to reshape regions of the complex plane by conformal mapping, and map solutions of Laplace's equation in one region into a solution in the re-shaped region. To see this suppose that

$$z = g(w) \tag{19.19}$$

is a holomorphic mapping. Then we have, if $f(z)$ is given,

$$\frac{\partial^2 f}{\partial w \partial \bar{w}} = |g'(w)|^2 \frac{\partial^2 f}{\partial z \partial \bar{z}} \tag{19.20}$$

which means, using Eq. (19.18), that if f is a solution of Laplace's equation in terms of x, y, it is also a solution of

$$\frac{\partial^2 f}{\partial u^2} + \frac{\partial^2 f}{\partial v^2} = 0 \tag{19.21}$$

where $w = u + i v$.

A common technique is to use a conformal mapping to transform the region of interest into a disc or half-plane. (Quite how this is done in general is an issue in itself – the Riemann mapping theorem, discussed in Chapter 21 guarantees that it can be done under very general conditions, but does not tell you how!) There are then very explicit solutions for the Dirichlet problem for the cases of the disc and the half-plane, which we shall discuss presently. For now we need to introduce some terminology related to boundary conditions.

The *Dirichlet* problem is one where the value of ϕ is prescribed on the boundary, whereas the *Neumann* problem is one where the normal derivative (i.e. in the direction of the outward normal to the region) is specified on the boundary. For the Dirichlet problem for the Laplace equation, the solution is unique, whereas for the Neumann problem it is unique up to a constant. (See Exercise 19.3). Furthermore, the problems where either the function, or its normal derivative, is zero on a given boundary are invariant under conformal mapping. For the first case, if the function is zero and the coordinates are changed, it clearly remains zero. For the latter observation, the conformality property guarantees that the differentiation direction remains normal to the boundary under the conformal mapping, and hence it follows that if the normal derivative is zero in one set of coordinates, it remains zero after a conformal mapping has been applied.

Note that such zero boundary conditions are typically only applied on the exterior of a given region – otherwise, for interior solutions, the solution must be zero or constant inside if it is non-singular inside, by the uniqueness result.

■ Additional observations

There is some additional structure arising from the Cauchy–Riemann equations. The relationship between ϕ and ψ is of some signifiance. In the case of fluid theory the relationship is of particular importance. Suppose that we identify ϕ as the potential for the velocity field $\underline{v}$:

$$\underline{v} = \left(\frac{\partial \phi}{\partial x}, \frac{\partial \phi}{\partial y}\right) \tag{19.22}$$

Then by the Cauchy–Riemann equations

$$\underline{v} = \left(\frac{\partial \psi}{\partial y}, -\frac{\partial \psi}{\partial x}\right) \tag{19.23}$$

and so if we consider the dot product of the velocity field with the gradient of ψ, we note that

$$\underline{v}.\underline{\nabla}\psi = \left(\frac{\partial \psi}{\partial y}, -\frac{\partial \psi}{\partial x}\right).\left(\frac{\partial \psi}{\partial x}, \frac{\partial \psi}{\partial y}\right) = 0 \tag{19.24}$$

so that ψ is constant along the velocity field. We call ψ the 'stream function', and the curves ψ = constant are literally the streamlines of the flow. This is a powerful visualization tool, and we see that both parts of the holomorphic function are useful. For non-fluid applications the curves ϕ = constant are called equipotentials, and the curves ψ = constant may be interpreted as 'force field' or 'flux' lines, but for fluids there is additional signficance in that these latter curves are the paths followed by fluid particles. This is *not* the case, for example, for a charged test-particle (i.e. one for which we ignore its own field) in an electrostatic or gravitational field. In the case of heat flow, these two sets of curves are sometimes called *isothermal* lines and flux lines.

19.3 Integral formulae for the half-plane and disk

There are special results for the Dirichlet problem when the boundary is a circle centred on the origin, or the upper half-plane. Both of these formulae follow from Cauchy's integral formula. First we give the result for the half-plane under straightforward assumptions.

Let u be the real part of the holomorphic function $f(z)$, and suppose that $u(x, 0)$ is known along the entire real axis. Suppose further that f tends to zero as $|z| \to \infty$ in the upper half-plane. Then

$$u(x, y) = \frac{1}{\pi} \int_{-\infty}^{\infty} \frac{y\, u(q, 0)}{(x-q)^2 + y^2}\, dq \tag{19.25}$$

Before proving this, let's take a look at the result.

■ Example of the half-plane result applied

Suppose that we have a function u that satisfies Laplace's equation in the upper half-plane, and that when $y = 0$, then $u = 0$, except between $x = -a$ and $x = a$, where it is unity. What is the solution? All we need to do is to evaluate the integral. We supply a suitable set of **Assumptions**:

```
Integrate[y / ((x - q) ^2 + y^2) / Pi, {q, -a, a},
 Assumptions → {a > 0, y > 0, Im[x] == 0}]
```

$$\frac{\tan^{-1}\left(\frac{a-x}{y}\right) + \tan^{-1}\left(\frac{a+x}{y}\right)}{\pi}$$

Let's take a look at the solution with $a = 1$:

```
Plot3D[% /. a -> 1, {x, -4, 4},
 {y, 0.0001, 4}, PlotRange -> All,
 PlotPoints -> 40, ViewPoint -> {1, 2, 1}]
```

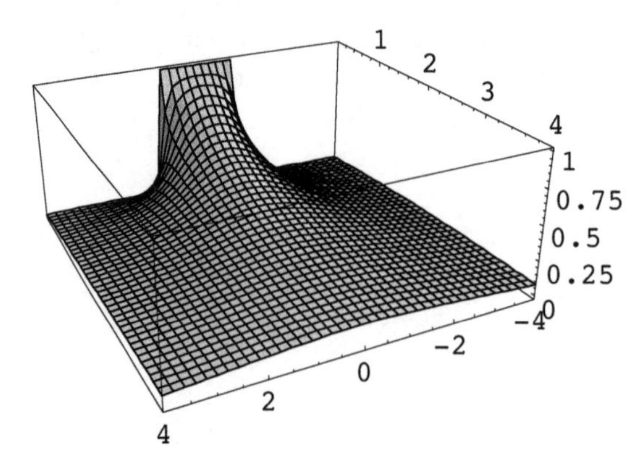

Does the solution satisfy Laplace's equation?

```
Simplify[D[%%, {x, 2}] + D[%%, {y, 2}]]
```

0

Some other examples are given in Exercises 19.4–19.6.

■ Proof of the half-plane result

This is a simple re-arrangment of the Cauchy Integral Formula. Let C be the boundary of a semicircular region D, where $z \in D$, with C consisting of two pieces: C_1, the real interval $-R < x < R$, and a semicircular arc C_2 radius R, centred on the origin, and in the upper half-plane. The integral formula gives

$$f(z) = \frac{1}{2\pi i} \int_C \frac{f(w)}{w - z} \, dw \tag{19.26}$$

and Cauchy's theorem gives

$$0 = \frac{1}{2\pi i} \int_C \frac{f(w)}{w - \bar{z}} \, dw \tag{19.27}$$

since $\bar{z}$, the image of z in the real axis, lies outside D. Now we subtract these two results, to obtain

$$f(z) = \frac{1}{2\pi i} \int_C f(w) \left(\frac{1}{w - z} - \frac{1}{w - \bar{z}} \right) dw = \frac{1}{2\pi i} \int_C \frac{f(w)(z - \bar{z})}{(w - z)(w - \bar{z})} \, dw \tag{19.28}$$

Now suppose that f is bounded in the sense that $|f| \leqslant M$, for large z, where $M \to 0$, as $z \to \infty$. Then the contribution from the integral over C_2 vanishes in the limit $R \to \infty$, leaving us with the contribution from Eq. (19.28) along the real axis. This gives us

$$f(x + i\, y) = \frac{1}{\pi} \int_{-\infty}^{\infty} \frac{f(q + 0i)\, y}{(x - q)^2 + y^2} \, dq \tag{19.29}$$

Then, if $f = u(x, y) + i\, v(x, y)$

$$\begin{aligned} u(x, y) &= \frac{1}{\pi} \int_{-\infty}^{\infty} \frac{u(q, 0) y}{(x - q)^2 + y^2} \, dq \\ v(x, y) &= \frac{1}{\pi} \int_{-\infty}^{\infty} \frac{v(q, 0) y}{(x - q)^2 + y^2} \, dq \end{aligned} \tag{19.30}$$

so that we obtain the half-plane result for both the real and the imaginary part of f. One point to note is that one cannot just put $y = 0$ in this formula – you must evaluate the integral before taking the limit of zero y.

■ Dispersion relations

Another more subtle point, which we shall not pursue, is that a similar argument applied to Eq. (19.26) alone allows you to extract the real part from the imaginary part integrated along the real axis, and vice versa. This observation leads to the notion of Hilbert transforms and dispersion relations in mathematical physics, the general idea being that the notion of holomorphicity allows information about one process, characterizing the real part, to be extracted from another that is related to the imaginary part. For an illuminating discussion of this, see Jackson (1975), Chapter 7. Exercise 19.7 gives a hint of what is happening. This concept has many applications.

■ Green's function version

If you have taken a course in potential theory based on Green's function methods, you may know a different way of deriving this result based on the divergence theorem applied to the function under consideration together with the Green's function. You might like to contrast the two methods – but note below how we use the method of images to derive the fundamental solution for a half-plane with Dirichlet boundary conditions.

■ Poisson's formula for the disk

The result for a disk is very similar, but is written out in polar coordinates (r, θ). Suppose that $u(R, \phi)$ is specified on the disk $r = R$. Then for $r < R$, we have

$$u(r, \theta) = \frac{1}{2\pi} \int_0^{2\pi} \frac{(R^2 - r^2)\, u(R, \phi)}{r^2 + R^2 - 2Rr\cos(\theta - \phi)}\, d\phi \tag{19.31}$$

The proof of this follows exactly the same route as the half-plane case, but where, if z lies inside the boundary circle, the analogue of the image point is the point

$$\hat{z} = \frac{R^2}{\bar{z}} \tag{19.32}$$

The details are left to you as Exercise 19.8. An example of the application of this result is given in Exercise 19.9.

19.4 Fundamental solutions

Many solutions of interest can be built from very simple solutions and careful use of the geometry. We are particularly interested in solutions corresponding to point-like sources, which become singular at isolated points. We are also interested in simple solutions corresponding to uniform fields. We shall now explore a few of these, and see how to build more exotic solutions from them. As we shall see in the next section, this is all 'done with mirrors'.

■ Uniform fields

An important simple case is the elementary function

$$f(z) = U_0\, e^{-i\alpha}\, z \tag{19.33}$$

where U_0 is a real constant. Let's find the real and imaginary parts of this function

```
ComplexExpand[Exp[-I α] (x + I y)]
```

$$x\cos(\alpha) + y\sin(\alpha) + i\,(y\cos(\alpha) - x\sin(\alpha))$$

The real part is therefore

$$\phi = U_0\,(x\cos(\alpha) + y\sin(\alpha)) \tag{19.34}$$

and the imaginary part is

$$U_0\,(y\cos(\alpha) - x\sin(\alpha)) \tag{19.35}$$

With this and the simple functions developed here, we shall take the time to interpret each one in terms of fluid flow. If your particular interest lies in another application, you should consider what these functions mean in that context. So if Eq. (19.34) gives the potential for a fluid velocity field, the vector field giving the velocity is the gradient, which is just

$$U_0\,\{\cos(\alpha),\, \sin(\alpha)\} \tag{19.36}$$

which is manifestly a constant vector of magnitude U_0 in a direction at an angle $+\alpha$ to the positive real axis. Note that there is a minus sign in the exponential of Eq. (19.33).

■ Sources and sinks

A source of strength $k > 0$ at the point $z = a$ corresponds to the function

$$f(z) = k\log(z - a) \tag{19.37}$$

Similarly, a sink of strength $k > 0$ at the point $z = a$ corresponds to the function

$$f(z) = -k\log(z - a) \tag{19.38}$$

Suppose that $a = 0$, so that the real part of f is just

$$\phi = \pm k\log(r) \tag{19.39}$$

and the velocity field is therefore of magnitude

$$v = \frac{\pm k}{r} \tag{19.40}$$

in a radial direction away $(+)$ or towards $(-)$ the origin.

Vortices

We can build elementary swirling solutions corresponding to a point vortex at $z = a$. This is just

$$f(z) = -ik\log(z - a) \tag{19.41}$$

This case is important as it includes the multi-valued effect we remarked on in the discussion of magnetostatic fields. Suppose we write

$$z = a + re^{i\theta} \tag{19.42}$$

so that we are using polar coordinates centred on the vortex. Then

$$f(z) = -ik\log(re^{i\theta}) = k\theta - ik\log(r) \tag{19.43}$$

whose real part is then

$$\phi = k\theta \tag{19.44}$$

The gradient of this is then the vector whose magnitude is

$$\frac{k}{r} \tag{19.45}$$

and which points in a direction tangent to a circle radius r centred at a. This is manifestly a swirling flow centred on a. Note that the potential is multi-valued. In the magnetostatic case this is the potential for magnetic field induced by a wire passing through a, normal to the complex plane, and carrying a current that is essentially (i.e. up to some electromagnetic constants that do not concern us) k.

Mathematica visualizations of elementary flows

Mathematica has a number of interesting visualization tools to help you understand the structure of the fields. One of the more interesting ones is the family of **PlotField** functions, which in particular allow a vector field to be plotted given only its scalar potential – this is particularly appropriate for the current application. The package for two dimensions is loaded as follows:

```
Needs["Graphics`PlotField`"]
```

The functions that have been loaded are given by:

The function that has been loaded that we want is **PlotGradientField**:

```
?PlotGradientField
```

```
PlotGradientField[f, {x, x0, x1, (xu)}, {y, y0, y1, (yu)}, (options)]
  produces a vector field plot of the gradient vector field of the
  scalar function f by calculating its derivatives analytically.
```

Let's look at the uniform flow for this case:

```
ComplexExpand[Exp[-I Pi / 6] (x + I y)]
```

$$\frac{\sqrt{3}\,x}{2} + \frac{y}{2} + i\left(\frac{\sqrt{3}\,y}{2} - \frac{x}{2}\right)$$

```
PlotGradientField[
 (Sqrt[3] x + y) / 2, {x, -2, 2}, {y, -2, 2}]
```

Here is the source field – note that some warning messages about the behaviour at the origin have been suppressed in the printed text:

```
PlotGradientField[
 Log[x^2 + y^2] / 2, {x, -3, 3}, {y, -3, 3}]
```

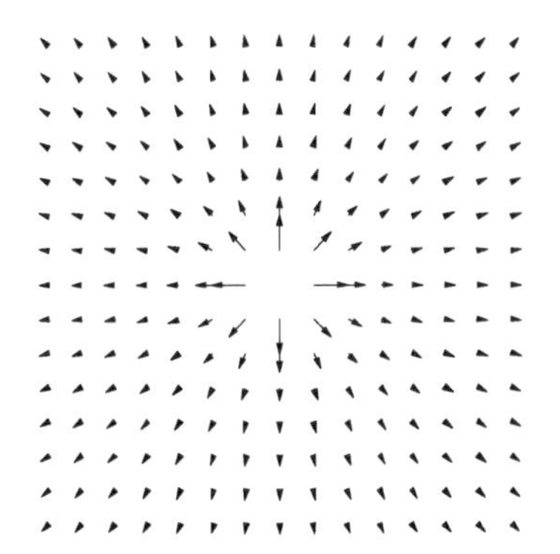

To treat the vortex, we need to make an explicit conversion to polar coordinates – this is best done using the two-argument form of the arctangent function:

```
PlotGradientField[ArcTan[x, y], {x, -3, 3}, {y, -3, 3}]
```

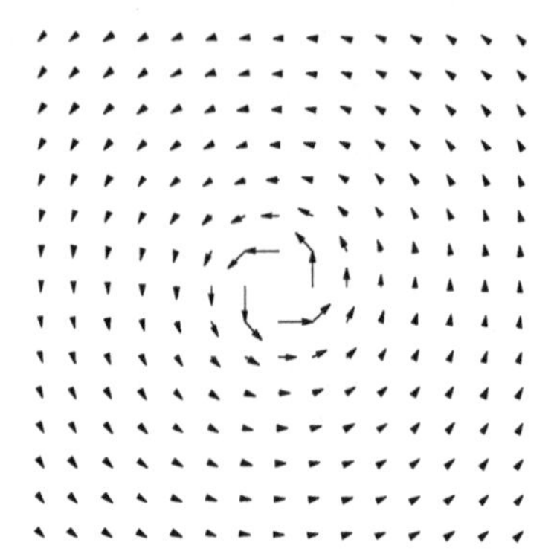

Some of these functions can also be interpreted as giving translation-independent fields in three dimensions – for example, the vortex flow is equivalent to the magnetic field around a straight current-carrying wire. To view the functions in this form is easy – you just use a related package:

```
Needs["Graphics`PlotField3D`"]
```

The function that we want is **PlotGradientField3D,** applied as follows:

```
PlotGradientField3D[ArcTan[x, y],
 {x, -3, 3, 0.5}, {y, -3, 3, 0.5},
 {z, -3, 3, 0.5}, ViewPoint -> {1, 2, 10}]
```

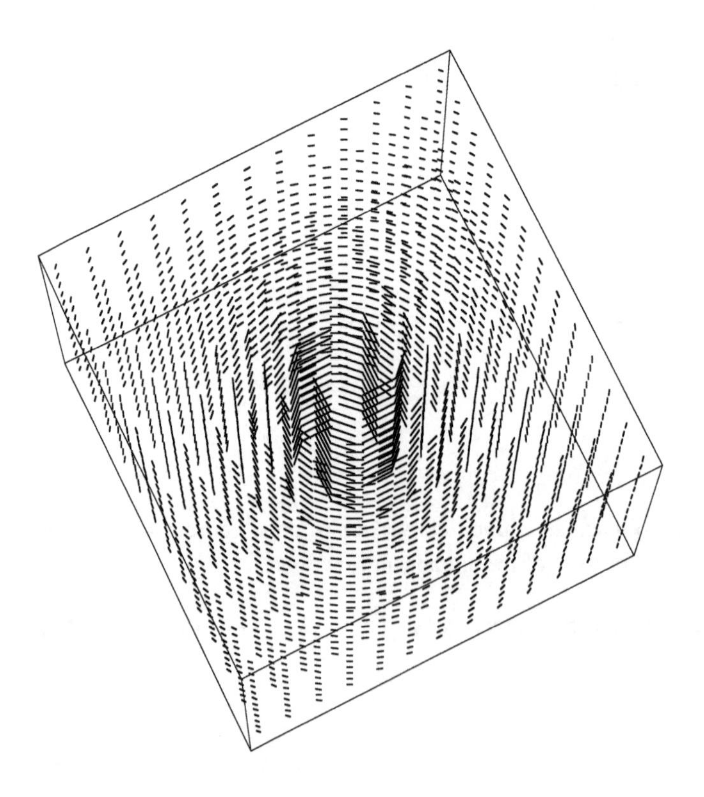

19.5 The method of images

The extension of the simple solutions just described, to more complicated regions, depends on how complicated the region's shape is. An important approach is the 'method of images'. In this approach boundary conditions of two simple types can be satisfied on several classes of boundary. The boundary conditions of interest are:

(1) Dirichlet zero, i.e.$\phi = 0$ on the boundary;
(2) Neumann zero, i.e. $\partial\phi/\partial n = 0$ on the boundary, where n is the normal direction.

The regions where this can be carried out particularly straightforwardly are:

(i) a half-plane;
(ii) a quadrant;
(iii) a wedge of interior angle $2\pi/k$ for some integer k.

We have used the terms 'Dirichlet' and 'Neumann' as these mathematical characterizations of the two types of boundary condition are general. You may find other terms in use to describe particular physical situations. In the electrostatic case, the Dirichlet problem may be described by saying that the boundary is 'earthed'. In the case of steady heat flow, the Neumann problem corresponds to an insulated boundary.

You should also appreciate that the method of images is a very powerful technique that can be applied in many other situations beyond those can can be modelled by two-dimensional potential theory and complex variables. For example, in applied electromagnetic analysis, the applications of the technique (to the Helmholtz equation in three dimensions) span a whole range of applications, from simple tutorial applications through to the construction of Green's functions for scattering from complex layered geometries. See, for example, Jackson (1975) and Lindell (1995). The text by Lindell gives some applications where a distribution of *complex* images is used to satisfy boundary conditions on dielectric media boundaries. In financial analysis, the method can be applied to solve option pricing problems where the value of a financial instrument becomes zero on one or more 'barrier' lines (see Shaw, 1998). However, here our goals are rather more modest, and we begin by exploring case (i), where we wish to satisfy boundary conditions on a half-plane.

■ ❄ Half-plane boundary conditions

We now consider the case of a half-plane characterized by $\text{Im}(z) > 0$, and a Dirichlet zero boundary condition imposed along the line $y = 0$. In the absence of the boundary, the complex function characterizing a source of strength k at a, with $\text{Im}(a) > 0$, is just $f(z) = k\log(z - a)$. We can satisfy the zero boundary condition on the real part by adding a corresponding sink at the conjugate point $\overline{a}$. In total this gives us a new complex function

$$g(z) = k\log(z-a) - k\log(z-\overline{a}) = k\log\left(\frac{z-a}{z-\overline{a}}\right) \tag{19.46}$$

Let us write $a = \alpha + i\beta$, $z = x + iy$. Then the real part of this function is just

$$\phi(x,\ y) = k\log\left|\left(\frac{x+iy-\alpha-i\beta}{x+iy-\alpha+i\beta}\right)\right| = \frac{1}{2}k\log\left(\frac{(x-\alpha)^2+(y-\beta)^2}{(x-\alpha)^2+(y+\beta)^2}\right) \tag{19.47}$$

This manifestly vanishes when $y = 0$. For example, consider a source of strength unity at (0, 1). Here is a contour plot of the resulting potential:

```
ContourPlot[1/2*Log[(x^2+(y-1)^2)/(x^2+(y+1)^2)],
{x,-2,2},{y,0,4}, PlotPoints -> 30]
```

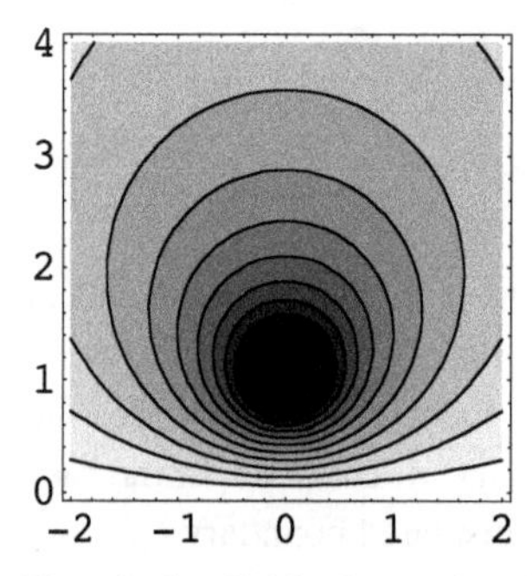

Here is the field of associated gradients:

```
PlotGradientField[1/2*Log[(x^2+(y-1)^2)/(x^2+(y+
1)^2)],{x,-2,2}, {y,0,4}]
```

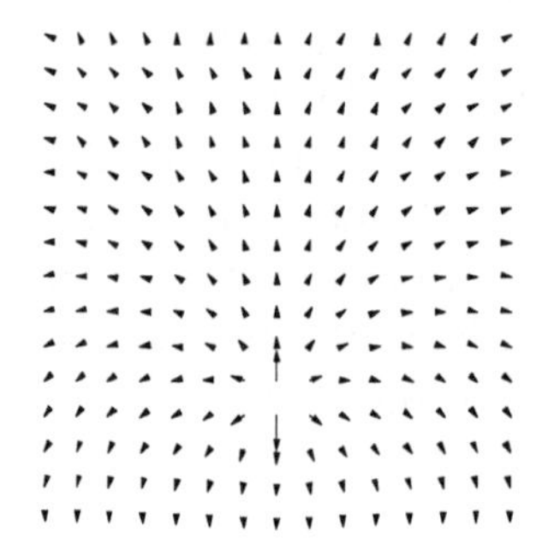

The corresponding Neumann zero problem is solved by adding a source of idential strength at the image point. This balances the flows so that there is zero flow through the boundary. In this case

$$\phi(x,\ y) = \frac{1}{2}k\log(((x-\alpha)^2+(y-\beta)^2)((x-\alpha)^2+(y+\beta)^2)) \tag{19.48}$$

Consider again a source of strength unity at (0, 1). Here is a contour plot of the resulting potential:

```
ContourPlot[1/2*Log[(x^2+(y-1)^2)*(x^2+(y + 1)^2)],
{x,-2,2},{y,0,4}, PlotPoints -> 30]
```

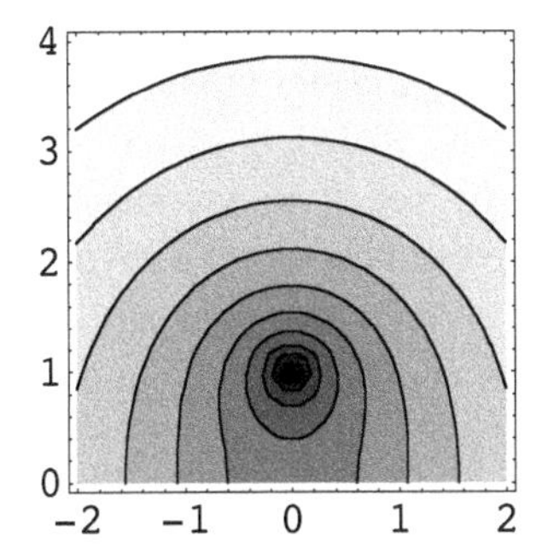

Here is the field of associated gradients – note that the flow is horizontal at $y = 0$:

```
PlotGradientField[1/2*Log[(x^2 + (y - 1)^2)*(x^2 + (y
+ 1)^2)], {x, -2, 2}, {y, 0, 4}]
```

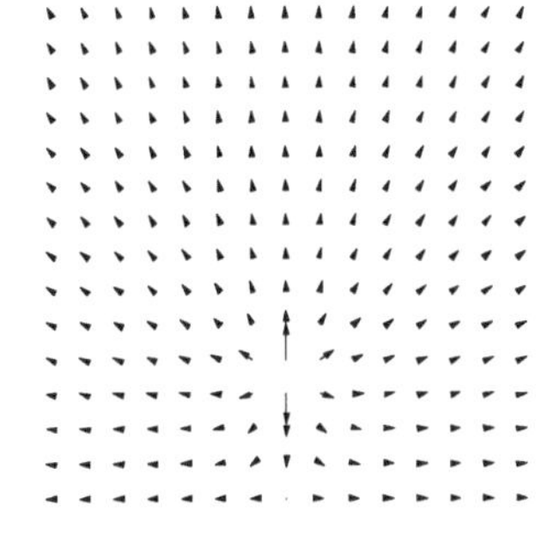

■ Quadrant and wedge boundary conditions

The approach just used for the half-plane can be extended to treat quadrants, and, under certain circumstances wedges. For example, the Nuemann zero problem in a quadrant can be managed by adding three image points in each of the other three quadrants with the same strength as the original source. Details are left for you to analyse in Exercise 19.10. The Dirichlet problem for a quadrant is discussed in Exercise 19.11, and the wedge in Exercise 19.12.

19.6 Further applications to fluid dynamics

We have mentioned the application to fluid flow already. A motion of a fluid that is both incompressible and irrotational, and two-dimensional, can be characterized in terms of the complex potential

$$f(z) = \phi(x, y) + i\psi(x, y) \tag{19.49}$$

where ϕ is the ordinary real potential, such that the velocity $\underline{v}$ of the fluid is given by $\underline{v} = \underline{\nabla}\phi$. The function ψ is the stream function, and the velocity field is tangent to the any curve ψ = constant.

If we introduce solid boundary surfaces into the geometry under consideration, there can be no flow normal to the boundary. This means that as a matter of course we are considering a Neumann problem for ϕ. In particular, we have already solved this problem where the geometry consists of a half-plane and (if you have investigated Exercises 19.11-19.12 sufficiently) quadrants and wedges. We can now plot the stream-lines of these flows as well, by just using the imaginary parts of the functions already considered. For the case of the half-plane Neumann problem the complex potential is

$$f(z) = k\log(z-a) + k\log(z-\overline{a}) = k\log((z-a)(z-\overline{a})) \tag{19.50}$$

and the imaginary part of this is

$$\psi = k\,\mathrm{Arg}((z-a)(z-\overline{a})) \tag{19.51}$$

Now consider the case of a unit source at (0, 1). We need to evaluate the argument of

```
Expand[(x + I (y - 1)) * (x + I (y + 1))]
```

$x^2 + 2\,i\,y\,x - y^2 + 1$

and this is best done using the **ArcTan** function. Here is the resulting plot, with some options set to produce a reasonably clean picture (the thick line extending vertically is an artifact):

```
ContourPlot[ArcTan[1 + x^2 - y^2, 2 x y],
 {x, -2, 2}, {y, 0, 2},
 ContourShading -> False, PlotPoints -> 400,
 ContourStyle → AbsoluteThickness[0.1],
 Contours -> Table[k / 4, {k, -15, 15, 1 / 2}],
 AspectRatio -> 1 / 2]
```

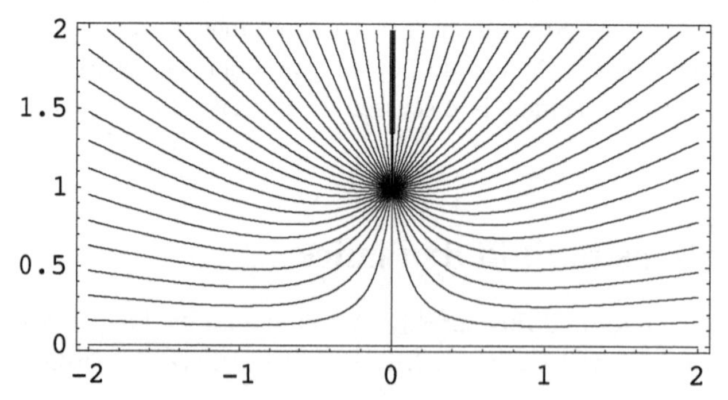

This makes the role of the boundary very clear. You might like to explore the streamlines for point sources within quadrants and wedges in the same way. Other flows within wedges can be obtained by elementary methods (see Exercise 19.13).

Boundary conditions on a plane and circle for general flows

Thus far we have considered some simple boundary conditions associated with simple flows arising from a point source. Suppose we have a general complex potential $f(z)$ that is associated with a fluid flow in the absence of a boundary, and we add a half-plane boundary. We can also ask about the introduction of a circular boundary. What is the appropriate new complex potential?

Consider the planar boundary first, and suppose that $f(z)$ only has singularities in the upper half-plane. The only sources and sinks are in the upper half-plane. We introduce a plane boundary at $y = 0$ and impose a Neumann boundary condition there. Then the appropriate new complex potential is just

$$g(z) = f(z) + \overline{f(\bar{z})} \tag{19.52}$$

Why is this? Consider first the boundary $y = 0$. There

$$g(x) = f(x) + \overline{f(x)} \in \mathbb{R} \tag{19.53}$$

and so $\psi = 0$ along this boundary. That is, this boundary is a streamline! The Neumann boundary condition on ϕ is then guaranteed by the orthogonality relations between ϕ and ψ. Furthermore, the function we have added only has singularities in the lower half-plane and so does not introduce any new fictitious sources in the upper half-plane. Using this method it is easy to generate new solutions in a half-plane geometry (see e.g. Exercise 19.14).

A similar trick can be used to calculate flows with boundary conditions on a circle. Suppose that we wish to create a solid boundary on the circle $|z| = a$, with a real. Suppose further that $f(z)$ is the complex potential for a flow in the entire complex plane, with any singularities located outside the circle $|z| = a$. Then the new complex potential for the corresponding flow with a solid boundary on the circle is just

$$g(z) = f(z) + \overline{f\left(\frac{a^2}{\bar{z}}\right)} \tag{19.54}$$

Why this works is similar to the argument for a half-plane (see Exercise 19.15).

Flow past a cylinder

A useful application of these methods is to consider the modification of a uniform flow by the introduction of a cylinder (i.e. a circular boundary). In the absence of the circle the flow is that we have already seen, and we write Eq. (19.33) in the form:

$$f(z) = (U - iV)\, z$$

In the presence of the circle, this now becomes

$$g(z) = (U - iV)\, z + (U + iV)\, \frac{a^2}{z} \tag{19.55}$$

If the flow is in the x-direction at infinity, this becomes just

$$g(z) = U\left(z + \frac{a^2}{z}\right) \tag{19.56}$$

In polar coordinates the corresponding potential and stream function are then given by

$$\phi = U\cos(\theta)\left(\frac{a^2}{r} + r\right) \tag{19.57}$$

$$\psi = U\sin(\theta)\left(r - \frac{a^2}{r}\right) \tag{19.58}$$

which makes it obvious that $r = a$ is indeed a streamline (see Exercise 19.16). In Cartesian coordinates, the stream function is

$$\psi = Uy\left(1 - \frac{a^2}{x^2 + y^2}\right) \tag{19.59}$$

We can now visualize this flow very easily, using the standard **ContourPlot** routine to see the streamlines. Note that they are also shown for the interior of the circle, but the solution we want is just for the exterior region.

```
ContourPlot[y * (1 - 1 / (x^2 + y^2)),
 {x, -4, 4}, {y, -2, 2}, AspectRatio -> 1 / 2,
 ContourShading -> False, PlotPoints -> 400,
 Contours -> 41, ContourStyle → AbsoluteThickness[0.1]]
```

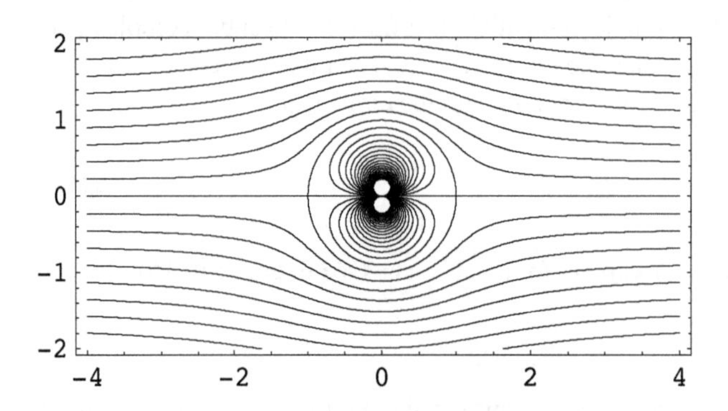

▪ Flow past a cylinder with circulation

The fact that the boundary is a streamline is not altered by the addition of a vortex term centred at the origin. Now we get the complex potential

$$g(z) = (U - iV)\,z + (U + iV)\,\frac{a^2}{z} - \frac{ik}{2\pi}\log(z) \tag{19.60}$$

If the flow is in the x-direction at infinity, this becomes just

$$g(z) = U\left(z + \frac{a^2}{z}\right) - \frac{ik}{2\pi}\log(z) \tag{19.61}$$

In polar coordinates, the corresponding potential and stream function are then,

$$\phi = U\cos(\theta)\left(\frac{a^2}{r} + r\right) + \frac{k\theta}{2\pi} \tag{19.62}$$

$$\psi = U\sin(\theta)\left(r - \frac{a^2}{r}\right) - \frac{k}{2\pi}\log(r) \tag{19.63}$$

and it is clear that $r = a$ remains a streamline. In Cartesian coordinates, the stream function is now

$$\psi = U\,y\left(1 - \frac{a^2}{x^2 + y^2}\right) - \frac{k}{4\pi}\log(x^2 + y^2) \tag{19.64}$$

We can also visualize this flow as before. Here we set $k = -5$:

```
ContourPlot[
 y * (1 - 1 / (x^2 + y^2)) + 5 / (4 Pi) Log[x^2 + y^2],
 {x, -4, 4}, {y, -2, 2}, AspectRatio -> 1 / 2,
    ContourShading -> False, PlotPoints -> 400,
 Contours -> Table[k / 4, {k, -15, 15, 1}],
 ContourStyle → AbsoluteThickness[0.1]]
```

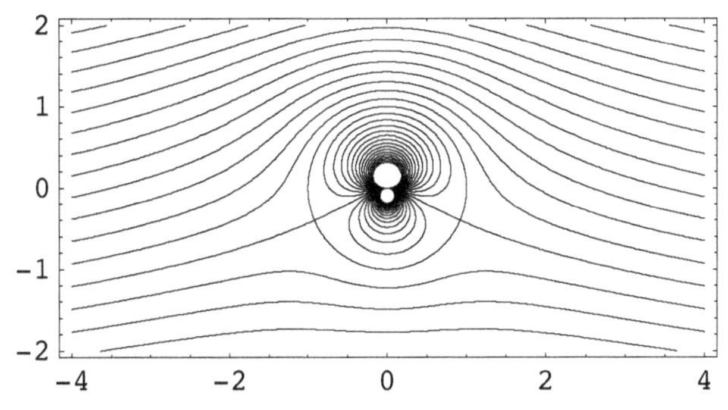

The points where the streamlines hit the cylinder are an example of *stagnation points*. You might like to explore what happens to these as the strength of the circulation term is increased.

■ Blasius' theorem

We shall not prove this here (it is an application of Bernoulli's theorem in fluid dynamics), but there is an interesting result due to Blasius that gives the force exerted on an object as a function of the complex potential $g(z)$. Suppose the fluid has density ρ and the force (per unit length in the third dimension) is given by the complex constant

$$F = F_x - iF_y \tag{19.65}$$

Then the theorem states that

$$F = \frac{i\rho}{2} \int_C \left(\frac{\partial g(z)}{\partial z}\right)^2 dz \tag{19.66}$$

where C is the boundary of the body. Note that since the integrand is holomorphic, C can be deformed in general, and the calculus of residues used to evaluate the integral. Let's get *Mathematica* to work out the force on the cylinder in the flow with circulation that we have considered:

```
g[z_] = (U - I*V)*z + (U + I*V)*a^2/z -
(I*k)/(2*Pi)*Log[z];
```

```
Collect[Expand[(D[g[z], z])^2], z]
```

$$U^2 - 2\,i\,V\,U - V^2 + \frac{-\frac{i\,k\,U}{\pi} - \frac{k\,V}{\pi}}{z} + \frac{-\frac{k^2}{4\pi^2} - 2\,a^2\,U^2 - 2\,a^2\,V^2}{z^2} + \frac{\frac{i\,a^2\,k\,U}{\pi} - \frac{a^2\,k\,V}{\pi}}{z^3} + \frac{U^2\,a^4 - V^2\,a^4 + 2\,i\,U\,V\,a^4}{z^4}$$

```
(I * ρ / 2) * (2 Pi I) * Residue[%, {z, 0}]
```

$$i\,(k\,U - i\,k\,V)\,\rho$$

```
Expand[%]
```

$$i\,k\,U\,\rho + k\,V\,\rho$$

So the force in the x-direction is $F_x = \rho\,k\,V$; that in the y-direction is $F_y = -\rho\,k\,U$. When $V = 0$, so that the flow at infinity is horizontal, we have $F_x = 0$. This calculation shows that when $k < 0$ we get a net upwards lift on the cylinder.

This calculation illustrates two further points. First, the lift is generated by the presence of circulation around the cylinder. Without it there is no lift. Second, there is no horizontal force in the absence of circulation, which is rather unphysical – there is no 'air resistance' This is a consquence of the introduction of the idealization of potential flow, which is not totally realistic.

■ ✵ Conformal mappings on the cylinder

In Chapter 16 we briefly investigated the Joukowski mapping. This transforms the region exterior to a circle of radius a to that exterior to an aerofoil shape. The mapping is $\zeta \longrightarrow z$, with

$$z = e^{i\theta}\left(\frac{\lambda^2}{w + \zeta} + w + \zeta\right) \tag{19.67}$$

With λ general this can be used to transform circles into ellipses. For example, here we generate a rotated ellipse from a unit circle:

```
h[ξ_, λ_, w_, θ_] := Exp[I θ] * (λ^2 / (w + ξ) + w + ξ)

ParametricPlot[{Re[h[1 * Exp[I ϕ], 1.5, 0, Pi / 4]],
  Im[h[1 * Exp[I ϕ], 1.5, 0, Pi / 4]]}, {ϕ, 0, 2 Pi}]
```

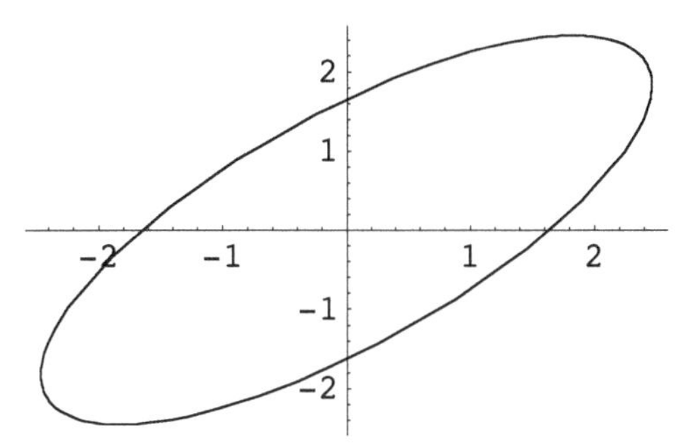

If, additionally, the constraint of Chapter 16 is to be applied, we would impose

$$\lambda = \sqrt{a^2 - \mathrm{Im}(w)^2} - \mathrm{Re}(w); \tag{19.68}$$

This generates the aerofoil family we considered earlier. We shall need the inverse of this mapping with the right behaviour at infinity. Solving the associated quadratic equation gives

$$\zeta = -w + \frac{1}{2}\, e^{-i\theta}\, z + \frac{1}{2}\sqrt{z^2 e^{-2i\theta} - 4\lambda^2} \tag{19.69}$$

The mapping $z \longrightarrow \zeta$ transforms the region exterior to an aerofoil to that exterior to a circle. The complex potential for a flow with circulation is then just

$$g(\zeta) = U\left(\zeta + \frac{a^2}{\zeta}\right) - \frac{i\,k}{2\pi}\log(\zeta) \tag{19.70}$$

Let's look at the cases of ellipses first. Here is the complex potential. Note that *Mathematica* does not always manage the branch points in the right way, so we write the inverse mapping in this particular way to try to force the issue.

```
JoukowskiFlow[z_, θ_, a_, w_, U_, V_, k_, λ_] :=
    Module[{ξ},
    ξ = -w + Exp[-I * θ] * z / 2 *
      (1 + Sqrt[1 - 4 λ^2 Exp[2 I θ] / z^2]);
        (U - I * V) ξ + (U + I * V) a^2 / ξ -
   (Log[ξ] * I * k) / (2 * Pi)]
```

We now extract the real potential and stream function by taking real and imaginary parts:

```
JoukowskiPhi[x_, y_, θ_,
  a_, α_, β_, U_, V_, k_, λ_] :=
    Re[JoukowskiFlow[x + I y, θ, a, α + I β, U, V, k, λ]]
```

```
JoukowskiPsi[x_, y_, θ_,
  a_, α_, β_, U_, V_, k_, λ_] :=
    Im[JoukowskiFlow[x + I y, θ, a, α + I β, U, V, k, λ]]
```

As our first example, we can now visualize the flow around an ellipse given by

$$\frac{x^2}{A^2} + \frac{y^2}{B^2} = 1 \tag{19.71}$$

and as an example we take $A = 2$, $B = 1$. The Joukowski mapping takes the ellipse to the exterior of a circle of radius $(A + B)/2$, which is $3/2$ here, with $\lambda = \sqrt{a^2 - b^2}/2$, which here is $\sqrt{3}/2$. Here we show the streamlines associated with this ellipse in a flow with velocity at infinity (1, 0.5) and $k = -4$. We make a separate copy of the ellipse in order to superimpose the boundary clearly.

```
EllipseGraphic[A_, B_] :=
 ParametricPlot[{A Cos[s], B Sin[s]},
  {s, 0, 2 Pi}, PlotStyle -> Thickness[0.008]]
```

```
ell = EllipseGraphic[2, 1]
```

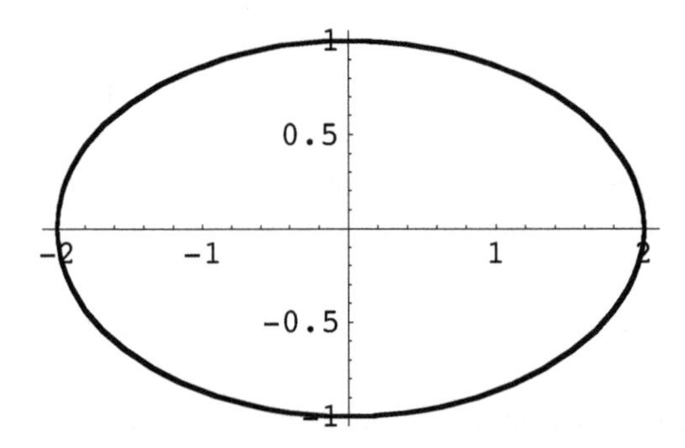

In this and the other examples in this chapter, the value of **PlotPoints** has been set quite high (400) to obtain a clean graphic – on an older computer you might wish to lower the value.

```
myplot = ContourPlot[JoukowskiPsi[
   x, y, 0, 3 / 2, 0, 0, 1, 0.5, -4, Sqrt[3] / 2],
  {x, -4, 4}, {y, -2, 2}, ContourShading -> False,
  AspectRatio -> 1 / 2, PlotPoints -> 300,
  Contours -> Table[k / 4, {k, -20, 20, 1}]]
```

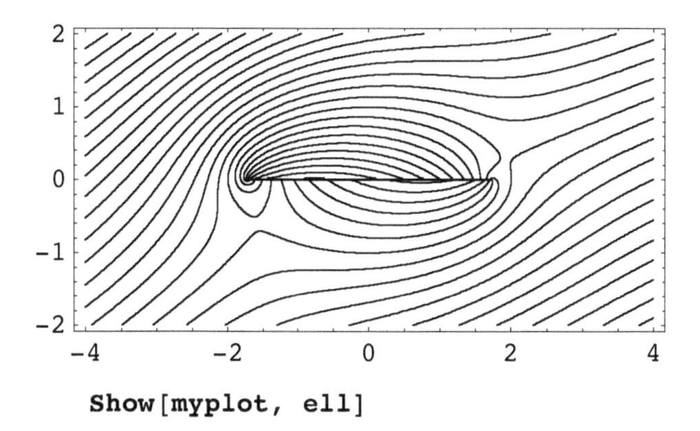


```
Show[myplot, ell]
```

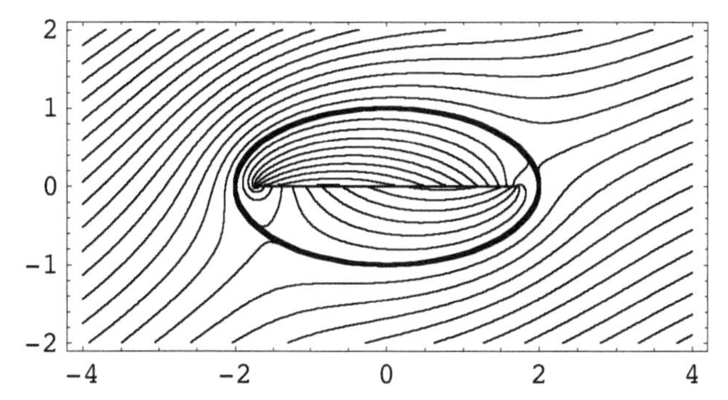

If we keep $A = 2$ but set $B = 0$ the ellipse collapses to a flat plate, as shown in the following streamline plot:

```
ellb = EllipseGraphic[2, 0]
```

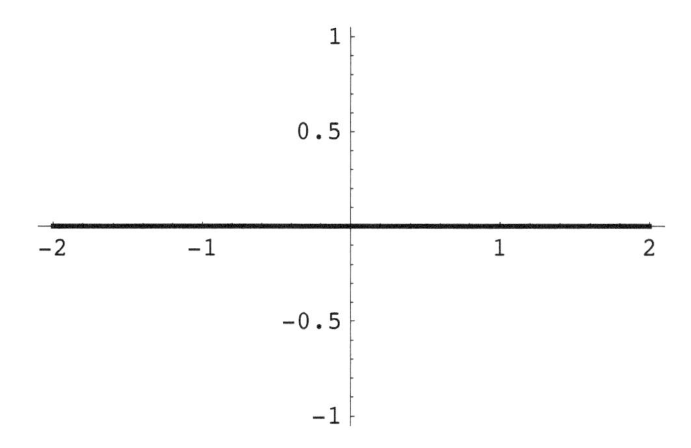

```
myplotb = ContourPlot[
  JoukowskiPsi[x, y, 0, 1, 0, 0, 1, 0.5, -4, 1],
  {x, -4, 4}, {y, -2, 2}, ContourShading -> False,
  AspectRatio -> 1 / 2, PlotPoints -> 400,
  Contours -> Table[k / 4, {k, -20, 20, 1}]]
```

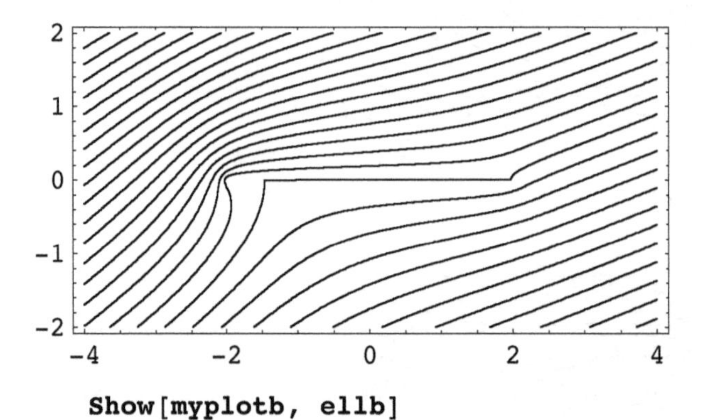

```
Show[myplotb, ellb]
```

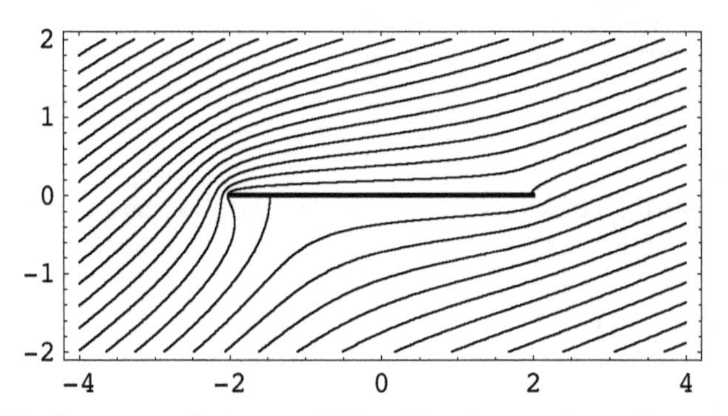

Finally, we recall that aerofoils with pointy ends can be created using the same transformation but with λ constrained according to Eq. (19.68). Here is one example of how to implement it – you are encouraged to explore this further in Exercise 19.17.

```
AeroJoukowskiFlow[z_, θ_, a_, w_, U_, V_, k_] :=
   Module[{λ = Sqrt[a^2 - Im[w]^2] - Re[w], ζ},
   ζ = -w + Exp[-I*θ]*z/2*
     (1 + Sqrt[1 - 4 λ^2 Exp[2 I θ] / z^2]) ;
        (U - I*V) ζ + (U + I*V) a^2 / ζ -
  (Log[ζ]*I*k) / (2*Pi)]

AeroJoukowskiPhi[x_, y_,
  θ_, a_, α_, β_, U_, V_, k_] :=
   Re[AeroJoukowskiFlow[x + I y, θ, a, α + I β, U, V, k]]
```

```
AeroJoukowskiPsi[x_, y_,
  θ_, a_, α_, β_, U_, V_, k_] :=
    Im[AeroJoukowskiFlow[x + I y, θ, a, α + I β, U, V, k]]
```

Here is a symmetrical aerofoil:

```
ContourPlot[
 AeroJoukowskiPsi[x, y, 0, 3 / 2, 0.5, 0.0, -1, 0.0, 0],
 {x, -4, 4}, {y, -2, 2}, ContourShading -> False,
 AspectRatio -> 1 / 2, PlotPoints -> 200,
 Contours -> Table[k / 4, {k, -20, 20, 1}]]
```

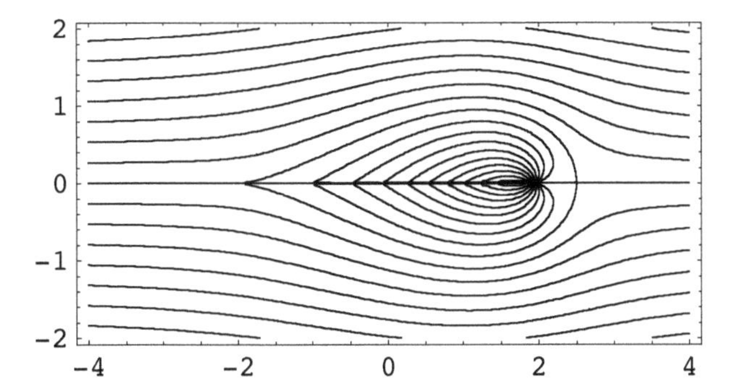

19.7 The Navier–Stokes equations and viscous flow

The use of complex functions in fluid dynamics is not at all confined to potential flow. There are many beautiful applications to other types of fluid flow that are known, and indeed there appear to be many unresolved issues. The author is grateful to Dr John Ockendon, F.R.S. for pointing out some of the basic work in this area, and to Professor K.B. Ranger for supplying several of his papers. The following discussion is intended to be a brief introduction only. For an introduction to viscous flow and the biharmonic equation at a student level within the context of fluid theory, see Ockendon and Ockendon (1995). The book by Ockendon *et al.* (2003) provides a discussion of complex solutions of the biharmonic equation within the general context of PDEs, and also gives a short introduction to the work of Ranger (see, for example, Ranger (1991, 1994) and references therein), which develops develops solution techniques for the complex representation of the 2-D steady Navier–Stokes equations originally given by Legendre (1949). Some particular applications of complex variables appear in many articles. Notable examples include Garabedian (1966), who discusses free boundary flows, and Finn and Cox (2001), who discuss a method of images for Stokes flow. This section will require some familiarity with basic identities of vector calculus.

■ Fluid flow in more generality

A large class of fluids can be characterized by their density ρ, a scalar field not initially presumed constant, and their dynamic viscosity μ. The velocity vector field is $\underline{v}$, and the pressure is the scalar field p. In general the mass conservation equation

$$\frac{\partial \rho}{\partial t} + \underline{\nabla}.(\rho \underline{v}) = 0 \tag{19.72}$$

holds. The conservation of momentum is expressed by the Navier–Stokes equations in the form

$$\rho\left(\frac{\partial \underline{v}}{\partial t} + (\underline{v}.\underline{\nabla})\, \underline{v}\right) = -\, \underline{\nabla} p \, + \, \mu \, \nabla^2 \, \underline{v} \tag{19.73}$$

As before, if the fluid is incompressible in the sense that ρ must be independent of both position and time, we deduce that

$$\underline{\nabla}.\underline{v} = 0 \tag{19.74}$$

In general, fluid mechanics involves the solution of the Navier–Stokes equations. Rather than assuming, as before, zero vorticity, we now define the vorticity as

$$\underline{\omega} = \underline{\nabla} \times \underline{v} \tag{19.75}$$

In what follows we *retain incompressibility* but allow for *non-zero vorticity*. We now need to use some vector identities. We define the scalar field H as

$$H = p + \frac{\rho}{2} \underline{v}.\underline{v} \; = \; p + \frac{1}{2} \rho \, \underline{v}^2 \tag{19.76}$$

and note that if $\underline{\nabla}.\underline{v} = 0$, then $\nabla^2 \, \underline{v} \; = -\underline{\nabla} \times \underline{\omega}$. We also recall the vector calculus identity

$$\nabla\left(\frac{1}{2} \underline{v}.\underline{v}\right) \; = \; (\underline{v}.\underline{\nabla})\underline{v} + \underline{v} \times \underline{\omega} \tag{19.77}$$

Putting this all together allows us to recast the Navier–Stokes equations as

$$\rho\left(\frac{\partial \underline{v}}{\partial t} - \underline{v} \times \underline{\omega}\right) + \underline{\nabla} H = - \; \mu \, \underline{\nabla} \times \underline{\omega} \tag{19.78}$$

Taking the curl of this, and noting that the divergence of the viscosity is identially zero, we get the *vorticity equation*

$$\rho\left(\frac{\partial \underline{\omega}}{\partial t} + (\underline{v}.\underline{\nabla})\, \underline{\omega} - (\underline{\omega}.\underline{\nabla})\, \underline{v}\right) = \nu \nabla^2 \, \underline{\omega} \tag{19.79}$$

where the *kinematic viscosity* $\nu \; = \; \mu / \rho$. We note that setting the vorticity to zero is consistent with the vorticity equation, leading us back to potential flow.

■ Steady planar flows and the stream function

Previously we have regarded the *stream function* as the imaginary part of the complex potential for steady potential flow. The concept of a stream function has wider applicability as we shall now see. In what follows I employ the sign conventions of Ranger (1994). We let $\psi(x, y)$ be chosen so that

$$\underline{v} = \underline{\nabla} \times (-\underline{k}\,\psi) = \left(-\frac{\partial \psi}{\partial y}, \frac{\partial \psi}{\partial x}, 0\right) \tag{19.80}$$

In the time-independent or steady case, the vorticity equation (18.79) then reduces to the scalar equation

$$\nabla^4 \psi = \frac{1}{\nu} \frac{\partial(\psi, \nabla^2 \psi)}{\partial(x, y)} \tag{19.81}$$

where the right hand side is the usual Jacobian. Note that it is consistent to demand that $\nabla^2 \psi = 0$ in this equation, taking us back to potential flow, but now there are many other interesting cases to consider. We also note that in the various forms of the momentum (Navier–Stokes) or vorticity equation (18.79) or the 2-D stream form (18.81), and for problems where one can identify a *unique* length scale L and speed scale U that characterizes the problem at hand, it makes sense to perform a further non-dimensionalization of the variables and introduce the *Reynolds number*

$$R_e = \frac{U\,L}{\nu} = \frac{\rho\,U\,L}{\mu} \tag{19.82}$$

so that, for example, Eq. (19.81) becomes, after suitable rescalings of all the variables,

$$\nabla^4 \psi = R_e \frac{\partial(\psi, \nabla^2 \psi)}{\partial(x, y)} \tag{19.83}$$

■ The Stokes flow limit in complex form

The approximation of Stokes flow is the limit where the Reynolds number becomes zero. In this case we have

$$\nabla^4 \psi = 0 \tag{19.84}$$

which is the *biharmonic* equation, also well known for its applications to elasticity. This is easily written in terms of complex variables. As usual, we let, $z = x + i y$. Using the notation introduced in Section 10.9

$$\frac{\partial}{\partial z} = \frac{1}{2}\left(\frac{\partial}{\partial x} - i\frac{\partial}{\partial y}\right) \tag{19.85}$$

$$\frac{\partial}{\partial \bar{z}} = \frac{1}{2}\left(\frac{\partial}{\partial x} + i\frac{\partial}{\partial y}\right) \tag{19.86}$$

we note that

$$\nabla^2 = 4\frac{\partial^2}{\partial z \partial \bar{z}}; \quad \nabla^4 = 16\frac{\partial^4}{\partial z^2 \partial \bar{z}^2} \tag{19.87}$$

So the biharmonic equation is the condition

$$\frac{\partial^4 \psi}{\partial z^2 \partial \bar{z}^2} = 0 \tag{19.88}$$

It is simple to integrate this equation to obtain, for ψ real,

$$\psi = \mathrm{Re}(\bar{z} f(z) + g(z)) \tag{19.89}$$

where f and g are holomorphic. The potential flow case arises when $f = 0$.

■ The general case in complex form

In the case where the kinematic viscosity is finite, then the stream function equation (19.83) becomes, instead (it is a useful exercise to check this):

$$i\frac{\partial^4 \psi}{\partial z^2 \partial \bar{z}^2} - \frac{R_e}{2}\left(\frac{\partial \psi}{\partial z}\frac{\partial^3 \psi}{\partial z \partial \bar{z}^2} - \frac{\partial \psi}{\partial \bar{z}}\frac{\partial^3 \psi}{\partial z^2 \partial \bar{z}}\right) = 0 \tag{19.90}$$

This can be reorganized as follows. We first note that Eq. (19.90) can be written as

$$\frac{\partial^2}{\partial \bar{z}^2}\left(i\frac{\partial^2 \psi}{\partial z^2} - \frac{R_e}{2}\left(\frac{\partial \psi}{\partial z}\right)^2\right) + \frac{\partial^2}{\partial z^2}\left(i\frac{\partial^2 \psi}{\partial \bar{z}^2} + \frac{R_e}{2}\left(\frac{\partial \psi}{\partial \bar{z}}\right)^2\right) = 0 \tag{19.91}$$

An auxiliary real function ϕ is introduced so as to satisfy the equation

$$\frac{\partial^2 \phi}{\partial z^2} = i\frac{\partial^2 \psi}{\partial z^2} - \frac{R_e}{2}\left(\frac{\partial \psi}{\partial z}\right)^2 \tag{19.92}$$

and its complex conjugate

$$\frac{\partial^2 \phi}{\partial \bar{z}^2} = -i\frac{\partial^2 \psi}{\partial \bar{z}^2} - \frac{R_e}{2}\left(\frac{\partial \psi}{\partial \bar{z}}\right)^2 \tag{19.93}$$

Then elimination of ϕ from Eq. (19.92) and Eq. (19.93) by differentiation leads to Eq. (19.91). This pair of complex equations is the representation of Legendre (1949), and is the basis for the complex solution techniques developed by Ranger (1991, 1994).

■ Projects

A general investigation of the complex equations for general viscous or Stokes flow is rather outside the scope of this text. There are many examples in the references below that merit detailed investigation, using *Mathematica* as a visualization tool, and you are encouraged to pursue this. As an example to illustrate this approach, the following example from Finn and Cox (2001) is interesting in that it makes use of the method of images with respect to a circular boundary. The following stream function is that for a "stokeslet" placed inside a circle of radius a. The strength of the stokeslet is $A + i B$, and the image is at $\sigma = a^2 / s$. Finn and Cox calculate the resulting stream function (we let **zb** denote $\bar{z}$) and in *Mathematica* input form it is given by

```
Ψ[z_, zb_, a_, A_, B_, σ_, s_] =
  1 / 4 (A (z + zb) - I B (z - zb))
    Log[σ / s Abs[z - s] ^ 2 / Abs[z - σ] ^ 2] + A / 4 / s
    (Abs[z] ^ 2 - a ^ 2) ((z + s) / (z - σ) + (zb + s) / (zb - σ)) +
   I B (z - zb) (σ - s) (Abs[z] ^ 2 - a ^ 2) / (4 s Abs[z - σ] ^ 2);
```

That is, Ψ is given by

```
Ψ[z, z̄, a, A, B, σ, s]
```

$$\frac{i B (\sigma - s) (|z|^2 - a^2) (z - \bar{z})}{4 s |z - \sigma|^2} + \frac{1}{4} \log\left(\frac{\sigma |z - s|^2}{s |z - \sigma|^2}\right) (A (z + \bar{z}) - i B (z - \bar{z})) + \frac{A (|z|^2 - a^2) (\frac{s+z}{z-\sigma} + \frac{s+\bar{z}}{\bar{z}-\sigma})}{4 s}$$

Here is the resulting collection of streamlines as we let $s = 1/100, 1/7, 3/7, 5/7$. The boundary circle is shown in bold and the streamlines outside of it are ignorable. In the following code, in order to set the same options for four similar graphics, we first use **SetOptions** to ensure the consistent operation of **ContourPlot**.

```
SetOptions[ContourPlot, PlotPoints → 200, Contours → 50,
  ContourShading → False, PlotRange → All,
  Epilog → {Thickness[0.02], Circle[{0, 0}, 1]},
  DisplayFunction → Identity];

plota = ContourPlot[Re[Ψ[x + I y, x - I y, 1, 0.5,
     0.0, 100, 1 / 100]], {x, -1, 1}, {y, -1, 1}];
plotb = ContourPlot[Re[Ψ[x + I y, x - I y, 1, 0.5,
     0.0, 7, 1 / 7]], {x, -1, 1}, {y, -1, 1}];
plotc = ContourPlot[Re[Ψ[x + I y, x - I y, 1, 0.5,
     0.0, 7 / 3, 3 / 7]], {x, -1, 1}, {y, -1, 1}];
plotd = ContourPlot[Re[Ψ[x + I y, x - I y, 1, 0.5,
     0.0, 7 / 5, 5 / 7]], {x, -1, 1}, {y, -1, 1} ];
```

```
Show[GraphicsArray[{{plota, plotb}, {plotc, plotd}}]]
```

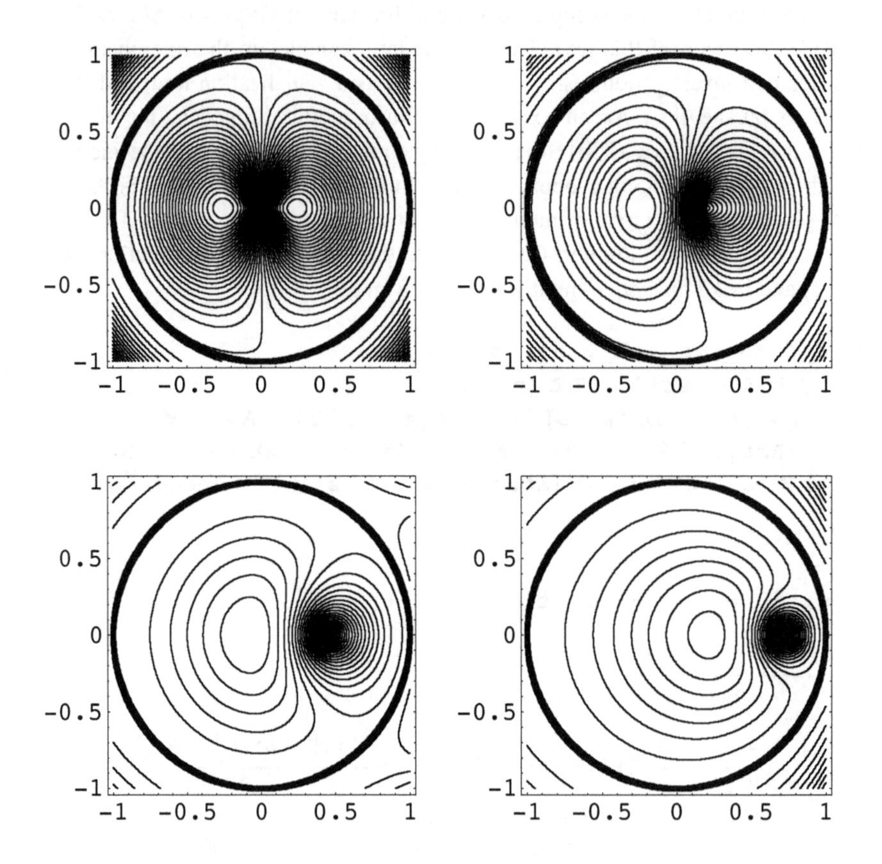

You are encouraged to pursue the work of Finn and Cox further, as the basic object described here is then built up to describe a more complex geometry. Another interesting source of examples, this time of *free boundary problems*, is the paper by Garabedian (1966).

Exercises

19.1 Show that the Cauchy–Riemann equations imply that both the real and imaginary parts of a holomorphic function satisfy Laplace's equation.

19.2 What are the analogues of the Cauchy–Riemann equations for a differentiable function of $\bar{z}$? Show that both the real and imaginary parts of such an anti-holomorphic function satisfy the Laplace equation.

19.3 Suppose that ϕ_1 and ϕ_2 are each solutions of Laplace's equation in a region D, and that on the boundary of D,

$$\phi_1 = \phi_2 = g$$

where g is a specified function defined on the boundary. Consider $h = \phi_1 - \phi_2$. What is the value of h on the boundary? Does h satisfy Laplace's equation? Now consider the identity

$$\underline{\nabla}.(h\,\underline{\nabla}\,h) = \nabla^2 h + |\underline{\nabla} h|^2$$

Integrate this identity over D and apply the divergence theorem. What does this tell you about

$$\int |\underline{\nabla}\,h|^2\,d\underline{x}$$

Deduce that h must be identically zero and that $\phi_1 = \phi_2$. What happens if a Neumann boundary condition is applied instead?

Note: in Exercises 19.4 and 19.5, you may work out the answer using pen and paper or *Mathematica* (preferably both, to check).

19.4 [❁] Find the solution of Laplace's equation in the upper half-plane that, along the real axis, takes the values $+1$ if $x > 0$ and -1 if $x < 0$.

19.5 [❁] Find the solution of Laplace's equation in the upper half-plane that, along the real axis, takes the values $+1$ if $x < -1$, 0 if $-1 < x < 1$ and -1 if $x > 1$.

19.6 ❁ Use *Mathematica* to visualize the solutions to Exercises 19.4 and 19.5.

19.7 Apply the integral formula

$$f(z) = \frac{1}{2\pi i}\int_C \frac{f(w)}{w - z}\,dw$$

directly to the semicircular region considered in Section 19.4, and, by taking real and imaginary parts, show how the real and imaginary parts of f can be computed from each other.

19.8 Using Cauchy's integral formula applied to a circle, and subtracting Cauchy's theorem as applied to the image point in Eq. (19.32), derive Poisson's formula as given by Eq. (19.31).

19.9 [❁] Work out the solution u of Laplace's equation on the interior of the unit disk if $u(1, \phi) = 1$ for $0 < \phi < \pi$ and $u(1, \phi) = -1$, for $\pi < \phi < 2\pi$. Visualize the answer using *Mathematica*.

19.10 [❁] A source of strength k is located in the quadrant $x > 0$, $y > 0$ at the point (α, β) and Neumann boundary conditions apply along the boundaries $x = 0$, $y = 0$. Use the method of images to find the appropriate complex function for this case, and check your answer by using the *Mathematica* functions **`ContourPlot`** and **`PlotGradient-Field`** to visualize the results.

19.11 [✲] A source of strength k is located in the quadrant $x > 0,\ y > 0$ at the point $(\alpha,\ \beta)$ and *Dirichlet* boundary conditions apply along the boundaries $x = 0,\ y = 0$. Use the method of images to find the appropriate complex function for this case, (be careful with the signs of the image sources, i.e. are they sources or sinks), and check your answer by using the *Mathematica* functions **`ContourPlot`** and **`PlotGradient-Field`** to visualize the results.

19.12 [✲] A Wedge of interior angle $2\pi/n$ with vertex at the origin has Neumann boundary conditions imposed along its edges. A source is located inside the wedge at the point $(\alpha,\ \beta)$. What distribution of image sources will guarantee that the Neumann condition is satisfied? Find the resulting potential, and check your answer by using the *Mathematica* functions **`ContourPlot`** and **`PlotGradientField`** to visualize the results. For what values of n is the corresponding Dirichlet problem easily solved in this way?

19.13 ✲ Use *Mathematica* to investigate the fluid flows given by the complex potential $f(z) = z^n$ for various choices of n. (You may find the cases $n = 1/2,\ 2/3,\ 1,\ 2,\ 3$ interesting).

19.14 By defining a vortex using Eq. (19.41), find the flow field corresponding to a vortex at $(0,\ 1)$, subject to a solid half-plane at $y = 0$.

19.15 Explain why Eq. (19.54) gives the complex potential for the flow outside a circle. (Hint: Do this in two stages: first explain why the circle is a streamline; second, locate the singularities of the new flow.)

19.16 Show that by setting $z = r\,e^{i\theta}$, the complex potential given by Eq. (19.56) implies the potential and stream function given by Eq. (19.57) and Eq. (19.58).

19.17 ✲ Explore the streamlines produced by the function **`AeroJoukowskiPsi`**. Consider varying at least:

(1) the flow direction;
(2) the amount of circulation;
(3) the camber.

Write a function to draw in the aerofoil outline (for example, using the methods of Chapter 16), and superimpose this on your plots (as with the examples we have given, there may not be a contour that conveniently outlines the shape of interest, and it is helpful to distinguish the exterior and interior clearly).

20 Numerical transform techniques

Introduction

In real-world applications if is often the case that a purely analytical approach to transform calculus is insufficient. This chapter explores two techniques that extend the utility of transform methods by allowing a purely numerical treatment. We can explore numerical methods for both Fourier and Laplace transforms – in principle any method could be applied to either type of transform, since one is a rotation of the other. However, in practice, two types of problem appear to be of most importance:

(1) An essentially numerical approach to forward and backward Fourier transforms;
(2) The inversion of a complicated Laplace transform given in analytical form, for which there is no known analytical inverse.

An entire book could be written about the first topic, which is at the heart of many problems in applied mathematics, physics and engineering. It is of particular importance for signal and image processing. We shall briefly explore *Mathematica*'s **`Fourier`** and **`InverseFourier`** functions. It should also be clear than any Fourier transform could be worked out numerically by the use of **`NIntegrate`**. In this chapter this approach will be explored in detail only for Laplace transforms.

The second topic is particularly important for generating solutions to partial differential equations. As we have seen in Chapters 18 and 19, certain partial differential equations may be simplified by transforming with respect to one or more of the independent variables, leading to a solvable algebraic or ordinary differential equation. It is necessary to then invert the solution of the transformed problem, and this is where difficulties may arise. We shall explore this issue, and, as an example, apply *Mathematica*'s numerical capabilities to some contemporary problems in financial engineering. We shall see that *Mathematica*'s easy management of complex contour integration rather simplifies the matter.

20.1 ❄ The discrete Fourier transform

The discrete Fourier transform acts on a list of complex numbers. This list may or may not be realized as the discrete values of a function whose continuous Fourier transform we wish to approximate. If the list has length n the discrete transform is based on the principal nth root of unity, which we denote by ω_n. Note that

$$\omega_n = e^{2\pi i/n} \tag{20.1}$$

Now suppose that we have a list a_r, for $1 \leqslant r \leqslant n$. Its discrete Fourier transform is given as a new list b_s, given by (with default settings)

$$b_s = \frac{1}{\sqrt{n}} \sum_{r=1}^{n} \omega_n^{(s-1)(r-1)} a_r \tag{20.2}$$

The corresponding inverse is given by

$$a_r = \frac{1}{\sqrt{n}} \sum_{s=1}^{n} \omega_n^{-(s-1)(r-1)} b_s \tag{20.3}$$

the proof of which relies on the identity (see Exercise 20.4):

$$1 + \omega_n + \omega_n^2 + ... + \omega_n^{n-1} = 0 \tag{20.4}$$

These are perhaps most easily thought of as being distinct operations from their continuous counterparts, though it can be shown that the continuous Fourier transform does indeed result from the discrete transform when a suitable limiting process is performed. The implementation in *Mathematica* is very straightforward.

```
? Fourier
```

```
Fourier[list] finds the discrete
  Fourier transform of a list of complex numbers.
```

```
? InverseFourier
```

```
InverseFourier[list] finds the discrete
  inverse Fourier transform of a list of complex numbers.
```

For example, we have

```
Fourier[{1, 1, 1}]
```

{1.73205 + 0. *i*, 0. + 0. *i*, 0. + 0. *i*}

```
InverseFourier[%]
```

{1., 1., 1.}

Note that it is *not* ncessary that the length of the list be a power of 2. It is particularly efficient when this *is* the case.

■ Parameter settings

With the setting **FourierParameters → {a, b}** the discrete Fourier transform computed by **Fourier** is

$$\frac{1}{n^{(1-a)/2}} \sum_{r=1}^{n} u_r \, e^{2\pi i b(r-1)(s-1)/n} \tag{20.5}$$

20.2 Applying the discrete Fourier transform in one dimension

Perhaps the most basic use of the discrete transform is to implement filters to either pass or bar the high or low frequency components of some signal. It is easy to carry out this type of modelling using the functions built in to *Mathematica*. Let's invent some noisy test data, and remind ourselves how big the data set is:

```
noisy = Table[Cos[t] + Sin[3 t] + 0.5 (Random[] - 0.5),
   {t, 0, 10, 0.01}];
```

A plot reveals the structure of signal + noise:

```
noisyplot =
 ListPlot[noisy, PlotStyle → PointSize[0.005]]
```

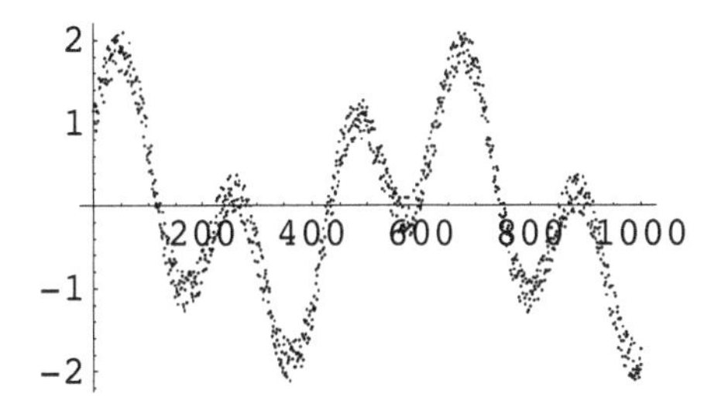

Now let's transform it, and plot the result – it is convenient to rotate the data around to reveal the peaks:

```
noisytfm = Fourier[noisy];
ListPlot[Abs[RotateRight[noisytfm, 500]],
 PlotRange -> All, PlotJoined -> True]
```

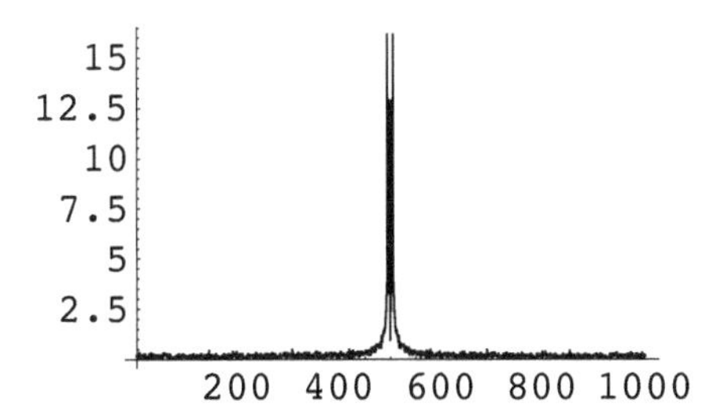

Let's get rid of what we think is the garbage, and have another look at it. Bearing in mind the vertical scale of this last plot, we shall cut out all transformed data less than unity in scale:

```
filtertfm = Chop[noisytfm, 1.0];

ListPlot[Abs[RotateRight[filtertfm, 500]],
 PlotRange -> All, PlotJoined -> True]
```

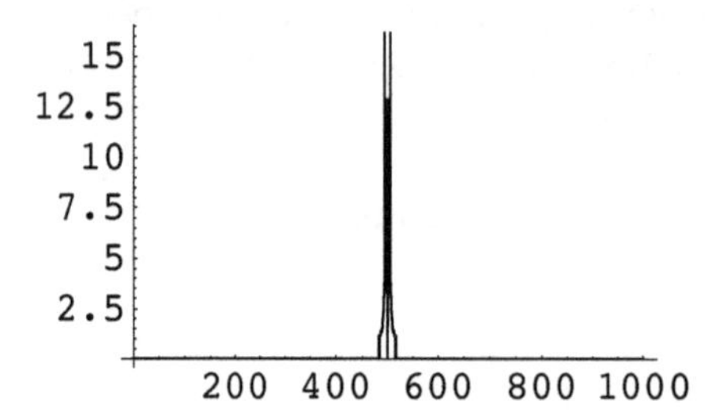

Now we invert and plot what we hope is the cleaned up transformed data:

```
cleanplot = ListPlot[
  Chop[InverseFourier[filtertfm]], PlotJoined -> True]
```

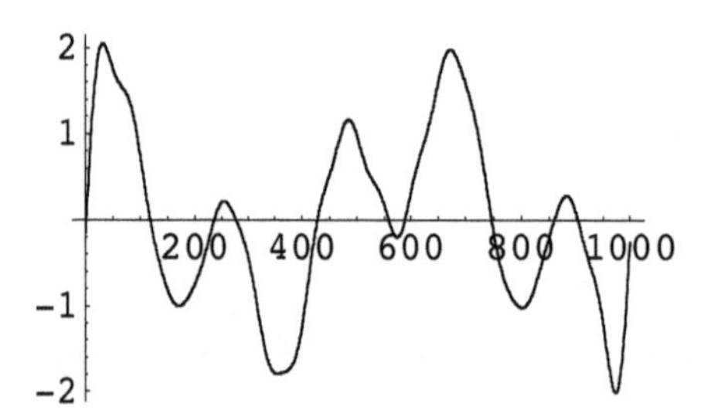


```
Show[cleanplot, noisyplot]
```

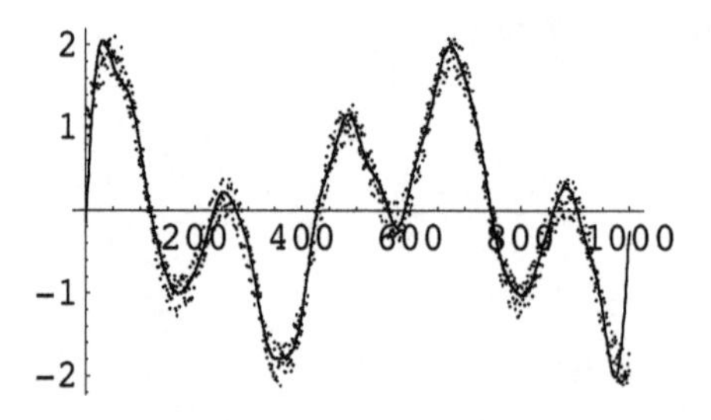

So this has clearly removed the noise. Note also that we have used **Chop** twice in slightly different ways. First we used it as the filter to clean up the transform. Then we used it to remove any small imaginary parts from the inverted data.

20.3 Applying the discrete Fourier transform in two dimensions

This follows the same pattern as in one dimension. Our goal here is simply to make you aware of the ability to apply the Fourier function to square arrays as well as one-dimensional lists. As in one dimension, it is very convenient to perform some list rotations in association with the operation of the transform. We define, for a square data set, the following

```
CentredFourier[x_List] :=
    Module[{len = Floor[Length[x] / 2]},
        Transpose[RotateRight[
     Transpose[RotateRight[Fourier[x], len]], len]]];

CentredInverse[x_List] :=
Module[{len = Floor[Length[x] / 2]},
    InverseFourier[    Transpose[
     RotateLeft[Transpose[RotateLeft[x, len]], len]]]];
```

■ Inventing Some Data

We shall invent some data that resembles the letter ‘E’ as follows:

```
chare[x_, y_] := Which[-30 < x < -20 && -40 < y < 40,
  0, -20 ≤ x ≤ 30 && -40 < y < -30,
  0, -20 ≤ x ≤ 30 && -5 < y < 5, 0,
  -20 ≤ x ≤ 30 && 30 < y < 40, 0, True, 1]

edata = Table[N[chare[x, y]],
   {y, -50, 50, 100 / 127}, {x, -50, 50, 100 / 127}];
ListDensityPlot[edata, Mesh -> False, PlotRange -> All]
```

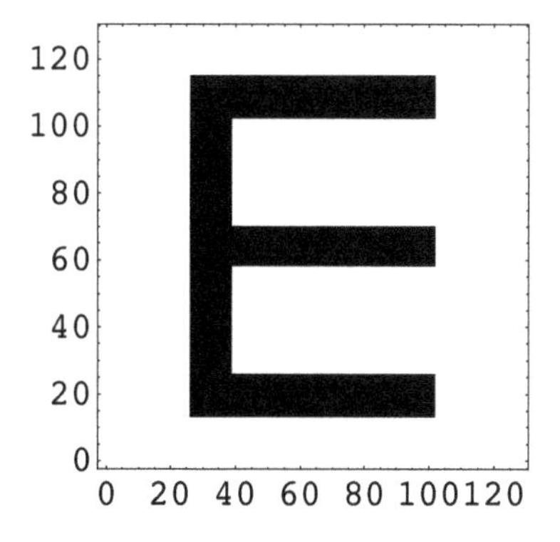

```
fourierddata = CentredFourier[edata];
ListDensityPlot[Abs[fourierddata],
 Mesh -> False, PlotRange -> {0, 0.5}]
```

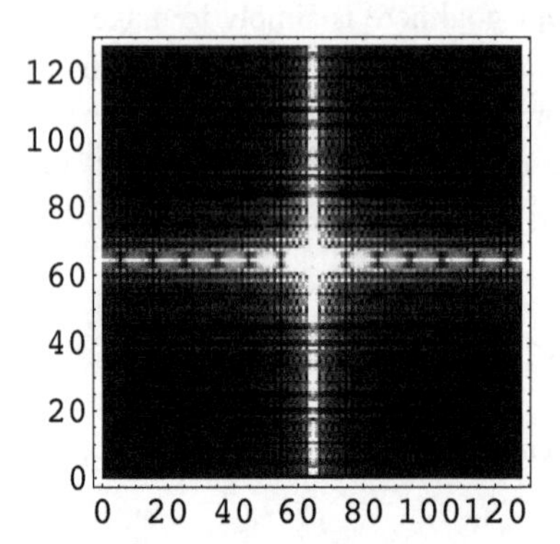

■ The Gaussian blur

We can construct a systematic blurring mechanism by using the convolution theorem. We build a blurring function first as a table, and then centre it:

```
blur = Table[Exp[-(x^2 + y^2) / 10],
   {x, -63.5, 63.5}, {y, -63.5, 63.5}];
blur = Map[RotateRight[#, 64] &, RotateRight[blur, 64]];
```

We wish to blur the ‘E’ by convolving it with this function. We use the convolution theorem – the observed blurred ‘E’ is the inverse of the product of the transform of the sharp ‘E’ with the transform of the blurring function:

```
observede = Chop[CentredInverse[
    CentredFourier[blur] * CentredFourier[edata]]];
ListDensityPlot[observede, Mesh -> False,
 PlotRange -> All]
```

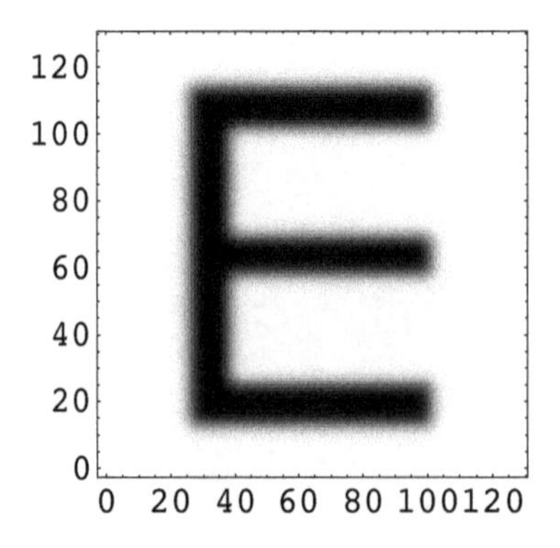

If we were to define the blurring function differently, we can increase the blurring effect:

```
blur = Table[Exp[-(x^2 + y^2) / 100],
   {x, -63.5, 63.5}, {y, -63.5, 63.5}];
blur = Map[RotateRight[#, 64] &,
   RotateRight[blur, 64]];
observede = Chop[CentredInverse[
    CentredFourier[blur] * CentredFourier[edata]]];
ListDensityPlot[observede, Mesh -> False,
 PlotRange -> All]
```

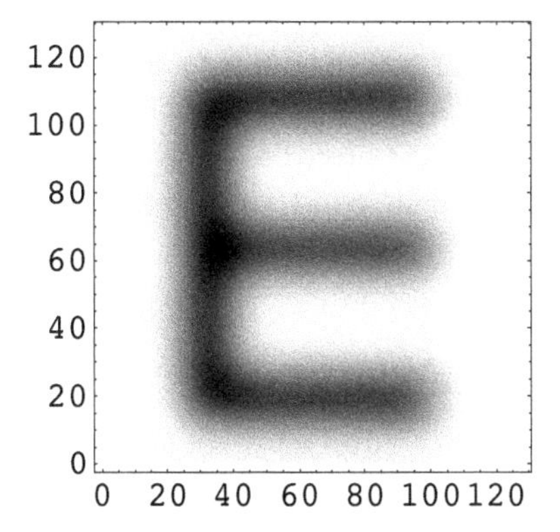

What we have shown here is how to implement the so-called 'Gaussian blur' with two different radii. This is one of a host of filters now commonplace in affordable commercial image processing programs, and available for operation on full colour imagery. Fourier transforms play a role in many of these filters. A delightful exercise on the relative importance of amplitude and phase in the tansform, in determining the shapes of letters, was given by Skotton (1994). This article also contains some handy tips for generating data associated with characters in a more elegant fashion than was considered here.

20.4 Numerical methods for Laplace transform inversion

The topic of numerical inversion of Laplace transforms has been actively studied for many years. An excellent survey of techniques in place up to the mid 1970s was given by Davies and Martin (1979). A family of *Mathematica* implementations of a subset of methods surveyed by Davies and Martin was given quite recently in the *Mathematica* journal (Cheng *et al.* 1994). The package produced in association with that article, **`NLaplaceInversion.m`**, is an excellent implementation of five of the older key numerical inversion techniques. You should check the *MathSource* electronic library for this package.

Work published by Talbot in 1979 has lead to renewed interest in numerical Laplace transform inversion. A good entry point to recent developments is a paper by

Rizzardi (1995), who extends Talbot's method in a very useful way. Talbot's original approach was to use the inversion formula over a cleverly chosen contour in the complex plane. Rizzardi has made the point that Talbot's method has the drawback that the contour chosen depends on the value of t of interest, so that recomputation is required for new values of t. He has extended the method so that the same contour can be used for an entire range of values of t. (It would be a very interesting exercise to use *Mathematica* to compare Rizzardi's methods, the work by Cheng *et al*, and the direct method introduced below.)

However, we have to confront the issue that any method that relies on a particular type of contour deformation may become useless if the transform under consideration has a structure that renders such a deformation impossible. For example, branch cuts extending to infinity along the imaginary axis can wreak havoc. At the same time, we wish to exploit the capabilities built in to *Mathematica* for numerical integration in the complex plane. So the approach taken here will be incredibly simple. We shall perform numerical inversion of the Laplace transform by using **NIntegrate** along the Bromwich contour! So we let the cunning integration routines built in to **NIntegrate** take care of the problems.

20.5 Inversion of an elementary transform

Let's introduce an elementary test problem with which to illustrate the method. Users of *Mathematica* version 3 or earlier will need to first enter the following:

```
Needs["Calculus`LaplaceTransform`"];
```

Most users can proceed straight to asking about the key inversion function.

```
? InverseLaplaceTransform
```

```
"InverseLaplaceTransform[expr, s, t] gives the inverse Laplace transform
  of expr. InverseLaplaceTransform[expr, {s1, s2, ... }, {t1, t2, ... }]
  gives the multidimensional inverse Laplace transform of expr.
```

```
FullSimplify[
 InverseLaplaceTransform[1 / ((s + 2) ^2 + 1), s, t]]
```

$e^{-2t}\sin(t)$

```
fninverse[t_] := Exp[-2 t] Sin[t]
```

Now we define the direct numerical inversion formula. In the following expression the intersection of the Bromwich contour with the real axis is given by the parameter **contour**. The truncation of the integration is given by the parameter **range**.

```
NBromwich[func_, t_, contour_, range_, options___] :=
(Exp[contour * t] / (2 Pi)) *
  NIntegrate[func[contour + I p] * Exp[I p t],
    {p, -range, range}, MaxRecursion -> 12, options]
```

```
NBromwich[(1 / ((# + 2)^2 + 1)) &, 1, 0, 500]
```

$0.113881 - 2.20974 \times 10^{-17}\, i$

The small imaginary part can best be managed by applying:

```
Chop[%]
```

0.113881

Here is the exact answer to six d.p.:

```
N[fninverse[1], 6]
```

0.113881

20.6 Two applications to 'rocket science'

■ A brief introduction to financial derivatives

You do not need to know anything about financial derivatives to appreciate the Laplace transform inversion issues that we are about to discuss, but the matter may be much clearer if we take a moment to explain a little about the pricing of options on assets. For full details in a *Mathematica* setting, see Shaw (1998). A *call option* is a financial instrument that confers on the holder the right (but not the obligation) to buy a certain amount of an underlying asset (such as a company share, or foreign currency), for a fixed price. If this right may be exercised at just one point in the future, after which the option becomes worthless, the option is said to be of *European* style. If the right can be exercised at any point in time up to a certain date, the option is said to be *American* in style.

Such options may be valued in a variety of ways, but for our purposes it will be sufficient to note that under certain assumptions, a general European option on a single asset can be valued by solving a certain partial differential equation, the *Black–Scholes differential equation*. This takes the form

$$\frac{\partial V}{\partial t} + S\,(r-q)\,\frac{\partial V}{\partial S} + \frac{1}{2}\,\sigma^2\,S^2\,\frac{\partial^2 V}{\partial S^2} - r\,V = 0 \tag{20.6}$$

where V is the value of the option, S is the value of the underlying asset (the price of one share), σ is the volatility of the asset, q is a measure of the dividend yield from the asset, and r is the interest rate. We shall assume that σ, r and q are constants. The differential equation is essntially the time-reversed diffusion equation with some extra terms, and it is to be solved (in the case of a *call* option) with a final condition (being a backwards diffusion this takes the place of an initial condition) that at time $t = T$

$$V = \text{Max}[S - K, 0] \tag{20.7}$$

If no other boundaries or variables are introduced, the solution to this equation is well-known – it is one example of the celebrated Black–Scholes formula:

$$\begin{aligned} V &= S\, e^{-qt}\, \Phi(d_1) - K\, e^{-rt}\, \Phi(d_2) \\ d_1 &= \frac{\log(S/K) + (r - q + \sigma^2/2)(T - t)}{\sigma\sqrt{T-t}} \\ d_2 &= \frac{\log(S/K) + (r - q - \sigma^2/2)(T - t)}{\sigma\sqrt{T-t}} \end{aligned} \tag{20.8}$$

where $\Phi(x)$ is the cumulative normal distribution function. This is easily implemented in *Mathematica*. In this definition t is the time remaining to expiry of the option ($T - t$ above):

```
Φ[x_] := 1/2*(1 + Erf[x/Sqrt[2]]);
done[s_, σ_, k_, t_, r_, q_] :=
((r - q)*t + Log[s/k])/(σ*Sqrt[t]) + (σ*Sqrt[t])/2;
dtwo[s_, σ_, k_, t_, r_, q_] :=
done[s, σ, k, t, r, q] - σ*Sqrt[t];

BlackScholesCall[s_, k_, σ_, r_, q_, t_] :=
s*Exp[-q*t]*Φ[done[s, σ, k, t, r, q]] -
k*Exp[-r*t]*Φ[dtwo[s, σ, k, t, r, q]];
```

A great many generalizations can be given where the final conditions are modified, or where additional boundaries are introduced. We shall look at two cases that have been considered recently from the point of view of Laplace transforms:

(i) A call option that becomes worthless if the underlying price reaches either of two values (a double-barrier option).
(ii) A call option where the final condition has the asset price replaced by its average value between $t = 0$ and $t = T$ (the continuously and arithmetically averaged Asian option).

■ The double-barrier call option

A double-barrier option is characterized by a strike, K, and two barriers, L, the lower barrier, and U, the upper barrier, with

$$L < K < U \tag{20.9}$$

The option becomes worthless if the stock price hits either barrier between the initiation and expiry of the option. Otherwise the payoff at $t = T$ is given by Eq. (20.7). This problem can be solved by series methods, and a paper by Kunitomo and Ikeda (1992) describes such a method that they argue converges very rapidly.

An alternative method was published recently. Geman and Yor (1996) have shown how to employ the Laplace transform method to double barrier options, and they demonstrated good agreement with the series method of Kunitomo and Ikeda. We shall explore and implement this method here.

■ The Geman–Yor model for double barrier options

Consider a double-barrier call option whose value is C. We write

$$C = C_{\mathrm{BS}} - e^{-r\,t}\,S\,\phi \tag{20.10}$$

where C_{BS} is the standard Black–Scholes value, given above, and ϕ is a normalized measure of the *decrease* in the value due to the double barrier. The interest rate is r and the stock price at $t = 0$ is S. The barrier term ϕ can be written as a function of the time to expiry T and the scaled variables:

$$h = \frac{K}{S}; \qquad m = \frac{L}{S}; \qquad M = \frac{U}{S} \tag{20.11}$$

The barrier term is Laplace-transformed to define

$$\psi(\lambda) = \int_0^\infty e^{-\lambda\,\tau}\,\phi(\tau, h, m, M)\,d\tau \tag{20.12}$$

Further scaling arguments can be used to write

$$\psi(\lambda) = \frac{\Psi(\frac{\lambda}{\sigma^2})}{\sigma^2} \tag{20.13}$$

Geman and Yor (1996) have found an exact closed-form solution for $\Psi(\theta)$. First, we define (positive) quantities a and b by

$$m = e^{-a} \qquad M = e^{b} \tag{20.14}$$

We also introduce a variable ν as follows:

$$\nu = \frac{r - q - \frac{\sigma^2}{2}}{\sigma^2} \tag{20.15}$$

Finally, we need a variable μ given in terms of the argument of the Laplace transform by

$$\mu = \sqrt{\nu^2 + 2\,\theta} \tag{20.16}$$

We also introduce some auxiliary functions f and g:

$$f(a, h, \mu, \nu) = \frac{h^{-\mu+\nu+1}\,e^{-\mu\,a}}{\mu\,(\mu - \nu)\,(\mu - \nu - 1)} \tag{20.17}$$

$$g(b, h, \mu, \nu) = \frac{2\,e^{b\,(\nu+1)}}{\mu^2 - (\nu + 1)^2} - \frac{2\,h\,e^{b\,\nu}}{\mu^2 - \nu^2} + \frac{h^{\mu+\nu+1}\,e^{-\mu\,b}}{\mu\,(\mu + \nu)\,(\mu + \nu + 1)} \tag{20.18}$$

The Laplace transform Ψ is given by

$$\Psi(\theta) = \frac{g(b, h, \mu, \nu)\sinh(\mu a)}{\sinh(\mu(a+b))} + \frac{f(a, h, \mu, \nu)\sinh(\mu b)}{\sinh(\mu(a+b))} \tag{20.19}$$

A closed-form inversion of this is not at present known, but one can use a variety of numerical methods to invert it. You might also like to consider the methods of Cheng *et al.* (1994) applied to the following.

■ *Mathematica* implementation of the Geman–Yor double-barrier model

First we just enter the basic changes of variable employed:

```
h[S_, K_] := K/S;
a[S_, L_] := Log[S/L];
b[S_, U_] := Log[U/S];
ν[σ_, r_, q_] := (r - q - σ^2/2)/σ^2;
μ[ν_, θ_] := Sqrt[ν^2 + 2*θ]
```

We define the auxiliary functions f and g given by

```
                           h^(-μ+ν+1) Exp[-μ a]
f[a_, h_, μ_, ν_] := ------------------------
                        μ (μ - ν) (μ - ν - 1)

g[b_, h_, μ_, ν_] :=
  2 Exp[b (ν + 1)]     2 h Exp[b ν]     h^(μ+ν+1) Exp[-μ b]
  ---------------- - ------------- + ----------------------
   μ^2 - (ν + 1)^2      μ^2 - ν^2      μ (μ + ν) (μ + ν + 1)
```

Then the Laplace transform Ψ is given by

```
Ψ[S_, K_, L_, U_, σ_, r_, q_, θ_] :=
 Module[{A = a[S, L], B = b[S, U],
   H = h[S, K], n = ν[σ, r, q], m}, m = μ[n, θ];
   g[B, H, m, n] Sinh[m A] + f[A, H, m, n] Sinh[m B]
   -------------------------------------------------]
                    Sinh[m (A + B)]
```

We now need to invert this transform, evaluated at $\sigma^2 T$, and subtract its discounted and normalized value from the Black–Scholes value. Now we define the Laplace inversion by integration along a vertical line in the complex plane (the Bromwich integral), with real part given by the parameter **contour**. The integration works, but *Mathematica* may warn us about the oscillatory nature of the integration. If you get irritating messages, you might like to turn it off, using

```
Off[NIntegrate::slwcon]
```

However, if you choose parameters that result in the integration yielding strange answers, you should not turn off such messages, and possibly explore the options to

NIntegrate. Now we define the valuation by the use of the Bromwich integral. In the following function we compute the Black–Scholes value, the correction for a unit stock price, and the full value:

```
DoubleBarrierCall[S_, K_, L_, U_, σ_, r_,
  q_, t_, contour_, range_, options___] :=
Module[{bs, dim},
    bs = BlackScholesCall[S, K, σ, r, q, t];
    dim = Re[Exp[-r*t] / (2 Pi) * NIntegrate[
          Ψ[S, K, L, U, σ, r, q, contour + I p] *
       Exp[(contour + I p) * σ^2 * t],
    {p, -range, range}, MaxRecursion -> 10,
      options]];     bs - S * dim]
```

The range {{contour − 1500 ∗ *I*}, {contour + 1500 ∗ *I*}} has been found to be useful for several practical cases. The truncation at 1500 may need to be modified for other situations.

■ Examples

The following numerical examples were considered by Geman and Yor (1996). If you consult their paper you can check that our model agrees with their numerical simulations:

```
DoubleBarrierCall[2, 2, 1.5,
 2.5, 0.2, 0.02, 0, 1, 10, 1500]
```

0.0410889

```
DoubleBarrierCall[2, 2, 1.5,
 3.0, 0.5, 0.05, 0, 1, 10, 1500]
```

0.0178568

```
DoubleBarrierCall[2, 1.75,
 1.0, 3.0, 0.5, 0.05, 0, 1, 10, 1500]
```

0.0761723

■ Comments on ‘hedging’ the double barrier call

One of the advantages of having a reasonably simple closed form for the Laplace transform is that any of the partial derivatives can be computed by inverting the partial derivatives of the transform. Of particular interest is the quantity

$$\Delta = \frac{\partial V}{\partial S} \tag{20.20}$$

which is related to the amount of shares to be held to properly hedge the option. The computation of this quantity and its derivative, called Γ, is given in the exercises.

■ Arithmetically averaged Asian call options

This is a much more subtle problem involving a modification to the Black–Scholes equation that depends on a *function of the path* taken by the underlying price variable - the average $< S >$ over the time to expiry of the option is used here, and the final value is

$$\text{Max}[0, \ < S > - K] \tag{20.21}$$

This section gives a brief summary of the exact solution for the Asian call given by Geman and Yor (1993), 'GY', also discussed by Eydeland and Geman (1995). Suppose that the current time is t, and that the option matures at a time $T > t$. The averaging is arithmetic, continuous, and began at a time $t_0 \leqslant t$. Suppose that the known average value of the underlying over the time interval $[t_0, \ t]$ is ES. GY define the following changes of variables:

$$\tau = \frac{1}{4} \sigma^2 (T - t) \tag{20.22}$$

$$\nu = \frac{2(r - q)}{\sigma^2} - 1 \tag{20.23}$$

$$\alpha = \frac{\sigma^2 (K (T - t_0) - (t - t_0) \text{ES})}{4 S} \tag{20.24}$$

They also define a function of a transform variable p as

$$\mu(p) = \sqrt{\nu^2 + 2 p} \tag{20.25}$$

The value of the average price option is then given by

$$\frac{e^{-r (T-t)} 4 S C(\tau, \ \nu, \alpha)}{(T - t_0) \sigma^2} \tag{20.26}$$

The remaining function $C[\nu, \ \tau, \ \alpha]$ is not given explicitly, but GY give its Laplace transform,

$$U(p, \nu, \alpha) = \int_0^{\infty} C(\tau, \ \nu, \alpha) e^{-p \tau} d\tau \tag{20.27}$$

as an integral:

$$U(p, \nu, \alpha) = \frac{\int_0^{\frac{1}{2\alpha}} x^{\frac{\mu-\nu}{2}-2} (1 - 2 \alpha x)^{\frac{\mu+\nu}{2}+1} e^{-x} dx}{p (p - 2 \nu - 2) \Gamma(\frac{\mu-\nu}{2} - 1)} \tag{20.28}$$

where μ is a function of p and ν, as given above. Geman and Yor develop a series description of the transform and show how it can be inverted. We shall now explore how this can be managed and simplified in *Mathematica*.

■ *Mathematica* implementation of exact arithmetical Asian

The first part of the translation to software is obvious – first we enter the definitions of the various basic functions:

```
τ[T_, t_, σ_] := σ^2 (T - t) / 4;
ν[r_, q_, σ_] := 2 (r - q) / σ^2 - 1;
α[S_, ES_, K_, σ_, T_, t_, to_] :=
  σ^2 / (4 * S) (K * (T - to) - (t - to) * ES);
μ[ν_, p_] := Sqrt[ν^2 + 2 * p]
```

Now we enter the definition of the integral that is part of the transform, and request immediate evaluation:

$$\texttt{F[p\_, μ\_, ν\_, α\_]} = \int_0^{\frac{1}{2\alpha}} x^{\frac{\mu-\nu}{2}-2}\,(1 - 2\,\alpha\,x)^{\frac{\mu+\nu}{2}+1}\,\texttt{Exp}[-x]\,dx$$

$$\text{If}\Big[\text{Re}(\mu-\nu) > 2 \wedge \text{Re}(\mu+\nu) > -4 \wedge \alpha > 0,\ \frac{1}{\Gamma(\mu+1)}\Big(2^{\frac{1}{2}(-\mu+\nu+2)}\,\alpha^{\frac{1}{2}(-\mu+\nu+2)}\,\Gamma\Big(\frac{1}{2}(\mu-\nu-2)\Big)\,\Gamma\Big(\frac{1}{2}(\mu+\nu+4)\Big)\,{}_1F_1\Big(\frac{1}{2}(\mu-\nu-2);\,\mu+1;\,-\frac{1}{2\alpha}\Big)\Big),$$
$$\text{Integrate}\Big[e^{-x}\,x^{\frac{1}{2}(\mu-\nu-4)}\,(1-2\,x\,\alpha)^{\frac{1}{2}(\mu+\nu+2)},\ \Big\{x,\,0,\,\frac{1}{2\alpha}\Big\},\ \text{Assumptions} \to \alpha \le 0 \vee \text{Re}(\mu-\nu) \le 2 \vee \text{Re}(\mu+\nu) \le -4\Big]\Big]$$

Mathematica has evaluated the integral in terms of hypergeometric functions – we extract exactly the piece we need by supplying the necessary **`Assumptions`** as follows:

```
G[p_, μ_, ν_, α_] = Integrate[x^((μ - ν)/2 - 2)*
    (1 - 2*α*x)^((μ + ν)/2 + 1)*Exp[-x], {x, 0,
1/(2*α)}, Assumptions -> {α>0, Re[μ+ν]>-4,
Re[ν]+2<Re[μ]}]
```

$$\frac{1}{\Gamma(\mu+1)}\Big(2^{\frac{1}{2}(-\mu+\nu+2)}\,\alpha^{\frac{1}{2}(-\mu+\nu+2)}\,\Gamma\Big(\frac{1}{2}(\mu-\nu-2)\Big)\,\Gamma\Big(\frac{1}{2}(\mu+\nu+4)\Big)\,{}_1F_1\Big(\frac{1}{2}(\mu-\nu-2);\,\mu+1;\,-\frac{1}{2\alpha}\Big)\Big)$$

There are further cancellations when we insert the other terms that make up the transform:

$$\texttt{U[p\_, μ\_, ν\_, α\_] = Simplify}\Big[\frac{\texttt{G[p, μ, ν, α]}}{\texttt{p (p - 2 ν - 2) Gamma}[\frac{\mu-\nu}{2} - 1]}\Big];$$

In standard mathematical notation, the transform is just (the output style has been converted here)

```
TraditionalForm[U[p, μ, ν, α]]
```

$$\frac{2^{\frac{1}{2}(-\mu+\nu+2)}\,\alpha^{\frac{1}{2}(-\mu+\nu+2)}\,\Gamma(\frac{1}{2}(\mu+\nu+4))\,{}_1F_1(\frac{1}{2}(\mu-\nu-2);\mu+1;-\frac{1}{2\alpha})}{p\,(p-2\,(\nu+1))\,\Gamma(\mu+1)} \tag{20.29}$$

We now have the ingredients to build the *Mathematica* model of the arithmetical average price Asian call. In the following the Laplace transform inversion is again done by direct numerical integration along the truncated Bromwich contour, whose location is determined in terms of the parameter ν. An explicit truncation parameters is supplied, and we also exploit some symmetry to do the integration over half the contour:

```
Off[NIntegrate::slwcon]

AriAsianPriceCall[S_, ES_, K_,
  r_, q_, σ_, T_, t_, to_, range_] :=
    Module[{ti = τ[T, t, σ], n = ν[r, q, σ],
   a = α[S, ES, K, σ, T, t, to], contour},
    contour = 2 n + 3;
    ac = Re[ 1 / (Pi) *
     NIntegrate[U[(contour + I p), μ[n, (contour + I p)],
        n, a] * Exp[(contour + I p) * ti],
      {p, 0, range}, MaxRecursion -> 11]];
 Exp[-r * (T - t)] * 4 * S / ((T - to) * σ^2) * ac]
```

Here is an example to show it working. The asset price is currently 1.9 currency units, the strike is at 2.0 units, and so on:

```
AriAsianPriceCall[1.9, 0, 2, 0.05, 0, 0.5, 1, 0, 0, 1500]
```

0.193174

This example shows the power of *Mathematica* very nicely. In lower level programming, in C/C++ in particular, one has to work *extremely hard* to work out such contour integrals. In using this result, one should note that the integration has to be taken over a larger value of the **range** variable if the volatility σ or the time to maturity $T-t$ is small. When this is done carefully this method can be used as a benchmark calculational tool for pricing this type of option, and has been used as such. See, for example, Vecer (2002).

Exercises

20.1 ✵ The square wave function is given by (evaluate to see the shape)

```
fsquare[x_] = Which[x < -0.5, 0, x > 0.5, 0, True, 1];
Plot[fsquare[x], {x, -1, 1}, AspectRatio -> 0.5];
```

Sample this function over 1000 points and investigate the effect of applying **Chop** with various parameters to the transformed data.

20.2 ✲ Explore the operation of **Chop** on the Fourier transform of the letter 'E'.

20.3 ✲ Explore the operation of convolutions based on non-Gaussian filters, making use of color to plot the results. The following is a useful template, using

$$J_1(x^2 + y^2)$$

instead of a Gaussian:

```
blur = Table[BesselJ[1, (x^2 + y^2)],
    {x, -63.5, 63.5}, {y, -63.5, 63.5}];;
blur = Map[RotateRight[#, 64] &,
   RotateRight[blur, 64]];
observedd = Chop[CentredInverse[
    CentredFourier[blur] * CentredFourier[edata]]];
ListDensityPlot[observedd, Mesh -> False,
  PlotRange -> All, ColorFunction -> Hue];
```

20.4 Prove the discrete Fourier inversion theorem. (Hint: Use Eq. (20.24).)

20.5 Work out the first and second partial derivatives with respect to S of the double-barrier transform function Ψ, and hence compute

$$\Delta = \frac{\partial V}{\partial S}$$

$$\Gamma = \frac{\partial^2 V}{\partial S^2}$$

for the double barrier call option. (Don't forget to include the derivatives of the ordinary Black–Scholes piece!)

20.6 ✲ Obtain the **NLaplaceInversion.m** package from *MathSource* and compare the results obtained on the transform function with the various implementations therein, against the direct method developed here, for the function

$$\frac{1}{(s+2)^2 + 1}$$

20.7 ✲ Look up Rizzardi's (1995) paper and implement the Talbot and modified Talbot methods in *Mathematica* – how well do these methods work on the double barrier call option inversion problem?

20.8 Identify the two poles in the transform of the solution of the Asian call option, and compute the residues there. Show that the contribution from these two poles alone to the value of the option is given by

$$e^{-r(T-t)} \left(\frac{\text{ES}\,(t - t_0)}{T - t_0} + \frac{(e^{(r-q)(T-t)} - 1)\, S}{(r - q)\,(T - t_0)} - K \right)$$

Compare the results of this calculation with the two examples given at the very end of Section 20.4, for larger values of S.

20.9 ✵ Work out a formula for the Δ of the Asian call option, and show that when

$S = 2$; ES $= 0$; $K = 2$;
$r = 0.05$; $q = 0$; $\sigma = 0.5$;
$t = 0$; $t_0 = 0$; $T = 1$;

the value of the option is about 0.25 and Δ is approximately 0.56.

20.10 Where are the branch points of the transform function involved in the Asian call option? Why does this make inversion awkward when some of the standard approximate methods are used?

21 Conformal mapping II: the Schwarz–Christoffel mapping

Introduction

In Chapter 16 we explored some of the geometrical properties of holomorphic functions, and in particular looked at the behaviour of the Möbius transformation. The key geometrical feature was that the mapping is conformal (where the derivative is non-zero) in the sense that it is locally angle-preserving. In Chapter 19 we highlighted the importance of conformal mapping to the solution of Laplace's equation in two dimensions. We produced several types of solution to Laplace's equation, including several examples where the region of interest was bounded by a circle or a line in the complex plane.

A question that arise naturally is how to manage matters when the region is not a half-plane or interior/exterior of a circle. Here we must draw a sharp distinction between issues of general principle and practicalities of implementation. We shall begin by stating without proof an important, but non-constructive, theorem that addresses the general principle. Then we shall introduce the Schwarz–Christoffel (SC) mapping that gives an explicit construction for polygonal regions.

There are very few examples of the SC mapping that can be worked out in closed-form in terms of 'simple' functions. A novel use of *Mathematica* is to use its advanced built-in special-function capabilities, and their linkage to the symbolic integrator, to give evaluations of several expressions usually left as complicated integrals in more traditional treatments. We can use such evaluations to facilitate the visualization of the mappings, and hence to confirm the correctness of the answer. As such this chapter contains an informal introduction to hypergeometric functions and elliptic functions, which arise naturally from SC mappings associated with triangles and rectangles. *Mathematica*'s symbolic integrator also allows the results for triangles to be extended to *regular n-gons*, again in terms of hypergeometric functions. In more recent versions of *Mathematica* the implementation of hypergeometric functions of two variables permits a still larger category of mappings to be explored. It should be emphasized that this is a purely analytical treatment, best used as a teaching tool.

For full professional use there is an extensive and highly developed numerical treatment already available. We shall also comment briefly at the numerical implementation of the SC mapping. Gratitude is expressed to Professor L. N. Trefethen F.R.S. for supplying background material on this topic. Readers interested in pursing SC mapping in more detail, whether analytical or numerical, are encouraged to see Section 2.17 and to consult the definitive and up-to-date text on the matter: Driscoll and Trefethen (2002). This chapter is best regarded as an educational bridge between the older and very limited analytical discussions in textbooks and the full numerical approach developed by Trefethen, Driscoll and co-workers. Our bridge is constructed using the advanced analytical tools in *Mathematica*.

21.1 The Riemann mapping theorem

This theorem is about simply connected domains D in $\mathbb{C}$ that are not the whole of $\mathbb{C}$. Given such a domain the theorem states that *there is a* 1-1 *holomorphic function* $w = f(z)$ *that maps* D *onto the interior of the unit disk, i.e. to the set of complex numbers* w *such that* $|w| < 1$. If we further specify that a given point z_0 satisfies $0 = f(z_0)$, and that a specified direction at z_0 is mapped into a specified direction at 0 (for example the direction along the positive real axis), the mapping is unique.

The proof of this result is rather beyond the scope of this book. A reasonably accessible treatment is given by Dettman (1965). Note that the result could just as easily have been stated about the existence of a mapping into the upper half-plane, by composing f with an appropriate Möbius transform.

The drawback of this theorem is that it is purely about existence – it tells us nothing about how to explicitly construct the mapping.

21.2 The Schwarz–Christoffel transformation

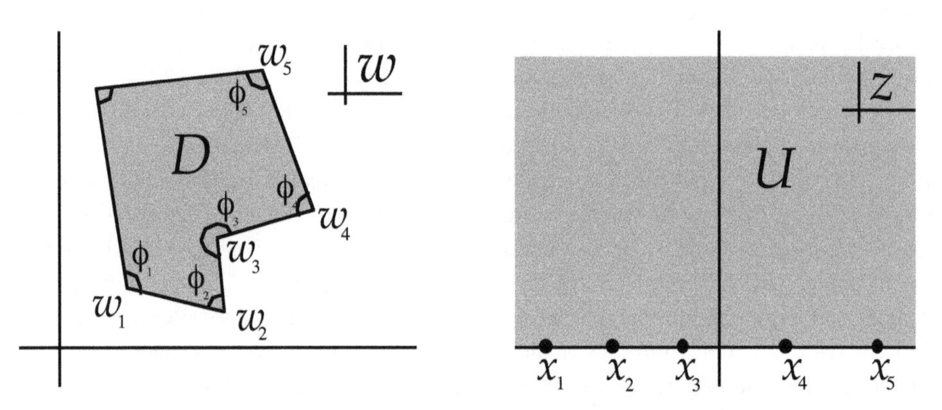

This is a very explicit constructive method for finding mappings that will cope with regions with polygonal boundaries. We can work with various base regions, such as the upper half-plane or interior of a circle, and to state the basic results we shall consider a mapping from the upper half-plane.

Suppose that we have a polygon in the w-plane with vertices at $w_1, w_2, \ldots, w_n$ and interior angles at those vertices $\phi_1, \phi_2, \ldots, \phi_n$. Suppose that these points map onto points $x_1, x_2, \ldots, x_n$ on the real axis of the z-plane. Let D denote the interior of the polygon and U the interior of the upper half-plane. Then a transformation that maps U onto D is given by the Schwarz–Christoffel (SC) formula:

$$w = A \int (z - x_1)^{(\phi_1/\pi)-1} (z - x_2)^{(\phi_2/\pi)-1} \ldots (z - x_n)^{(\phi_n/\pi)-1} \, dz + B \tag{21.1}$$

In differential form this is just

$$\frac{\partial w}{\partial z} = A\,(z - x_1)^{(\phi_1/\pi)-1}\,(z - x_2)^{(\phi_2/\pi)-1}\ldots(z - x_n)^{(\phi_n/\pi)-1} \tag{21.2}$$

■ Informal justification of the SC formula

The simplest way of understanding why the formula works is to consider a family of infinitesimal increments dz along the real axis in the z plane, and their images dw in the w plane. The key is to explore the arguments (or phases) of these infinitesimal changes. Now Eq. (21.2) tells us that:

$$\begin{aligned}&\mathrm{Arg}(dw)\\ &= \mathrm{Arg}(dz) + \mathrm{Arg}(A) + \left(\frac{\phi_1}{\pi} - 1\right)\mathrm{Arg}(z - x_1) + \ldots + \left(\frac{\phi_n}{\pi} - 1\right)\mathrm{Arg}(z - x_n)\end{aligned} \tag{21.3}$$

As z moves along the real axis in the positive direction, when z lies strictly between two of the x_i, (here i is an integer subscript labelling a point in the sequence) all the terms in Eq. (21.3) are constant, so that w moves in a *straight line*. This is why the image in the w plane is polygonal. Now we need to see what happens at the corners. When z passes through x_i, all the terms except those involving x_i remain constant, but the factor

$$\mathrm{Arg}(z - x_i) \tag{21.4}$$

jumps from a value of π, when $z < x_i$, to a value of 0, when $x > x_i$. The term

$$\left(\frac{\phi_i}{\pi} - 1\right)\mathrm{Arg}(z - x_i) \tag{21.5}$$

therefore jumps from the value $\phi_i - \pi$ to the value 0. The direction in which w is travelling therefore rotates in a positive sense (i.e. anti-clockwise), by $\pi - \phi_i$. A change in direction by this amount corresponds to the introduction of a corner in the polygon with interior angle ϕ_i.

What this discussion establishes is that the real axis maps to the boundary of a polygonal region. This will be sufficient for our purposes. One can go further and establish that the mapping is one-to-one and indeed takes the *upper* half-plane to the *interior* of a polygon. We can always check this latter point by working out sample values of particular mappings, such as the image of i.

■ The point at infinity: a simplification

We can assume that one of the points on the real axis, say x_n is at infinity, with it then being dropped from the list of points in the transformation. To see this, we write, with x_n finite,

$$A = \frac{A'}{(-x_n)^{(\phi_n/\pi)-1}} \tag{21.6}$$

$$\frac{\partial w}{\partial z} = A'\,(z - x_1)^{(\phi_1/\pi)-1}\,(z - x_2)^{(\phi_2/\pi)-1}\ldots\left(\frac{x_n - z}{x_n}\right)^{(\phi_n/\pi)-1} \tag{21.7}$$

Now we take the limit $x_n \to \infty$, keeping A' finite and the last factor drops out, leaving us with

$$\frac{\partial w}{\partial z} = A' \, (z - x_1)^{(\phi_1/\pi)-1} \, (z - x_2)^{(\phi_2/\pi)-1} \ldots (z - x_{n-1})^{(\phi_{n-1}/\pi)-1} \tag{21.8}$$

■ The exponents and other comments

We can make a simple geometrical argument that puts one constraint on the exponents. Note that the sum of the exterior angles of any closed polygon is 2π. So we can state that

$$(\pi - \phi_1) + (\pi - \phi_2) + \ldots + (\pi - \phi_n) = 2\pi \tag{21.9}$$

If we divide this by $-\pi$ we obtain

$$\left(\frac{\phi_1}{\pi} - 1\right) + \left(\frac{\phi_2}{\pi} - 1\right) + \ldots + \left(\frac{\phi_n}{\pi} - 1\right) = -2 \tag{21.10}$$

so that the sum of the exponents in the SC formula is -2. This is a constraint on what we can write down, and will also turn out to be useful in relating the SC formula for a half-plane to the corresponding result for a disk. There is also some freedom in the choice of the x_n. Note first that the constants A and B merely adjust the size and position of the polygon generated by the mapping. So if we set $A = 1$ and $B = 0$ we create a polygon, say P', that is *similar* to the one, say P, that is desired, but it is not the right size and in the right location. Let's examine the freedom in the x_i within this scheme. We shall assume that the specialization $x_n = \infty$ has not yet been made. Under these circumstances, given that the exterior angles are fixed, there are still constraints amongst the x_i. For P' to be similar to P, given that the angles are fixed (so that the remaining two side lengths are implied) requires that any $n - 2$ connected sides of P' must have lengths that are in a common constant ratio to corresponding sides of P. This implies $n - 3$ constraints on the vertices, which are themselves determined by the n variables x_i.

This means that before we make the specialization $x_n = \infty$, three of the x_i can be chosen at will. If this specialization is then made, two of the remaining x_i can be chosen. This freedom can also be understood in terms of the requirements within the Riemann mapping theorem needed to make the mapping unique – we must specify the image of one complex point (two real conditions) and a direction (one further real condition) to ensure uniqueness.

21.3 Analytical examples with two vertices

To get a firm grip on the SC formula, we now consider some simple analytical examples. It will be evident that the matter becomes rather more complicated once there are more than three vertices, since then we have to solve the problem of determining the x_i for all but three of the values of i. So we consider first some cases with just two vertices, which can be done analytically.

■ The semi-infinite vertical strip

Perhaps the simplest case is the semi-infinite vertical strip given by

$$-a \leqslant \mathrm{Re}(w) \leqslant a; \quad \mathrm{Im}(w) \geqslant 0 \tag{21.11}$$

In this case we have

$$\begin{aligned} w_1 &= -a; \quad \phi_1 = \frac{\pi}{2} \\ w_2 &= a; \quad \phi_2 = \frac{\pi}{2} \end{aligned} \tag{21.12}$$

and we can make the simple choice $x_1 = -1$, $x_2 = +1$. The SC formula gives us

$$\frac{\partial w}{\partial z} = A\,(z-1)^{\frac{\pi/2}{\pi}-1}\,(z+1)^{\frac{\pi/2}{\pi}-1} = \frac{A}{\sqrt{z^2-1}} = \frac{C}{\sqrt{1-z^2}} \tag{21.13}$$

where the shift of constant from A to C makes it rather obvious how to integrate this result in terms of inverse trigonometric functions – we have, immediately

$$w = B + C\sin^{-1}(z) \tag{21.14}$$

Now the constants B and C can be fixed by looking at the locations of the corners. Considering the first vertex, this gives us

$$-a = B + C\sin^{-1}(-1) = B - \frac{C\pi}{2} \tag{21.15}$$

The other vertex gives us

$$a = B + C\sin^{-1}(1) = B + \frac{C\pi}{2} \tag{21.16}$$

from which it follows that $B = 0$, and $K = 2a/\pi$, i.e.

$$w = \frac{2a\sin^{-1}(z)}{\pi}; \quad z = \sin\left(\frac{\pi w}{2a}\right) \tag{21.17}$$

This is one of the very few simple analytical outcomes of the SC formula. We can have a more detailed look at what this mapping does using the visualization tools we have already introduced.

■ ❄ Recall of visualization tools

As in Chapter 16, we shall use the following functions:

```
Needs["Graphics`ComplexMap`"];
```

```
CartesianConformal[func_,
   xrange_, yrange_, options___] :=
    Show[GraphicsArray[{
            CartesianMap[# &, xrange, yrange,
     options, DisplayFunction -> Identity],
            CartesianMap[func, xrange, yrange,
     options, DisplayFunction -> Identity]
       }], DisplayFunction -> $DisplayFunction];
```

Let's look at this first example with $a = 1$. Recall that **CartesianConformal** shows the images of the real and imaginary coordinate lines under the mapping.

```
CartesianConformal[(2 / Pi ArcSin[#] &),
 {-16, 16}, {10^(-8), 16}, Lines -> 20,
 PlotPoints -> 200, PlotRange -> All]
```

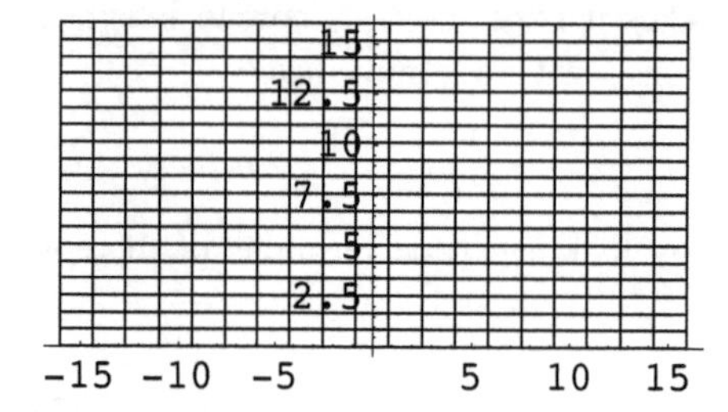

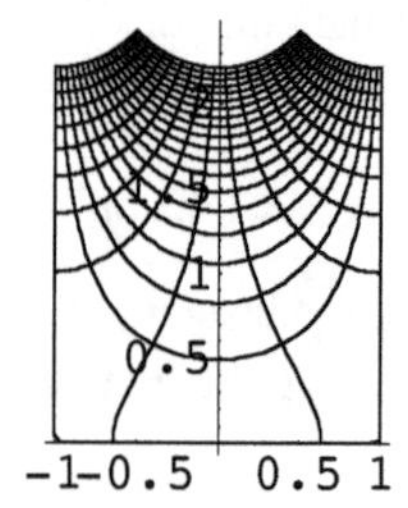

■ The step

There are other examples similar in style to this one, where there are just two vertices to be considered. An important case, left for you to investigate in detail in Exercise 21.1, is where the polygon has two right-angled turns of opposite sense – this leads to a mapping from the upper half-plane to one with a step-shaped boundary.

21.4 Triangular and rectangular boundaries

There are a class of problems where we can get 'formulae' for the answers, but where the formulae remain expressible only as an integral or, as we shall now see using *Mathematica*, as a 'special function'. Some interesting cases can be generated by considering the creation of triangular and rectangular polygons.

For the case of a triangle, consider the following picture (the parameters in the plot routine, which is shown only to show you how this figure was drawn, are of no particular signifiance), where we assume that the point A is at the origin and B is at $w = 1$. The interior angles at these two vertices are α and β respectively.

```
Show[Graphics[{{GrayLevel[0.6], Polygon[{{0, 0}, {1, 0}, {0.3, 1}}]},
        {PointSize[0.03], Point[{0, 0}]},
    {PointSize[0.03], Point[{1, 0}]},
        {PointSize[0.03], Point[{0.3, 1}]},
        Line[{{-1/2, 0}, {1.5, 0}}], Line[{{0, -1/2}, {0, 1.5}}],
        Text["A", {0.1, -0.1}],
    Text["B", {1.1, -0.1}], Text["C", {0.3, 1.2}],
        Text["α", {0.1, 0.1}], Text["β", {0.85, 0.1}]}]];
```

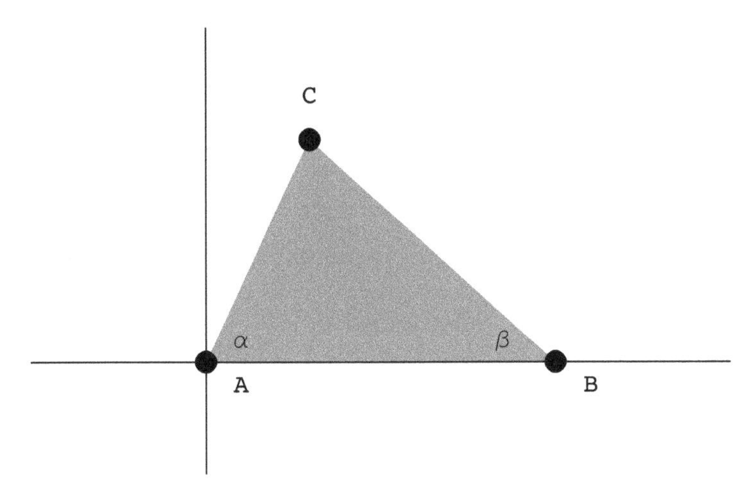

We wish to build a mapping $w(z)$ such that this triangle is the image of the upper half-plane, with A the image of the origin $z = 0$, and B is the image of $z = 1$. Here we shall make use of the fact that C can be set to be the image of infinity, so that the SC formula still only involves two terms:

$$\frac{\partial w}{\partial z} = A' \, z^{(\alpha/\pi)-1} \, (z-1)^{(\beta/\pi)-1} = C \, z^{(\alpha/\pi)-1} \, (1-z)^{(\beta/\pi)-1} \tag{21.18}$$

So the mapping is given by integrating this result:

$$w = C \int_0^z \zeta^{(\alpha/\pi)-1} \, (1-\zeta)^{(\beta/\pi)-1} \, d\zeta + B \tag{21.19}$$

Now, when $z = 0$, we also want $w = 0$, so we set $B = 0$. Furthermore, since $w = 1$ when $z = 1$, we must choose C to set

$$1 = C \int_0^1 \zeta^{(\alpha/\pi)-1} \, (1-\zeta)^{(\beta/\pi)-1} \, d\zeta \tag{21.20}$$

Now we can get *Mathematica* to do some work (which can be done by pen and paper if you know about Gamma and Beta functions):

```
Integrate[ζ^(α/Pi - 1) (1 - ζ)^(β/Pi - 1),
 {ζ, 0, 1}, Assumptions → {α > 0, β > 0}]
```

$$\frac{\Gamma(\frac{\alpha}{\pi})\,\Gamma(\frac{\beta}{\pi})}{\Gamma(\frac{\alpha+\beta}{\pi})}$$

We can assert that C is given by the reciprocal of this integral, i.e.

$$C = \frac{\Gamma(\frac{\alpha+\beta}{\pi})}{\Gamma(\frac{\alpha}{\pi})\,\Gamma(\frac{\beta}{\pi})} \tag{21.21}$$

The mapping is then given by

$$w = \frac{\Gamma(\frac{\alpha+\beta}{\pi})}{\Gamma(\frac{\alpha}{\pi})\,\Gamma(\frac{\beta}{\pi})} \int_0^z \zeta^{(\alpha/\pi)-1}\,(1-\zeta)^{(\beta/\pi)-1}\,d\zeta \tag{21.22}$$

■ Analysis with *Mathematica*

In traditional texts, this is usually where the discussion ends. However, armed with *Mathematica*, or a very good book of integrals, we can probe further into the structure of this mapping – let's ask *Mathematica* to do the integral:

```
Integrate[ζ^(α/Pi - 1) (1 - ζ)^(β/Pi - 1),
 {ζ, 0, z}, Assumptions → {α > 0, β > 0}]
```

$$\text{If}\Big[\text{Im}(z) \neq 0 \lor \text{Re}(z) \leq 1,\ \text{B}_z\Big(\frac{\alpha}{\pi},\ \frac{\beta}{\pi}\Big),$$
$$\text{Integrate}\big[(1-\zeta)^{\frac{\beta}{\pi}-1}\,\zeta^{\frac{\alpha}{\pi}-1},\ \{\zeta, 0, z\},\ \text{Assumptions} \to z > 1\big]\Big]$$

Clearly for a range values of z, the answer is given by the middle term of this expression multiplied by C. We augment the **Assumptions** accordingly:

```
w = Γ((α+β)/π) / (Γ(α/π) Γ(β/π)) *
  Integrate[ζ^(α/Pi - 1) (1 - ζ)^(β/Pi - 1), {ζ, 0, z},
   Assumptions → {Im[z] ≠ 0 || Re[z] ≤ 1, α > 0, β > 0}]
```

$$\frac{\text{B}_z(\frac{\alpha}{\pi}, \frac{\beta}{\pi})\,\Gamma(\frac{\alpha+\beta}{\pi})}{\Gamma(\frac{\alpha}{\pi})\,\Gamma(\frac{\beta}{\pi})}$$

To explain this output, we present the full form of the input that generates it:

```
TraditionalForm[Beta[z, a, b]]
```

$\text{B}_z(a, b)$

Now we ask about this function.

```
? Beta
```

```
Beta[a, b] gives the Euler beta function B(a, b).
  Beta[z, a, b] gives the incomplete beta function Bz(a, b).
```

Mathematica has now given us a result in terms of the *incomplete beta function*. The theory of *analytic continuation* then can be used to argue that this holds everywhere. A proper discussion of this is rather beyond the scope of this text, but, basically, if two holomorphic functions agree on a suitable region, they are equal everywhere, so we can use the formula in terms of a beta function, which is holomorphic, for other values of z besides those suggested by the output from the integrator. In any case, we can check this formula by using our visualization tool. We consider an equilateral triangle, with

$$\alpha = \beta = \frac{\pi}{3} \tag{21.23}$$

The formula for w is then:

```
w /. {α -> Pi / 3, β -> Pi / 3}
```

$$\frac{B_z(\frac{1}{3}, \frac{1}{3})\,\Gamma(\frac{2}{3})}{\Gamma(\frac{1}{3})^2}$$

We define a *Mathematica* formula by

```
g[z_] = Simplify[%];
```

Now let us take a look at the result (this needs a fast machine to compute reasonably quickly):

```
CartesianConformal[(g[#] &), {-8, 8},
 {10^(-12), 8}, Lines -> 30, PlotPoints -> 1000,
 PlotRange -> All, PlotStyle → AbsoluteThickness[0.01]]
```

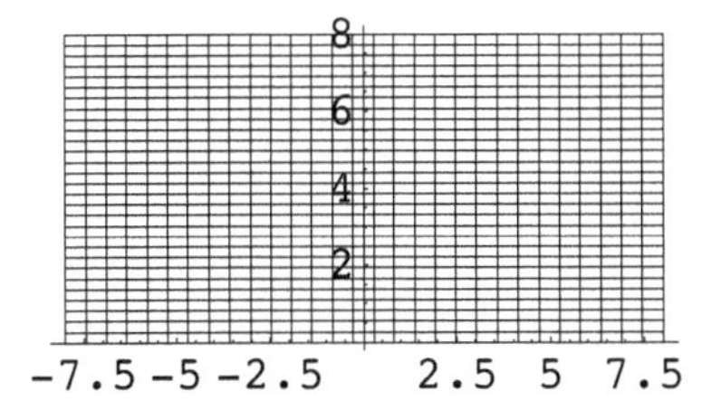

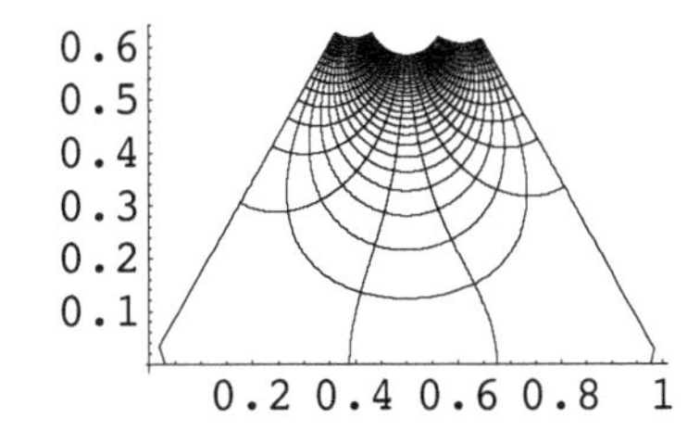

Slight numerical plotting effects at end-points aside, this confirms the validity of the closed-form answer.

Another interesting exercise is to consider the region in the upper half plane above a related triangle and the real axis – this is left for you to explore in Exercise 21.2.

■ Hypergeometric and beta functions

The incomplete beta function is a special case of the hypergeometric function. You can just ask the kernel for some basic information about such functions

```
?Hypergeometric2F1
```

```
Hypergeometric2F1[a, b, c, z]
  is the hypergeometric function 2F1(a, b; c; z).
```

More useful is the series expansion, which extrapolates the pattern shown here for the first few terms:

```
Series[Hypergeometric2F1[a, b, c, z], {z, 0, 3}]
```

$$1+\frac{a\,b\,z}{c}+\frac{a\,(a+1)\,b\,(b+1)\,z^2}{2\,c\,(c+1)}+\frac{a\,(a+1)\,(a+2)\,b\,(b+1)\,(b+2)\,z^3}{6\,c\,(c+1)\,(c+2)}+O(z^4)$$

The beta function arises from the following special case where $c = a + 1$:

```
FullSimplify[
 (z^a/a) Hypergeometric2F1[a, 1 - b, a + 1, z]]
```

$$\mathrm{B}_z(a,\,b)$$

The reader may well ask how the author knew how to dream up this particular form of the hypergeometric function! Traditionally this sort of identity would have been found in one of the classic books of results about special functions. A more modern approach is to consult an on-line database, such as `http://functions.wolfram.com`. This is a *Mathematica* based collection of results, approaching 100,000 formulae at the time this book was edited. It is quote common for SC integrals to evaluate to a hypergeometric function – we shall see some more in later sections. With any of these results it may be useful to check the integral by differentiating it. Here, for example, we obtain

```
D[Beta[z, a, b], z]
```

$$(1-z)^{b-1}\,z^{a-1}$$

■ Rectangles and elliptic functions

Now consider the polygon as a rectangle. As in the case of the vertical strip, each $\phi_i = \pi/2$. There are now four vertices, so strictly we might consider that this requires a search procedure to determine the fourth value of x_i. However, in this case we can appeal to *symmetry* to specify the four pre-image points as ± 1 and $\pm\alpha$ for some α, and investigate the family of shapes that result. We shall assume $\alpha > 1$ in the following example. In this case the differential SC mapping takes the form

$$\begin{aligned}\frac{\partial w}{\partial z} &= A\,(z+\alpha)^{-1/2}(z+1)^{-1/2}(z-1)^{-1/2}(z-\alpha)^{-1/2} \\ &= -\frac{A}{\sqrt{1-z^2}\,\sqrt{\alpha^2-z^2}}\end{aligned} \tag{21.24}$$

It is a little more convient to write this in terms of $a = 1/\alpha$ and a new overall constant A' as

$$\frac{\partial w}{\partial z} = \frac{A'}{\sqrt{1-z^2}\,\sqrt{1-a^2\,z^2}} \tag{21.25}$$

The integral of this is only moderately exotic, in that it defines one of the well known family of elliptic functions. We now set $A' = 1$ and evaluate the integral with some suitable **Assumptions**:

```
Integrate[1/(Sqrt[1 - z^2] * Sqrt[1 - a^2 z^2]),
 {z, 0, ζ}, Assumptions → {0 < a < 1, Im[ζ] > 0}]
```

$F(\sin^{-1}(\zeta) \mid a^2)$

To figure out what that output really means, we ask for it in **InputForm**:

```
InputForm[%]
```

```
EllipticF[ArcSin[ζ], a^2]
```

So we define:

```
SCRect[z_, a_] := EllipticF[ArcSin[z], a^2];
```

and check the derivative is indeed what we want:

```
D[SCRect[z, a], z]
```

$$\frac{1}{\sqrt{1-z^2}\,\sqrt{1-a^2\,z^2}}$$

What can the kernel tell us about this function?

```
?EllipticF
```

```
EllipticF[phi, m] gives the elliptic integral of the first kind F(phi|m).
```

Further information is provided in Section 3.2.11 of *The Mathematica Book* (Wolfram 2003), on elliptic integrals and elliptic functions. Let's set $\alpha = 2$, so that $a = 1/2$, and plot the mapping, to get a feel for what is happening:

```
CartesianConformal[(SCRect[#, 1 / 2] &),
 {-16, 16}, {10^(-8), 16}, Lines -> 40,
 PlotPoints -> 200, PlotRange -> All]
```

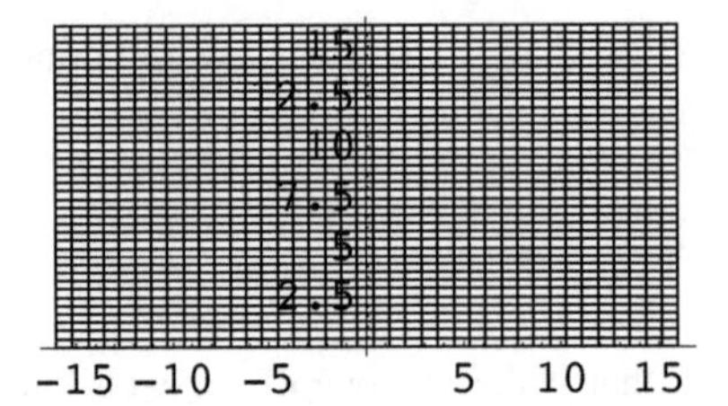

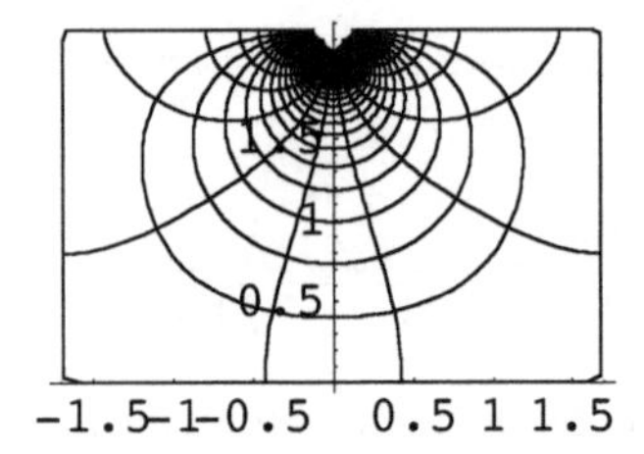

We can see very clearly that the **EllipticF** function is effecting the mapping we want. The four vertices are a the points $\pm b$, $\pm b + i\, c$, where, as functions of the parameter $a = 1/\alpha$, b is given by integrating the mapping from 0 to 1:

```
b [a_] = SCRect[1, a]
```

$$K(a^2)$$

where the *Mathematica* function just output is given by

```
InputForm[%]

 EllipticK[a^2]
```

For the case we have plotted, the value of this is:

```
b[1 / 2] // N
```

1.68575

To extract the right value of c we need to be slightly careful with the branching properties (also see Wolfram (1993)) of the **EllipticF** function – to this end we define:

```
c[a_, ϵ_] = SCRect[1 / a + ϵ I, a] - b[a]
```

$$F\left(\sin^{-1}\left(i\,\epsilon + \frac{1}{a}\right) \middle|\, a^2\right) - K(a^2)$$

For the example at hand, the most obvious approach does not quite work:

```
c[1 / 2, 0] // N
```

$$-4.00238 \times 10^{-9} - 2.15652\, i$$

If we nudge the point slightly into the upper half-plane, which is, after all, where we are working, we get:

```
c[1 / 2, 10^(-12)] // N
```

$-8.16425 \times 10^{-7} + 2.15651\, i$

```
Im[%]
```

2.15651

which is the right value, and consistent with our plot.

21.5 Higher-order hypergeometric mappings

These analytical considerations may be taken further, to deal with more general quadrilaterals. The calculation of the relevant two-variable hypergeometric functions that is needed is also quite computationally demanding and a GHz-class machine is recommened. You will also need version 4 or later of *Mathematica*. Let's add a further pre-vertex at $g > 1$ and consider the following

$$\frac{\partial w}{\partial z} = C\, z^{(\alpha/\pi)-1}\, (1 - z)^{(\beta/\pi)-1}\, (g - z)^{(\gamma/\pi)-1} \tag{21.26}$$

Let's look at the image of unity first, to get the normalization:

```
normfactor = Integrate[
  ζ^(α / Pi - 1) (1 - ζ)^(β / Pi - 1) (g - ζ)^(γ / Pi - 1),
  {ζ, 0, 1}, Assumptions → {g > 1, α > 0, β > 0}]
```

$$\frac{g^{\frac{\gamma}{\pi}-1}\, \Gamma(\frac{\alpha}{\pi})\, \Gamma(\frac{\beta}{\pi})\, {}_2F_1\left(\frac{\alpha}{\pi}, 1 - \frac{\gamma}{\pi}; \frac{\alpha+\beta}{\pi}; \frac{1}{g}\right)}{\Gamma(\frac{\alpha+\beta}{\pi})}$$

We will restrict attention to the situation where unity maps to unity, as in the case of the triangle, so we assert that C is given by the reciprocal of the central term, i.e.

```
norm = 1 / normfactor
```

$$\frac{g^{1-\frac{\gamma}{\pi}}\, \Gamma(\frac{\alpha+\beta}{\pi})}{\Gamma(\frac{\alpha}{\pi})\, \Gamma(\frac{\beta}{\pi})\, {}_2F_1\left(\frac{\alpha}{\pi}, 1 - \frac{\gamma}{\pi}; \frac{\alpha+\beta}{\pi}; \frac{1}{g}\right)}$$

So this mapping is given by

$$w = \frac{\left(\Gamma(\frac{\alpha}{\pi})\Gamma(\frac{\beta}{\pi})\right)^{-1} g^{1-(\gamma/\pi)}}{{}_2\tilde{F}_1\left(\frac{\alpha}{\pi}, 1 - \frac{\gamma}{\pi}; \frac{\alpha+\beta}{\pi}; \frac{1}{g}\right)} \int_0^z \zeta^{(\alpha/\pi)-1} (1 - \zeta)^{(\beta/\pi)-1} (g - \zeta)^{(\gamma/\pi)-1}\, d\zeta \tag{21.27}$$

What about the remaining integral? This is best described by introducing another hypergeometric function. Consider the following *Mathematica* function:

```
?AppellF1
```

```
"AppellF1[a, b1, b2, c, x, y] is the Appell
  hypergeometric function of two variables F1(a; b1, b2; c; x, y).
```

We will set:

```
new =
 (g^c*z^(1+a)*AppellF1[1+a, -b, -c, 2+a, z, z/g])/
  (1+a)
```

$$\frac{g^c\, z^{a+1}\, F_1\left(a+1; -b, -c; a+2; z, \frac{z}{g}\right)}{a+1}$$

Here is the observation we need:

```
FullSimplify[D[new, z]]
```

$$g^c\,(1-z)^b\, z^a \left(1-\frac{z}{g}\right)^c$$

We now define the mapping in full:

```
h[z_, α_, β_, γ_, g_] = norm*
   (new /. {a → α/Pi - 1, b → β/Pi - 1, c → γ/Pi - 1});
AppellMap[z_, α_, β_, γ_, g_] := h[z, α, β, γ, g];
TraditionalForm[AppellMap[z, α, β, γ, g]]
```

$$\frac{\pi\, z^{\frac{\alpha}{\pi}}\, F_1\left(\frac{\alpha}{\pi}; 1-\frac{\beta}{\pi}, 1-\frac{\gamma}{\pi}; \frac{\alpha}{\pi}+1; z, \frac{z}{g}\right)\Gamma\left(\frac{\alpha+\beta}{\pi}\right)}{\alpha\,\Gamma\left(\frac{\alpha}{\pi}\right)\Gamma\left(\frac{\beta}{\pi}\right)\,{}_2F_1\left(\frac{\alpha}{\pi}, 1-\frac{\gamma}{\pi}; \frac{\alpha+\beta}{\pi}; \frac{1}{g}\right)}$$

We can now work with more general quadrilaterals. The following takes some considerable time to compute, but nicely illustrates the construction of a trapezoidal region:

```
CartesianConformal[
 (AppellMap[#, Pi/2, Pi/2, 5*Pi/8, 2] &),
 {-6, 6}, {0.05, 8}, Lines -> 10,
 PlotPoints -> 30, PlotRange -> All]
```

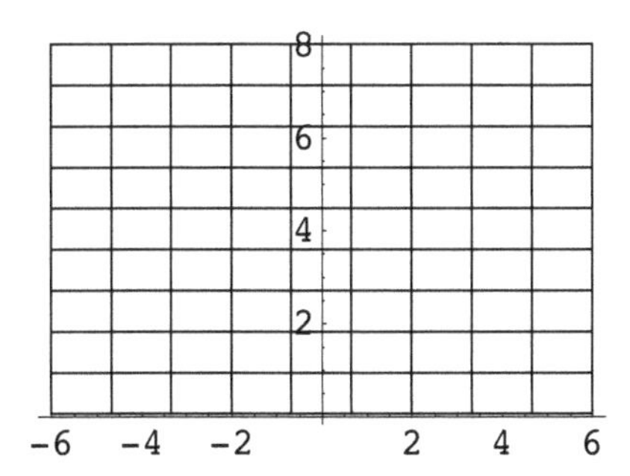

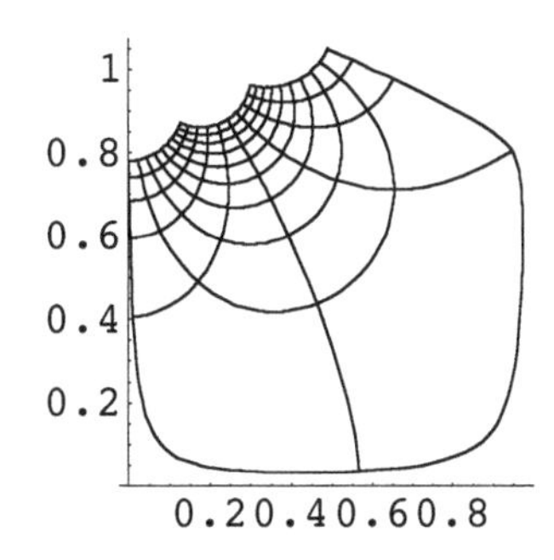

In principle, the Appell function could be made the basis of a semi-analytical treatment of general quadrilaterals. Note that the value of g has been specified in the example given. In general, one would need to find g given a specification of the target shape. This is an example of the 'pre-vertex' problem.

Hypergeometric mappings with symmetry

In the same spirit as we treated the rectangular regions, we can also treat some *six-vertex* regions where the domain is treated symmetrically about the origin. Consider the SC-mapping defined by

$$\frac{\partial w}{\partial z} = C\, z^{(\alpha/\pi)-1}\,(1-z^2)^{(\beta/\pi)-1}\,(g-z^2)^{(\gamma/\pi)-1} \tag{21.28}$$

Mathematica can integrate this directly reasonably quickly.

```
Integrate[
 ζ^(α/Pi - 1) (1 - ζ^2)^(β/Pi - 1) (g - ζ^2)^(γ/Pi - 1),
 {ζ, 0, 1}, Assumptions → {g > 1, α > 0, β > 0}]
```

$$\frac{1}{2}\, g^{\frac{\gamma}{\pi}-1}\, \Gamma\!\left(\frac{\alpha}{2\pi}\right)\Gamma\!\left(\frac{\beta}{\pi}\right)\, {}_2\tilde{F}_1\!\left(\frac{\alpha}{2\pi},\, 1-\frac{\gamma}{\pi};\, \frac{\alpha+2\beta}{2\pi};\, \frac{1}{g}\right)$$

Project

Explore the use of the Appell and ${}_2F_1$ hypergeometric function for these mappings. Show how to produce a mapping of the upper half-plane into a (symmetric) L-shaped region, by taking $\alpha = \beta = \gamma = \pi/2$ and $g = 2$.

21.6 Circle mappings and regular polygons

The SC formula may also be written in terms of a mapping from the unit circle to a polygon – the resulting mapping turns out to be no more complicated than the half-plane SC formula. When the polygon is a regular polygon, the integral simplifies considerably and can be evaluated by *Mathematica* in terms of hypergeometric functions.

■ The SC formula for the unit circle

Recall from Chapter 16 that the mapping

$$p(z) = \frac{z - i}{z + i} \tag{21.29}$$

takes the upper half-plane to the unit circle. The inverse of this mapping is just

$$\frac{i\,(1 + p)}{1 - p} \tag{21.30}$$

We can actually construct a very simple mapping from the interior of the unit circle to a polygon by composing this Möbius mapping with the SC formula. Suppose that the points x_i map into points p_i on the unit circle. Then, for each i, we have

$$z - x_i = \frac{i\,(p + 1)}{1 - p} - \frac{i\,(p_i + 1)}{1 - p_i} = \frac{2\,i\,(p - p_i)}{(1 - p)\,(1 - p_i)} \tag{21.31}$$

and the Jacobian of the transformation is

$$\frac{\partial z}{\partial p} = \frac{2\,i}{(1 - p)^2} \tag{21.32}$$

We now subsitute all of this into the SC formula (21.1) and make use of the exponent formula (21.10), which simplifies the result to

$$w = \tilde{A} \int (p - p_1)^{(\phi_1/\pi)-1}\,(p - p_2)^{(\phi_{12}/\pi)-1} \ldots (p - p_n)^{(\phi_n/\pi)-1}\, dp + B \tag{21.33}$$

where A is a new arbitrary constant. This result is of course structurally identical to the half-plane result, but now the p_n are points on the unit circle.

■ Regular polygons

It is evident from Eq. (21.29) that matters simplify if all the exponents are identical. Furthermore, if we exploit symmetry to consider points p_i that are evenly spaced around the unit circle, we can go further still. We take the points w_k to be the n nth roots of unity, in the form

$$p_k = \omega_n^k \tag{21.34}$$

$$\omega_n = e^{2\pi i/n} \tag{21.35}$$

The product

$$(p - p_1)(p - p_2) \ldots (p - p_n) \tag{21.36}$$

then simplifies to just

$$p^n - 1 \tag{21.37}$$

With our target a regular n-gon, the interior angles are all equal and given by:

$$\phi_k = \pi\left(1 - \frac{2}{n}\right) \tag{21.38}$$

and it follows that the exponents are all equal to $-2/n$. Therefore the desired mapping takes the form

$$w = \tilde{A}\int \frac{1}{(p^n - 1)^{n/2}}\, dp + B \tag{21.39}$$

In fact, slightly cleaner results will be obtained from the integration, which will run from $p = 0$ to $p = z$, by redefining A so that this is written as a linear transformation of

$$w = \int_0^z \frac{1}{(1 - p^n)^{n/2}}\, dp \tag{21.40}$$

▪ Example: the hexagon

A hexagon is obtained by setting $n = 6$. We now get the integrator to finish the calculation for us:

```
Integrate[1 / (1 - ζ^6) ^ (1 / 3), {ζ, 0, z}]
```

$$z\,\text{If}\Big[-1 \le \text{Re}(z) \le 1 \vee \text{Im}(z) \ne 0,\ {}_2F_1\left(\frac{1}{6}, \frac{1}{3}; \frac{7}{6}; z^6\right), \text{Integrate}\Big[\frac{1}{\sqrt[3]{1 - \zeta^6 z^6}},$$
$$\{\zeta, 0, 1\}, \text{Assumptions} \to \neg(-1 \le \text{Re}(z) \le 1 \vee \text{Im}(z) \ne 0)\Big]\Big]$$

To check what this function is doing we plot the result. To visualize this mapping from the circle we now recall, from Chapter 16, the polar version of our mapping viewer:

```
PolarConformal[func_, radial_, polar_, options___] :=
    Show[GraphicsArray[{
            PolarMap[# &, radial, polar,
      options, DisplayFunction -> Identity],
            PolarMap[func, radial, polar,
      options, DisplayFunction -> Identity]
       }], DisplayFunction -> $DisplayFunction];
```

```
PolarConformal[
 (# * Hypergeometric2F1[1 / 6, 1 / 3, 7 / 6, #^6] &),
 {0, 1}, {0, 2 Pi}, Lines -> 20,
 PlotPoints -> 40, PlotRange -> All]
```

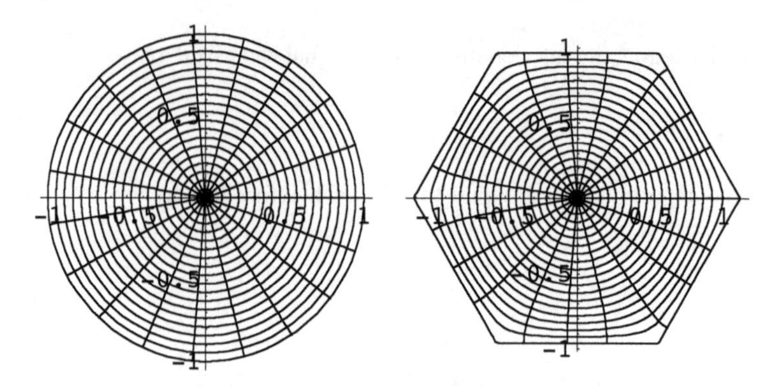

■ The general regular polygon

Mathematica is also quite happy dealing with a polygon with n vertices:

```
PolyMap[z_, n_] = Integrate[
  1 / (1 - ξ^n) ^ (2 / n), {ξ, 0, z}, Assumptions → n > 0]
```

$$z\,{}_2F_1\left(\frac{1}{n}, \frac{2}{n}; 1+\frac{1}{n}; z^n\right)$$

Let's take a look at this for $n = 3,\ 5$:

```
PolarConformal[(PolyMap[#, 3] &), {0, 1}, {0, 2 Pi},
 Lines -> 20, PlotPoints -> 320, PlotRange -> All]
```

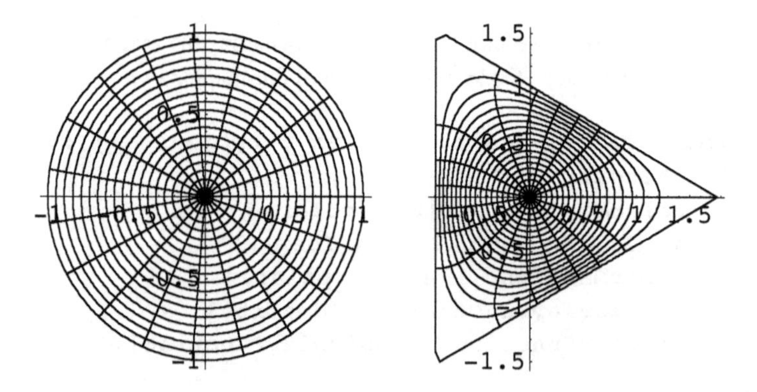


```
PolarConformal[(PolyMap[#, 5] &), {0, 1}, {0, 2 Pi},
 Lines -> 20, PlotPoints -> 80, PlotRange -> All]
```

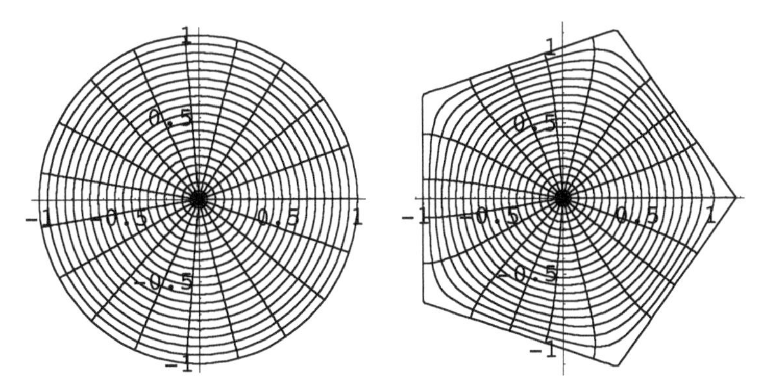

Of course, any of these mappings can be composed with the half-plane to circle Mobius mapping to show the effect of the mapping when applied to the coordinate lines within the upper half-plane. This gives less coverage of the target region for any given finite rectangular subset of the upper half-plane. For example, for the pentagon and square, we have:

```
CartesianConformal[(PolyMap[(# - I) / (# + I), 5] &),
 {-5, 5}, {0.0001, 5}, Lines -> 20,
 PlotPoints -> 40, PlotRange -> All]
```

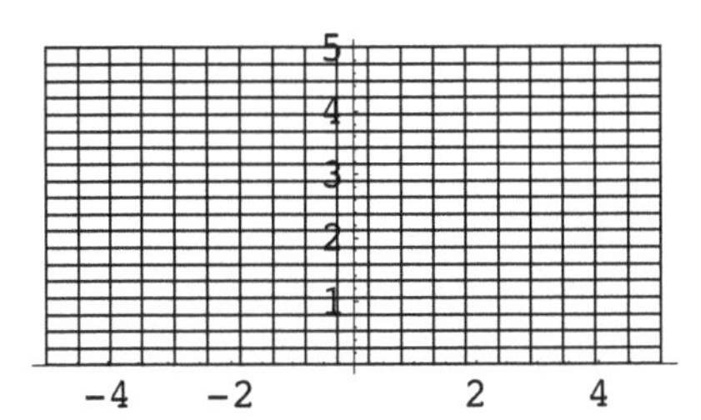

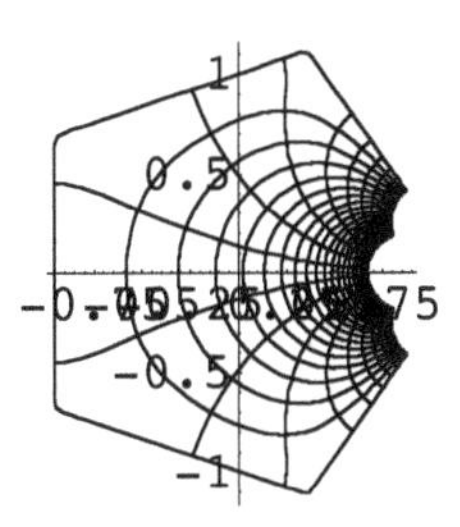

```
CartesianConformal[(PolyMap[(# - I) / (# + I), 4] &),
 {-5, 5}, {0.0001, 5}, Lines -> 20,
 PlotPoints -> 40, PlotRange -> All]
```

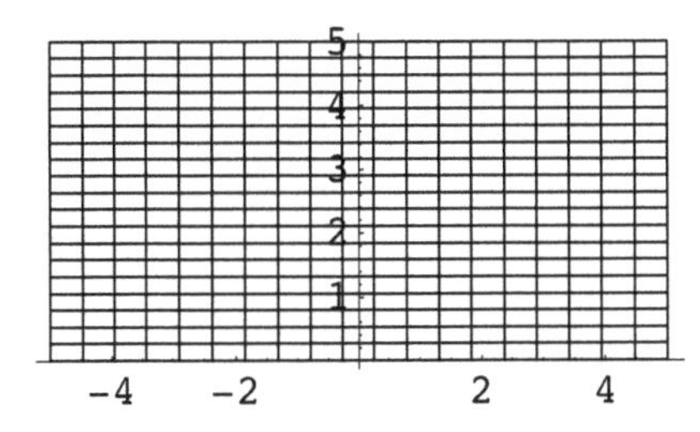

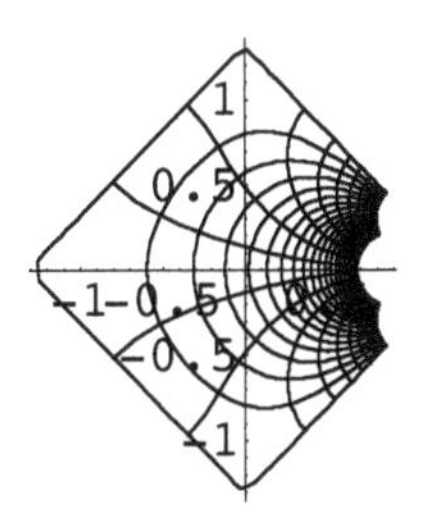

Note that these rather exotic mappings are the composition of a Mobius transform with a carefully chosen hypergeometric mapping! See Driscoll and Trefethen (2002, Section 4.4) for an approach to regions *exterior* to a polgon, also with hypergeometric functions.

21.7 Detailed numerical treatments

It is evident that the management of polygonal regions with many vertices and no regular symmetry takes us outside the domain of tractable analytical closed-form solutions, even when armed with advanced special-function libraries such as those built in to *Mathematica*. In addition to computational efficiency issues, there are two points of difficulty:

(1) determination of the x_i to generate a mapping corresponding to a prescribed polygon;
(2) efficient integration to compute the mapping (also needed within (1)).

The first part of this is called the pre-vertex problem. I have resisted the temptation to implement some of this in *Mathematica*, as there are very efficient and mature implementations already available in FORTRAN and MATLAB. Indeed, *Mathematica*'s strength in this area is probably in the routine management of the simpler examples we have cited using advanced analytics.

Much of the recent work in this area has been done by L.N. Trefethen and co-workers. A FORTRAN implementation, SCPACK, suitable for a class of problems was given by Trefethen (1980), and its successor, the Schwarz–Christoffel Toolbox (Driscoll, 1996), is now available for use within the MATLAB system, and contains many enhancements. This is available for download over the World-Wide Web at: `http://www.math.udel.edu/~driscoll/SC/`. Some key references are as follows. An extensive survey of developments up to about 1985 is given by Trefethen (1986). Some particular approaches for treating elongated structures are given by Howell and Trefethen (1990). A more recent but compact survey is given by Trefethen and Driscoll (1998). The text 'Schwarz-Christoffel Mapping' (Driscoll and Trefethen, 2002) is probably the best entry point.

The numerical approach involves many interesting features. The solution of the pre-vertex problem involves some cunning transformations to eliminate the constraints associated with ordering of the x_i. The integration also requires an approach that takes care of the singularities of the integrand at the pre-images of the vertices. This requires an enhancement of the Gauss–Jacobi method. Often the x_i crowd together in an awkward fashion, and this must also be managed. Those interested should see the references for information on the management of these matters. A supply of other *analytical* problems set in the interesting context of fluid dynamics is given by Milne-Thomson (1938).

Exercises

21.1 ❀ Consider the polygon in the w plane consisting of the negative real axis, the line from 0 to $i\,a$, and the horizontal line from $i\,a$ to $i\,a + \infty$. Use the SC formula to show that this can be obtained from the upper half-plane in the z plane by the mapping

$$w = \frac{2\,a\,i}{\pi}\left(\sin^{-1}\left(\sqrt{z}\right) + \sqrt{z\,(1 - z)}\right)$$

Confirm the validity of this formula by using the **CartesianConformal** graphics function to visualize the mapping with $a = 1$ and $a = 2$. (Hint: You will need to use the pure function

```
((2 * I) / Pi * (ArcSin[Sqrt[#]] + Sqrt[# * (1 - #)]) &)
```

over a suitable domain.)

21.2 Consider the polygon given by the shaded region below:

```
Show[Graphics[
  {{GrayLevel[0.6], Polygon[{{-2, 0}, {-1, 0}, {0, 1}, {1, 0}, {2, 0}, {2, 2}, {-2, 2}}]},
          {PointSize[0.03], Point[{-1, 0}]}, {PointSize[0.03], Point[{1, 0}]},
          {PointSize[0.03], Point[{0.0, 1.0}]},
          Line[{{-2, 0}, {2, 0}}], Line[{{0, -1/20}, {0, 2}}],
          Text["A:w=-b", {-1, -0.2}],
   Text["B:w=+b", {1.1, -0.2}], Text["C:w=ia", {0.4, 1}],
          Text["α", {-0.8, 0.1}], Text["α", {0.8, 0.1}]}]];
```

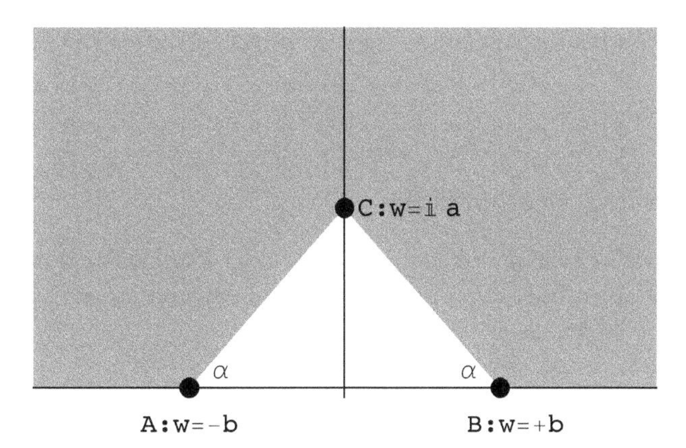

Construct an SC mapping with the properties that:

(i) A ($-b$) is the image of -1;
(ii) B ($+b$) is the image of $+1$;
(iii) C ($+i\,a$) is the image of 0.

and express your answer in terms of an integral. What constraint relates a, b and α? Using the *Mathematica* integrator, determine the arbitrary constants A and B. By considering the *Mathematica* evaluation of the expression

```
Integrate[ξ^(2 α / Pi) / (1 - ξ^2) ^ (α / Pi), {ξ, 0, z}]
```

find a closed form for the mapping. Now consider the form of your answer with

$$\alpha = \frac{\pi}{4};\ \beta = \frac{\pi}{4};\ a = b = 1$$

By making a construction similar to that given for the interior of the triangle, build the analogue of the function **g[z]** in the text, and visualize the results. You should be able to obtain a plot identical to the following, with the given parameters:

```
CartesianConformal[(g[#] &),
 {-1.5, 1.5}, {0.00001, 1.5}, Lines -> 30,
 PlotPoints -> 40, PlotRange -> All]
```

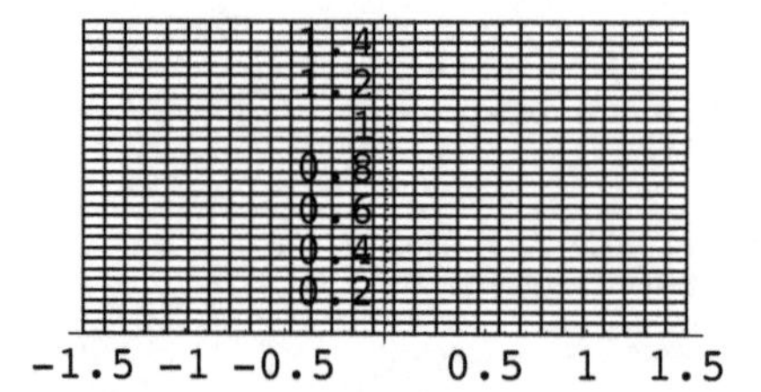

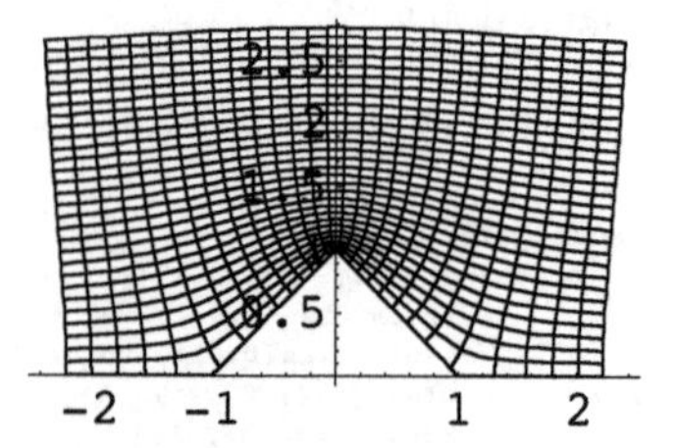

21.3 Using the SC formula for a square, establish a relationship between a special case of the hypergeometric function and a special case of an elliptic function.

21.4 ❁ What is the length of the size of the regular hexagon generated by the mapping

```
PolyMap[z_, n_] =
 Integrate[1 / (1 - ζ^n) ^ (2 / n), {ζ, 0, z}]
```

with $n = 6$? (Hint: Calculate the image of $z = 1$ and of $z = e^{2\pi i/6}$.)

21.5 ❁ The images of very small circles in the unit disk look at lot like circles in the image n-gon, in the examples given in Section 21.6. Explain this by considering the power series for **PolyMap**. You might like to consider the form of

```
Series[PolyMap[z, 10], {z, 0, m}]
```

to get you going, for various values of integers **m**.

21.6 ❁ (Reverse engineering) For each of the following mappings, use the **CartesianConformal** function to establish the image region, and then explain how to derive the mapping from the SC formula:

$$\frac{1}{\pi}\cosh^{-1}(z)\,;\quad \log(z)\,;\quad z^{1/3}$$

22 Tiling the Euclidean and hyperbolic planes

Introduction

In our studies so far we have been concerned with the complex plane interpreted as a two-dimensional Euclidean plane – when the concept of distance has been needed, it has always been the standard Euclidean notion expressed by Pythagoras' theorem. There are concepts of distance other than the standard Euclidean one. Indeed, this notion is at the heart of modern geometrical physics, and finds expression in both the non-positive-definite notions of distance of special relativity, and the non-flat metrics of general relativity.

In this chapter we shall meet the hyperbolic plane, which is perhaps the simplest non-Euclidean geometry. We shall not be able, in one chapter, to do full justice to this concept – indeed, excellent entire books have already been written about it (Coxeter, 1965; Stahl, 1993). What we shall do is explore a little of the geometry through the process of tiling the hyperbolic plane (see the Bibliography for papers by Coxeter and Levy on this particular matter also).

This chapter is based substantially (the sections on triangles and the 'ghosts and birdies' tiling) on a project carried out by a former colleague, V. Thomas, for the BBC Open University Production Centre. Gratitude is expressed to A.M. Gallen and, latterly, the Open University for permission to use this material, to Professor R. Penrose F.R.S. ('R. Penrose' for short) for several helpful suggestions, and to V. Thomas for permission to use her work in this text. Later sections (22.6–7) on the projective representations and tilings by squares were inspired by some suggestions by R. Penrose in connection with illustrations for his beautiful book (Penrose, 2004). Gratitude is again expressed to R. Penrose for his guidance on the development of some of the algorithms described here.

This chapter develops techniques for a process that, when given any suitable triangle, will tile the Euclidean plane using that shape. This then leads to a procedure for adapting such triangular shapes to produce Escher-like tilings of the Euclidean plane. These methods are then extended to hyperbolic tilings, represented on the Euclidean plane by the Poincaré disc, more properly known as Beltrami's conformal representation. This is illustrated with specific triangles and an Escher-like pattern. Then a look is taken at Beltrami's projective representation, and tilings by hyperbolic squares in both the conformal and projective representations. An extension of the technique provides a hyperbolic tiling using heptagons. This chapter also illustrates yet another role for the Möbius transformation!

22.1 Background

The familiar geometry of the plane, Euclidean geometry, is only one of many different geometries that describe the various intrinsic structures of different surfaces. Most of these geometries do not have as much symmetry as Euclidean geometry, but there are two geometries that do: the geometry of the surface of a sphere, and another geometry called *hyperbolic* geometry. Hyperbolic geometry shares many of the properties of Euclidean geometry but differs in certain significant respects. Most importantly, the ordinary Euclidean notion of parallel lines does not hold. (In Euclidean geometry there is a unique line through a given point parallel to a given line.) Also the angles in a hyperbolic triangle do not add up to 180 degrees but always to something less than 180, the difference being a measure of the area of the triangle. This is a geometry that cannot be directly represented by any diagram we could draw on paper. However, we can perform a transformation and obtain a picture of how the geometry would look if we were to embed it into a Euclidean world in which the whole of the infinite hyperbolic plane is mapped onto an open unit disc in the Euclidean plane. This particular transformation can be achieved so that the angular measurements made on this 'Euclidean' picture of the hyperbolic plane (called the Poincaré disc) will be the same as the angles measured in the actual hyperbolic plane. The Dutch artist M. C. Escher made several woodcuts illustrating this representation of hyperbolic geometry.

For a Euclidean space we intuitively know what it would be like to move around in a space defined by that geometry, and can define transformations that describe these movements. If we think of the Euclidean plane as the complex plane then these transformations can be described in terms of functions applied to the complex number z as

$$\begin{aligned} &z \to a z + b \\ &|a| = 1 \end{aligned} \tag{22.1}$$

if orientation is preserved (i.e. translation and rotation), or, if orientated is reversed, we use the non-holomorphic mapping:

$$\begin{aligned} &z \to a \bar{z} + b \\ &|a| = 1 \end{aligned} \tag{22.2}$$

In each case a and b are complex numbers.

For the hyperbolic plane the relevant transformations are a subset of the Möbius transformations and their non-holomorphic equivalents. As a procedure for enabling us to manipulate the structures of this hyperbolic world, it is useful to think of the transformations taking the hyperbolic geometry into itself, as represented on the Poincaré disc. Taking the Euclidean plane as the complex plane, we look at the transformations that will take the open unit disc (representing the hyperbolic plane) into itself. The unit disc is described by the set of complex z with $|z| < 1$. The transformations that take this disc into itself are of the form

$$z \to \frac{\alpha z + \beta}{\gamma z + \delta} \tag{22.3}$$

where

$$\begin{aligned} &\alpha \bar{\alpha} = \delta \bar{\delta} > \beta \bar{\beta} = \gamma \bar{\gamma} \\ &\alpha \bar{\gamma} = \beta \bar{\delta} \end{aligned} \tag{22.4}$$

These are the self-transformations of the disc that preserve orientation. There are also orientation-reversing transformations. These are the same as above but composed with $z \to \bar{z}$, so we can write them as

$$z \to \frac{\alpha\,\bar{z} + \beta}{\gamma\,\bar{z} + \delta} \tag{22.5}$$

with the same relations amongst α, β, γ and δ.

22.2 Tiling the Euclidean plane with triangles

The procedure for creating a tiling that will cover the whole of a surface with a single (or many) shapes is achieved by defining the transformations of polygonal shapes that define their positions on a particular plane. One way of defining such a tiling is to reflect a single shape in its edges repeatedly to produce a pattern built up of many copies of the same shape. To be able to do this we need to define a reflection in a Euclidean plane. The reflection of a point c in the line joining the points a and b, given as points in the complex plane, is the point c' described by

$$c' = \frac{(\overline{a}\,b - \overline{b}\,a) + (a - b)\,\overline{c}}{\overline{a} - \overline{b}} \tag{22.6}$$

We can use this to produce a general function that will reflect the point r, in the line joining points p1 and p2. In the following function r, p1, p2 are all lists of length two representing coordinates in the plane:

```
reflection[{r_, p1_, p2_}] :=
 Module[{cr, cp1, cp2, newcr, nr},
        {cr, cp1, cp2} =
   Map[(First[#] + I Last[#]) &, {r, p1, p2}];
newcr = (((cp1 - cp2) Conjugate[cr] +
       (Conjugate[cp1] cp2 - cp1 Conjugate[cp2])) /
     (Conjugate[cp1] - Conjugate[cp2]));
nr = N[{Re[newcr], Im[newcr]}]]
```

To show this reflection in action, we introduce the function **showReflection**, which will take any three points and reflect the first in the line joining the other two and illustrate this procedure. The function **pointLabels** is employed only to ensure that the labels associated with each point illustrated occur in an appropriate position:

```
pointLabels[{a_, la_}, {b_, lb_}] :=
Module[{theta, disp},
theta = Arg[(First[#] + I Last[#]) & [a - b]];
disp = 0.25 * {Cos[theta - Pi / 2], Sin[theta - Pi / 2]};
{Text[FontForm[la, {"Times", 14}], a + disp],
Text[FontForm[lb, {"Times", 14}], b + disp]}]
```

```
showReflection[r_, p1_, p2_] :=
Module[{X},
    Show[Graphics[
    Join[
    {PointSize[0.015],
        Map[Point,
       {r, p1, p2, X = reflection[{r, p1, p2}]}],
        Line[{p1, p2}]},
    pointLabels[{p1, a}, {p2, b}],
    pointLabels[{X, x'}, {r, x}]]
    ],
    AspectRatio -> Automatic]]
```

```
showReflection[{4, 1}, {5, 4}, {1, 1}]
```

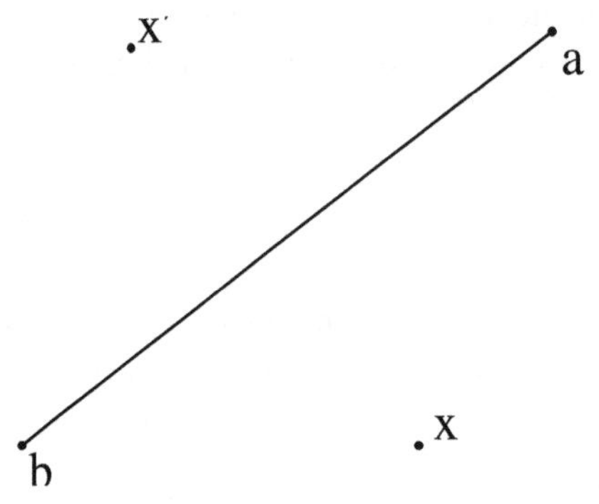

Now we can extend this to produce the reflection of a triangle in one of its sides, as illustrated. For this it will be useful to have a standard function defined that will join the three points of a triangle and also show the vertices as points. Now we introduce the general reference **triangle**, which carries the information about the vertices and the colour of the triangle:

```
triangleANDpoints[
triangle[color_: GrayLevel[0], {r_, p1_, p2_}]] :=
{{Line[{p2, r, p1, p2}]},
{PointSize[0.015], Point[r],
Point[p1],
Point[p2]}}
```

To illustrate the reflection of a triangle *abc* in one of its sides *bc* we can write the function **reflectedtriangle** and apply that to some set of three vertices, the labels only being added for clarity:

```
reflectedTriangle[{a_, b_, c_}] :=
Module[{A},
Show[Graphics[
Join[
{triangleANDpoints[triangle[{a, b, c}]],
triangleANDpoints[
      triangle[{A = reflection[{a, b, c}], b, c}]]},
pointLabels[{A, "a'"}, {a, "a"}],
pointLabels[{b, "b"}, {c, "c"}]]],
AspectRatio -> Automatic]]

reflectedTriangle[{{6, 1}, {0, 0}, {5, 6}}]
```

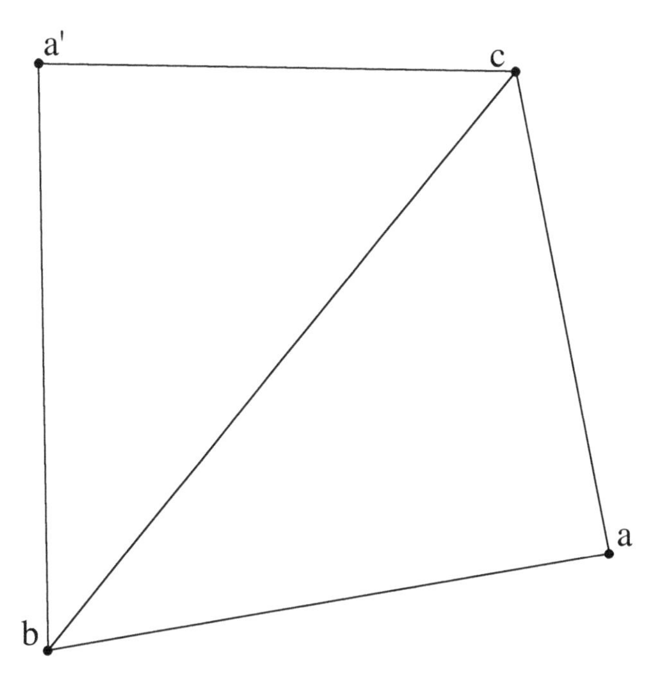

However, instead of only reflecting one of the points of the triangle in one of the sides, to tile the plane quickly with these triangular tiles it will be useful to be able to reflect all three points of the triangle in their opposite sides. Here we will want to apply the function reflection to the cyclic permutation of the points of the triangle:

```
cyclicPermutations[x_List] :=
    NestList[RotateLeft, x, Length[x] - 1]
```

The function **reflectTriangles** will return the set of three triangles that are the reflections of the original triangle in each of its sides and the colour of these triangles will have been changed:

```
reflectTriangles[
    triangle[color_: GrayLevel[0], {a_, b_, c_}]] :=
Map[triangle[
    If[Head[color] == Hue,
        ReplacePart[#,
            Switch[
         #[[1]], 0.15, 0.72, 0.72, 0.15],
            1], #, #] &[color], #] &,
    MapThread[({a, b, c} /. #2 -> reflection[#1]) &,
    {cyclicPermutations[{a, b, c}],
    {a, b, c}}]
    ]
```

Now we can obtain the first step towards tiling the Euclidean plane by applying this reflecting function to our base triangle once:

```
Show[Graphics[
    Map[triangleANDpoints,
        reflectTriangles[
        triangle[{{6, 1}, {0, 0}, {5, 6}}]]]],
AspectRatio -> Automatic]]
```

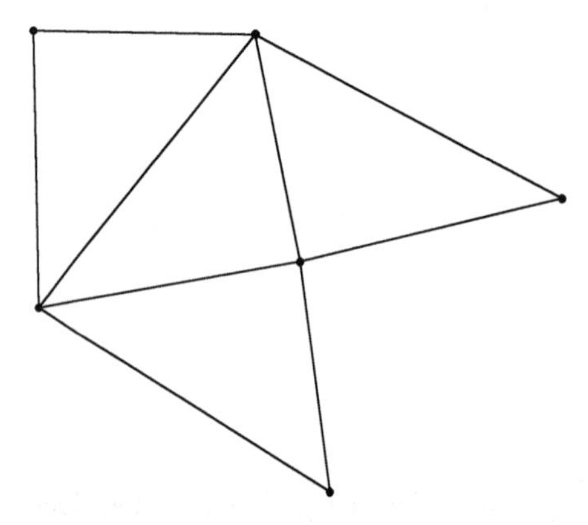

The next stage in this process will be to reflect each of the three new triangles in each of their three sides. As we come to deal with larger and larger lists of triangles we will want **reflectTriangles** to perform this same procedure on each triangle that makes up that list. We therefore need it to have the attribute of being **Listable**:

```
SetAttributes[reflectTriangles, Listable]
```

In order for this process to produce a larger and larger grouping of triangles it will be useful to be able to apply the **reflectTriangles** function over and over again to the results obtained at each stage. For this we can use the **NestList** function, which will give the set containing the triangles produced after applying the reflecting function a certain number of times.

In our function **triangleTilingTry**, which will perform this process a certain number of times, we therefore combine these procedures to repeatedly apply the **reflectTriangles** function and then produce the graphic using **triangleAND-points**:

```
triangleTilingTry[triangle_, n_] :=
Show[Graphics[
    Map[triangleANDpoints,
    Flatten[NestList[
        reflectTriangles,
        triangle,
        n]]]], AspectRatio -> Automatic]
```

Let's see this process carried out for an example where we reflect from a base triangle twice:

```
triangleTilingTry[triangle[{{6, 1}, {0, 0}, {5, 6}}], 2]
```

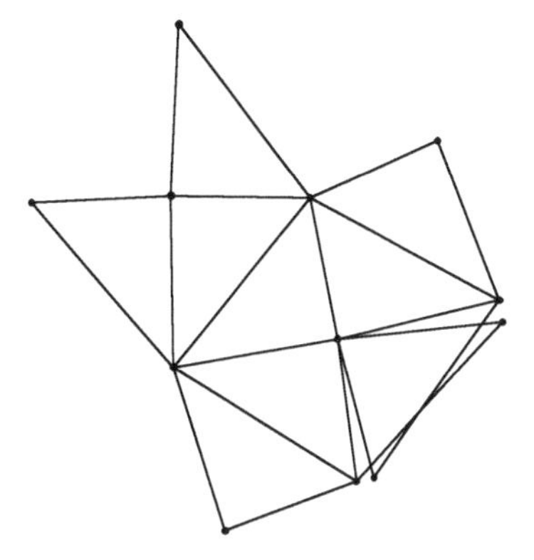

At this stage of the process one of the main problems of the whole procedure begins to come to light. It will get more complicated the more iterations are performed. Each set of reflections of a single triangle will produce the triangles surrounding the original triangle. This means, first, at any given level there is likely to be some overlapping of triangles and, second, that triangles already produced at an earlier level in the process are all going to be re-obtained (this is obvious as in our original reflection of the point c in *ab*, reflecting *c′* in *ab* will get us back to *c*). There is a great deal of duplication of triangles taking place.

Rather than being a problem, the fact that triangles produced at an earlier stage are re-obtained when the next reflection is performed means that it is not necessary to carry around information relating to all the triangles of the pattern but only the last two set of triangles obtained by the iterative process. It is therefore only necessary to use the **Nest** function and add in a holder that will keep the last set of triangles acted upon stored. This can then be joined onto the current set produced for production of the graphic. This is done using the following code (this is not for actual evaluation – it is to illustrate the structure):

```
Join[Nest[
        (temp = #; refelction procedure [#]) &,
        starting triangle,
        number of iterations ],
    temp]
```

The removal of duplicates within any single level of reflections can, at least at this stage, be done by just asking for duplicates to be removed using the **Union** function. Later we will need to take more care over this process. So, the reflections can be performed and duplicates removed using the following procedure:

```
Flatten[Union[reflectTriangles[#]]] &
```

The function **Flatten** is added so that **triangleANDpoints** can be used for displaying the result.

Before we go on to apply all of this and tile a larger area we should take note of something about the properties of triangles that will successfully tile the Euclidean plane. The triangle with vertices {{6, 1},{0, 0},{5, 6}}, that has been used so far has given us a problem after the second iteration, by not closing the pattern. The two tiles on the bottom right of the diagram are not coinciding but overlapping. It is always the case that only tiles whose internal angles are sub-multiples of π will tile the Euclidean plane successfully. So we can now improve our function to become **triangleTiling** and for a simple example of a triangular shape that will tile the Euclidean plane we use an equilateral triangle:

```
triangleTiling[triangle_, n_] :=
Module[{temp},
  Show[Graphics[
    Map[triangleANDpoints,
    Join[Nest[(temp = #;
        Flatten[Union[reflectTriangles[#]]]) &,
        triangle, n], temp]],
    AspectRatio -> Automatic]]];
```

```
triangleTiling[
triangle[{{-1, 0}, {1, 0}, {0, N[Sqrt[3]]}}], 6]
```

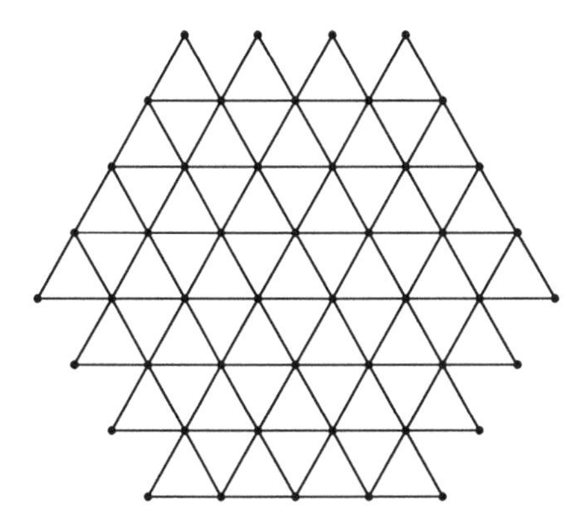

And this can continue as far as you want!

22.3 Tiling the Euclidean plane with other shapes

Now what we can do is to try to associate some different shape with the three points of the triangle, instead of the simple straight lines. This can then be used to produce an Escher-like pattern where we are using a simple tile shape to tile the Euclidean plane.

Before proceeding with this, it is necessary to point out that the triangles of our patterns divide naturally into two classes, which we shall call **B** (black) and **W** (white), depending on whether the triangle is obtained from the original one by an even or odd number of reflections. Thus the **B** class consists of triangles whose orientations are the same as the original one, and the **W** class, those whose orientations are opposite from it. In modifying the triangles to provide an Escher-like tiling it is conceptually easiest if we allow the **B** shapes to be different from the **W** shapes. Then we can specialize afterwards to the particular cases where these two shapes are congruent to each other. We can deform the boundary lines of the **B** triangles in any way we choose (within certain limitations, so that unwanted overlaps are avoided). Since the orientations of the **W** triangles are oppposite to those of the **B** triangles, the **W** shapes defined by the boundary lines will be 'inside out' with respect to the **B** shapes.

To make the two shapes fit, as in the Escher-like design that we shall be describing, we can make the deformation on one of the edges the same as (actually the reverse of) that on one of the others, whilst requiring that that on the third edge be rotationally symmetric about its central point. In this way we can ensure that the 'inside out' **W** shape is actually congruent to the **B** shape. These requirements can be met by deforming the straight lines of the triangle into the curves shown below.

The curves are defined by giving a set of points that will join the points -1 and 1 (in the Argand plane) by following the prescribed curve. The ones we want (after some experimentation with graph paper – highly recommended!) are:

```
Clear[w];
Show[Graphics[Line[
   pointsBC = {1, 0.7 - 0.1*I, 0.34 - 0.24*I, -0.4*I,
0, 0.4*I, -0.34 + 0.24*I, -0.7 + 0.1*I, -1} /.
     w_?NumberQ -> {Re[w], Im[w]}]],
AspectRatio -> Automatic]
```

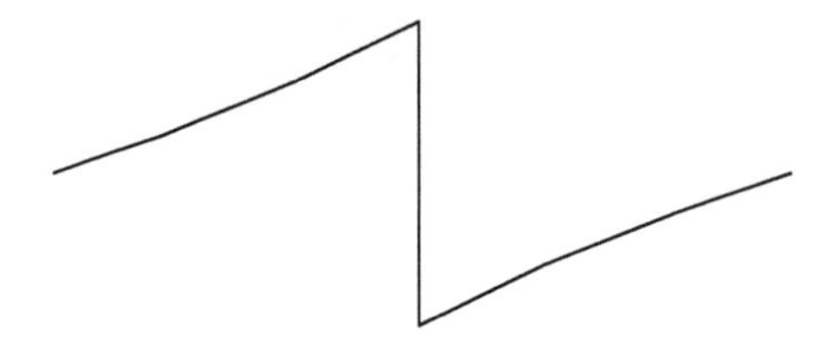

and

```
Show[Graphics[Line[pointsAB =
    {-1, -0.8 - 0.08*I, -0.6 - 0.15*I,
    -0.4 - 0.19*I, -0.3 - 0.21*I,
    -0.3 - 0.03*I, -0.29 + 0.14*I,
    -0.26 + 0.28*I, -0.18 + 0.4*I,
    -0.09 + 0.3*I, 0.04 + 0.14*I, 0.14,
    0.24 - 0.19*I, 0.5 - 0.13*I,
    0.66 - 0.1*I, 0.82 - 0.07*I,
    1} /. w_?NumberQ -> {Re[w], Im[w]}]],
AspectRatio -> Automatic]
```

When these two line shapes are used for particular sides of a triangle, they form a shape that at least somewhat resembles a bird. However, to get these lines to join any three points that form a triangle, we need to be able to transform the lines that join the points -1 and 1 to lines that will join any two points, say a and b. The transformation

$$z \to \frac{a+b}{2} + \frac{1}{2}(a-b)\,z \tag{22.7}$$

indeed takes the line $[-1, 1]$ into $[a, b]$, where a and b are the complex numbers representing the points a and b. So now we can define a function that will join any pair of general points by a line taking the form of one of our curves, employing the transformation defined in Eq. (22.7):

```
birdLine1[{x_, y_}] := Module[{a, b, w, pointsBC},
{a, b} = Map[(First[#] + I Last[#]) &, {x, y}];
pointsBC = {-1, -0.7 + 0.08 * I, -0.34 + 0.16 * I, 0.2 * I,
    0, -0.2 * I, 0.34 - 0.16 * I, 0.7 - 0.08 * I, 1};
 Map[(# (b - a) / 2 + (a + b) / 2) &,
    pointsBC] /. w_?NumberQ -> {Re[w], Im[w]}]
```

We can define a similar function for the other set of points but this time we must take care over the two forms it will appear in. Though we reflect to change black triangles to white triangles and vice-versa, we don't actually reflect the curves. Theses are simply transported by (non-reflective) Euclidean motions (or, later, by hyperbolic motions) to the places that they need to be in order to join the required points.

So the second line we are going to use is equivalent to drawing the curve from b to a, which is why we reverse the order of **pointsAB**, but we must then remember to adjust the transformation function to be one that takes the interval $[1, -1]$ into $[a, b]$. This is all done by using the marker type to take the value 1 or -1:

```
birdLine2[type_, {x_, y_}] :=
 Module[{a, b, z, w, pointsAB},
    {a, b} = Map[(First[#] + I Last[#]) &, {x, y}];
pointsAB = {-1, -0.8 - 0.08 * I, -0.6 - 0.15 * I,
-0.4 - 0.19 * I, -0.3 - 0.21 * I, -0.3 - 0.03 * I,
    -0.29 + 0.14 * I, -0.26 + 0.28 * I, -0.18 + 0.4 * I,
    -0.09 + 0.3 * I, 0.04 + 0.14 * I, 0.14, 0.24 - 0.19 * I,
    0.5 - 0.13 * I, 0.66 - 0.1 * I, 0.82 - 0.07 * I, 1};
    If[type == -1, pointsAB = Reverse[pointsAB]];
    Map[(# * type * (b - a) / 2 + (a + b) / 2) &,
        pointsAB] /. w_?NumberQ -> {Re[w], Im[w]}]
```

A function that will take any three points and join them to form a bird will therefore take the form:

```
birdyTriangleLine[triangle[color_, {a_, b_, c_}]] :=
{color,
  Line[Join[birdLine2[1, {a, b}], birdLine1[{b, c}],
birdLine2[-1, {c, a}]]]}
```

This can be shown using an equilateral triangle as the underlying base:

```
Show[Graphics[birdyTriangleLine[
   triangle[RGBColor[1, 0, 0],
    {{-1, 0}, {1, 0}, {0, N[Sqrt[3]]}}]
]], AspectRatio -> Automatic]
```

These line birds can be changed to coloured birds by swapping **Polygon** for **Line** in the definition of **birdyTriangle**:

```
birdyTriangle[triangle[color_, {a_, b_, c_}]] :=
{color,
Polygon[Join[birdLine2[1, {a, b}], birdLine1[{b, c}],
birdLine2[-1, {c, a}]]]}
```

So now we are ready to combine these two processes. First, create the series of points that give the vertices of the shapes that tile the plane and then join them using the bird pattern:

```
birdTiling[shape_, n_] :=
Module[{temp}, Show[Graphics[
Map[birdyTriangle, Join[
Nest[(temp = #;
          Flatten[Union[reflectTriangles[#]]]) &,
    {shape}, n], temp]]
], AspectRatio -> Automatic]]

birdTiling[triangle[Hue[0.72],
{{-1, 0}, {1, 0}, {0, N[Sqrt[3]]}}], 6]
```

We are almost in a position to be able to do a similar kind of tiling that, instead of working in a Euclidean plane, applies to a hyperbolic plane and produces a tiling that will fit on the Poincaré disc. This will mean a change in the formulae used to perform the reflections and also in the way in which lines will be drawn that join the vertices of the hyperbolic triangles.

22.4 Triangle tilings of the Poincaré disc

Having developed the procedure for tiling a Euclidean plane, we can take the same procedure and adapt that for use on the Poincaré disc, where we need to redefine functions that are going to reflect a point in a line, and need to explain how to join points by a hyperbolic line. For a point r that is to be reflected in the hyperbolic line joining the points a and b, the reflection is defined by the transformation rule:

$$r \to \frac{b\,\overline{a} - a\,\overline{b} + (a - b + b\,\overline{b}\,a - a\,b\,\overline{a})\,\overline{r}}{\overline{a} - \overline{b} + b\,\overline{b}\,\overline{a} - a\,\overline{a}\,\overline{b} + (a\,\overline{b} - \overline{a}\,b)\,\overline{r}} \tag{22.8}$$

This is implemented in the function **hyperbolicReflection** where we are reflecting the point r in the hyperbolic line joining p1 and p2. :

```
hyperbolicReflection[{r_, p1_, p2_}] :=
Module[{cr, cp1, cp2, newcr, nr},
{cr, cp1, cp2} =
   Map[(First[#] + I Last[#]) &, {r, p1, p2}];
newcr = ((Conjugate[cp1] cp2 - Conjugate[cp2] cp1) +
    Conjugate[cr] (cp1 - cp2 + cp1 cp2 Conjugate[cp2] -
        cp2 cp1 Conjugate[cp1])) /
    ((Conjugate[cp1] - Conjugate[cp2] +
    Conjugate[cp1] cp2 Conjugate[cp2] -
    Conjugate[cp2] cp1 Conjugate[cp1]) +
 Conjugate[cr]
       (Conjugate[cp2] cp1 - Conjugate[cp1] cp2));
nr = N[{Re[newcr], Im[newcr]}]]
```

Similarly, we need to define the form of the hyperbolic line that joins any two points to be able to construct the geometric shapes that will make up the tiling of the hyperbolic plane. The hyperbolic lines in the Poincaré disc are represented by Euclidean circles that meet the bounding unit circle at right angles, and where, as a limiting case, the diameters of that unit circle also represent hyperbolic straight lines. A way of obtaining the curves representing the hyperbolic straight line joining two points, let us say a and b, is to think of the Euclidean straight line running from -1 to 1 (in the complex plane) and define a transformation that will take that line to the hyperbolic line joining a and b. Such a transformation takes the form

$$z \to \frac{z(a - \lambda_{ab} b) - (a + \lambda_{ab} b)}{z(1 - \lambda_{ab}) - (1 + \lambda_{ab})} \tag{22.9}$$

where

$$\lambda_{ab} = \frac{(1 - a\overline{b})}{|1 - a\overline{b}|} \sqrt{\frac{1 - a\overline{a}}{1 - b\overline{b}}} \tag{22.10}$$

This technique can be used to define the function **hyperbolicLine**, which will join any two points, a and b, by transforming the set of points on the straight line running from -1 to 1, under the transformation as defined above:

```
hyperbolicLine[{a_List, b_List}] :=
Module[{aC, bC, points, lamdaAB, pointsAB},
{aC, bC} = Map[(First[#] + I Last[#]) &, {a, b}];
points = Range[-1, 1, 0.1];
lamdaAB =
((1 - aC Conjugate[bC]) / Abs[1 - aC Conjugate[bC]]) *
    Sqrt[(1 - aC Conjugate[aC]) / (1 - bC Conjugate[bC])];
pointsAB = Map[
    ((# (aC - lamdaAB bC) - (aC + lamdaAB bC)) /
        (# (1 - lamdaAB) - (1 + lamdaAB))) &,
points] /. x_?NumberQ -> {Re[x], Im[x]}]
```

To allow us to follow the same technique as used for the Euclidean tiling we need to define a function to produce a graphic form of, let us say, a hyperbolic triangle, using the function that will join any two points with a hyperbolic line. As the function **hyperbolicLine** returns a list of points making up the line, we can join the different lists that define the three lines of the triangle to obtain a sequence of points that define the perimeter and create a **Polygon** of that shape:

```
hyperbolicTriangle[
triangle[color_: GrayLevel[0], {a_, b_, c_}]] :=
Join[{color},
    {Polygon[
        Apply[Join,
            Map[hyperbolicLine,
                Partition[{a, b, c, a}, 2, 1]
            ]]]}]
```

The function **reflectHyptTri** defines the function that will produce the next level of triangles obtained by reflecting a single triangle in each of its sides:

```
reflectHypTri[triangle[color_, {a_, b_, c_}]] :=
Map[triangle[
If[Head[color] == Hue,
    ReplacePart[#,
    Switch[#[[1]], 0.15, 0.72, 0.72, 0.15], 1] &[
      color]
], #] &,
 MapThread[
   ({a, b, c} /. #2 -> hyperbolicReflection[#1]) &,
    {cyclicPermutations[{a, b, c}],
    {a, b, c}}]]
```

This function similarly alters the colouring of the triangles after each reflection. This function also needs to have the attribute **Listable**:

```
SetAttributes[reflectHypTri, Listable]
```

Finally, to produce the hyperbolic tiling, just as with the Euclidean case, we can define a function **hypTriangleTiling1** (this is provisional – hence the '1' affix) and apply that to a set triangle:

```
hypTriangleTiling1[triangle_, n_] :=
Module[{temp},
Show[Graphics[
Map[hyperbolicTriangle,
Join[
Nest[(temp = #; Flatten[Union[reflectHypTri[#]]]) &,
 triangle, n], temp]]
], AspectRatio -> Automatic]]

hypTriangleTiling1[triangle[Hue[0.15],
  {{0, 0}, {0.266077, 0}, {0, 0.140626}}], 8]
```

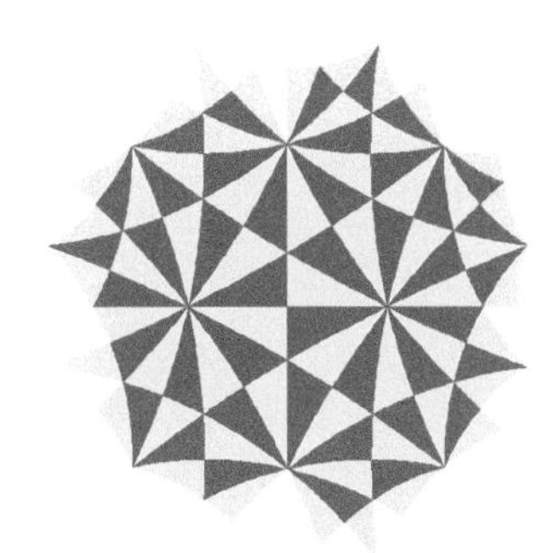

An example of one method by which one calculates the details of the starting shape is given in Section 22.6.

■ Accuracy and duplication management

One property of a tiling of the hyperbolic plane is that as we approach the boundary circle we move out to hyperbolic infinity. This means that as the tiling approaches the boundary circle the triangles are going to get smaller and smaller as more and more of them are squeezed into a limited space. The vertices of these small triangles need to be calculated to a high enough accuracy to ensure that the triangles match neatly. For this reason we start by taking the vertices of the original triangles to an accuracy of ten decimal places, which should be sufficient for our needs. As an aside, we note that this idea is a very general and powerful one. There are many cases where some form of iteration may be used to define a sequence, possibly of functions, but the iteration may even be unstable and produce growing errors. One of the useful features of *Mathematica* is the ability to do arbitrary-precision arithmetic. Hence, by writing down very high precision starting values, unstable iteration becomes practical!

In calculating the vertices of the reflected triangles some rounding error is going to occur. This is not a problem in the first few iterations of the reflecting procedure, but does become apparent as the process is continued. The problem that this then produces is that two triangles which for our purposes are identical will not be recognized as being so, since their vertices will differ in some decimal place. The most straightforward way to remove these repeated triangles is to check whether their vertices differ by less than a particular amount (we are choosing 10^{-15}), and if they do, remove one of the triangles. (This is also needed for the Euclidean tiling when a large number of iterations are performed.) This is implemented in **removeDuplicates**:

```
removeDuplicates[list_List] :=
Module[{removelist, listnew, i, n, diff},
listnew = Sort[Flatten[list]];
For[n = 1, n <= Length[listnew], n++,
removelist = {};
    For[m = n + 1, m <= Length[listnew], m++,
    If[Min[Flatten[diff = Map[
        Flatten[Map[Abs, (listnew[[n, 2]] - #)]] &,
            Join[#, Map[Reverse, #]] &[
            cyclicPermutations[listnew[[m, 2]]]
          ]]]] > 0.001(*0.1, for heptagons*),
        Break[],
    For[l = 1, l <= Length[diff], l++,
    If[Apply[Plus, diff[[l]]] < 10^-8,
        If[
        ! MemberQ[Map[(# < 10^-9) &, diff[[l]]],
    False], AppendTo[removelist, listnew[[m]]];
            Break[]
        ]]]]];
    listnew = Complement[listnew, removelist]
]; listnew]
```

You will notice that the list of structures to be checked for duplicates is first sorted and then, within the loops, which pick their way through pairs of triangles (or whatever shape is being compared), the value of **diff** is calculated, which holds the differences between each of the coordinates defining the shape. If the minimum value here is greater than 0.001, then we can safely say that these shapes are far enough apart that we would have moved away from the region containing shapes near to the shape we are checking, and can move on to check the next shape. Otherwise the process for actually checking whether two shapes are the same continues and any repeated shapes are removed as necessary.

Having this function now allows us to repeat the reflection process for many iterations and approach the point where we can fill the unit disc. To do this we can define a general function that, given any hyperbolic triangle, will perform the specified number of reflections to produce the tiling.

```
hypTriangleTiling[
  shape_?((MatchQ[Head[#], triangle]) &), n_] :=
Module[{temp},
Show[Graphics[
{Circle[{0, 0}, 1],
Map[hyperbolicTriangle,
Join[
Nest[(temp = #; Print[Length[temp]];
    Flatten[
           removeDuplicates[reflectHypTri[#]]]) &,
     {shape}, n],
    temp]]
    }
],
AspectRatio -> Automatic,
PlotRange -> {{-1, 1}, {-1, 1}}]]
```

Here we have carried out this process for a many reflections ($n = 40$). You should check that this works on your system and get a feel for timings by trying, for example, $n = 20$ first. The $n = 20$ test calculation takes under a minute to caclulate and render on a Power Macintosh G4 at 1.4 GHz. Note that running the following calculation will take rather longer! It will also produce progress output that is suppressed here.

```
hypTriangleTiling[triangle[Hue[0.15],
{{0, 0}, {0.2660772453, 0}, {0, 0.14062593}}], 40]
```

22.5 Ghosts and birdies tiling of the Poincaré disc

Now we have all the tools necessary to be able to produce a tiling of the hyperbolic plane similar to those of M. C. Escher. The shapes we shall use are the same as those for the Euclidean tiling and we shall just add some more details to characterize two different types of element.

To define the shape of the bird, the functions **birdLine1** and **birdLine2** used to produce the Euclidean drawing are employed again, with the alteration that instead of using a linear transformation to define the lines it is replaced with the corresponding hyperbolic transformation. For example,

```
(# (b - a) / 2 + (a + b) / 2) &
```

goes to

```
((# (a - lamdaAB b) - (a + lamdaAB b))
    / (# (1 - lamdaAB) - (1 + lamdaAB))) &
```

In addition, we can add some lines that will allow us to create more characterful shapes. Here we add lines to characterize the basic shape into either a ghost or a raven (time to get the graph paper out again!):

```
ghost[{x_, y_}] :=
Module[{a, b, lamdaAB, face1},
{a, b} = Map[(First[#] + I Last[#]) &, {x, y}];
face1 = {{0.23, 0.21 + 0.03 * I, 0.17 + 0.04 * I, 0.14,
     0.15 - 0.04 * I, 0.205 - 0.045 * I, 0.24},
{0.09, 0.07 + 0.03 * I, 0.03 + 0.04 * I, 0.,
     0.01 - 0.04 * I, 0.065 - 0.045 * I, 0.09},
{-0.1 + 0.08 * I,
      -0.08 + 0.12 * I, -0.04 + 0.16 * I, 0.19 * I,
     0.06 + 0.2 * I, 0.1 + 0.2 * I, 0.16 + 0.19 * I,
     0.2 + 0.15 * I}};
{GrayLevel[0], Thickness[0.006 * Abs[a - b]],
lamdaAB =
((1 - a Conjugate[b]) / Abs[1 - a Conjugate[b]]) *
    Sqrt[(1 - a Conjugate[a]) / (1 - b Conjugate[b])];
Map[Line, Map[((# (a - lamdaAB b) - (a + lamdaAB b)) /
          (# (1 - lamdaAB) - (1 + lamdaAB))) &,
face1] /. w_?NumberQ -> {Re[w], Im[w]}]}]
```

The raven not only has a face, but tails and wing feathers. These are defined separately as each set of characteristics is connected with one of the three lines of the triangle that forms the basis of our raven shape:

```
birdLineHyp1[{x_, y_}] :=
Module[{a, b, lamdaAB, pointsBC},
{a, b} = Map[(First[#] + I Last[#]) &, {x, y}];
pointsBC =
    {-1, -0.7 + 0.08 * I, -0.34 + 0.16 * I, 0.2 * I, 0,
    -0.2 * I, 0.34 - 0.16 * I, 0.7 - 0.08 * I, 1};
lamdaAB =
((1 - a Conjugate[b]) / Abs[1 - a Conjugate[b]]) *
    Sqrt[(1 - a Conjugate[a]) / (1 - b Conjugate[b])];
Map[((# (a - lamdaAB b) - (a + lamdaAB b)) /
        (# (1 - lamdaAB) - (1 + lamdaAB))) &,
    pointsBC] /. w_?NumberQ -> {Re[w], Im[w]}
]
```

```
birdLineHyp2[type_, {x_, y_}] :=
Module[{a, b, z, w, pointsAB},
    {a, b} = Map[(First[#] + I Last[#]) &, {x, y}];
    pointsAB = {-1, -0.8 - 0.08 * I, -0.6 - 0.15 * I,
        -0.4 - 0.19 * I, -0.3 - 0.21 * I,
        -0.3 - 0.03 * I, -0.29 + 0.14 * I,
        -0.26 + 0.28 * I, -0.18 + 0.4 * I,
        -0.09 + 0.3 * I, 0.04 + 0.14 * I,
        0.14, 0.24 - 0.19 * I, 0.5 - 0.13 * I,
        0.66 - 0.1 * I, 0.82 - 0.07 * I, 1};
If[type == -1, pointsAB = Reverse[pointsAB]];
lamdaAB =
((1 - a Conjugate[b]) / Abs[1 - a Conjugate[b]]) *
    Sqrt[(1 - a Conjugate[a]) / (1 - b Conjugate[b])];
Map[((type * # * (a - lamdaAB b) - (a + lamdaAB b)) /
      (type * # * (1 - lamdaAB) - (1 + lamdaAB))) &,
       pointsAB] /. w_?NumberQ -> {Re[w], Im[w]}]

raven[{x_, y_}] :=
Module[{a, b, lamdaAB, face2},
{a, b} = Map[(First[#] + I Last[#]) &, {x, y}];
face2 = {{-0.12 + 0.3 * I, -0.11 + 0.28 * I, -0.1 + 0.2 * I,
-0.1 + 0.04 * I,
 -0.16 + 0.09 * I, -0.2 + 0.12 * I, -0.28 + 0.14 * I},
 {-0.04 + 0.21 * I, -0.04 + 0.2 * I, -0.06 + 0.14 * I,
     -0.07 + 0.1 * I, -0.1 + 0.04 * I},
 {-0.2 - 0.09 * I, -0.18 - 0.06 * I, -0.14 - 0.06 * I,
 -0.12 - 0.09 * I, -0.13 - 0.13 * I, -0.16 - 0.14 * I,
 -0.19 - 0.13 * I, -0.2 - 0.09 * I},
 {-0.28 - 0.34 * I, -0.285 - 0.3 * I, -0.3 - 0.22 * I},
 {0.22 - 0.2 * I, 0.2 - 0.21 * I, 0.16 - 0.22 * I,
  0.1 - 0.24 * I}};
{GrayLevel[0], Thickness[0.006 * Abs[a - b]],
lamdaAB =
((1 - a Conjugate[b]) / Abs[1 - a Conjugate[b]]) *
    Sqrt[(1 - a Conjugate[a]) / (1 - b Conjugate[b])];
Map[Line, Map[((# (a - lamdaAB b) - (a + lamdaAB b)) /
          (# (1 - lamdaAB) - (1 + lamdaAB))) &,
face2] /. w_?NumberQ -> {Re[w], Im[w]}]}];


rightFeather[{x_, y_}] :=
Module[{a, b, lamdaAB, feathers1},
{a, b} = Map[(First[#] + I Last[#]) &, {x, y}];
feathers1 = {{-0.13 - 0.24 * I, -0.1 - 0.23 * I,
-0.02 - 0.2 * I},
```

```
 {0.01 - 0.3 * I, -0.21 * I},
 {-0.18 - 0.16 * I, -0.1 - 0.14 * I, -0.02 - 0.11 * I},
 {-0.26 - 0.06 * I, -0.2 - 0.04 * I, -0.1 - 0.01 * I,
 -0.02 + 0.02 * I}, {-0.36 + 0.03 * I, -0.26 + 0.06 * I,
 -0.1 + 0.1 * I, -0.02 + 0.12 * I}};
{GrayLevel[0], Thickness[0.006 * Abs[a - b]],
lamdaAB =
((1 - a Conjugate[b]) / Abs[1 - a Conjugate[b]]) *
    Sqrt[(1 - a Conjugate[a]) / (1 - b Conjugate[b])];
Map[Line, Map[((# (a - lamdaAB b) - (a + lamdaAB b)) /
          (# (1 - lamdaAB) - (1 + lamdaAB))) &,
feathers1] /. w_?NumberQ -> {Re[w], Im[w]}]}];

leftFeather[{x_, y_}] :=
Module[{a, b, lamdaAB, feathers2},
{a, b} = Map[(First[#] + I Last[#]) &, {x, y}];
feathers2 = {{0.68, 0.56 + 0.05 * I, 0.42 + 0.08 * I,
0.31 + 0.1 * I},
 {0.56 - 0.12 * I, 0.48 - 0.09 * I, 0.4 - 0.07 * I,
 0.31 - 0.06 * I}, {0.47 - 0.24 * I, 0.38 - 0.21 * I,
 0.3 - 0.2 * I}, {0.36 - 0.34 * I, 0.26 - 0.3 * I},
 {0.29 - 0.44 * I, 0.2 - 0.4 * I},
 {-0.37 + 0.135 * I,
      -0.325 + 0.06 * I, -0.22 - 0.1 * I,
 -0.14 - 0.2 * I}, {-0.575 + 0.1 * I, -0.56 + 0.06 * I,
 -0.48 - 0.04 * I, -0.38 - 0.16 * I, -0.28 - 0.275 * I},
 {-0.78 + 0.06 * I, -0.72,
      -0.66 - 0.06 * I, -0.6 - 0.12 * I}};
{GrayLevel[0], Thickness[0.006 * Abs[a - b]],
lamdaAB =
((1 - a Conjugate[b]) / Abs[1 - a Conjugate[b]]) *
    Sqrt[(1 - a Conjugate[a]) / (1 - b Conjugate[b])];
Map[Line, Map[((# (a - lamdaAB b) - (a + lamdaAB b)) /
          (# (1 - lamdaAB) - (1 + lamdaAB))) &,
feathers2] /. w_?NumberQ -> {Re[w], Im[w]}]}]
Null
```

The function **ghostsANDravens** will produce the two basic tiles with which we will fill the Poincaré disc. To show the difference between the two shapes we are employing, we will change the function **reflectHypTri** so that under reflection not only the colour of a shape is reversed but also the type from ghost to raven. So we define the function **ghostsANDravens** to act differently depending on whether the argument is a ghost or a raven:

```
ghostsANDravens[ghost[color_, {a_, b_, c_}]] :=
{{color,
Polygon[Join[birdLineHyp2[1, {a, b}],
birdLineHyp1[{b, c}],
birdLineHyp2[-1, {c, a}]]]},
ghost[{b, a}]}

ghostsANDravens[raven[color_, {a_, b_, c_}]] :=
{{color,
Polygon[Join[birdLineHyp2[1, {a, b}],
birdLineHyp1[{b, c}],
birdLineHyp2[-1, {c, a}]]]},
raven[{a, c}],
rightFeather[{c, b}],
leftFeather[{b, a}]}
```

To see how these look:

```
Show[Graphics[ghostsANDravens[
ghost[Hue[0.15], {{0, 0}, {0.448288, 0.258819},
                 {0.448288, -0.258819}}]
]],
AspectRatio -> Automatic]
```

Before checking the graphic for the raven, let us define a new reflection function that will swap between the raven and the ghost at each reflection:

```
reflectHypShape[
    shape_?((MatchQ[#, raven] ||
        MatchQ[#, ghost]) &)[
   color_, {a_, b_, c_}]] :=
Map[Switch[shape, ghost, raven, raven, ghost][
If[Head[color] == Hue,
    ReplacePart[#,
    Switch[#[[1]], 0.15, 0.72, 0.72, 0.15], 1] &[
      color]
], #] &,
 MapThread[
   ({a, b, c} /. #2 -> hyperbolicReflection[#1]) &,
    {cyclicPermutations[{a, b, c}],
    {a, b, c}}]
    ]

SetAttributes[reflectHypShape, Listable]
```

Now with the raven we must take care. The outline of this shape is the same as the original ghost shape, but the triangle is given with the opposite orientation. This, as you will recall, was the reason for using the specific lines for **birdline1** and **birdline2.** However now we have introduced to the shapes elements that do not have these properties we must take care how they are used. If we are to start the tiling with the ghost then the shape as displayed is quite acceptable. The raven outlines will appear as the gaps between the ghosts, but unreflected, whereas the triangles themselves have been obtained by reflection. This is taken care of by the symmetry of one curve and the repetition of the other. We must take care that the orientation of the edges is consistent so that the two shapes emerge as they are supposed to.

```
Show[Graphics[ghostsANDravens[reflectHypShape[
ghost[Hue[0.15], {{0, 0}, {0.448288, 0.258819},
{0.448288, -0.258819}}]][[1]]]],
AspectRatio -> Automatic]
```

If we had looked at the raven shape in its unreflected form, it would have looked a little odd, since the ordering of the vertices defining the shape determine whether the shading of the polygon will occur on one side or the other of any one line defining our shape.

Now we have all the tools to define the function that will produce a hyperbolic tiling of some specified shape up to a certain number of reflections:

```
hyperbolShapes[shape_, n_] :=
Module[{temp},
Show[Graphics[{Circle[{0, 0}, 1],
Map[ghostsANDravens,
Join[Nest[(temp = #; Print[Length[#]];
Flatten[
            removeDuplicates[reflectHypShape[#]]]) &,
 {shape}, n], temp]]}
], AspectRatio -> Automatic]]
```

Try the following, which should take just a few seconds on a GHz-class machine:

```
hyperbolShapes[
ghost[Hue[0.15], {{0, 0}, {0.4482877361, 0.2588190451},
           {0.4482877361, -0.2588190451}}], 8]
```

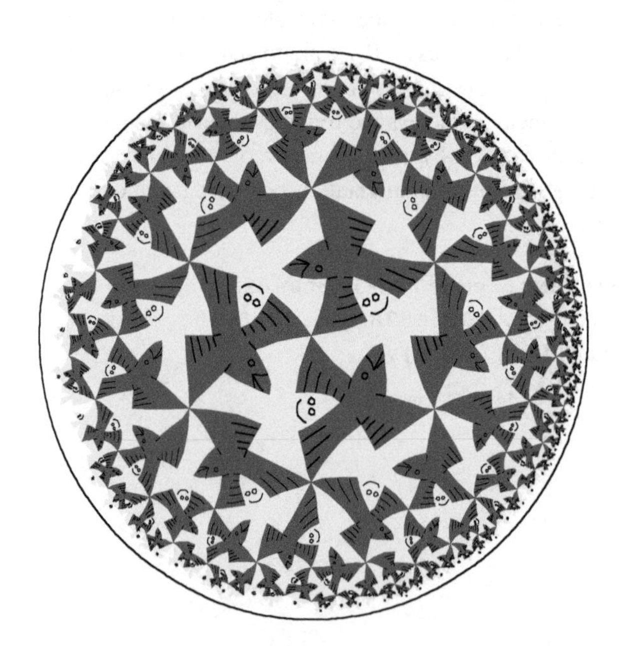

The following takes a few minutes to compute and render on a GHz-class machine:

```
hyperbolShapes[
ghost[Hue[0.15], {{0, 0}, {0.4482877361, 0.2588190451},
           {0.4482877361, -0.2588190451}}], 14]
```

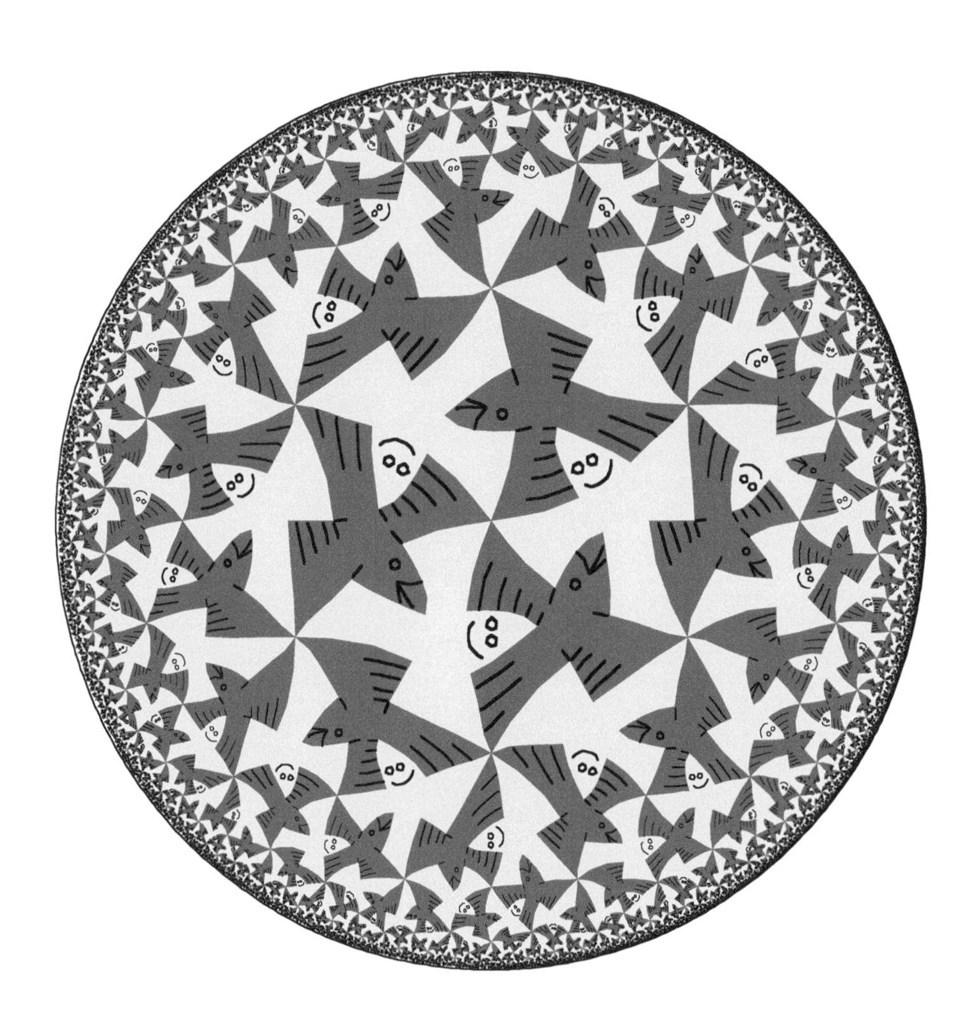

22.6 The projective representation

The geometric representation we have considered so far is not the only one of interest. A detailed discussion of the history and geometry relevant to these variations is given by Penrose (2004). The various possibilities appear to have first been realized by the Italian geometer E. Beltrami. The representation we have discussed so far, often known as the Poincaré disc, is better known as *Beltrami's conformal representation*. There is also a *Beltrami projective representation*. When working with respect to a unit radius disc, the mapping from conformal to projective is obtained by the mapping taking the point (x, y) to that given by the following function:

```
Projective[{x_, y_}] :=
 Module[{rsq = x^2 + y^2}, {x, y} * 2 / (1 + rsq)]
```

That is,

```
Projective[{x, y}]
```

$$\left\{\frac{2x}{x^2+y^2+1}, \frac{2y}{x^2+y^2+1}\right\}$$

We could re-engineer all of our tilings based on the conformal representation, but it is easier just to insert the function **Projective** at various places in our existing *Mathematica* functions. For the triangles, all we need to do is the following:

```
PhyperbolicTriangle[
triangle[color_: GrayLevel[0], {a_, b_, c_}]] :=
Join[{color},
    {Polygon[Map[Projective[#] &,
        Apply[Join,
            Map[hyperbolicLine,
                Partition[{a, b, c, a}, 2, 1]
            ]]]]}]

PhypTriangleTiling[
  shape_? ((MatchQ[Head[#], triangle]) &), n_] :=
Module[{temp},
Show[Graphics[
Map[PhyperbolicTriangle ,
Join[
Nest[(temp = #;
    Flatten[
         removeDuplicates[reflectHypTri[#]]]) &,
     {shape}, n],
    temp]]
],
AspectRatio -> Automatic,
PlotRange -> {{-1, 1}, {-1, 1}}]]
```

We do not need quite so many iterations to appear to fill out the geometry to the edge of the circle:

```
PhypTriangleTiling[triangle[Hue[0.15],
{{0, 0}, {0.2660772453, 0}, {0, 0.14062593}}], 20]
```

This figure illustrates the key difference between the conformal and projective representations. The latter conveys the impression of being drawn on a sphere. The circles implicit in the tiling in the conformal representation are now straight lines in the projective representation. Another beautiful example of this, based on the work of M.C. Escher in his woodcut *Circle Limit I*, is given in Sections 2.4 and 2.5 of Penrose (2004).

22.7 Tiling the Poincaré disc with hyperbolic squares

The methods developed above for triangles can be readily extended to squares. The basic functions are the same as for triangles, as follows, except that we simplify the colouring scheme to black and white.

```
hyperbolicSquare[
square[color_: GrayLevel[0], {a_, b_, c_, d_}]] :=
Join[{color},
    {Polygon[
        Apply[Join,
            Map[hyperbolicLine,
                Partition[{a, b, c, d, a}, 2, 1]
    ]]]}]

PhyperbolicSquare[
    square[color_: GrayLevel[0],
        {a_, b_, c_, d_}]] :=
    Join[{color},
        {Polygon[Map[Projective[#] &,
            Apply[Join,
                Map[hyperbolicLine,
                Partition[{a, b, c, d, a}, 2, 1]
                ]]]]}]

reflectHypSqu[
    square[color_: GrayLevel[0],
            {a_, b_, c_, d_}]] :=
MapThread[square,
{Table[color, {4}],
lines = Partition[{a, b, c, d, a}, 2, 1];
refl[{x__}] := (# :> hyperbolicReflection[{#, x}]) &;
MapThread[{a, b, c, d} /. (Map[refl[#1], #2]) &,
{#, Map[Complement[{a, b, c, d}, #] &, #]}
] &[lines]}]

SetAttributes[reflectHypSqu, Listable];
SetAttributes[hyperbolicSquare, Listable];
SetAttributes[PhyperbolicSquare, Listable]
```

Now we ask ourselves two questions:

(1) how many hyperbolic squares will meet at a vertex?
(2) how do we figure out in detail the coordinates of the initial hyperbolic square?

Note how odd it is that we can even ask the first question! This would not make any sense in the usual Euclidean plane. These two questions are related. We need to write down equations governing the intersection (α, α) of two circles centred at points $(0, \beta)$ and $(\beta, 0)$, each of which interesects the unit circle at a right angle, and which intersect each other at an angle of $2\pi/n$, where n is prescribed. If you do this (I leave this as a worthwhile exercise!) you find that α is governed by the quartic (effectively quadratic) equation

$$\alpha^2 + \frac{1}{4\alpha^2} = \sec\left(\frac{2\pi}{n}\right) \tag{22.11}$$

```
n = 6;
Solve[α^2 + 1 / 4 / α^2 == 1 / Cos[2 Pi / n], α]
```

$$\left\{\left\{\alpha \to \frac{1}{2}\left(-1-\sqrt{3}\right)\right\}, \left\{\alpha \to \frac{1}{2}\left(1-\sqrt{3}\right)\right\}, \left\{\alpha \to \frac{1}{2}\left(-1+\sqrt{3}\right)\right\}, \left\{\alpha \to \frac{1}{2}\left(1+\sqrt{3}\right)\right\}\right\}$$

```
N[%, 20]
```

$\{\{\alpha \to -1.3660254037844386468\}, \{\alpha \to -0.36602540378443864676\}, \{\alpha \to 0.36602540378443864676\}, \{\alpha \to 1.3660254037844386468\}\}$

The value we want is the one that is real and positive and less than one, i.e.

```
size = N[(Sqrt[3] - 1) / 2, 20]
```

0.36602540378443864676

Having worked out our starting value, we can have a look at the results. This will take some time.

```
Show[Graphics[{Circle[{0, 0}, 1],
    Map[hyperbolicSquare,
    Nest[
    Flatten[removeDuplicates[reflectHypSqu[#]]] &,
{square[{{size, size}, {-size, size},
        {-size, -size}, {size, -size}}]}, 8]]}
], AspectRatio -> Automatic]
```

Note that we can move to the projective representation just as before. This time the impact is not quite so dramatic!

```
Show[Graphics[{Circle[{0, 0}, 1],
    Map[PhyperbolicSquare,
    Nest[
    Flatten[removeDuplicates[reflectHypSqu[#]]] &,
{square[{{size, size}, {-size, size},
        {-size, -size}, {size, -size}}]}, 4]
    ]}
], AspectRatio -> Automatic]
```

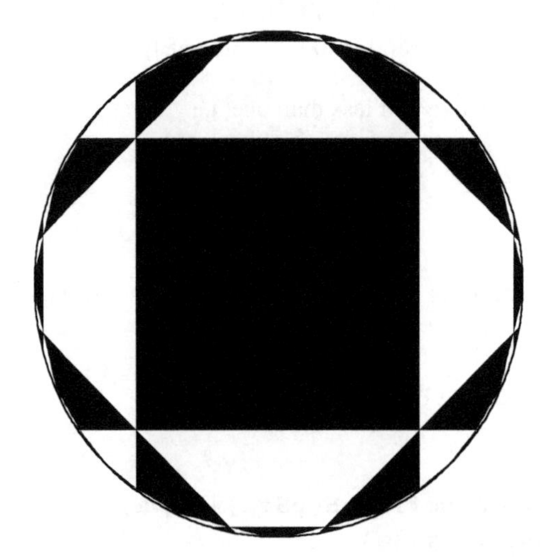

Here we have employed a two-colour (or rather black and white) representation. This does require that n is of course even. With a bit of tidying up we can semi-automate the production of graphics for other even n. Let's look at $n = 8$:

```
n = 8;
size =
  N[Select[α /. Solve[α^2 + 1 / 4 / α^2 == 1 / Cos[2 Pi / n],
       α], 0 < # < 1 &], 20][[1]];
Show[Graphics[{Circle[{0, 0}, 1],
    Map[hyperbolicSquare,
    Nest[
    Flatten[removeDuplicates[reflectHypSqu[#]]] &,
{square[{{size, size}, {-size, size},
        {-size, -size}, {size, -size}}]}, 8]
    ]}
], AspectRatio -> Automatic]
```

```
Show[Graphics[{Circle[{0, 0}, 1],
    Map[PhyperbolicSquare,
    Nest[
    Flatten[removeDuplicates[reflectHypSqu[#]]] &,
{square[{{size, size}, {-size, size},
        {-size, -size}, {size, -size}}]}, 4]
    ]}
], AspectRatio -> Automatic]
```

With a larger value of n we need to do even more reflections just to get the right basic picture at the vertices of the original square. This one, with $n = 14$, takes a while but is worth the wait!

```
n = 14;
size =
  N[Select[α /. Solve[α^2 + 1 / 4 / α^2 == 1 / Cos[2 Pi / n],
      α], 0 < # < 1 &], 20][[1]];
Show[Graphics[{Circle[{0, 0}, 1],
    Map[hyperbolicSquare,
    Nest[
    Flatten[removeDuplicates[reflectHypSqu[#]]] &,
{square[{{size, size}, {-size, size},
        {-size, -size}, {size, -size}}]}, 8]
    ]}], AspectRatio -> Automatic]
```

For larger values of n the corresponding projective picture approaches that of a square inscribed in the circle with very little edge detail. The following does not try to fill in all the edge detail as it is not visible on this scale, but you can just see the meeting of 14 squares at the vertices of the central square:

```
n = 14;
size =
  N[Select[α /. Solve[α^2 + 1 / 4 / α^2 == 1 / Cos[2 Pi / n],
       α], 0 < # < 1 &], 20][[1]];
Show[Graphics[{Circle[{0, 0}, 1],
    Map[PhyperbolicSquare,
    Nest[
    Flatten[removeDuplicates[reflectHypSqu[#]]] &,
{square[{{size, size}, {-size, size},
      {-size, -size}, {size, -size}}]}, 6]
    ]}], AspectRatio -> Automatic]
```

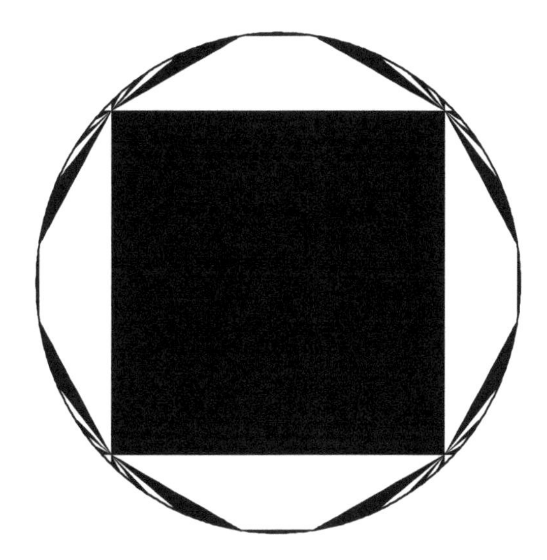

■ Working with unfilled polygons

If n is odd, we cannot use a simple two-colour scheme. A simple approach is to just display lines. In many ways the case $n = 5$ is the most interesting, especially in the projective representation.

```
hyperbolicSquareL[
square[color_: GrayLevel[0], {a_, b_, c_, d_}]] :=
Join[{color},
    {Line[
        Apply[Join,
            Map[hyperbolicLine,
                Partition[{a, b, c, d, a}, 2, 1]
    ]]]}]
```

```
PhyperbolicSquareL[
    square[color_: GrayLevel[0],
        {a_, b_, c_, d_}]] :=
    Join[{color},
        {Line[Map[Projective[#] &,
            Apply[Join,
                Map[hyperbolicLine,
                Partition[{a, b, c, d, a}, 2, 1]
                ]]]]}]


SetAttributes[hyperbolicSquareL, Listable];
SetAttributes[PhyperbolicSquareL, Listable]

n = 5;
size =
  N[Select[α /. Solve[α^2 + 1 / 4 / α^2 == 1 / Cos[2 Pi / n],
       α], 0 < # < 1 &], 20][[1]];
Show[Graphics[{Circle[{0, 0}, 1],
    Map[hyperbolicSquareL,
    Nest[
    Flatten[removeDuplicates[reflectHypSqu[#]]] &,
{square[{{size, size}, {-size, size},
        {-size, -size}, {size, -size}}]}, 7]
    ]}
], AspectRatio -> Automatic]
```

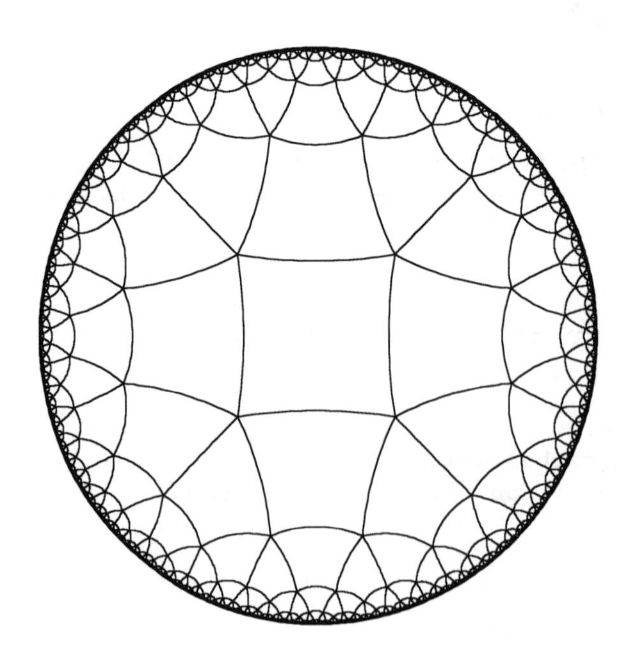

```
n = 5;
size =
  N[Select[α /. Solve[α^2 + 1 / 4 / α^2 == 1 / Cos[2 Pi / n],
       α], 0 < # < 1 &], 20][[1]];
Show[Graphics[{Circle[{0, 0}, 1],
    Map[PhyperbolicSquareL,
    Nest[
    Flatten[removeDuplicates[reflectHypSqu[#]]] &,
{square[{{size, size}, {-size, size},
        {-size, -size}, {size, -size}}]}, 6]
    ]}
], AspectRatio -> Automatic]
```

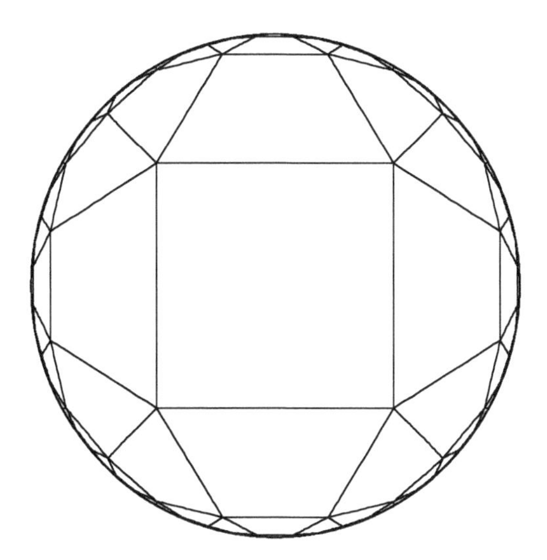

22.8 Heptagon tilings

As well as the simple tilings we have already created exactly the same techniques, can be employed to tile either the Euclidean or Hyperbolic planes using suitable shapes. As a further example, we can also tile the hyperbolic plane using regular heptagons, something that is possible on the hyperbolic plane, but not on the Euclidean plane. In the same way as we have in the previous examples we define a function that will produce a **Graphics** object from the **heptagon** object. Here we are only using an outline of a heptagon, rather than shading, and consider a case where three heptagons meet at a vertex:

```
hyperbolicHeptagon[
    heptagon[color_: GrayLevel[0],
        {a_, b_, c_, d_, e_, f_, g_}]] :=
    Join[{color},
        {Line[
            Apply[Join,
                Map[hyperbolicLine,
                Partition[
        {a, b, c, d, e, f, g, a}, 2, 1]
                ]]]}]
```

```
PhyperbolicHeptagon[
    heptagon[color_: GrayLevel[0],
        {a_, b_, c_, d_, e_, f_, g_}]] :=
    Join[{color},
        {Line[Map[Projective[#] &,
            Apply[Join,
                Map[hyperbolicLine,
                Partition[
        {a, b, c, d, e, f, g, a}, 2, 1]
                ]]]]}]
```

We also define a function that will reflect a given heptagon in each of its sides and return the resulting set of seven heptagons:

```
reflectHypHept[
    heptagon[color_: GrayLevel[0],
            {a_, b_, c_, d_, e_, f_, g_}]] :=
MapThread[heptagon,
{Table[color, {7}],
lines = Partition[{a, b, c, d, e, f, g, a}, 2, 1];
refl[{x__}] := (# :> hyperbolicReflection[{#, x}]) &;
MapThread[{a, b, c, d, e, f, g} /. (Map[refl[#1], #2]) &,
{#,
Map[Complement[{a, b, c, d, e, f, g}, #] &, #]}
] &[lines]}
]
```

Now each of these functions needs to be **Listable** in order for the procedure of continually reflecting and then the drawing process to work:

```
SetAttributes[reflectHypHept, Listable];
SetAttributes[hyperbolicHeptagon, Listable];
SetAttributes[PhyperbolicHeptagon, Listable]
```

The same technique for removing duplicated shapes is used as for removing duplicated triangles. The only change is that checking for the duplicates only stops when individual elements are more than 0.1 units apart rather than the previous 0.001 units. This arises since the ordering of vertices of these heptagons is not regular and so ordering sets of them does not mean that all shapes within a certain region will be as close to each other in the ordered set as was the case with triangles. Now we can produce a graphic that will start with a central hyperbolic heptagon (the coordinates being calculated using the same technique as for the hyperbolic triangles) and by reflection fill the whole of the unit circle:

```
Show[Graphics[{Circle[{0, 0}, 1],
    Map[hyperbolicHeptagon,
    Nest[
    Flatten[removeDuplicates[reflectHypHept[#]]] &,
    {heptagon[
    {{0.2709597367419336349, 0.1304873319336850037},
    {0.0669215284039119685 8, 0.2932023733985011144},
    {-0.1875099557726562724, 0.2351300474557987219},
    {-0.3007426187463786622, 0.},
    {-0.1875099557726562724,
        -0.2351300474557987219},
    {0.0669215284039119685 8, -0.2932023733985011144},
    {0.2709597367419336348, -0.1304873319336850037}
    }]}, 5]]}], AspectRatio -> Automatic]
```

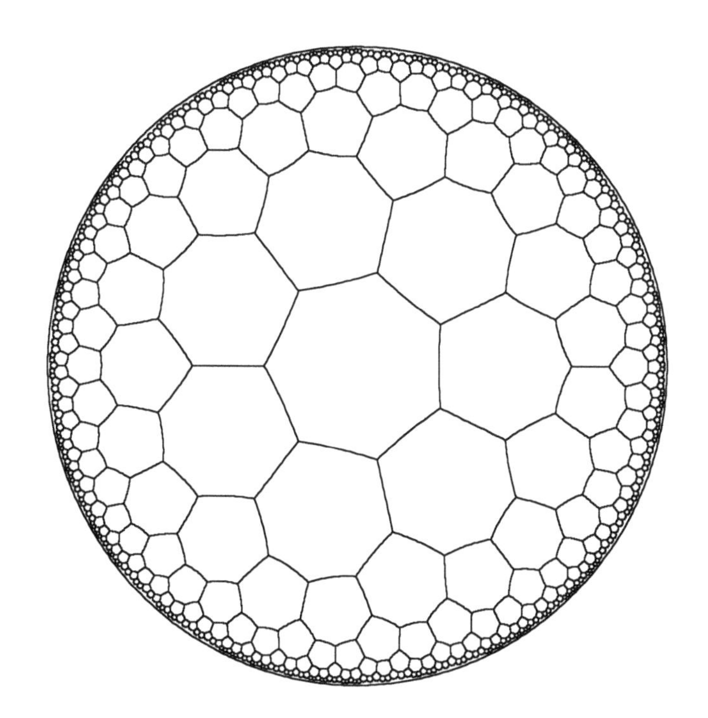

```
Show[Graphics[{Circle[{0, 0}, 1],
   Map[PhyperbolicHeptagon,
   Nest[
   Flatten[removeDuplicates[reflectHypHept[#]]] &,
   {heptagon[
   {{0.2709597367419336349, 0.1304873319336850037},
   {0.06692152840391196858, 0.2932023733985011144},
   {-0.1875099557726562724, 0.2351300474557987219},
   {-0.3007426187463786622, 0.},
   {-0.1875099557726562724,
        -0.2351300474557987219},
   {0.06692152840391196858, -0.2932023733985011144},
   {0.2709597367419336348, -0.1304873319336850037}
   }]}, 4]]}], AspectRatio -> Automatic]
```

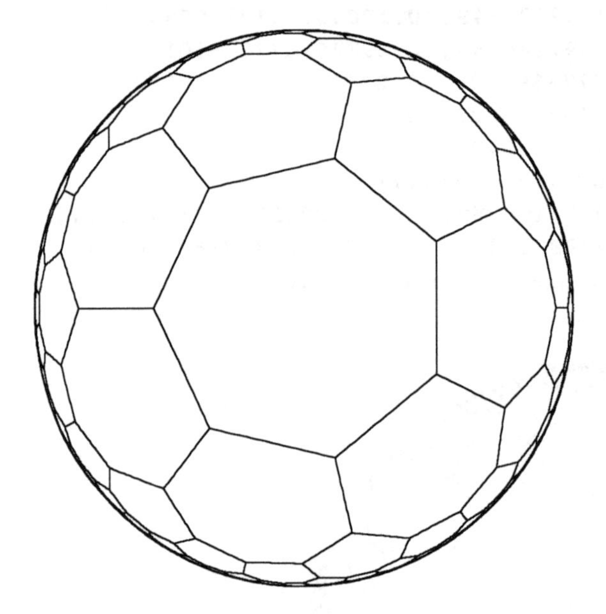

It is possible to colour this figure so that the colouring scheme is consistent across the whole figure using nine different colours. This has not been implemented algorithmically at the time of publication and any ideas on elegant methods for this would be most welcome!

22.9 ✵ The upper half-plane representation

We already know how to map the upper half-plane into the interior of the unit circle. This is just the mapping

$$w = \frac{z - i}{z + i} \tag{22.12}$$

Writing $w = x + i\,y$, we can invert this as

$$z = \frac{i\,(x + i\,y + 1)}{1 - x - i\,y} \tag{22.13}$$

We can define an analogue of the map **Projective**, which we call **ToUHP**, as follows:

```
re = ComplexExpand[Re[I (1 + x + I y) / (1 - x - I y)]];
im = Simplify[
   ComplexExpand[Im[I (1 + x + I y) / (1 - x - I y)]]];

g[x_, y_] = {re, im};
ToUHP[{x_, y_}] := g[x, y]

ToUHP[{x, y}]
```

$$\left\{-\frac{2\,y}{(1-x)^2 + y^2},\ -\frac{x^2 + y^2 - 1}{x^2 - 2\,x + y^2 + 1}\right\}$$

We now insert this function in place of **Projective**, as follows:

```
UHPhyperbolicTriangle[
triangle[color_: GrayLevel[0], {a_, b_, c_}]] :=
Join[{color},
    {Polygon[Map[ToUHP[#] &,
        Apply[Join,
            Map[hyperbolicLine,
                Partition[{a, b, c, a}, 2, 1]
            ]]]]}]

UHPhypTriangleTiling[
  shape_?((MatchQ[Head[#], triangle]) &), n_] :=
Module[{temp},
Show[Graphics[{Line[{{-5, 0}, {5, 0}}],
     Line[{{0, 0}, {0, 10}}],
Map[UHPhyperbolicTriangle ,
Join[
Nest[(temp = #;
    Flatten[
           removeDuplicates[reflectHypTri[#]]]) &,
     {shape}, n], temp]]} ], AspectRatio -> Automatic,
PlotRange → {{-5, 5}, {0, 12}}]]
```

We can now obtain another famous representation, best shown by example, with the lower boundary also shown.

```
UHPhypTriangleTiling[triangle[Hue[0.15],
{{0, 0}, {0.2660772453, 0}, {0, 0.14062593}}], 40]
```

Exercises

22.1 ✵ Re-visualize all these tilings by mapping the tilings of the unit disc into the upper half-plane, using a suitable Möbius transformation.

22.2 ✵ Investigate the transformation of other tilings into the projective representation.

22.3 ✵ Investigate the tiling of the unit disc by n-gons for values of $n > 5$.

Comment: the author recognizes that his own contributions here are somewhat ad hoc! Anyone interested in developing a systematic and optimized package to do these kinds of graphics, for general n-gons, unfilled or filled, general colouring schemes and conformal, projective or half-plane representations, is encouraged to do so and contact the author!

23 Physics in three and four dimensions I

Introduction

When we come to explore real dimensions greater than two, matters become considerably more interesting. Indeed, in my own undergraduate studies, the question as to how to solve Laplace's equation in three or more dimensions using methods analogous to those presented here went unanswered, and remains unanswered in most undergraduate curricula the world over. I did not see the answer until my postgraduate studies, studying twistor methods in Oxford, and did not fully understand many of the geometrical aspects until my own post-doctoral work on twistor descriptions of minimal surfaces and strings. However, this author at least is convinced that many of the concepts are easily understood using only the elementary complex analysis already presented here, and this chapter is in part an attempt to get the message across in such a fashion. Another goal of this chapter and the subsequent one is to persuade some of you that, as well as being a basis for research in fundamental theoretical physics, there are also some interesting problems in basic and very applied mathematics that might well be solved with such methods, if only more people worked on it!

In some ways the presentation is easier if we make the jump straight to four dimensions, and treat the relativistic case. Results for three dimensions can then be obtained by constraining matters to a hyperplane $t = 0$. This is also more elegant as our starting point is then a detailed investigation of the 'celestial spehere', and the interpretation of Möbius transforms as relativistic Lorentz transformations, which was promised earlier, then emerges naturally. So we shall dwell at length on a relativistic interpretation of stereographic projection to begin with. This is all very necessary, as clearly it is no longer possible to associate real and complex spaces through simple associations such as

$$\omega = x + iy \tag{23.1}$$

as there as now a third space coordinate, which we shall term z, and a time coordinate, t. The relativistic interpretation of stereographic projection is in fact the key to understanding how to make the jump to four dimensions. In this chapter we shall avoid the use of z as denoting a complex number, for obvious reasons! One might wonder if further extensions of the number system are needed to deal with three and four dimensions, but it will turn out (just as for dealing with polynomials) that complex numbers are enough!

The discussion here is intended to be relatively informal and accessible to anyone who has progressed thorugh the basic material on complex analysis discussed previously in this text. In particular, familiarity with the material of Chapter 16 is assumed. However, there is a large body of advanced work in mathematical physics and pure mathematics related to these topics. A good place to go for a thorough discussion and a detailed exposition of spinor theory is that of Penrose and Rindler (1984a, b), the material of which is the basis of a large portion of this chapter.

23.1 Minkowski space and the celestial sphere

We shall consider Minkowski space, $\mathbb{M}$, as the vector space represented by four-dimensional position vectors

$$X^\mu = \{ct, x, y, z\} \tag{23.2}$$

where $\mu = 0, 1, 2, 3$, and c is the speed of light, equipped with an inner product, or 'metric' $\eta_{\mu\nu} = \text{diag}(+1, -1, -1, -1)$, so that the length-squared of a vector, be it a position vector or a displacement vector, is,

$$X^2 = \eta_{\mu\nu} X^\mu X^\nu = c^2 t^2 - x^2 - y^2 - z^2 \tag{23.3}$$

We use the summation convention introduced by Einstein, which states that the appearance of a *repeated* index implies *summation* over its possible values. A vector X is said to be *timelike* if has positive length, *null* if it has zero length, and *spacelike* if it has negative length. This convention, that ordinary Euclidean three-vectors have negative length, may seem rather bizarre to a beginner in relativity, but it is just a convention, and does avoid many irritating factors of -1 later. The null vectors are especially important, as they represent the directions of light rays. If we fix an origin O in $\mathbb{M}$, the set of null position vectors with origin at O is given by the family of points satisfying

$$\underline{r}^2 = x^2 + y^2 + z^2 = c^2 t^2 \tag{23.4}$$

which represent the locus of points moving at speed c towards or away from the origin. If we fix the time-scale (so that we consider only the directions of vectors based at the origin) so that

$$|ct| = 1 \tag{23.5}$$

we get two copies of the unit sphere characterized by

$$x^2 + y^2 + z^2 = 1 \tag{23.6}$$

The sphere with $ct = +1$ is denoted S^+ and that with $ct = -1$ is denoted S^-. These two spheres represent the directions of light rays emanating from the origin, in the case of S^+, and arriving at the origin, in the case of S^-. For this reason S^- is often called the *celestial sphere*. It represents the field of vision for an observer located at O – everything that he or she sees is represented in its direction by a point on S^-. This is very important for our understanding of how our observer's view of the world shifts when they acquire a velocity.

23.2 Stereographic projection revisited

The key to using complex variables in a relativistic setting is to associate the spheres S^+ and S^- with the Riemann sphere and the Argand plane, by stereographic projection. We shall not go into detail on the geometrical aspects of this, as an algebraic treatment will be more immediately useful. Consider first S^+. We let, for points on S^+ only,

$$w = \frac{x + iy}{1 - z} \tag{23.7}$$

The modulus-squared of w is given by

$$w\overline{w} = \frac{x^2 + y^2}{(1 - z)^2} = \frac{1 - z^2}{(1 - z)^2} = \frac{1 + z}{1 - z} \tag{23.8}$$

We can now invert the relationship to obtain

$$x = \frac{w + \overline{w}}{1 + w\overline{w}}; \quad y = \frac{w - \overline{w}}{i(1 + w\overline{w})}; \quad z = \frac{w\overline{w} - 1}{1 + w\overline{w}} \tag{23.9}$$

Note that the point at infinity is given by the North pole ($z = 1$) of the sphere S^+. We can also use polar coordinates in the form

$$w = e^{i\phi} \cot\left(\frac{\theta}{2}\right) \tag{23.10}$$

What do we do about S^-? There is a natural mapping between S^+ and S^- known as the antipodal mapping. A light ray passing through O and arriving at the point (x, y, z) on S^+ must have passed through the point $(-x, -y, -z)$, on S^- so that the natural mapping is to associate antipodal points on the two spheres. On the complex coordinates, the antipodal map is given by

$$w \longrightarrow -\frac{1}{\overline{w}} \tag{23.11}$$

so that the polar coordinate representation is

$$w = -e^{i\phi} \tan\left(\frac{\theta}{2}\right) \tag{23.12}$$

23.3 Projective coordinates

There is a certain awkwardness about the coordinate w introduced in the previous section, as we need to allow for the point at infinity. This can be managed more elegantly by introducing coordinates (ζ, η) related to w by

$$w = \frac{\zeta}{\eta} \tag{23.13}$$

This pair are *projective* or *homogeneous* coordinates for the sphere(s) – the pairs (ζ, η) and $(\lambda\zeta, \lambda\eta)$ represent the same point, provided $\lambda \in \mathbb{C}$ and $\lambda \neq 0$. The point at infinity is then given by setting $\eta = 0$. These projective coordinates also play another role. To see this most straightforwardly, we first make some adjustments to the scaling of the relations characterizing stereographic projection. If we let $b = \zeta\overline{\zeta} + \eta\overline{\eta}$, then Eq. (23.9) can be recast as

$$x = \frac{\zeta\overline{\eta} + \eta\overline{\zeta}}{b}; \quad y = \frac{\zeta\overline{\eta} - \eta\overline{\zeta}}{ib}; \quad z = \frac{\zeta\overline{\zeta} - \eta\overline{\eta}}{b}; \tag{23.14}$$

The persistent appearance of b is an inconvenience, and so, rather than working with a sphere of radius unity, we shall work with a sphere of radius $b/\sqrt{2}$. This gives us the four relations

$$\begin{aligned} ct &= \frac{1}{\sqrt{2}}(\zeta\overline{\zeta} + \eta\overline{\eta}); \; x = \frac{1}{\sqrt{2}}(\zeta\overline{\eta} + \eta\overline{\zeta}); \\ y &= \frac{1}{i\sqrt{2}}(\zeta\overline{\eta} - \eta\overline{\zeta}); \quad z = \frac{1}{\sqrt{2}}(\zeta\overline{\zeta} - \eta\overline{\eta}); \end{aligned} \tag{23.15}$$

This leads directly to an elegant matrix characterization of this relationship, in the form

$$\frac{1}{\sqrt{2}}\begin{pmatrix} ct+z & x+iy \\ x-iy & ct-z \end{pmatrix} = \begin{pmatrix} \zeta\overline{\zeta} & \zeta\overline{\eta} \\ \eta\overline{\zeta} & \eta\overline{\eta} \end{pmatrix} = \begin{pmatrix} \zeta \\ \eta \end{pmatrix}(\overline{\zeta}, \overline{\eta}) = S^A \overline{S}^{A'} \tag{23.16}$$

where S^A is the two-component complex 'vector'

$$S^A = \begin{pmatrix} \zeta \\ \eta \end{pmatrix} \tag{23.17}$$

and the index A, labelling the rows, runs from zero to one. (This is just a convention linked to the fact that the time coordinate is often the zero component of the four-dimensional vector X^μ.) The index A' is attached to the complex conjugate vector and labels the columns of the matrices.

These two-component vectors play a very special role in relativistic physics and are given a special name – they are called *spinors*. (Physicists often refer to *four-component*, or *Dirac*, spinors that are made up of a pair of the more fundamental ones discussed here.) What we have done is to show that spinors, up to a scale factor, characterize the (real) null directions emanating from a point. When we attach the scale factor, a two-component spinor characterizes a null (real) four-vector.

We have assumed that the coordinates (ct, x, y, z) are themselves real. If we allow them to be complex we can still use these algebraic characterizations of null four-vectors. Inspection of the matrix

$$X^{AA'} = \frac{1}{\sqrt{2}}\begin{pmatrix} ct+z & x+iy \\ x-iy & ct-z \end{pmatrix} \tag{23.18}$$

shows that it corresponds to a zero-length vector if and only if the determinant of $X^{AA'}$ is zero. This occurs if and only if the rows are linearly dependent, and, as there are only two rows, this means that this matrix can be factorized in the form

$$X^{AA'} = S^A \pi^{A'} \tag{23.19}$$

for some pair of spinors S and π. In the special case when X corresponds to a real vector π is the complex conjugate of S, but in the complex setting this is relaxed.

You may well ask, what about the case when X^μ is not zero-length? Not all points arise as points separated from the origin by a null direction. We shall return to this presently. But already we have enough mathematics at our disposal to investigate some very interesting effects to do with the appearance of objects to an observer that is moving rapidly, and the link between Möbius transforms and relativistic physics. We begin with the latter.

23.4 Möbius and Lorentz transformations

The importance of spinors and projective coordinates for the sphere emerges when we consider how to interpret the set of linear transformations that act on spinors. Consider the linear transformation (remember that summation over $B = 0, 1$ is implicit)

$$S^A \to \tilde{S}^A = \Lambda^A_B S^B \tag{23.20}$$

where the matrix Λ is given by four complex numbers:

$$\Lambda = \begin{pmatrix} a & b \\ c & d \end{pmatrix} \tag{23.21}$$

If we write this our in coordinates, we have

$$\begin{aligned} \zeta \to \tilde{\zeta} &= a\zeta + b\eta \\ \eta \to \tilde{\eta} &= c\zeta + d\eta \end{aligned} \tag{23.22}$$

and on the single non-projective coordinate w this induces the mapping

$$w \to \tilde{w} = \frac{aw + b}{cw + d} \tag{23.23}$$

which is our old friend, the Möbius transformation. Let's explore this new matrix interpretation of such mappings. Note first that if we rescale all four coordinates (a, b, c, d), the same Möbius transformation is obtained. So we shall rescale Λ so that $\det \Lambda = 1$, i.e. the *unimodular* condition applies:

$$ad - bc = 1 \tag{23.24}$$

The set of complex matrices of this form is given a name, it is $SL(2, \mathbb{C})$, and algebraically gives a multiplicative group. So far we have thought about it only in terms of its action on the two-component spinors. But it also acts on matrices of the form given by Eq. (23.18), and indeed this action will have the same form whether or not $X^{AA'}$ is a null vector representable as a product of spinors. In terms of matrices, the mapping induced on X is

$$X \to \tilde{X} = \Lambda X \Lambda^* \tag{23.25}$$

where Λ^* is the transposed complex conjugate. From this relation we can deduce the action on the determinants of X:

$$\det X \to \tilde{\det} X = \det(\Lambda X \Lambda^*) = \det \Lambda \det X \det \Lambda^* = \det X \tag{23.26}$$

So the determinant is invariant. For a general X this determinant is, from Eq. (23.18), one half of

$$c^2 t^2 - x^2 - y^2 - z^2 \tag{23.27}$$

which is precisely the relativistic distance function. This, by definition, is a Lorentz transformation. (It is in fact a special form of Lorentz transformation that preserves the sense of time – these are called *restricted* Lorentz transformations.)

We have established that unimodular Möbius transformations correspond to elements of SL(2, $\mathbb{C}$), each of which generate a restricted Lorentz transformations. The relationship is not quite 1:1, because both Λ and $-\Lambda$ give the same Lorentz transformation. This correspondence is at the heart of a complex-analytic treatment of relativistic physics. One immediate question is to ask: What does our knowledge of Möbius transformations tells us about relativity?

23.5 The invisibility of the Lorentz contraction

One of the most immediate results from this analysis concerns how rapidly moving obsevers perceive the world compared to observers who are stationary. Many elementary discussions of relativity contain a confusing description of the 'Lorentz Contraction' , with objects appearing to be squashed in the direction of motion of the observer. A favourite construction is to show what Einstein saw on his tram ride to work, with a tram moving close to the speed of light, and the houses on the side of the street looking rather narrower than usual! Such discussions completely confuse the business of going out and measuring obects, usually with 'rods' and 'clocks', with the calculation of what happens to light rays arriving in the eye of an observer. The latter is a rather more subtle affair.

That there is something rather incomplete with the simple contraction picture can be seen by recalling what happens when you drive (preferably rather slowly) into an area of falling snow. If you are stationary, the snow appears to fall from directly above. When you drive at some speed, the snow appears to arrive on your windscreen at a non-zero angle to the vertical. It is possible, as this author recalls, to drive fast enough that the snow appears to be travelling horizontally, though I must emphatically discourage others from trying to reproduce this effect. What this crude physical analogy is telling us is that objects *appear to be moved to in front of us,* but it does not tell us how their apparent shape changes. To get a grip on this we need to exploit the concept of the celestial sphere, and the effect of Lorentz transformations on the points on this sphere.

This issue was not properly appreciated until quite recently – see Terrell (1959) and Penrose (1959). In particular, we shall look at what happens to our perceptions of spheres. To a stationary observer, a stationary sphere will present a circular outline on that observer's celestial sphere. Let's explore this and see what happens when that observer accelerates.

The first observation we need is that if an object presents a circular outline on $S^{\pm}$, then its outline corresponds also to a circle in the Argand plane. To see this, note that a circle outlined on $S^{\pm}$ corresponds to the intersection of a plane with $S^{\pm}$. So let the equation of the plane be characterized by the direction cosines (l, m, n) of its unit normal. Then the intersection of $S^{\pm}$ with this plane is given by those points on $S^{\pm}$ satisfying

$$lx + my + nz = d \tag{23.28}$$

with $d^2 < 1$. If we scale, for example, the sphere S^{+} so that $ct = 1$, then we can parametrize points on it by Eq. (23.9), to obtain

$$l(w + \overline{w}) - im(w - \overline{w}) + n(w\overline{w} - 1) = d(w\overline{w} + 1) \tag{23.29}$$

and on reorganization this gives

$$(n - d)w\overline{w} + (l - im)w + (l + im)\overline{w} = \mu + d \tag{23.30}$$

We have already studied this equation – provided $n \neq d$ it is the equation of a circle, otherwise it is the equation of a straight line. Similar equations are obtained on S^{-} by the use of the antipodal map. We already know also that under a Möbius transformation, the family of equations given by Eq. (23.30) are transformed into the same family. Since S^{-} has the interpretation as the field of view of an observer at the origin, objects with a circular outline, such as a sphere, will still appear spherical to an observer travelling rapidly but instantaneously coincident with the origin, i.e. spheres do *not* appear to be squashed in the direction of motion.

This analysis can be taken one step further. We already know that Möbius transformations are conformal maps of the Argand plane. It is, in fact, also the case that stereographic projection between the Riemann sphere and the Argand plane is conformal. So if we compose these observations, if follows that the shape of a *small* object (characterized by the local angles between elements of its outline on S^{-}) is also left invariant by a Lorentz transformation.

To appreciate that stereographic projection is conformal requires a basic appreciation of the relationship between metrics and angles – we have not spelled this out before now. First of all, the notion of distance in Euclidean three-space is given by the simple Pythagorean metric

$$ds^2 = dx^2 + dy^2 + dz^2 \tag{23.31}$$

We need to know what form this takes when restricted to the Riemann sphere, for example S^{+}. It is straightforward to show (see Exercise 23.1) that if we use the parametrization given by Eq. (23.9), then

$$ds^2 = \phi(w, \overline{w})\, dw d\overline{w} \tag{23.32}$$

where

$$\phi = \frac{4}{(1 + w\overline{w})^2} \tag{23.33}$$

So the distance function on S^+ is a multiple of the distance function on the Argand plane, the latter being just $dwd\overline{w}$. That is, stereographic projection effects a locally *isotropic* expansion of the neighbourhood of points, and hence is angle-preserving.

Thus far we have established two important observations. For all observers instantaneously co-located

(1) the apparent shape of small objects is invariant under changes of observer velocity;
(2) the circular outline of an object of arbitrary size is also invariant under changes of observer velocity.

These two results, due, respectively, to Terrell (1959) and Penrose (1959), constitute what is known as the 'invisibility of the Lorentz contraction'. As we have seen, these results follow directly from the conformality and circle-preserving properties of the Möbius (Lorentz) transformation. So now we know that objects do *not* get anisotropically squashed. Next we need to get a better grip on the detail of what does happen.

23.6 Outline classification of Lorentz transformations

We now know that a Lorentz transformation is one of the family of mappings

$$w \to \tilde{w} = M(w) = \frac{aw + b}{cw + d}$$

with the property that $ad - bc = 1$. We have already discussed some of the properties of these mappings in Chapter 16. We shall look again now with the fresh eye of relativity. The most straightforward way of analysing the mappings is to classify them on the basis of their fixed points. If we set $\tilde{w} = w$ in this last equation, we obtain a quadratic equation for w, which in general has two, possibly coincident, roots. So we can consider these two cases separately.

■ Distinct Roots

It is immensely helpful to analyse this case by picking appropriate coordinates. We are assuming that there are two distinct complex numbers w_1 and w_2 that are left invariant by the Möbius transformation under consideration. These correspond to two distinct points on each of the spheres S^+ or S^-. As discussed in Section 16.8, we can construct a second Möbius transformation, say G, to send each of these two points to any given new pair of points. We shall choose the new pair of points to be zero and infinity. So let

$$\xi = G(w); \quad \tilde{\xi} = G(\tilde{w}) \tag{23.34}$$

and it follows that

$$\tilde{\xi} = G(\tilde{w}) = G(M(w)) = G(M(G^{-1}(\xi))) \tag{23.35}$$

It also follows that the mapping

$$\xi \to \tilde{\xi} = N(\xi) = G \circ M \circ G^{-1}(\xi) \tag{23.36}$$

is a Möbius transformation, as the transformations form a compositional group, and it has fixed points at zero and infinity. What is the nature of such a transformation, expressed by N? Let us now suppose that we have written this new mapping in the form

$$N(\xi) = \frac{A\xi + B}{C\xi + D}$$

Inspection of the formula for the fixed points reveals that there is a fixed point at infinity if and only if $C = 0$. So the possession of a fixed point at infinity is equivalent to being able to write N in the form

$$N(\xi) = A'\xi + B'$$

for complex contants A', B'. Clearly the origin is then also a fixed point if and only if $B' = 0$. Thus we obtain the following result. A Möbius/Lorentz transformation with a pair of distinct fixed points is equivalent (under a change of coordinates expressed by Eq. (23.35)) to the simple complex rescaling:

$$\xi \to \tilde{\xi} = A'\xi \tag{23.37}$$

■ Equal Roots

We have also done most of the work to analyse the case of a unique fixed point. We apply a transformation of the form of Eq. (23.35) to place the single fixed point at infinity. So the transformation must be equivalent to

$$\xi \to \tilde{\xi} = A'\xi + B' \tag{23.38}$$

Now how do we ensure that there are no other fixed points? If we write down the equation

$$\xi = A'\xi + B' \tag{23.39}$$

we see that there is a second finite solution unless $A' = 1$. So the case of equal roots is equivalent to

$$\xi \to \tilde{\xi} = \xi + B' \tag{23.40}$$

which is a simple complex translation. This classification will be sufficient for our needs here. A much more detailed exposition is given by Needham (1997).

■ Physical interpretations

What we have shown is that any Möbius/Lorentz transformation is equivalent to either a multiplication of the form of Eq. (23.37) or a translation of the form of Eq. (23.40). What do these observation mean physically? We need to do two things: (a) interpret the imposition of infinity, and optionally, zero, as the fixed points, and (b) interpret the transformations then given by Eqs. (23.37) and (23.40).

Suppose first that we impose zero and infinity as fixed points. This means that both B and C are zero. The matrix of the transformation with respect to these coordinates is then just

$$\Lambda = \begin{pmatrix} \lambda & 0 \\ 0 & \lambda^{-1} \end{pmatrix} \tag{23.41}$$

where $A' = \lambda^2$, $\lambda \in \mathbb{C}$, and we observe that acting on the corresponding spatial coordinates, this induces a mapping via Eq. (23.25), which is particularly simple. The directions given by $ct + z$ and $ct - z$ are left invariant, and there is no mixing of these coordinates with the x, y coordinates. In detail, we have, if

$$\lambda = \rho e^{i\psi} \tag{23.42}$$

then (see Exercise 23.2 for the details)

$$\begin{aligned} c\tilde{t} + \tilde{z} &= \rho^2 (ct + z) \\ c\tilde{t} - \tilde{z} &= \frac{1}{\rho^2}(ct - z) \\ \tilde{x} + i\tilde{y} &= e^{2i\psi}(x + iy) \end{aligned} \tag{23.43}$$

So the use of zero and infinity as fixed points corresponds to making the choice of $ct \pm z$ as an invariant pair of directions (they are not absolutely invariant, being rescaled by reciprocal factors). The form of the set of equations (23.43) suggests that we split this case into two further special cases. The first is where $\rho = 1$. The t, z pair are then left invariant, and we observe that the effect of the transformation is a rotation of the x, y plane by an angle 2ψ. The second case is perhaps a little more interesting. We set ψ to zero. The x, y pair are then left invariant, and we have new formulae expressing the mapping from the pair z, t to the new pair z', t'. This is most easily expressed by introducing new and rather suggestive parameters as follows. First we set

$$\rho = e^{\mu} \tag{23.44}$$

and define $\gamma \geqslant 1$ by

$$\gamma = \cosh(2\mu) \tag{23.45}$$

We now introduce a speed parameter v defined up to a $\pm$ sign by the equation

$$\gamma = \frac{1}{\sqrt{1 - \frac{v^2}{c^2}}} \tag{23.46}$$

and we fix its sign by requiring that

$$\sinh(2\mu) = \frac{\gamma v}{c} \tag{23.47}$$

It then follows from Eq. (23.43) (see Exercise 23.3 for details) that

$$\begin{aligned} \tilde{t} &= \gamma\left(t + \frac{zv}{c^2}\right) \\ \tilde{z} &= \gamma(z + vt) \end{aligned} \tag{23.48}$$

This is perhaps the most famous form of the *Lorentz transformation*. The plane given by setting $\tilde{z} = 0$ is equivalent to the constraint $z = -vt$. The effects of time dilation and the Lorentz contraction can now be derived with care from these relations. However, our interest is in the *appearance of objects*. This is most easily expressed by use of the Möbius transformation

$$\xi \to \tilde{\xi} = \rho^2\, \xi \tag{23.49}$$

What is ρ^2 expressed in terms of v? This is now easy. We have

$$A' = \rho^2 = e^{2\mu} = \cosh(2\mu) + \sinh(2\mu) = \gamma\left(\frac{v}{c} + 1\right) = \left(\frac{1 + v/c}{1 - v/c}\right)^{1/2} \tag{23.50}$$

Now all of this analysis has been worked out on coordinates adapted to S^+. The corresponding action on S^- can be deduced by applying the antipodal map to Eq. (23.49), which gives the effect as the real rescaling factor

$$w \to \tilde{w} = \frac{w}{\rho^2} \tag{23.51}$$

Now we use the polar coordinate representation for S^-, i.e.

$$w = -e^{i\phi} \tan\left(\frac{\theta}{2}\right) \tag{23.52}$$

to deduce that the effect of the transformation is just

$$\tan\left(\frac{\theta'}{2}\right) = \left(\frac{1 - v/c}{1 + v/c}\right)^{1/2} \tan\left(\frac{\theta}{2}\right) \tag{23.53}$$

This is a very important formula. It tells us how the apparent direction of a light ray changes as we adjust the velocity of the observer at the origin. It is sometimes called the *angular abberation formula*. This relation also helps us to fix the sense of v in our minds – if a stationary obsever perceives a star or other object at an angle θ to the North pole, then an obsever heading *to* the North pole with speed $v > 0$ perceives that same object to be located at an angle θ' to the North pole, at that instant when he/she is instantaneously co-located with the stationary observer.

The second case – that of one fixed point at infinity – is left as an exercise for you to explore the details. For now we wish to explore the effects of Eq. (23.53) in a little more detail.

23.7 Warping with *Mathematica*

In order to implement this, we shall simplify the graphics somewhat, and represent all objects by collections of points. This makes the drawing a little inefficient, but makes the implementation of the conformal map very trivial. We shall consider an imaginary race of aliens called The Grob, who travel around in cubical spaceships. We want to find out what their spaceship looks like if we view it from an observer that is initially stationary with respect to the Grob ship, and then from an observer travelling at 'Warp Factor' 9. With due deference to certain other cultural interpretations of this term, for the purposes of this discussion Warp Factor 9 means 0.9 times the speed of light. (The management of observers equipped with trans-warp drive is beyond the scope of this discussion.) We can also consider more boring speeds such as Warp Factor 5.

First of all we shall build our cubes:

```
xside[n_] := Table[{k / n, 0, 0}, {k, 0, n}];
yside[n_] := Table[{0, k / n, 0}, {k, 0, n}];
zside[n_] := Table[{0, 0, k / n}, {k, 0, n}];

move[origin_, points_] := Map[(origin + #) &, points]

PointCube[origin_List, n_] :=
    Module[{xpts = xside[n],
            ypts = yside[n],
            zpts = zside[n]},
            Flatten[{move[origin, xpts],
                move[origin + {0, 1, 0}, xpts],
                move[origin + {0, 0, 1}, xpts],
                move[origin + {0, 1, 1}, xpts],
                move[ origin, ypts],
                move[ origin + {1, 0, 0}, ypts],
                move[ origin + {1, 0, 1}, ypts],
                move[ origin + {0, 0, 1}, ypts],
                move[origin, zpts],
          move[origin + {1, 0, 0}, zpts],
                move[origin + {0, 1, 0}, zpts],
                move[origin + {1, 1, 0}, zpts]
             }, 1]]
```

Let's have two cubes, in different locations:

```
testa = PointCube[1.1 * {1, 1, 1.5}, 40];
testb = PointCube[1.1 * {-1, -1, 0}, 40];
```

Now we define a function that projects the points onto the unit sphere. The idea is that our observer will sit in the centre of the nose cone of a spaceship, and the cone has an observation window that is the cap of a sphere. The observer therefore sees the outside world entirely in terms of a 'painting' on his window – this window forms a section of his or her celestial sphere. We will draw the window also as part of a sphere – we actually do the painting slightly inside. The following function takes points in space and paints them onto the inside surface of the observation window:

```
ProjectToUnitSphere[{x_, y_, z_}] :=
 Module[{r = Sqrt[x^2 + y^2 + z^2], θ, ϕ},
       θ = ArcCos[z / r];
       ϕ = ArcCos[x / (r * Sin[θ])];
0.999 * {Sin[θ] Cos[ϕ], Sin[θ] Sin[ϕ], Cos[θ]}]
```

Here is the painting carried out with the angular abberation factor derived above:

```
AbberateProjectToUnitSphere[{x_, y_, z_, warp_}] :=
 Module[{r = Sqrt[x^2 + y^2 + z^2], θ, ϕ, newtheta,
   abber = Sqrt[((1 - warp / 10) / (1 + warp / 10))]},
       θ = ArcCos[z / r];
       ϕ = ArcCos[x / (r * Sin[θ])];
       newtheta = 2 * ArcTan[abber * Tan[θ / 2]];
0.999 * {Sin[newtheta] Cos[ϕ],
    Sin[newtheta] Sin[ϕ], Cos[newtheta]}]

newcubea = Map[ProjectToUnitSphere[#] &, testa];

newcubeb = Map[ProjectToUnitSphere[#] &, testb];

abbernewcubea[warp_] :=
  Map[AbberateProjectToUnitSphere[Append[#, warp]] &,
   testa];

abbernewcubeb[warp_] :=
  Map[AbberateProjectToUnitSphere[Append[#, warp]] &,
   testb];
```

In order to calculate the graphics effectively, we shall load a package that makes the computation of scaled coordinates slightly easier. This is from Wickham-Jones (1994). We shall exhibit the relevant settings, however – you do not need to get this package, but it is very useful to have, and is available from `http://library.wolfram.com` (if you get a warning about a **CrossProduct** object on loading this package ignore it).

```
Needs["ExtendGraphics`View3D`"]
```

```
?ViewPointFromUser
```

```
ViewPointFromUser[ pt, plotrange, boxratios]
  returns the ViewPoint that corresponds  to the point in
  user coordinates pt, with PlotRange prng and BoxRatios box.
```

We shall deal with a cap that is part of the hemisphere:

```
MyPlotRange = {{-1.1, 1.1},{-1.1, 1.1},{0.4, 1.1}};
zwidth = MyPlotRange[[3,2]] - MyPlotRange[[3,1]];
MyBoxRatios = {2.2, 2.2, zwidth};
MyViewPoint = ViewPointFromUser[{0,0,0.0},
MyPlotRange, MyBoxRatios];
```

Here is the actual data:

```
{MyBoxRatios, MyViewPoint}
```

$$\begin{pmatrix} 2.2 & 2.2 & 0.7 \\ 0. & 0. & -0.340909 \end{pmatrix}$$

Let's draw the observation window with these settings:

```
HemiSphere = ParametricPlot3D[
  1.1 * {Sin[θ] Cos[ϕ], Sin[θ] Sin[ϕ], Cos[θ]},
  {θ, 0, Pi / 2}, {ϕ, 0, 2 Pi},
        Boxed -> False, BoxRatios -> MyBoxRatios,
  Axes -> False, Shading -> False,
        PlotRange -> MyPlotRange,
  ViewPoint -> MyViewPoint,
        PlotPoints -> {10, 30}]
```

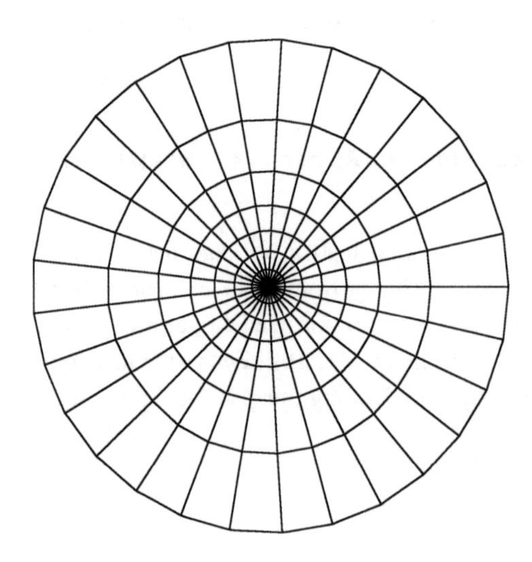

Now we shall view this window painted with the external scene:

```
baseplota = Show[Graphics3D[Map[Point, newcubea],
   Boxed -> False,  BoxRatios -> MyBoxRatios,
   Axes -> False, Shading -> False,
   PlotRange -> MyPlotRange, ViewPoint -> MyViewPoint,
            DisplayFunction -> Identity]]

baseplotb = Show[Graphics3D[Map[Point, newcubeb],
   Boxed -> False,  BoxRatios -> MyBoxRatios,
   Axes -> False, Shading -> False,
   PlotRange -> MyPlotRange, ViewPoint -> MyViewPoint,
            DisplayFunction -> Identity]]
```

Here is the zero velocity scene. There are limitations in the way *Mathematica* projects this, but you certainly perceive the cubes as having straight sides. This will change when we boost the observer's speed.

```
Show[HemiSphere, baseplota, baseplotb]
```

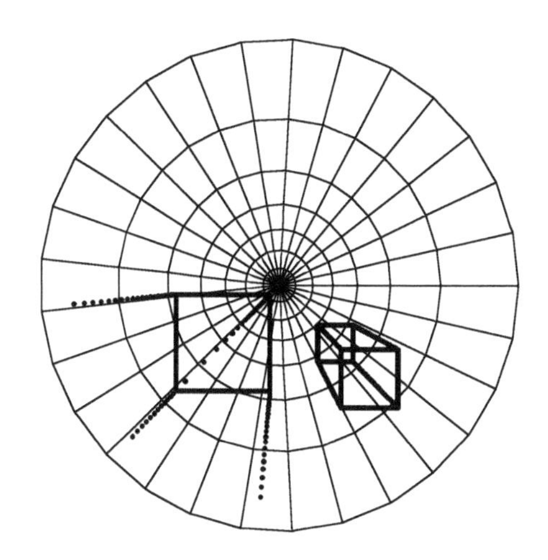

```
abberplota[warp_] :=
 Show[Graphics3D[Map[Point, abbernewcubea[warp]],
   Boxed -> False,  BoxRatios -> MyBoxRatios,
   Axes -> False, Shading -> False,
   PlotRange -> MyPlotRange, ViewPoint -> MyViewPoint,
   DisplayFunction -> Identity]]

abberplotb[warp_] :=
 Show[Graphics3D[Map[Point, abbernewcubeb[warp]],
   Boxed -> False,  BoxRatios -> MyBoxRatios,
   Axes -> False, Shading -> False,
   PlotRange -> MyPlotRange, ViewPoint -> MyViewPoint,
   DisplayFunction -> Identity]]
```

Now let's look at the same scene from an observer moving at half the speed of light:

```
Show[HemiSphere, abberplota[5], abberplotb[5]]
```

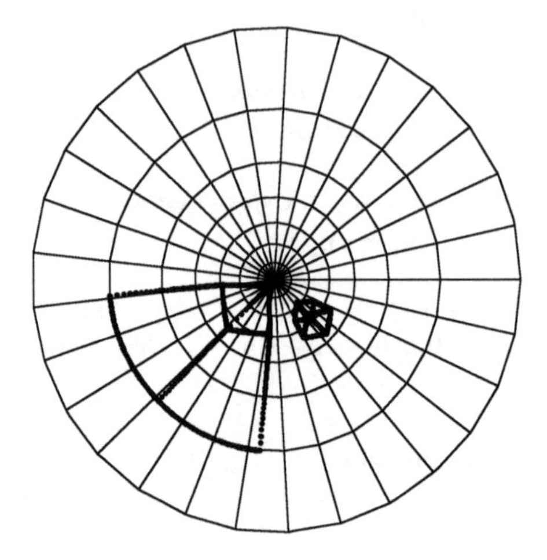

At Warp Factor 9, i.e. 90 per cent of *c* the effect is even more dramatic:

```
Show[HemiSphere, abberplota[9], abberplotb[9]]
```

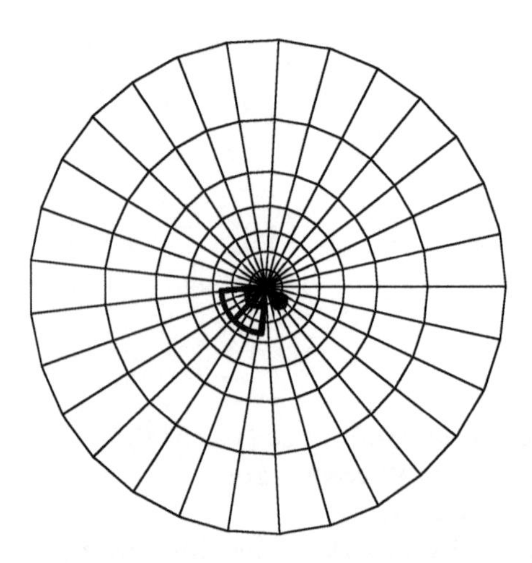

You might like to make a movie of the process of 'going to warp' – here is one to try (see also the exercises – the output is not shown here):

```
Do[Show[HemiSphere, abberplota[m], abberplotb[m]],
  {m, 0, 9, 1}]
```

Finally, if you have, or can obtain, the add-on application *Dynamic Visualizer*, try using this to obtain still more realistic results.

23.8 From null directions to points: twistors

We now have a very firm grip on what happens to the field of observations of observers under Lorentz transformations. But to start doing physics in three and four dimensions, we shall need rather more. In particular, we shall need a way of representing points in space and space-time. All we have so far is an understanding of the properties of points that are obtained by passing a light ray through an origin.

In order to describe general points via a suitable complex geometry, we will need to introduce the concept of a twistor, due to Penrose (see for example, Penrose and Rindler, 1984b). The following is intended to be an informal discussion, but will contain the key ingredients. We need first to extend the structures asociated with spinor algebra. We have already encountered the concept of a spinor as a two-component object:

$$S^A = \begin{pmatrix} \zeta \\ \eta \end{pmatrix}$$

We also have their complex conjugates $\overline{S}^{A'}$. Furthermore, we can define corresponding elements of the dual space, say λ_A, and elements of the dual conjugate space $\pi_{A'}$. As usual, the operation of contraction (summing over repeated indices), gives us a complex number – for example:

$$S^A \lambda_A = S^0 \lambda_0 + S^1 \lambda_1 \in \mathbb{C} \tag{23.54}$$

and similarly for their complex conjugates. The most obvious question from a linear algebra point of view is to ask whether there is a natural mapping from a spinor to a dual spinor. This as usual relies on the existence of a natural inner product on spinors. There is indeed one, but it takes a rather unusual form that is induced by the relativistic concept of distance embodied in Eq. (23.3). We have already observed that this norm on vectors corresponds to the operation of taking determinants of matrices of the form given by Eq. (23.18), which were

$$X^{AA'} = \frac{1}{\sqrt{2}} \begin{pmatrix} ct + z & x + iy \\ x - iy & ct - z \end{pmatrix}$$

Such determinants can also be calculated by the introduction of suitable alternating symbols. We define the *spinor metric* ϵ_{AB} to be $+1$ if $A = 0,\ B = 1$; -1 if $A = 1,\ B = 0$ and zero otherwise. The complex conjugate is identical. Now consider the expression (summation convention applies as usual):

$$\begin{aligned} &\epsilon_{AB}\, \epsilon_{A'B'} X^{AA'}\, X^{BB'} \\ &= \epsilon_{01}\epsilon_{0'1'} X^{00'} X^{11'} + \epsilon_{10}\epsilon_{1'0'} X^{11'} X^{00'} \\ &\quad + \epsilon_{01}\epsilon_{1'0'} X^{01'} X^{10'} + \epsilon_{10}\epsilon_{0'1'} X^{10'} X^{01'} \\ &= 2\epsilon_{01}\epsilon_{0'1'} X^{00'} X^{11'} + 2\, \epsilon_{01}\epsilon_{1'0'} X^{01'} X^{10'} \\ &= c^2 t^2 - x^2 - y^2 - z^2 \end{aligned} \tag{23.55}$$

So then this strange alternating metric on spinors is naturally compatible with the relativistic concept of distance. Thus we do have a natural inner product on spinors:

$$<\lambda, \mu> = \epsilon_{AB}\, \lambda^A \mu^B \tag{23.56}$$

and the inner product of any spinor with itself is identically zero. (This takes some getting used to.) Furthermore, we can use this ϵ_{AB} quantity to define dual spinors from spinors:

$$\lambda_B = \lambda^A\, \epsilon_{AB} \tag{23.57}$$

and the corresponding inverse mapping ϵ^{AB} exists and we can extract spinors from dual spinors thus:

$$\lambda^A = \epsilon^{AB} \lambda_B \tag{23.58}$$

We shall need these structural observatiosn in the following discussion.

A twistor, in its simplest guise, is a pair consisting of a spinor and an element of the complex conjugate dual space:

$$Z^\alpha = \{\omega^A, \pi_{A'}\} \tag{23.59}$$

The new index α runs from 0 to 3. It is not to be confused with the space-time index which also runs from 0 to 3. The introduction of such objects gives us enough structure to properly define a relationship between general points in Minkowski space and the associated complex space known as twistor space. This relationship is that of incidence between a space-time point and a twistor, and in Minkowski space it is given by the relationship

$$\omega^A = i\, X^{A A'}\, \pi_{A'} \tag{23.60}$$

The nature of this relationship is extraordinarily subtle. First, we note that it only involves the coordinates Z^α in a projective sense – there is no scaling implicit in this relationship. Second, given a space-time point x expressed by the matrix $X^{A A'}$, we can obtain a two-dimensional family of twistors (or, projectively, a one-dimensional family) that are incident with x. Finally, given a twistor Z^α there are an entire family of space-time points that are incident with it. Indeed, if x_0 is any one solution, so will that given by

$$X^{AA'} = X_0^{A A'} + \lambda^A \pi^{A'} \tag{23.61}$$

for all values of the spinor λ. This, geometrically, represents a null 2-plane in (complex) Minkowski space.

So how do we define a point in Minkowski space in twistor terms? Since a point corresponds to a one-dimensional family in projective terms – a line in projective twistor space, it is now straightforward to specify a point as corresponding to a pair of twistors (two points on the line). Algebraically this should come as no surprise. The incidence relation (23.60) corresponds to two equations. To specify the four coordinates embedded in $X^{AA'}$ we need four equations – so incidence with two distinct twistors will suffice. So a point can be characterized by a pair of incidence relations:

$$\begin{aligned}\omega_1^A &= iX^{AA'}\pi_{1A'}\,;\\ \omega_2^A &= iX^{AA'}\pi_{2A'}\end{aligned} \tag{23.62}$$

and this has the solution

$$iX^{AA'} = (\omega_1^A \pi_2^{A'} - \omega_2^A \pi_1^{A'})/(\pi_{1C'}\pi_2^{C'}) \tag{23.63}$$

23.9 Minimal surfaces and null curves I: holomorphic parametrizations

We shall now consider our first application of basic twistor algebra. This is not, in fact, an immediate application to equations like Laplace's equation, but is an application to a construction that does *not* have an analogue in two dimensions – the construction of minimal surfaces (soap bubbles!). There is some beautiful geometry that we can explore that is based on the properties of complex holomorphic curves whose tangent is everywhere null. The visualization of the real expressions of these curves is naturally achieved within *Mathematica*. We shall also learn some useful facts about points and how they are represented in complex terms – this will also be useful in Chapter 24.

This is not a text about differential geometry, so we shall not indulge in an exhaustive discussion on minimal surfaces, but we do need to make the link between minimal surfaces and certain holomorphic null curves. This correspondence applies in general dimension d. We are interested in finding maps from $\mathbb{R}^2$ to $\mathbb{R}^d$, represented as $x^a(u, v)$, such that the area of an immersed two-surface is a minimum. Such surfaces arise very naturally in nature – in particular the surface formed when a wire frame is dunked in soapy water is a surface of least area such that its boundary is given by the wire frame.

This is a problem in the calculus of variations and in general leads to non-linear differential equations. These equations can be simplified considerably if one exploits the freedom to reparametrize the surface. We shall assume that this can be done in such a way that the surface is conformal to a plane in the (u, v) coordinates. In other words, the infinitesimal distance ds between two points is given in terms of their coordinate shifts, (du, dv), by

$$ds^2 = \phi(u, v)\,(du^2 + dv^2) \tag{23.64}$$

Such coordinates are often called *isothermal* coordinates. The differential equations governing the minimal surface then become the set

$$\left(\frac{\partial^2}{\partial u^2} + \frac{\partial^2}{\partial v^2}\right) x^a = 0 \tag{23.65}$$

but there are additional constraints associated with Eq. (23.3), which take the form that the tangent vectors to the surface along the u and v directions are of the same length and orthogonal, i.e. if the metric function is η_{ab} (not necessarily of the signature of Eq. (23.3), nor necessarily in four dimensions, but composed of a dimension-d diagonal matrix of ± 1 down the diagonal), then

$$\eta_{ab} \frac{\partial x^a}{\partial u} \frac{\partial x^b}{\partial u} = \eta_{ab} \frac{\partial x^a}{\partial v} \frac{\partial x^b}{\partial v} \tag{23.66}$$

$$\eta_{ab} \frac{\partial x^a}{\partial u} \frac{\partial x^b}{\partial v} = 0 \tag{23.67}$$

For a detailed discussion of the differential geometry associated with these surfaces, and on isothermal coordinates, see Chapter 23 of Gray (1993), which contains many applications of *Mathematica* to differential geometry, and the references therein.

■ Holomorphic representation via null curves

The point of discussing these equations in a complex analysis setting is that Eq. (23.65) can be solved by requiring that x^a is the real part of a complex vector field, i.e. a holomorphic function of $w = u + i v$:

$$x^a = \mathrm{Re}(X^a[w]) \tag{23.68}$$

The constraints (23.66), (23.67) are then equivalent to demanding that $X^a(\zeta)$ has a tangent that is null:

$$\eta_{ab} \frac{\partial X^a}{\partial w} \frac{\partial X^b}{\partial w} = 0 \tag{23.69}$$

So we see that the differential equation has been essentially eliminated, leaving us with a holomorphic system satisfying a simple geometrical condition given by Eq. (23.69).

This is a good point for a pause for thought on what has been accomplished. The equations characterizing a minimal surface constitute a set of non-linear differential equations. We have simplified these by introducing isothermal coordinates. The differential equation states that each component of the position vector characterizing a point on the surface must be a harmonic function of the isothermal coordinates, and so can be managed in terms of holomorphic functions. We are left with a nullity constraint.

■ The construction of null holomorphic curves in Minkowski space

We shall now go one step further and solve the nullity constraint completely. The work we have done setting up the theory of spinors and twistors makes dealing with this very easy indeed for the case of four dimensions and nullity condition being based on the Minkowski metric given by Eq. (23.3). The twistor geometry of this was first set out by the author (Shaw, 1985), though the formulae that result are *much* older, as we shall see presently. The analysis for some other signatures and dimensions is summarized by Shaw and Hughston (1990).

First of all, the nullity of the tangent vector implies the existence of a spinor field $\pi_{A'}(w)$ with the property that

$$\frac{\partial X^{AA'}}{\partial w}\pi_{A'} = 0 \tag{23.70}$$

This spinor field is only defined projectively by this equation. We define an associated twistor field by requiring incidence everywhere:

$$\omega^A(w) = iX^{AA'}(w)\pi_{A'}(w) \tag{23.71}$$

Now we differentiate both sides with respect to w, using a dot to denote differentiation. Since Eq. (23.70) holds we deduce that

$$\dot{\omega}^A(w) = iX^{AA'}(w)\dot{\pi}_{A'}(w) \tag{23.72}$$

so that the point is also incident with the derivative twistor. But now we can extract X from the twistor, as two incidence relations serve to define the point in general. That is,

$$iX^{AA'} = \left(\omega^A\dot{\pi}^{A'} - \dot{\omega}^A\pi^{A'}\right)/\left(\pi_{C'}\dot{\pi}^{C'}\right) \tag{23.73}$$

This gives us a formula for a null curve in terms of an arbitrary differentiable twistor curve. There is a great deal of freedom in this formula. First, we note that it is invariant under w-dependent rescalings of the twistor curve. (You might like to check this.) So only the proective components matter. We exploit this and choose the complex parameter such that

$$w = \pi_{0'}/\pi_{1'}\,;\quad \pi_{A'} = \pi_{1'}(w,\ 1)\,;\qquad \pi^{A'} = \pi_{1'}(1,\ -w) \tag{23.74}$$

We also introduce explicit coordinates for ω^A as:

$$\omega^A = \sqrt{2}\,i\pi_{1'}\{f(w),\ g(w)\} \tag{23.75}$$

It is now a matter of unravelling Eq. (23.73) using the variables in Eqs. (23.74) and (23.75). For example

$$iX^{00'} = \sqrt{2}\,i\,f'(w) \tag{23.76}$$

becomes the condition

$$ct(w) + z(w) = f'(w) \tag{23.77}$$

Unpacking the other conditions and reorganizing gives

$$\begin{aligned} ct &= f' + g - wg' \\ x &= g' + f - wf' \\ iy &= f - wf' - g' \\ z &= f' - g + wg' \end{aligned} \tag{23.78}$$

These formulae are actually rather old! The representation of null curves and minimal surfaces by these formulae were first given in a series of papers by Montcheuil (1905) and Eisenhart (1911, 1912). However, their understanding in terms of complex relativistic geometry emerged relatively recently (Shaw, 1985). The corresponding results for dimension three are older still – we shall look at this right away.

■ Reduction to three-dimensional Euclidean space

If we constrain Eqs. (23.78) so that the resulting curve lies in the hyperplane $t = 0$, we must impose the condition

$$f' = -g + w g' \tag{23.79}$$

This can be integrated. If we demand that $g = G'$ for some G, then

$$f = wG' - 2G \tag{23.80}$$

The remaining equations in (23.78) then boil down to

$$\begin{aligned} x &= -2G + 2wG' + (1 - w^2)G'' \\ iy &= -2G + 2wG' - (1 + w^2)G'' \\ z &= 2wG'' - 2G' \end{aligned} \tag{23.81}$$

This is a very important formula. It gives the general solution for a holomorphic null curve in complex Euclidean three-space. Its real part is a minimal surface in real Euclidean three-space. Note that neither this formula nor its four-dimensional counterpart contain any integration – we have solved the problem completely in terms of free holomorphic functions.

Note that the result for three dimensions can also be derived by an *intrinsically three-dimensional* argument. This has been given by Hitchin (1982), and gives the Weierstrass (1866a) parametrization for minimal surfaces, in its integral-free form. The paper by Weierstrass is available on the internet (Weierstrass 1866b).

■ Degeneration to points

Before we look at the surfaces generated by these formulae, let's take a quick look at degenerate situations. It will be evident that if we take f and g to be both linear functions of w, Eq. (23.78) will give constant values for (ct, x, y, z). Similarly, if G is quadratic, Eq. (23.81) will also give constant values for (x, y, z). Let's explore this a little, and suppose that

$$G = a + bw + cw^2 \tag{23.82}$$

It follows that

$$\begin{aligned} G' &= b + 2cw; \\ G'' &= 2c \end{aligned} \tag{23.83}$$

Application of Eq. (23.81) then gives us

$$x = 2(c - a);\ iy = -2(a + c);\ z = -2b \tag{23.84}$$

or inverting this

$$G = -\frac{1}{4}((x + iy) + 2zw - (x - iy)w^2) \tag{23.85}$$

This is a simple way of observing that points in Euclidean three-space correspond to quadratic functions of w. We shall need this in Chapter 24. Similarly, we can give an explicit coordinate representation of the fact that points in Minkowski space correspond to a pair of linear functions of w. What we need to observe for now is that adding a linear term to f and/or g, or adding a quadratic term to G, will not materially alter the form of the null curve or associated minimal surface – it will just translate it.

23.10 Minimal surfaces and null curves II: minimal surfaces and visualization in three dimensions

We now consider the business of generating minimal surfaces using *Mathematica*. Let us suppose that we always work in terms of a variable w and supply a function G whose complex argument is w. First we wish to extract the null curve. We do this by using a method similar to that introduced by Gray (1993), but avoiding any integration:

```
NullCurve[G_][w_] :=
Module[{ww, temp, one, two},
one = D[G[ww], ww];
two = D[G[ww], {ww, 2}];
temp = {-2*G[ww]+2*ww*one + (1-ww^2)*two,
I*(2*G[ww]-2*ww*one+(1+ww^2)*two),
2*ww*two - 2*one};
Simplify[temp /. ww->w]
]
```

Let's check it is working by giving it a trivial quadratic – we should get a point back:

```
NullCurve[-1/4((x+I y) + 2 z #-(x - I y)#^2)&][w]
```

$\{x, y, z\}$

Next we wish to write out the real part of these null curves, based on a convenient parametrization:

```
MinimalSurface[G_, w_, params_] :=
Module[{null},
null = NullCurve[G][w];
ComplexExpand[Re[null /. w-> params]]]
```

Next we wish to generate some automatic plotting routines. We also want the flexibility to use either Cartesian or Polar coodinates, or possibly a hybrid. Let's do a Cartesian version first:

```
CartesianMinimalSurface[G_, w_, {umin_, umax_},
{vmin_, vmax_}, options___] :=
Module[{ms, u, v},
ms = MinimalSurface[G, w, u + I v];
ParametricPlot3D[Evaluate[ms], {u, umin, umax}, {v,
vmin, vmax}, options]]
```

Here are some versions for other coordinates:

```
PolarMinimalSurface[G_, w_, {umin_, umax_}, {vmin_,
vmax_}, options___] :=
Module[{ms, u, v},
ms = MinimalSurface[G, w, u*Exp[I v]];
ParametricPlot3D[Evaluate[ms], {u, umin, umax}, {v,
vmin, vmax}, options]]

ExponentialMinimalSurface[G_, w_, {umin_, umax_},
{vmin_, vmax_}, options___] :=
Module[{ms, u, v},
ms = MinimalSurface[G, w, Exp[u+I v]];
ParametricPlot3D[Evaluate[ms], {u, umin, umax}, {v,
vmin, vmax}, options]]
```

Here, for example, we find *Enneper's surface* corresponds to the simple choice w^3:

```
CartesianMinimalSurface[#^3&, w, {-4, 4}, {-4, 4},
PlotPoints -> 30]
```

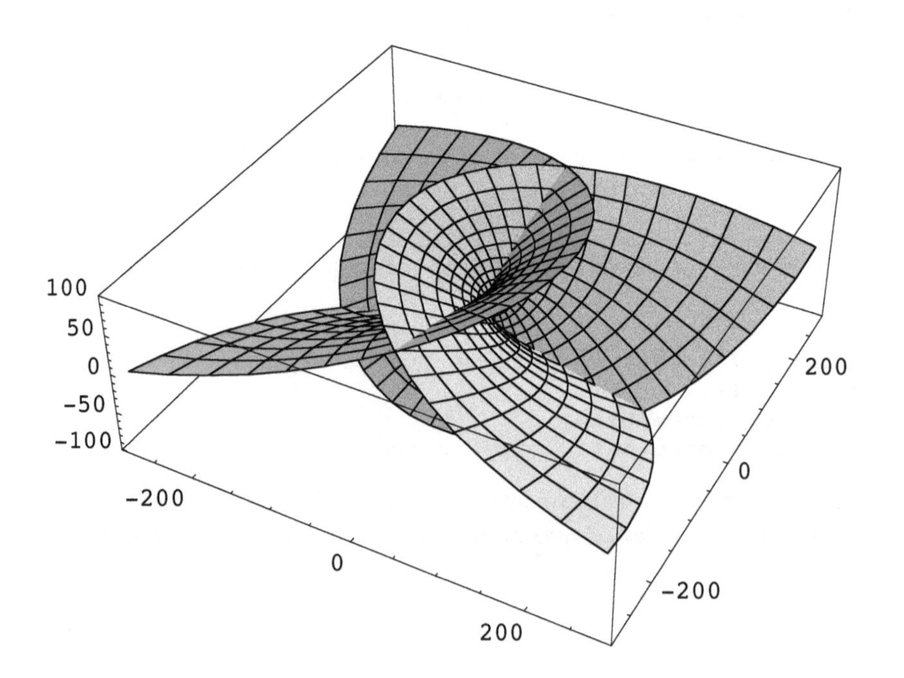

The catenoid is obtained by chooing $G = w\,log(w)$, but we need to consider carefully what coordinates to use:

```
CartesianMinimalSurface[#*Log[#]&, w, {-4, 4}, {-4,
4}, PlotPoints -> 30]
```

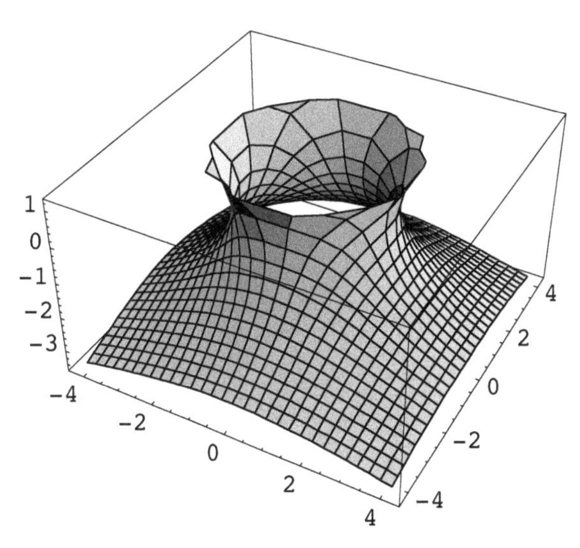


```
ExponentialMinimalSurface[#*Log[#]&, w, {-2, 2}, {0, 2
Pi}, PlotPoints -> 30]
```

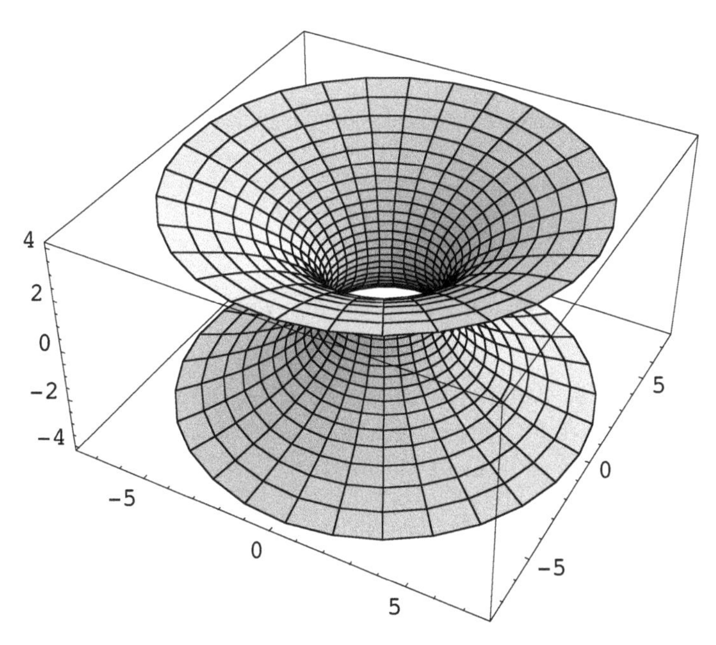

The exploration of these surfaces is left to you – some suggestions are given in Exercise 23.6. Here is one more to get you started:

```
CartesianMinimalSurface[1 / (10 - #^3)^(1 / 3) &,
 w, {-1, 1}, {-1, 1}, PlotPoints → 30]
```

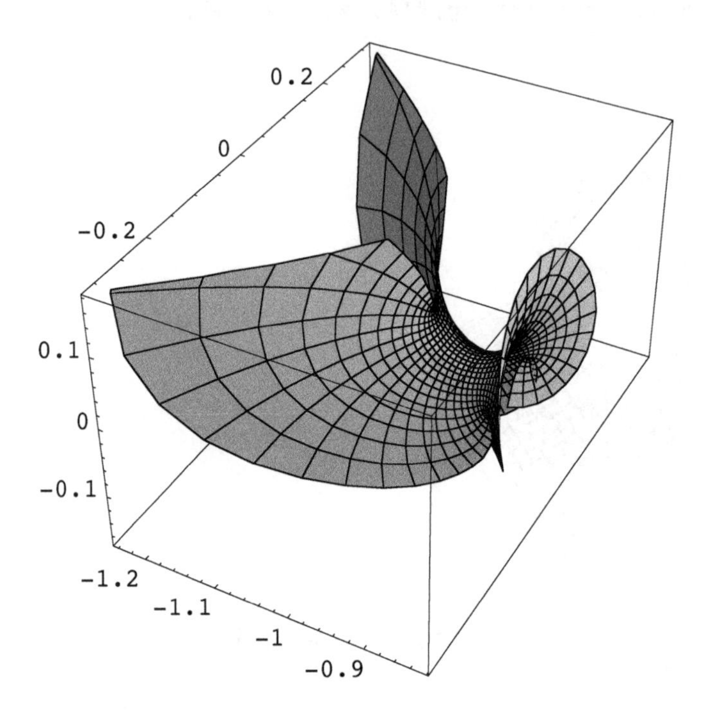

Exercises

23.1 [✵] Suppose that we have a curve on the sphere S^+ paramterized by $w(s)$ as s varies, with Eq. (23.9). Show that

$$\dot{x}^2 + \dot{y}^2 + \dot{z}^2 = \frac{4\,\dot{w}\,\dot{\overline{w}}}{(1 + w\,\overline{w})^2}$$

where the dot denotes differentiation with respect to s. Confirm your results by repeating the calculation using *Mathematica*.

23.2 Show that if the mapping defined by Eq. (23.25) in the special form of Eq. (23.41) is applied to a matrix of the form of Eq. (23.18), then the result is as expressed by Eq. (23.43).

23.3 Derive Eq. (23.48) from Eq. (23.43) by using the parametrizations expressed by Eqs. (23.45-23.47).

23.4 Elucidate the effect of a transformation such as that given by Eq. (23.40), with one fixed point at infinity, by deriving the equations analogous to Eq. (23.43).

23.5 ❁ There are many ways of extending the warping observer example given in Section 23.8. Try extending this to include:

(i) management of shaded polygonal objects.
(ii) recomputation of the scene as the observer actually moves through the field of alien ships.
(iii) more 'realistic' alien ships.
(iv) addition of colour and Doppler shift effects.

23.6 ❁ There are many issues in the theory of minimal surfaces that you can explore. Some suggestions are:

(i) Given a choice of G, what is the effect of replacing this by iG?
(ii) Explore what happens within **`ParametricPlot3D`** when the functions (such as log) contain branch points. Investigate how to fix this. (See also Maeder, 1994).
(iii) Try taking G to be any of *Mathematica*'s special functions, as well as interesting polynomial, rational and algebraic expressions.

23.7 Given the explicit formulae in Eq. (23.78) and Eq. (23.81), in each case work out the complex tangent vector to the curve, and verify that it has zero length with respect to the appropriate measure of distance.

24 Physics in three and four dimensions II

Introduction

The goal of this Chapter is to provide you with a very basic understanding of how Laplace's equation (in three space variables) and the scalar wave equation (in three space and one time dimension) can be solved using holomorphic methods. This chapter will build on the methods developed in Chapter 23 in a very direct way, and you are recommended to read that chapter now before proceeding further here.

The material on dimension three is, in part, a much simplified version of part of a series of lectures given by Nigel Hitchin (now Professor, F.R.S.) in Oxford in the early 1980s. The picture presented here will not give you anything like the full geometrical picture underlying the results, which we shall develop by elementary methods. Part of the theory of the intrinsic three-dimensional approach is developed in a paper published by Hitchin (1982). The four-dimensional picture is covered most comprehensively by its principle architect, Professor Sir Roger Penrose, F.R.S., in Penrose and Rindler (1984a, b).

24.1 Laplace's equation in dimension three

Perhaps the most natural place to start is with the solution of Laplace's equation in three variables. In Chapter 23 we found, in Eq. (23.85), a natural holomorphic representation of a point in (possibly complex) three-dimensional space, arising as a degenerate case of a holomorphic null curve. This representation is in terms of a special type of quadratic holomorphic function. Absorbing the irrelevant factor of $1/4$ into the definition of the function G, we have the formula

$$G = (x + i\,y) + 2\,z\,w - (x - i\,y)\,w^2 \tag{24.1}$$

This quadratic function of the complex variable w plays a very special role, and we shall give it a name: $\eta_{\underline{x}}(w)$, where $\underline{x} = (x,\, y,\, z)$. For a proper discussion of the geometrical significance of $\eta_{\underline{x}}(w)$ you should see Hitchin (1982). Now suppose that $f(\eta,\, w)$ is holomorphic in both of its arguments η and w. What can we say about $f(\eta_{\underline{x}}(w),\, w)$? As it is holomorphic, we can certainly work out its second derivatives:

$$\frac{\partial^2}{\partial x^2}\, f(\eta_x(w),\, w) = (1 - w^2)^2\, f_{\eta\eta} \tag{24.2}$$

$$\frac{\partial^2}{\partial y^2}\, f(\eta_x(w),\, w) = -(1 + w^2)^2\, f_{\eta\eta} \tag{24.3}$$

$$\frac{\partial^2}{\partial z^2}\, f(\eta_x(w),\, w) = 4w^2\, f_{\eta\eta} \tag{24.4}$$

It follows immediately that

$$\nabla^2 f(\eta_x(w), w) = 0 \tag{24.5}$$

by elementary addition of Eqs. (24.2–24.4). So for any choice of w and f, we have a solution of Laplace's equation. By the linearity of Laplace's equation, we can linearly superimpose such solutions for different choices of w, and we can express this through integration. Thus, given $f(\eta, w)$, we define a scalar field ϕ as follows:

$$\Phi(x, y, z) = \int_C f(x + iy + 2zw - (x - iy)w^2, w)\, dw \tag{24.6}$$

where, provided only f is holomorphic, ϕ satisfies Laplace's equation (apart from possible singular points) for a given choice of the contour C.

We should emphasize that such integral representations are not new and were derived without twistor methods many years ago. Whittaker (1903) wrote down a formula of a similar type to Eq. (24.6). It should also be appreciated that there is much geometric subtlety associated with Eq. (24.6) that we shall not be able to explore properly. The point is that there may be many choices of f yielding a given ϕ, each differing from the other by a function that integrates to zero under the contour. Cauchy's theorem guarantees that there will be many such functions. A proper study of this and how it is managed mathematically is outside the scope of this book. We can, nevertheless, do a great many interesting things with Eq. (24.6) and a naive approach.

24.2 Solutions with an axial symmetry

In order to get a grip on Eq. (24.6) we shall explore some simple special cases where the field ϕ possesses some symmetry. Perhaps the most interesting case is that of scalar fields ϕ that are axis-symmetric. That is, choosing the axis of symmetry to be the z-direction, we seek invariance under rotations within the x–y plane. In our classification of Lorentz transformations given in Chapter 23, in particular by the last of the Eqs. (23.43), we saw that the mapping (we divide ψ by two here, and use w as a spinor coordinate)

$$w \to e^{i\psi} w \tag{24.7}$$

induces the rotation

$$x + iy \to e^{i\psi}(x + iy) \tag{24.8}$$

It follows that

$$\eta_{\underline{x}}(w) = (x + iy) + 2zw - (x - iy)w^2 \to e^{i\psi} \eta_{\underline{x}}(w) \tag{24.9}$$

So the induced action on the η variable is also multiplication by $e^{i\psi}$. We also note that

$$dw \to e^{i\psi} dw \tag{24.10}$$

So to obtain axis-symmetric solutions, the function $f(\eta,\ w)$ must be of the form

$$f(\eta, w) = \frac{1}{w}\, g\left(\frac{\eta}{w}\right) \tag{24.11}$$

for some holomorphic g, and we obtain the formula for axis-symmetric solutions in the form (inserting a normalization factor that will clean the results up):

$$\Phi = \frac{1}{2\pi i}\int g\left(\frac{\eta_{\underline{x}}(w)}{w}\right)\frac{1}{w}\, dw \tag{24.12}$$

This formula is equivalent to that given by Whittaker and Watson (1984, note 18.3). Its geometrical characteristics in twistor terms have been uncovered in a series of articles by, amongst others, by Ward (1981, 1990) and Mason (1990). It was originally written down, in trigonometric form, by Whittaker (1903).

To elucidate the structure of this formula, we assume that g has a Laurent series about the origin, and the contour is deformable to a circle containing the origin. We can now consider components of g that are just powers of its argument. That is, we are interested in evaluating

$$\Phi_n = \frac{1}{2\pi i}\int\left(\frac{\eta_{\underline{x}}(w)}{w}\right)^n \frac{1}{w}\, dw = \frac{1}{2\pi i}\int\frac{\eta^n{}_{\underline{x}}(w)}{w^{n+1}}\, dw \tag{24.13}$$

where the integration is over a circle. To evaluate Eq. (24.13), let us set $y = 0$, $z = r\cos(\theta)$, $x = r\sin(\theta)$. Then Eq. (24.13) simplifies to

$$\frac{r^n}{2\pi i}\int\left(2\cos(\theta) + \left(\frac{1}{w} - w\right)\sin(\theta)\right)^n \frac{1}{w}\, dw \tag{24.14}$$

Now set $w = e^{i t}$ to parametrize the integral. We obtain

$$\frac{2^n r^n}{2\pi}\int_0^{2\pi}(\cos(\theta) - i\sin(t)\sin(\theta))^n\, dt \tag{24.15}$$

The evaluation of this integral is well known in terms of elementary functions. It is given, for example, by Gradshteyn and Ryzhik (1980, Eq. 3.611.3) in the form:

$$\begin{aligned}&\int_0^{\pi}(\cos(\theta) + i\cos(x)\sin(\theta))^n\, dx\\ &= \int_0^{\pi}(\cos(\theta) + i\cos(x)\sin(\theta))^{-n-1}\, dx\\ &= \pi P_n[\cos(t)]\end{aligned} \tag{24.16}$$

where P_n is the nth Legendre polynomial. Making some simple changes of variable, we obtain, for Eq. (24.15), if $n \geqslant 0$,

$$2^n r^n P_n[\cos(\theta)] \tag{24.17}$$

and if $n = -k - 1$, for $k = 1, 2, 3, \ldots$

$$\frac{P_k[\cos(\theta)]}{2^{k+1}\,r^{k+1}} \tag{24.18}$$

Thus we obtain the well-known family of solutions in terms of Legendre polynomials. This contour integral approach can also lead to other interesting representations of the Legendre polynomials (see exercise 24.1.) Exercise 24.2 invites you to explore similar solutions with the property that

$$\Phi \to e^{im\psi}\,\Phi \tag{24.19}$$

under a rotation about the z axis.

24.3 Translational quasi-symmetry

We can generate some other simple solutions by exploring solutions with a translational 'symmetry':

$$\Phi(x, y, a+z) = e^{\lambda a}\,\Phi(x, y, z) \tag{24.20}$$

We can express this equivalently as the condition that

$$\frac{\partial \Phi}{\partial z} = \lambda\,\Phi \tag{24.21}$$

and this will of course generate (some) solutions of the 2-D Helmholtz equation

$$\frac{\partial^2 \Phi}{\partial x^2} + \frac{\partial^2 \Phi}{\partial y^2} + \lambda^2\,\Phi = 0 \tag{24.22}$$

if we pull out the exponential function of z. Suppose we impose a condition analogous to Eq. (24.21) on the underlying complex function itself. This gives us

$$2\,w\frac{\partial f}{\partial \eta} = \lambda f \tag{24.23}$$

which has the solution

$$f = e^{\lambda\eta/(2w)}\,h(w) \tag{24.24}$$

where h is general (holomorphic apart from some poles). What types of field does this give us? We now need to evaluate expressions that boil down to contour integrals of the form

$$\frac{e^{\lambda z}}{2\pi i}\int e^{\frac{1}{2}\lambda((x+iy)/w-(x-iy)w)}\,h(w)\,dw \tag{24.25}$$

where h (we assume) has a Laurent series. Under this assumption, we may as well consider expressions of the form

$$\frac{1}{2\pi i}\int e^{\frac{1}{2}\lambda((x+iy)/w-(x-iy)w)}\,w^{-(n+1)}dw \tag{24.26}$$

for n an integer. To evaluate this we make use of the generating function for Bessel functions in the form

$$e^{\frac{1}{2}p\,(t-1/t)} = \sum_{k=-\infty}^{\infty} t^k J_k(p) \tag{24.27}$$

We set, with ϕ now the plane polar angle:

$$p = \lambda\rho = \lambda\sqrt{x^2+y^2} \tag{24.28}$$

$$t = -\frac{(x-iy)w}{\rho} = -e^{-i\phi}\,w \tag{24.29}$$

The integral becomes

$$\frac{1}{2\pi i}\int\sum_{k=-\infty}^{\infty} w^k\,(-1)^k\,e^{-ik\phi}\,J_k(\lambda\rho)w^{-(n+1)}dw \tag{24.30}$$

and this can be evaluated directly based on residues, to give

$$(-1)^n\,e^{-in\phi}\,J_n(\lambda\rho) \tag{24.31}$$

which is an important family of basic solutions to the 2-D Helmholtz equations.

24.4 From three to four dimensions and back again

As with the discussion of null curves given in Chapter 23, matters are slightly more straightforward in the full relativistic and four-dimensional picture. We can make direct use of the twistor-space-time incidence relation

$$\omega^A = i x^{AA'}\,\pi_{A'} \tag{24.32}$$

to construct solutions of the scalar wave equation in a natural holomorphic way. We are now interested in the interpretation of holomoprhic functions of twistors $f(Z^\alpha)$ in space-time terms. Now by splitting the argument up into its spinor pairs:

$$f(Z^\alpha) = f(\omega^A, \pi_{A'}) \tag{24.33}$$

it then becomes natural to consider space-time fields obtained by imposing the incidence relation thus:

$$f(i x^{AA'}\,\pi_{A'}, \pi_{A'}) \tag{24.34}$$

and carrying out a suitable integration over possible values of $\pi_{A'}$. Now because the incidence relation is really a relationship between space-time points and twistors considered projectively – the scale being irrelevant, we need to work this out in projective terms. In our standard coordinates, we have

$$\sqrt{2}\,\omega^0 = i(ct+z)\,\pi_{0'} + i(x+iy)\,\pi_{1'} \tag{24.35}$$

$$\sqrt{2}\,\omega^1 = i(x-iy)\,\pi_{0'} + i(ct-z)\,\pi_{1'} \tag{24.36}$$

Now we set

$$w = \frac{\pi_{0'}}{\pi_{1'}};\; p = \frac{-i\sqrt{2}\,\omega^0}{\pi_{1'}};\; q = \frac{-i\sqrt{2}\,\omega^1}{\pi_{1'}}; \tag{24.37}$$

and this gives us

$$p = (ct+z)w + (x+iy) \tag{24.38}$$

$$q = (x-iy)w + (ct-z) \tag{24.39}$$

Now working projectively with the scale factor $\pi_{1'}$ removed, we wish to consider contour integrals of the form

$$\int g(p,\,q,\,w)\,dw \tag{24.40}$$

or, in explicit coordinates:

$$\Psi(x,\,y,\,z,\,t) = \int g[w(ct+z)+(x+iy),\, w(x-iy)+(ct-z),\, w]\,dw \tag{24.41}$$

where g is holomorphic in each of its arguments but otherwise arbitrary. We now investigate this formula. But we note first that the geometric appreciation of this formula is due to Penrose – a full discussion is given by Penrose and Rindler (1984a, b), including the result for other types of massless fields including electromagnetic fields. However, results equivalent to this were first obtained by Bateman (1904, 1944) based on Whittaker's earlier analysis for the Laplace equation.

■ Solving the wave equation

We write the wave equation in the form

$$\frac{1}{c^2}\frac{\partial^2\Psi}{\partial t^2} - \frac{\partial^2\Psi}{\partial x^2} - \frac{\partial^2\Psi}{\partial y^2} - \frac{\partial^2\Psi}{\partial z^2} = 0 \tag{24.42}$$

The fact that Eq. (24.41) satisfies Eq.(24.42) can be hammered out by direct differentiation, but it is rather more transparent if more suitable coordinates are chosen. We introduce

$$u = ct - z;\; v = ct + z;\; \zeta = x + iy;\; \overline{\zeta} = x - iy \tag{24.43}$$

and note that (see Exercise 24.3) Eq. (24.42) is equivalent to the condition

$$4\left(\frac{\partial^2 \Psi}{\partial u \partial v} - \frac{\partial^2 \Psi}{\partial \zeta \partial \overline{\zeta}}\right) = 0 \tag{24.44}$$

Now note that in these coordinates, Eq. (24.41) gives us

$$\Psi = \int g(wv + \zeta, w\overline{\zeta} + u, w)\, dw \tag{24.45}$$

and satisfaction of Eq. (24.44) is immediate, since each term in Eq. (24.44) is the contour integral of w times the mixed second derivative of g with respect to its first and second arguments, and they cancel!

■ Solving the Helmholtz equation in three dimensions

Suppose that we now constrain Eq. (24.45) and demand that

$$\frac{\partial \Psi}{\partial t} = -i\omega\Psi \tag{24.46}$$

We can solve this by demaning that g itself satisfy the same condition when constrained to space-time points, i.e.

$$\frac{\partial}{\partial t} g[(x + iy) + w(ct + z), w(x - iy) + (ct - z), w] = -i\,\omega\, g \tag{24.47}$$

Now let, as before

$$\begin{aligned} p &= x + iy + w(ct + z) \\ q &= ct - z + w(x - iy) \end{aligned} \tag{24.48}$$

Then Eq. (24.47) becomes the condition

$$c\left(w\frac{\partial g}{\partial p} + \frac{\partial g}{\partial q}\right) = -i\omega g \tag{24.49}$$

As usual, we let $\omega = ck$ for some k, so that this is the condition

$$w\frac{\partial g}{\partial p} + \frac{\partial g}{\partial q} = -ikg \tag{24.50}$$

This first-order partial differential equation can be solved in a variety of ways. Perhaps the simplest is to remove the right-hand side by use of an integrating factor based on the second argument, q, of g. So we write

$$g(p, q, w) = e^{-ikq}\, h(p, q, w) \tag{24.51}$$

where

$$w\frac{\partial h}{\partial p} + \frac{\partial h}{\partial q} = 0 \tag{24.52}$$

we solve this PDE by inspection in the form

$$h(p, q, w) = H(p - wq, w) \tag{24.53}$$

for H some holomorphic function of two variables. Note further that

$$\begin{aligned} &p - wq \\ &= x + iy - w(w(x - iy) + (ct - z)) + w(ct + z) \\ &= (x + iy) + 2zw - w^2(x - iy) \\ &= \eta_x(w) \end{aligned} \tag{24.54}$$

Re-assembling all this, we have constructed a solution of the form

$$\Psi = e^{-i\omega t}\Phi(x, y, z) \tag{24.55}$$

where

$$\Phi = \int e^{-ik(w(x-iy)-z)} H(\eta_x(w), w) dw \tag{24.56}$$

and this satisfies the Helmholtz equation:

$$\frac{\partial^2 \Phi}{\partial x^2} + \frac{\partial^2 \Phi}{\partial y^2} + \frac{\partial^2 \Phi}{\partial z^2} + k^2\Phi = 0 \tag{24.57}$$

Note also that when $k = 0$ this formula also reduces naturally to that we have already given for the Laplace equation in 3-D. This is one of the ways in which the formulae for three dimensions emerge as special cases of the relativistic result. Another approach makes use of the freedom in the detailed choice of the integrating factor. If we balance it symmetrically between the two arguments of g we obtain an equivalent representation:

$$g(p, q, w) = e^{\frac{-ik}{2}(p/w+q)} H(p - wq, w) \tag{24.58}$$

This leads to the formula for solutions of the Helmholtz equation in the form

$$\Phi = \int e^{-\frac{1}{2}ik(w(x-iy)+(x+iy)/w)} H(\eta_x(w), w) dw \tag{24.59}$$

Exercise 24.44 guarantees that any solution of Eq. (24.50) can be written in this form.

■ Change of global coordinates

One observation that is sometimes useful is to appreciate what happens when we switch from the w-coordinate to its inverse, in order to analyse functions that are well-behaved at infinity in w-space. We make the replacement

$$w \longrightarrow \tilde{w} = \frac{1}{w} \tag{24.60}$$

Inspection of Eq. (24.37) shows that if we correspondingly re-scale the ω^A spinor by $\pi_{0'}$ instead of $\pi_{1'}$ we induce the rescalings

$$p \longrightarrow \tilde{p} = \frac{p}{w} = \tilde{w}\, p \tag{24.61}$$

$$q \longrightarrow \tilde{q} = \frac{q}{w} = \tilde{w}\, q \tag{24.62}$$

If we now follow through the argument above for time-independent fields, we find that the analogous formula for η is given (up to a sign) by

$$\tilde{\eta} = \tilde{q} - \tilde{p}\,\tilde{w} = (x - i y) - 2 z \tilde{w} - (x + i y)\tilde{w}^2 = -\frac{\eta}{w^2} \tag{24.63}$$

■ The differential geometry view

That Eq. (24.63) this is *exactly* the right formula to use (i.e. with the right sign) follows from a rather deeper geometrical argument, where η is interpreted as representing complex tangent vector fields (where the tangents are those to the Riemann sphere), in the sense of differential geometry. That is, a tangent vector is given by

$$\eta \frac{\partial}{\partial w} \tag{24.64}$$

If we wish to equate this to its corresponding form in reciprocal coordinates:

$$\eta \frac{\partial}{\partial w} = \tilde{\eta} \frac{\partial}{\partial \tilde{w}} \tag{24.65}$$

But, by Eq. (24.60), we have

$$\frac{\partial}{\partial \tilde{w}} = -w^2 \frac{\partial}{\partial w} \tag{24.66}$$

So we must have

$$\tilde{\eta} = -\frac{\eta}{w^2} \tag{24.67}$$

You might wish to explore some other related issues. For example, you might like to show that (case (2) is for the more adventurous)

(1) The form of the axis-symmetry constraint is preserved under such an inversion;
(2) The only vector fields that are holomorphic everywhere has η a quadratic in w.

24.5 Translation symmetry: reduction to 2-D

This section represents an informal presentation of an argument due to N. Hitchin.

So far so good. But are these *contour integral* expressions really the appropriate generalization of the simple formula for two dimensions? In two dimensions we usually just write down

$$\phi(x, y) = \text{Re}(f(w)); \quad w = x + i\,y \tag{24.68}$$

One way to explore this is to see what happens when we reduce dimensions from three to two, and seek solutions of Laplace's equation in three dimensions, but which satisfy

$$\Phi(x, y, z + u) = \Phi(x, y, z) \tag{24.69}$$

Now inspection of Eqs. (24.20–23) reveals that the procedure outlined there fails if $\lambda = 0$ We only get functions that are independent of η. So this is a matter of some subtlety. This is the point where we *must* get a better grip on the fact that our holomorphic functions represent spatial fields only once the contour integration has been carried out. We cannot just impose the condition

$$f(\eta + \tau\, w, w) = f(\eta, w) \tag{24.70}$$

and we must take better account of the contour integration. Let us simplify matters and assume that the contour of integration C is deformable to the unit circle. Suppose that instead of Eq. (24.62) we demand that

$$f(\eta + \tau w, w) = f(\eta, w) + g_0(\tau, \eta, w) - g_1(\tau, \eta, w) \tag{24.71}$$

where g_0 is holomorphic on a region containing all of the interior of the unit circle, and g_1 is holomorphic on all of the exterior region, including infinity. Remember that this means that we set $\zeta = 1/w$ and demand that g_1 is holomorphic at $\zeta = 0$. Now when we integrate Eq. (24.63) over C, we obtain, by Cauchy's Theorem, the result that

$$\int_C f(\eta + w\tau, w)\, dw = \int_C f(\eta, w)\, dw \tag{24.72}$$

which is what is actually required. We must take account of (63) when imposing translational symmetry. Now, differentiating Eq. (24.71) with respect to τ, and then setting $\tau = 0$, we obtain

$$w\,\frac{\partial f}{\partial \eta} = \dot{g}_0 - \dot{g}_1 = G_0[\eta, w] - G_1[\eta, w] \tag{24.73}$$

where the dot denotes τ-differentiation, and G_0 is holomorphic on the interior region, G_1 on the exterior region. Now we integrate this relation with respect to η to obtain

$$w f = H_0[\eta, w] - H_1[\eta, w] \tag{24.74}$$

where H_0 is holomorphic on the interior region, H_1 on the exterior region. Let us look at the contribution of H_0 to the remaining contour integration. The contribution is of the form

$$\int_C \frac{H_0[\eta_x(w), w]}{w}\, dw \tag{24.75}$$

but since H_0 is holomorphic on the interior of C we can evaluate this using the calculus of residues at the manifest simple pole at $w = 0$. The answer is just

$$2\pi i H_0[\eta_x(0), 0] = 2\pi i H_0[x + i y, 0] \tag{24.76}$$

which is just a holomorphic function of $x + iy$. A detailed analysis of the exterior peice requires some more analysis, based on the change of global coordinate given in the previous section. Inspection of Eq. (24.63) shows that the corresponding calculation leads to a contribution that is a holomorphic function of $\tilde{\eta}_x(0)$, i.e. $x - iy$. In other words, the contour integrals that are translation invariant reduce to expressions of the form

$$h_0(x + i y) + h_1(x - i y) \tag{24.77}$$

where each h_i is a holomorphic function of its argument. So the contour integration evaporates, leaving us with just the evaluation of functions. In this picture, dimension two is rather special – the use of contour integration is rather more general. Note that taking the real part of a holomorphic function of $x + iy$ is just a special case of Eq. (24.77). It is noteworthy that an essentially holomorphic 3-D picture reduces to a situation where both $w = x + iy$ and $\overline{w} = x - iy$ play a role. So *superficially non-holomorphic* structures in 2-D are really the projection of a larger and *essentially holomorphic* structure in higher dimensions. This is a very important observation. It has interesting consequences for applied mathematics. For example, in the solution of the biharmonic problem for fluid flow discussed in Section 19.7, in Eq. (19.89) in particular (where z represented $x + iy$), the solution looks non-holomorphic in a very essential way. Again, this turns out to be an illusion of the 2-D projection from a 3-D holomorphic structure, but a proper discussion of this is well outside the scope of this book.

24.6 Comments

Chapters 23 and 24 were intended to be only the briefest of introductions to what is possible in higher dimensions. You should be aware that this is just a rather blinkered glimpse at the full power of twistor methods. In this chapter we have just been discussing the Laplace equation, scalar Helmholtz equation, and scalar (massless) wave equation, and using very simple ideas about contour integration.

The pure mathematicians among you should appreciate that we have glossed over a proper treatment of the holomorphic functions and the contour integration. The right setting for all of this is within sheaf cohomology. A good place to start in understanding how to set this up is Chapter 6 of Penrose and Rindler (1984b).

The theoretical/mathematical physicists among you should be aware that the treatment of the scalar massless wave equation is a very special case. There are analogous contour integral solutions on Minkowski space-time of the source-free:

(1) Maxwell's equations;
(2) massless neutrino equations;
(3) linearized Einstein equations.

Indeed, massless fields of arbitrary helicity are easily handled by twistor methods, as is also discussed in Chapter 6 of Penrose and Rindler (1984b). The methods extend, rather remarkably, to Yang–Mills fields and at least part (so far) of the way to the full Einstein equations in General Relativity. These are essentially non-linear systems – in some ways the solution of the minimal surface problem discussed in Chapter 23 is a model for the management of such non-linear systems – Plateau's problem for minimal surfaces starts off as a hard non-linear problem. It only becomes easy when viewed in the right (holomorphic) way. Recent work includes clearer management of sources and currents. A collection of useful recent entry points to work in twistor theory including these issues is given, for example, by Bailey and Baston (1990), Mason and Hughston (1990), Mason *et al.* (1990).

Those of you interested in applied mathematics in more of an engineering context may wonder what these methods have to offer you. At the time of writing, this is less clear, but there are some fascinating unexplored aspects. The contour integral methods introduced here were first made widely known (albeit without a clear geometrical picture, the non-cohomological aspects of which have been presented here) by Whittaker and Watson (1984 in its current available form), in a text first published much earlier. They have not (yet) become part of the standard undergraduate presentation, probably because approaches based on repeated separation of variables (SOV) are easier to teach, and, more importantly, because the imposition of boundary conditions is easy within the SOV approach, provided the boundaries lie on coordinate surfaces. This is an important provision. For example, exact solutions for scattering of time-harmonic plane waves by objects are only available for simple structures such as spheres, cylinders and so on. (A good catalogue is given by Bowman *et al.* 1987.) The computation of scattering by general objects is now a huge industry, remains difficult, and is still well beyond the scope of numerical methods for a general two-dimensional boundary surface for frequency domains of common interest. Yet there are tantalising sporadic appearances of contour integral methods, from the 'Watson transformation' to the 'Sommerfeld integrals' (see Bowman *et al.*, 1987, for some examples). So far as this author is aware, there is no systematic appreciation of whether and how such occasional appearances are related to the general twistor formulae. Ph.D. project recommendation no. 1 is to sort this out (hard) – a lesser and more practical Ph.D. proposition would be to try to understand how specific boundary conditions influence the choice of twistor function and contour. The author would be delighted to hear of progress in this area, and in applications of twistor methods to fields such as fluid dynamics, as well as scalar and electromagnetic scattering.

Exercises

24.1 The fact that each Legendre polynomial corresponds to a certain power of η/w makes it easy to add up series of such polynomials using the formula for a geometric series. Using the calculus of residues and the integral representation for

$$2^n r^n P_n[\cos(\theta)]$$

to show that

$$\sum_{n=0}^{\infty} 2^n r^n P_n[\cos\theta] = \frac{1}{\sqrt{1 - 2(2r)\cos(\theta) + (2r)^2}}$$

24.2 (Project) Explore the fields corresponding to the holomorphic functions

$$G_{n,m}(\eta,\ w) = \left(\frac{\eta}{w}\right)^n w^{m-1}$$

for $n > 0$, $m \geqslant 0$, as follows:

(i) What is the value of the field ϕ if $m > n$?
(ii) How does the resulting field ϕ change under a rotation by ψ about the z-axis?
(iii) Evaluate the field along the axis $y = 0$ as far as you can.

Look up the definition of the associated Legendre functions and try to relate them to your expressions. See also Note 18.31 of Whittaker and Watson (1984).

24.3 Suppose that $u = ct - z$; $v = ct + z$; $\zeta = x + iy$; $\overline{\zeta} = x - iy$. Show, by using the chain rule for partial derivatives, that

$$\frac{1}{c^2}\frac{\partial^2 \Psi}{\partial t^2} - \frac{\partial^2 \Psi}{\partial x^2} - \frac{\partial^2 \Psi}{\partial y^2} - \frac{\partial^2 \Psi}{\partial z^2} = 0$$

is equivalent to:

$$4\left(\frac{\partial^2 \Psi}{\partial u \partial v} - \frac{\partial^2 \Psi}{\partial \zeta \partial \overline{\zeta}}\right) = 0$$

24.4 Consider the partial differential equation

$$w\frac{\partial g}{\partial p} + \frac{\partial g}{\partial q} + ikg = 0 \tag{24.78}$$

Let new coordinates be defined by

$$\alpha = p - wq$$
$$\beta = p + qw$$

Show, using the chain rule, that the PDE of Eq. (24.78) becomes

$$\frac{\partial g}{\partial \beta} = -\frac{ikg}{2w}$$

and deduce that the general solution is of the form

$$g = e^{-ik\beta/(2w)}\ h(\alpha,\ w)$$

Bibliography

V. Adamchik and M. Trott (1994). *Solving the Quintic*, poster and numerous web resources available at `http://library.wolfram.com/examples/quintic/`
L.V. Ahlfors (1953, 1979). *Complex Analysis: An Introduction to the Theory of Analytic Functions of One Complex Variable.* (First and Third ed.) McGraw-Hill.
J. D'Angelo (2002). *Inequalities in Complex Analysis.* Mathematical Monograph #28, Mathematical Assoication of America.
T.B. Bahder (1995). *Mathematica for Scientists and Engineers.* Addison-Wesley.
T.N. Bailey and R.J. Baston (ed.) (1990). *Twistors in Mathematics and Physics.* London Mathematical Society Lecture Note Series 16. Cambridge University Press.
H. Bateman (1904). The solution of partial differential equations by means of definite integrals. *Proceedings of the London Mathematical Society*, **1** (2), p. 451-8.
H. Bateman (1944). *Partial Differential Equations of Mathematical Physics.* Dover.
A.F. Beardon (1991). *Iteration of Rational Functions.* Springer Graduate Texts in Mathematics, no. 132. Springer.
R.P. Boas (1987). *Invitation to Complex Analysis.* Random House.
J.J. Bowman, T.B.A. Senior and P.L.E. Uslenghi (1987). *Electromagnetic and Acoustic Scattering by Simple Shapes*, revised edition. Hemisphere Publishing.
D.M. Burton (1995). *Burton's History of Mathematics, An Introduction,* 3rd edn. Wm. C. Brown Publishers.
G. Cardano (1993). *The Great Art, or the Rules of Algebra,* translated by Richard Witmer. Dover Reprint.
A. Cayley (1879). The Newton–Fourier imaginary problem. *American Journal of Mathematics,* **2**, p. 97.
H. Cartan (1961). *The elementary theory of analytic functions of one or several complex variables.* The original French version is 1961. The english translation dating from 1963 is available as a 1995 Dover reprint, where the relevant discussion can be found on p. 126.
A.H.-D. Cheng, P. Sidauruk and Y. Abousleiman (1994). Approximate inversion of the Laplace transform. *The Mathematica Journal*, **4** (2), p. 76-82.
H.S.M. Coxeter (1964). Regular compound tessellations of the hyperbolic plane. *Proceedings of the Royal Society,* **A278**, p. 147-167.
H.S.M. Coxeter (1965). *Non-Euclidean Geometry*, 5th ed.. University of Toronto Press.
B. Davies and B. Martin (1979). Numerical inversion of the Laplace transform: a survey and comparison of methods. *Journal of Computational Physiscs,* **33** (1), p. 1-32.
J.W. Dettman (1965). *Applied Complex Variables*, Dover reprint.
R.M. Dickau (1997). Compilation of iterative and list operations. *The Mathematica Journal*, **1** (1), p. 14-15.
T.A. Driscoll (1996). Algorithm 765: a MATLAB toolbox for Schwarz–Christoffel mapping. *ACM Transactions on Mathematical Software,* **22**, p. 168-186.
T.A. Driscoll and L.N. Trefethen (2002). *Schwarz–Chistoffel Mapping.* Cambridge University Press.

L.P. Eisenhart (1911). A fundamental parametric representation of space curves, *Annals of Mathematics* (Ser II), **XIII**, p. 17-35.
L.P. Eisenhart (1912). Minimal surfaces in Euclidean four-spaces, *American Journal of Mathematics*, **34**, p. 215-236.
A. Eydeland and H. Geman (1995). Asian options revisited: inverting the Laplace transform. *RISK Magazine*, March.
M. Field and M. Golubitsky (1992). *Symmetry in Chaos: A Search for Pattern in Mathematics, Art and Nature*. Oxford University Press. (Chapter 4). See also the web site at `http://nothung.math.uh.edu/~mike/`
M.D. Finn and S.M. Cox (2001). Stokes flow in a mixer with changing geometry. *Journal of Engineering Mathematics*, **41**, p. 75-99.
P.R. Garabedian (1966). Free boundary flows of a viscous liquid. *Communications on pure and applied mathematics*, **XIX** (4), p. 421-434.
H. Geman and M. Yor (1993). Bessel processes, Asian options, and perpetuities. *Mathematical Finance,* **3** (4), p. 349-375.
H. Geman and M. Yor (1996). Pricing and hedging double-barrier options: a probabilistic approach. *Mathematical Finance,* **6** (4), p. 365-378.
I.S. Gradshteyn and I.M. Ryzhik (1980). *Tables of integrals, series and products*, corrected and enlarged edition. Academic Press.
A. Gray (1993). *Modern differential geometry of curves and surfaces*. CRC Press.
T.W. Gray and J. Glynn (1991). *Exploring Mathematics with Mathematica*, Addison–Wesley.
Getting Started with Mathematica on (Windows, Macintosh, Linux etc.) Systems. Wolfram Research *Mathematica* Documentation Kit.
N.J. Hitchin (1982). Monopoles and geodesics. *Communications in Mathematical Physics*, **83**, p. 579-602.
L.H. Howell and L.N. Trefethen (1990). A modified Schwarz–Christoffel transformation for elongated regions. *SIAM Journal of Scientific and Statistical Computing,* **11**, p. 928-949.
R. R. Huilgol (1981). Relation of the conjugate harmonic functions to $f(z)$ in cylindrical polar coordinates. *Australian Mathematical Society Gazette,* **8** (1), p. 23-25.
J.D. Jackson (1975). *Classical Electrodynamics,* Second edition. Wiley.
N. Kunitomo and M. Ikeda (1992). Pricing options with curved boundaries. *Mathematical Finance,* **2** (4), p. 275-297.
E.V. Laitone (1977). Relation of the conjugate harmonic functions to $f(z)$. *The American Mathematical Monthly*, **84** (4), p. 281-283. (Available on-line at JSTOR).
R. Legendre (1949). Solutions plus complète du problème Blasius. *Comptes Rendus,* **228**, p. 2008-2010.
I.V. Lindell (1995). *Methods for Electromagnetic Field Analysis*. IEEE Press/Oxford University Press Series on Electromagnetic Wave Theory.
S. Levy (1993). Automatic generation of hyperbolic tilings. In *The Visual Mind*, M. Emmer (ed.), MIT Press.
S. Levy and T. Orloff (1990). Automatic Escher. *The Mathematica Journal,* **1** (1), p.34-35.
R. Maeder (1994). *The Mathematica Programmer*. AP Professional.

R. Maeder (1995). Function iteration and chaos. *The Mathematica Journal,* **5** (2), p. 28-40.
R. Maeder (1997). *Programming in Mathematica.* Third Edition. Addison Wesley.
L.J. Mason (1990). Sources and currents; relative cohomology and non-Hausdorff twistor space. In *Further Advances in twistor theory Vol. I: The Penrose transform and its Applications*. Pitman Research Notes in Mathematics 231. Longman.
L.J. Mason and L.P. Hughston (ed.) (1990). *Further advances in twistor theory, Vol. 1: The Penrose transform and its Applications.* Pitman Research Notes in Mathematics 231. Longman.
L.J. Mason, L.P. Hughston P.Z. and Kobak (ed.) (1990). *Further advances in twistor theory, Vol. 2: Integrable systems, conformal geometry and gravitation*. Pitman Research Notes in Mathematics 232. Longman.
Mathematica Applications Package – Dynamic Visualizer (1999). See the web site at: `http://www.wolfram.com/products/applications/visualizer/`
R. May (1976). Simple mathematical models with very complicated dynamics. *Nature,* **261**, p. 459-469.
L.M. Milne-Thomson (1937). On the relation of an analytic function of z to its real and imaginary parts. *Mathematical Gazette*, **244** (21), p. 228-229.
L.M. Milne-Thomson (1938). *Theoretical Hydrodynamics*, first edition. See also second edition (1949), third Edition (1955), fourth Edition (1962). The fifth edition remains available as a Dover reprint.
M. Montcheuil (1905). Résolution de l'équation $ds^2 = dx^2 + dy^2 + dz^2$, Bulletin de la Société Mathématique de France, **33**, p. 170-171.
T. Needham (1997). *Visual Complex Analysis*. Clarendon Press.
H. Ockendon and J.R. Ockendon (1995). *Viscous Flow.* Cambridge texts in applied mathematics, Cambridge University Press.
J.R. Ockendon, S.D. Howison, A. Lacey and A. Movchan (2003). *Applied Partial Differential Equations*, revised edition, Oxford University Press.
H-0. Peitgen and P.H. Richter (1986). *The Beauty of Fractals: images of complex dynamical systems*, Springer.
R. Penrose (1959). The apparent shape of a relativistically moving sphere. *Proceedings of the Cambridge Philosophical Society*, 55, p. 137-9.
R. Penrose (1999). *The Emperor's New Mind.* Oxford University Press.
R. Penrose (2004). *The Road to Reality*. Jonathan Cape.
R. Penrose and W. Rindler (1984a). *Spinors and space-time*, Vol. 1: *Two-spinor calculus and relativistic fields.* Cambridge University Press.
R. Penrose and W. Rindler (1984b). *Spinors and space-time*, Vol. 2: *Spinor and twistor methods in space-time geometry*. Cambridge University Press.
W.H. Press, S.A. Teukolsky, W. T. Vetterling and B.P. Flannery (1992). *Numerical Recipes in C, The Art of Scientific Computing, Second Edition*. Cambridge University Press.
A.P. Prudnikov, Yu.A. Brychkov and O.I. Marichev (1998). *Integrals and Series*, *Vol 4*: *Direct Laplace Transforms,* (second printing). Gordon and Breach.
A.P. Prudnikov, Yu.A. Brychkov and O.I. Marichev (2002). *Integrals and Series*, *Vol 5*: *Inverse Laplace Transforms,* (digital edition). Gordon and Breach.

K.B. Ranger (1991). A complex variable integration technique for the two-dimensional Navier–Stokes equations. *Quarterly of Applied Mathematics*, **XLIX** (2), p. 555-562.
K.B. Ranger (1994). Parametrization of general solutions for the Navier–Stokes equation., *Quarterly of Applied Mathematics*, **LII**, p. 335-41.
M. Rizzardi (1995). A modification of Talbot's method for the simultaneous approximation of several values of the inverse Laplace transform. *ACM Transactions on Mathematical Software*, **21** (4), p. 347-371.
W. Rudin (1976). *Principles of Mathematical Analysis*. McGraw-Hill.
W.T. Shaw (1985). Twistors, minimal surfaces and strings, *Classical and Quantum Gravity*, **2**, L113-119.
W.T. Shaw and J. Tigg (1993). *Applied Mathematica*. Addison-Wesley.
W.T. Shaw (1995). Symmetric chaos in the complex plane. *The Mathematica Journal,* **5** (3), p. 85-89.
W.T. Shaw (1995b). MathLive and the virtual dynamics lab. *The Mathematica Journal,* **5** (2), p. 14-20.
W.T. Shaw (1998). *Modelling Financial Derivatives with Mathematica*, Cambridge University Press.
W.T. Shaw (2004). Recovering holomorphic unctions from their real or imaginary parts without the Cauchy-Riemann equations. *SIAM Review*, **46** (4), p. 717-728.
W.T. Shaw and L.P. Hughston (1990). Twistors and strings. In *Twistors in mathematics and physics,* eds. T.N. Bailey and R. Baston. London Mathematical Society Lecture Notes, no. 156. Cambridge University Press.
B.C. Skotton (1994). On amplitude and phase in printed characters. *The Mathematica Journal*, **4** (2), p. 83-86.
M.R. Spiegel (1965). *Laplace Transforms*. Schaum's Outlines series, McGraw-Hill.
M.R. Spiegel (1981). *Theory and Problems of Complex Variables*. Schaum's outline series, McGraw-Hill.
S. Stahl (1993). *The Poincaré half-plane: a gateway to modern geometry*. Jones and Bartlett.
R.A. Struble (1979). Obtaining analytic functions and conjugate harmonic functions. *Quarterly of Applied Mathematics*, **37** p.79-81.
A. Talbot (1979). The accurate numerical inversion of Laplace transforms. *Journal of the Institute of Mathematics and its Applications*, **23**, p. 97-120.
J. Terrell (1959). Invisibility of the Lorentz contraction, *Physical Review D*, **116** (4), p. 1041-5.
L.N. Trefethen (1980). Numerical computation of the Schwarz–Christoffel transformation. *SIAM Journal of Scientific and Statistical Computing,* **1**, p. 82-102.
L.N. Trefethen (ed.) (1986). *Numerical Conformal Mapping*. North-Holland.
L.N. Trefethen and T.A. Driscoll (1998). *Schwarz–Christoffel mapping in the computer era*, Oxford University Computing Laboratory report 98/08.
J. Vecer (2002). Unified Asian pricing. *RISK Magazine*, June.
L.I. Volkovyskii, G.L. Lunts, I.G. Aramanovich (1960). (Russian original edition) *A Collection of Problems on Complex Analysis*. English translation (1965) Pergamon Press. Currently available as Dover reprint (1991).
S. Wagon (1991). *Mathematica in Action*. W.H. Freeman.

R.S. Ward, R.S (1981). Axissymmetric stationary fields. In Twistor Newsletter 11. Reprinted (1990) In *Further Advances in twistor theory Vol. I: The Penrose transform and its Applications*. Pitman Research Notes in Mathematics 231. Longman.
K. Weierstrass (1866a). Über die Flächen derren mittlere Krümmung überall gleich null ist. *Monatsberichte der Königlich Preußischen Akademie der Wissenschaften zu Berlin*, p. 612-625.
K. Weierstrass (1866b). (1866a via the world-wide web):
`http://bibliothek.bbaw.de/bibliothek-digital/digitalequellen/schriften/anzeige/index_html?band=09-mon/1866&seite:int=628`
E.T. Whittaker (1903). On the partial differential equations of mathematical physics. Mathematische Annalen, **57**, p. 333-355.
E.T. Whittaker and G.N. Watson (1984). *A course of modern analysis*. Cambridge University Press.
T. Wickham Jones (1994). *Mathematica Graphics, Techniques and Applications*. Springer.
S. Wolfram (2003). *The Mathematica Book*, 5th Edition. Wolfram Media.

Index

/., *Mathematica* concept of "given", 3
i, notation for imaginary number, 12
j, notation for imaginary number, 12
[[ ]], to refer to position in list, 3
[ ], to contain arguments of functions, 3
δ-function
  as Fourier transform of unity, 362
  as limit of sequence of functions, 360
  definition, 359, 360
  in *Mathematica*, 362
{ }, to denote lists, 3
`//N`, to denote numerical evaluation, 4
`==`, to denote equality within equation, 3
`=`, to denote assignment of one quantity to another, 3

3-D
  complex numbers for, 513, 540

4-D
  complex numbers for, 513, 540

`Abs`, modulus function, 18
absolute convergence, 195
  in Taylor's theorem, 270
aerofoil, 424
  by a conformal map, 345
Ahlfors, L.V., 229
Ahlfors–Struble theorem, 208
  finding harmonic conjugates with, 231
  history of, 228
  implementation with *Mathematica*, 226
  inverting `ComplexExpand`, 227
alternating series test for convergence, 198
analytic functions, 214
angle-preserving map, 347
animations
  making with `Do`, 52, 87
Appel hypergeometric function, 464
`AppellF1`, 464
`ArcCosh`, 170
`ArcCot`, 170
`ArcCoth`, 170
`ArcCsc`, 170
`ArcCsCh`, 170
`ArcSec`, 170
`ArcSech`, 170
`ArcSin`, 170
  branching of, 173
  visualization of, 181, 183
arcsine
  visualization of, 181
`ArcSinh`, 170
`ArcCos`, 170
`ArcTan`, 170
`ArcTanh`, 170
area-scaling induced by holomorphic function, 348
`Arg`, argument function, 17
Argand plane, 14
Argand, J.R., 14
argument
  Mathematica function `Arg`, 17
  ambiguity in, 16
  of complex number, 16

principal value, 17
principle of the, 328
Asian call option, 446
assignment, use of =, 3
`Assumptions`
use with `FourierTransform`, 358
use with `Integrate`, 312
attraction, basin of, 59
axis-symmetric solutions of Laplace's equation, 541

basin of attraction, 59
Beardon, A.F., 56
`Beta`, 458, 460
beta function, 458, 460
bifurcation
and symmetry generation, 100
diagrams of, 98
explained in terms of stable/unstable roots, 86, 88, 91
symmetry-increasing, 143
biharmonic equation, 427
complex variable solution, 428
bilinear transformation, 348
Black–Scholes equation, 441, 442
Blasius' theorem, 419
Boas, H., 208
Boas, R.P., 229
boundary condition
Dirichlet, 404
Neumann, 404
branch cut, 171
Mathematica conventions, 172
detailed investigation of arcsin, 173
branch point, 171

C code
use with *Mathematica*, 120, 148
calculus
fundamental theorem of, 241
call option, 441
with Asian feature, 446
with double barrier, 442
Cardano, G., 1, 41
`CartesianMap`, 177
catenoid, 537
Cauchy inequalities, 271
Cauchy integral formula, 263
Cauchy principal value, 313
Cauchy sequence, 196
Cauchy's theorem
converse to, 274
strong (Cauchy–Goursat) form, 250
weak form based on Green's theorem, 250
Cauchy–Goursat theorem
for a triangle, 250
for star-shaped set, 255
Cauchy–Riemann equations, 208, 212, 213, 403
and orthogonal curves, 214
harmonic functions, 215
Laplace's equation, 215
using to recover function from real part, 221
Cayley planet
for cubic, 74
for septic, 75
Cayley's problem
solution for a cubic, 62
solution for a simple septic, 73
solution for quadratic, 60
Cayley, A., 59
celestial sphere, 514
`Chop`
example of use, 174, 436
circle
complex equation of, 350
circline
defined, 351
explored with *Mathematica*, 351
closed set, 163
cobwebbing, 78
convergence of, 79
error behaviour, 79
plots of, 93, 94, 96
`Coeffcient`
applied to quartic equation, 48
`Collect`

applied to expression, 49
colour function examples, 146
colouring schemes for fractals, 64
comparison test for convergence, 197
complex conjugate
Mathematica function `Conjugate`, 18
of complex number, 14
complex number
n'th root of, 26
$x + iy$ notation, 12
$x + jy$ notation, 12
`x+Iy` notation, 13
algebraic properties, 13
Argand plane representation, 14
argument, 16
as a point in the plane, 14
complex conjugate of, 14
cube root of, 27
exponential form, 29
functions of, 159
imaginary part, 14
modulus of, 14
ordered pair notation, 12
polar representation, 15
real part, 14
square root of, 25
working in 3-D, 513, 540
working in 4-D, 513, 540
complex plane
extended, 175
complex roots of equations, 5
`ComplexExpand`, 161
applied to trigonmetric and hyperbolic functions, 168
as function to take real and imaginary parts of expressions, 19
inverting with *Mathematica* implemenation of Ahlfors–Struble theorem, 227
complexification
issues with, 224
`ComplexInequalityPlot`, 339, 354
`ComplexMap` package, 177, 338
conformal map, 338
and Laplace transforms, 395
and Laplace's equation, 404
on a cylinder, 420
visualization of, 341–345
conformal representation of hyperbolic plane, 498
conformality of holomorphic functions, 347
`Conjugate`, complex conjugate function, 18
connected set, 248, 286
continuity
of a function at a point, 164
of sequence of uniformly convergent functions, 199
under addition etc., 164
contour
circular, 305
defined as a piecewise smooth path, 238
deforming, 255
for treating branching integrands, 320, 321, 323
integral defined, 240
length of, 242
mousehole, 318
rectangular, 324
semicircular, 313, 316, 366
UHP or LHP?, 316, 367
zero theorem, 314
semicircular with indentation, 318
contour integration, 240
issues with older versions of *Mathematica*, 245
with *Mathematica*, 244
`ContourPlot`, 414, 416, 418, 419, 422, 423, 425, 429
convergence
absolute, 195
in Taylor's theorem, 270
Cauchy's condition for, 196
of monotone bounded real sequence, 196
of subsequence of bounded se-

quence, 196
radius of, 202
terms tending to zero, 196
tests for, 196
uniform, 195
in Taylor's theorem, 270
uniform and integration, 243
uniform behaviour of power series, 204
convergence test
alternating series, 198
comparison, 197
Dirichlet, 198
integral, 198
nth root, 197
ratio, 197
convergence-time algorithm for Newton–Raphson, 63
convergent subsequence in iterated map, 81
convolution theorem
for Fourier transforms, 364
for Laplace transforms, 390
generalized (Efros), 395
`Cos`, 170
power series, 29
`Cot`, 170
`Cosh`, 170
`Coth`, 170
`CountRoots`, 334
`Csc`, 170
`CsCh`, 170
cube root of unity, 27
role in solving cubic equation, 43
cubic equation
as iterated logistic map, 81
history, 41
resolvent for quartic, 48
solution, 44
solution with *Mathematica*, 42
solved via Newton–Raphson, 62, 67
cyclic group, 138
in images generated from polynomial maps, 142
deforming contours, 255
to a circle, 256
del Ferro, S., 41
delta-function
as Fourier transform of unity, 362
as limit of sequence of functions, 360
definition, 359, 360
in *Mathematica*, 362
derivative
complex, 212
financial, 441
differentiability
complex, 212
complex conjugates, 220
implies partial derivatives exist, 209
insufficiency of existence of partial derivatives, 209
of complex function considered as two real functions, 211
of function of two real variables, 209
of polynomial, 217
of power series, 218
other notations, 220
relation to Cauchy–Riemann equations, 213
differential equation
solved by Fourier transform, 373
solved by Laplace transform, 391, 392
solved with `DSolve`, 392, 393
differentiation
addition rule, 215
and Fourier transforms, 365
and Laplace transforms, 383
and sequences of uniformly convergent functions, 200
chain rule, 216
product rule, 216
quotient rule, 217
reciprocal rule, 216
diffusion equation, 401
solved by Fourier transform, 374

solved by Laplace transform, 393
dihedral group, 138
in images generated from non-polynomial maps, 142
in images generated from polynomial maps, 140
Dirichlet boundary condition, 404
and disk, 408
and half-plane, 406, 414
and quadrant, 415
and wedge, 415
Dirichlet's test for convergence, 198
disc
closed, 163
open, 163
punctured, 163
discrete Fourier transform, 433
applied to 1-D filtering, 435
applied to 2-D filtering, 437
distribution, 359
`Do`
for animated graphics, 52
domain, 159, 248
simply connected, 248
`DSolve`, 392, 393

Efros's theorem, 395
Einstein's equations, 550
Eisenhart, L.P.
minimal surfaces in four dimensions, 533
ellipse, fluid flow past, 422
`EllipticF`, 461
elliptic function, 461
Enneper's surface, 536
equality, use of == to denote equality in an equation, 3
`Erfc`, 178
visualization of, 179, 181, 184
error function
visualization of, 179, 181, 184
essential singularity, 288
other characterization of, 289
visualized, 292
`Exp`
exponential function, 13
power series, 29
`Expand`, 160
applied to quartic equation, 48
with `Trig -> True` option, 25
exponential function
Mathematica function `Exp`, 13
complex properties, 31
definition for complex variable, 165
related to trigonometric functions, 29
expression
imaginary part with `ComplexExpand`, 19
real part with `ComplexExpand`, 19
showing detailed stucture with `FullForm`, 21
extracting a holomorphic function from its real part
Ahlfors–Struble method, 221
Cauchy–Riemann method, 221

`Factor`, 6
applied to quartic equation, 47, 48
`FactorList`, 6
Ferrari, L., 41
solution of quartic equation, 47
Field, M., 138
filtering noise, 435
financial derivatives, 441
`FindRoot`
applied to solving equations, 59
`FixedPoint`, 63
exercises with, 76
for Newton–Raphson problems, 63
`FixedPointList`, 63
applied to Mandelbrot map, 119
exercises with, 76
fluid
conservation of mass, 402
detailed flows, 415
flow past a flat plate, 423
flow past an ellipse, 422

force on body, 420
incompressible, 402
irrotational flow, 403
potential flow past a cylinder, 417
with circulation, 418
source above a half-plane, 416
source outside a circle, 417
uniform flow, 409
viscous flow, 425
in 2-D, 427
with circulation, 418
zero vorticity condition, 403
Fontana, N., 41
force on a body in a fluid, 420
four dimensions
physics in, 513, 544
`Fourier`, 433–435, 437
Fourier transform, 357
and differential equations, 373
and heat/diffusion equation, 374
and Laplace's equation, 375
conventions, 358
convolution theorem, 364
defined, 358
differentiation theorem, 365
discrete, 433
in a complex setting, 372
in older *Mathematica* versions, 377
inversion theorem, 363
of Cauchy p.d.f., 369
of Gaussian function, 370
of sine function, 368
of unity, 362
scaling theorem, 365
shift theorem, 365
`FourierParameters`, 434
`FourierTransform`, 358
conventions, 358
options, 358
fractal
from iterated sine function, 189, 191
Newton–Raphson for cubics, 65, 67
planet, 73
fractional transformation, 348
`FullForm`, function to show detailed structure of expression, 21
function
analytic, 214
and conformality, 347
Appel hypergeometric two variable, 464
area-scaling induced by, 348
beta, 458, 460
checking for being harmonic with *Mathematica*, 230
defined by power series, 205
defined formally, 159
definition by series, 165
definition by use of real analogues, 165
domain of, 159
elliptic, 461
exponential, 165
harmonic, 215
holomorphic, 214
hyperbolic, 167
hypergeometric, 460, 463
integral over a contour, 240
multi-valued, 171
range of, 159
real and imaginary parts of, 160, 161
regular, 214
trigonometric, 166
visualization with *Mathematica*, 176, 183
fundamental theorem of algebra, 272
fundamental theorem of calculus, 241

`Gamma`
visualization of, 185
gamma function
visualization of, 185
Geman-Yor model
for double barrier options, 443
geometry
Euclidean, 474
hyperbolic, 474

ghosts and birdies tiling
of the Euclidean plane, 484
of the Poincaré disc, 490, 497
given, /. notation, 3
Golubitksy, M., 138
graphics using checkers and holes, 187
Green's theorem in the plane, 249
for complex function, 249
group
cyclic, 138
dihedral, 138
Möbius transforms, 349

harmonic conjugates
finding with Ahlfors–Struble theorem, 231
harmonic function
and Cauchy–Riemann equations, 215
checking with *Mathematica*, 230
`Head`, function to show type of nnumber or expression, 20
heat equation, 401
solved by Fourier transform, 374
solved by Laplace transform, 393
hedging
applied to double barrier call option, 445
Helmholtz equation
twistor solution, 546
Hitchin, N., 534, 540
holomorphic 3-D structures, 550
holomorphic functions, 214
hyperbolic functions
Mathematica expressions for, 170
definition for complex variable, 167
hyperbolic plane, 473
conformal representation, 498
projective representation, 498
UHP representation, 511
hypergeometric function, 460, 463
Appel two variable, 464
confluent, 447
in Asian option theory, 447
`Hypergeometric1F1`, 447
`Hypergeometric2F1`, 460, 463

`Im`, imaginary part function, 18, 161
image processing, 438
images
method of, 413
imaginary numbers, introduction of, 1
imaginary part
Mathematica function `Im`, 18
of complex number, 14
of expression, *Mathematica* function `ComplexExpand`, 19
ineqaulities
visualizing, 339
inequality
length, 242
value, 242
`InequalityGraphics`
Mathematica package, 339
`InequalityGraphicsPlot`, 354
infinity
adding to complex plane, 175
`InputForm`, 5
integral test for convergence, 198
`Integrate`
care with use, 302
use to check residue calculation, 308, 312, 316, 317, 323, 324, 326
use with `Assumptions`, 302, 312
integration
contour, 240
of branching functions, 321, 323
of function with periodic singularities, 324
of powers of z about the origin, 240
of sequence of uniformly convergent functions, 199
over arbitrary angle about simple pole, 318
over infinite range, 313, 316
trigonometric, 305
with *Mathematica*, 302
inverse function

for sinh, 169
for trigonometric and hyperbolic functions, 170
inverse hyperbolic functions
Mathematica expressions for, 170
inverse trigonometric functions
Mathematica expressions for, 170
`InverseFourier`, 433, 434, 436, 437
`InverseLaplaceTransform`, 387, 440
inversion, 348
isolated singularity, 287
isolated zeroes, 287
iterated map
implementation with `Nest`, 82

Jordan's lemma, 366
Joukowski flow, 421, 424
Joukowski map, 345
attack angle variation, 346
camber variation, 346
visualized, 345

Kummer's function, 447

Laitone, E.V., 229
Laplace transform, 381
and change of time variable, 397
and differential equations, 391, 392
and heat/diffusion equation, 393
convolution theorem, 390
definition, 381
differentiation theorem, 383
holomorphic property, 384
inversion, 387
of algebraic function, 389
of rational function, 388
with branch cut, 389
making tables with *Mathematica*, 386
numerical inversion, 439, 441, 445
scaling theorem, 383
shift theorem, 383
Laplace's equation, 401
and fluid flow, 402
and heat flow, 401
and holomorphic functions, 403
axis-symmetric, 541
from Cauchy–Riemann equations, 215
in three dimensions, 540
invariance under conformal maps, 404
solved by Fourier transform, 375
`LaplaceTransform`, 385
Laurent series, 278
uniquness of coefficients, 281
use of known power series, 283
Legendre polynomial, 542
length inequality, 242
limit
of a function at a point, 164
`Limit`
use to calculate residue, 296, 299
use with residue theorem, 308
line
complex equation of, 350
linear transformation, 348
Liouville's theorem, 271
list
generating with `Table`, 7
picking an element of a given position within, 3
`Listable`, 478
`Log`
branching of, 172
logarithm
ambiguity in, 32
as inverse to exponential, 32
branching of, 172
modulus-argument form, 32
why there is no global form, 241
logistic map, 78
cubic, 80, 81
quadratic, 80, 103
Lorentz contraction
invisibility of, 518
Lorentz transformation
angular abberation, 523
as Möbius map, 517
classification, 520

details of velocity change, 522
remapping light rays with *Mathematica*, 524

Mandelbrot map
definition of, 106
fixed points of, 107
Misiurewicz points, 113
periodic orbits of, 110
Mandelbrot set
escape-time algorithm, 114
pictures of, 119, 123, 127–129
problems in drawing it properly, 129
purist black and white pictures of, 131–133
purist vs pretty pictures, 134
what *precisely* is it?, 115
map
conformal, 338
logistic, 78
Mandelbrot, 106
Möbius, 338
Newton–Raphson, 56
non-linear with symmetry, 138
Schwarz–Christoffel, 451
`Map`
applied to solving an equation, 59
Margolis, B., 229
Mathematica older versions
Fourier transforms in, 377
MathLink
use in drawing fractals, 120
use in drawing images of symmetric chaos, 148
maximum modulus theorem, 275
Maxwell's equations, 550
May, R., 80, 104
Mean value theorem for the modulus, 237
mean-value theorem, 275
method of images, 413
for circle, 417
for half-plane, 414, 416
for quadrant, 415
for wedge, 415
metric
relativistic, 514
Milne-Thomson, L.M., 229
minimal surface, 531
and holomorphic null curves, 532
in four-space, 533
in three-space, 534
Montcheuil–Eisenhart formula, 533
visualized with *Mathematica*, 535
Weierstrass formula, 534
Minkowski space, 514
Mittag-Leffler theorem, 328
`Module`
use of in graphics program, 27
modulus
Mathematica function `Abs`, 18
of complex number, 14
Montcheuil, M.
minimal surfaces in four dimensions, 533
Morera's theorem, 274
movies
making with `Do`, 52, 87
multi-valued function, 171
Möbius map, 338
and Poincaré disc, 474
as a group, 349
as Lorentz trasnformation, 517
inverted with *Mathematica*, 61
link to Newton–Raphson and Cayley's problem, 60
simple components of, 348
simple example, 340
to map three points as specified, 352

`N`, to denote numerical evaluation, 4
Navier–Stokes equations, 425
complex 2-D form, 428
neighbourhood, 163
deleted, 163
`Nest`
applied to iterated map, 138
applied to logistic map, 82
`NestList`

applied to iterated map, 91, 138
Neumann boundary condition, 404
and half-plane, 414, 416, 417
and quadrant, 415
and wedge, 415
noise filtering with Fourier transform, 435
`NResidue`
use of package, 297, 308
`NSolve`, 332
applied to quartic equation, 52
nth root test for convergence, 197
null
holomorphic curve, 532
twistor construction, 533
vector, 514

Ockendon, J.R., 229
open set, 163
option pricing, 441
`Options`, 7
order of pole, 288

`ParametricPlot`, 421
path
closed, 237
defined, 237
drawing in *Mathematica*, 238
length of, 242
piecewise smooth, 238
simple, 237
smooth, 238
Penrose, R., 473, 529, 540
physics
in four dimensions, 513, 544
in three dimensions, 513, 540
in two dimensions, 401
relativistic, 545
plane
Euclidean, 473
hyperbolic, 473
plate
fluid flow past, 423
`Plot`, 9
`Plot3D`
applied to complex functions, 171
use of, 406
`PlotGradientField`, 410
`PlotGradientField3D`, 412
`PlotRange`, 9
polar representation of complex number, 15
`PolarMap`, 178
pole, 288
order of, 288
other characterization of, 289
residue at, 292
residue for simple case, 293
visualization of, 291
polynomial
complex differentiability, 217
polynomial equations
have all roots complex, 272
iteration solution, 56
numerical solution, 56
Possion's formula for the disk, 408
potential theory, 401
power series
behaviour on circle of convergence, 203
convergence of, 202
defining functions by, 205
uniform convergence of, 204
`PowerExpand`
applied to quartic equation, 50
`Prime`, 7
principal value, of the argument, 17
principle of the argument, 328
projective representation of hyperbolic plane, 498

quadratic equation
as motivation for i, 2
solution of, 2
solved by Newton–Raphson, 60
treated as iterated logistic map, 103
quartic equation
history, 41
solution by Ferrari's method, 47
solution with *Mathematica*, 46
solved by Newton–Raphson, 71

quintic equation
- issues with, 51
- numerical solution with `NSolve`, 52
- solved by Newton–Raphson, 71

range, 159
ratio test for convergence, 197
rational numbers, motivation from solving equations, 11
`Re`, real part function, 18, 160, 161
real numbers, motivation from solving equations, 10
real part
- *Mathematica* function `Re`, 18
- of complex number, 14
- of expression, *Mathematica* function `ComplexExpand`, 19

`Reduce`, 25
regular functions, 214
relativity
- and complex numbers, 515
- metric of, 514

removable singularity, 288
- Riemann's theorem, 288

residue
- defined from Laurent series, 282
- example calculations of, 293
- formula for pole, 292
- formula for simple pole, 293
- theorem for integrals, 302

`Residue`
- at essential singularities, 296
- examples of use, 294–296, 298

residue theorem
- and branching integrands, 320
- and indented semicircular contour, 318
- and rectangular contour, 324
- and semicircular contour, 313, 316
- and series summation, 326
- applications, 304
- trigonometric integrals, 305

resolvent cubic, 48
Reynolds number, 427
Riemann
- mapping theorem, 452
- sphere, 175
- zeta function
  - visualization of, 182

rocket science, 441
`RootIsolation`
- *Mathematica* package, 334

roots
- branching of, 171
- of complex numbers, 26
- of unity, 27

rotation, 348
Rouché's theorem, 330

scaling theorem
- for Fourier transforms, 365
- for Laplace transforms, 383

Schwarz–Christoffel map, 451
- advanced numerical methods, 470
- defined, 452
- for hexagon, 467, 468
- for pentagon, 469
- for rectangle, 461
- for regular polygon, 466
- for trapezoid, 465
- for triangle, 456, 468
- for vertical strip, 455
- from circle, 465, 466
- point at infinity, 453
- power of *Mathematica*, 451

`Sec`, 170
`Sech`, 170
semicircle integration, 314
- with complex exponentials, 366

septic equation
- solved by Newton–Raphson, 71

sequence
- of complex functions, 194
- of complex numbers, 194

series
- Laurent, 278
- of complex functions, 195
- power series, 202
- summation, 326

`Series`

function for generating power series, 29
in computing Laurent expansion, 283
set
closed, 163
connected, 248, 286
open, 163
star-shaped, 254
`SetAttributes`, 478
`SetOptions`, 429
shift theorem
for Fourier transforms, 365
for Laplace transforms, 383
silly face picture, 70
simple pole, 293
`Sin`, 170, 177
iterated, 189, 191
power series, 29
visualization of, 181, 184
sine
iterated to make a fractal, 189, 191
visualization of, 181, 184
singularity
classification of (isolated), 288
essential, 288
isolated, 287
pole, 288
removable, 288
`Sinh`, 170
sink, 409
soap bubble, 531
`Solve`, 332
applied to cubic equation, 42
applied to quadratic equation, 3
applied to quartic equation, 46
source, 409
spinor, 516
two-component, 516
`Sqrt`, 7
branching of, 171
square roots of complex numbers, 25
`StandardForm`, 5
stereographic projection, 175, 515
Stokes flow, 427
and cylinder, 429
stream function
for viscous flow, 427
stretching, 348
Struble, R.A., 229
summation of infinite series
by residue method, 326
with *Mathematica*, 327, 328
symmetric chaos, 138
high-resolution imagery, 150
symmetry in complex non-linear map, 138

`Table`, 7
`Tan`, 170
`Tanh`, 170
Tartaglia, 41
Taylor's theorem, 265
absolute convergence, 265, 270
differences from real case, 266
uniform convergence, 265, 270
tests for convergence – see convergence test, 196
Thomas, V., 473
three dimensions
physics in, 513, 540
reduction to two, 548
holomorphic properties, 550
tiling, 473
of Euclidean plane with ghosts and birdies, 484
of Euclidean plane with other shapes, 481
of Euclidean plane with triangles, 475
of Poincaré disc with ghosts and birdies, 490, 497
of Poincaré disc with heptagons
conformal representation, 509
projective representation, 510
of Poincaré disc with hyperbolic squares
conformal representation, 501, 503, 504, 506

projective representation, 502, 503, 505, 507
of the Poincaré disc with triangles, 485, 490
projective representation, 499
UHP representation, 512
`TraditionalForm`, 5
transform
Fourier, 357, 358
Laplace, 381
translation, 348
Trefethen, L.N., 451
triangle subdivision, 251
with *Mathematica*, 259
`TrigExpand`, 24, 168
`TrigFactor`, 25
trigonometric functions
Mathematica expressions for, 170
Mathematica function `Cos`, 24
Mathematica function `Sin`, 25
converting multiple angles to powers, 23
with `TrigExpand`, 24
converting powers to multiple angles, 23
with `TrigReduce`, 25
definition for complex variable, 166
factorizing with `TrigFactor`, 25
power series, 29
related to exponential function, 29
`TrigReduce`, 169
twistor
and minimal surfaces in four dimensions, 533
and minimal surfaces in three dimensions, 534
as pair of spinors, 530
theory, 529, 540, 544, 550
two-component spinor, 516

uniform convergence, 195
and continuity, 199
and differentiation, 200
and integration, 199, 243
continuity and integration counter-examples, 201
in Taylor's theorem, 270
M-test, 200
upper half-plane representation of hyperbolic plane, 511

value inequality, 242
vector
null, 514
spacelike, 514
timelike, 514
viscous fluids, 425
equation for stream function, 427
visualization
alternative 3D schemes, 185, 187
of aerofoil, 345
of angular abberation, 524
of arcsine function, 181, 183
of catenoid, 537
of complex functions, 176
of complex functions in three dimensons, 183
of conformal map, 341–345
of discontinuous function with partial derivatives, 210
of double pole, 291
of Enneper's surface, 536
of essential singularity, 292
of flow around a flat plate, 424
of flow around ellipse, 423
of fluid flow past a cylinder, 418
with circulation, 419
of fundamental theorem of algebra, 273
of gamma function, 185
of ghosts and birdies tiling of Poincaré disc, 497
of heptagon tiling of Poincaré disc
conformal representation, 509
projective representation, 510
of inequalities, 339, 354
of Joukowski map, 345
of Lorentz transformation, 524
of minimal surfaces, 535
of potential flows, 410

of power series, 205
of SC map
for hexagon, 468
for pentagon, 469
for rectangle, 462
for trapezoid, 465
for triangle, 459, 468
for vertical strip, 456
of simple pole, 291
of sin function, 177
of sine function, 181, 184
of singularities, 290
of source above half-plane, 416
of square tiling of Poincaré disc
conformal representation, 501, 503, 504, 506
projective representation, 502, 503, 505, 507
of Stokes flow in cylinder, 429
of triangle tiling of Poincaré disc, 490
projective representation, 499
UHP representation, 512
of vector field, 410, 412
using contour plots, 176
using surface plots, 183
using surfaces with holes, 187
with `CartesianMap`, 177
with `PolarMap`, 178
vortex, 410
vorticity equation, 426

wave equation
twistor solution, 545
Weierstrass, K.
M-test, 200
minimal surface formula, 534
Wessel, C., 14
Wessel–Argand plane, 14
winding number, 264

zeroes
isolation of, 287
locating with Rouché's theorem, 332
locating with *Mathematica*, 332
location of, 330
`Zeta`
visualization of, 182
zeta function
visualization of, 182

Lightning Source UK Ltd.
Milton Keynes UK
UKHW050109081122
411680UK00016BD/1